SKEW-ELLIPTICAL DISTRIBUTIONS AND THEIR APPLICATIONS

A JOURNEY BEYOND NORMALITY

SKEW-ELLIPTICAL DISTRIBUTIONS AND THEIR APPLICATIONS

A JOURNEY BEYOND NORMALITY

EDITED BY Marc G. Genton

CRC Press
Taylor & Francis Group
Boca Raton London New York

CRC Press is an imprint of the
Taylor & Francis Group, an **informa** business
A TAYLOR & FRANCIS BOOK

First published 2004 by Taylor & Francis Group

Published 2020 by CRC Press
Taylor & Francis Group
6000 Broken Sound Parkway NW, Suite 300
Boca Raton, FL 33487-2742

First issued in paperback 2020

ISBN 13: 978-0-367-57831-2 (pbk)
ISBN 13: 978-1-58488-431-6 (hbk)

Visit the Taylor & Francis Web site at
http://www.taylorandfrancis.com

and the CRC Press Web site at
http://www.crcpress.com

Library of Congress Cataloging-in-Publication Data

Skew-elliptical distibutions and their applications : a journey beyond normality / edited by Marc G. Genton.

p. cm.

Includes bibliographical references and index.

ISBN 1-58488-431-2 (alk. paper)

1. Distribution (Probability theory). 2. Skew fields. 3. Multivariate analysis. I. Genton, Marc G.

QA273.6.S54 2004
519.2′4—dc22 2004049449

Library of Congress Card Number 2004049449

Preface

The introduction of non-normal distributions can be traced back to the nineteenth century. Edgeworth (1886) examined the problem of fitting asymmetrical distributions to asymmetrical frequency data. A few years later, Karl Pearson investigated the dissection of an asymmetrical frequency curve into a mixture of two normal curves, and also defined a "generalized form of the normal curve of an asymmetrical character" in a letter to *Nature* (Pearson, 1893). It appears that Erastus De Forest derived independently such a "skew curve" in his article (De Forest, 1882, 1883). Edgeworth brought this work to the attention of Pearson who acknowledged De Forest's priority in a letter to *Nature* (Pearson, 1895). This topic of skew curves created a "competition" between Edgeworth and Pearson, whose correspondence is mostly reprinted in Stigler (1978). For further details on those historical developments, the reader is referred to the sections in Stigler (1986) on *Skew Curves* and *The Pearson Family of Curves.*

The twentieth century has also seen a great deal of interest in non-normal distributions. For instance, Tukey (1977) introduced the univariate g-and-h family of distributions. It encompasses a wide variety of distributional shapes and is defined in terms of quantile functions relative to the standard normal distribution. Another approach to introducing skewness is given by Azzalini (1985) with his skew-normal distribution, although this idea is suggested in earlier work, see, e.g., Birnbaum (1950), Nelson (1964), Weinstein (1964), Roberts (1966), and O'Hagan and Leonard (1976). It consists of modifying the normal probability density function by a multiplicative skewing function. During the past two decades, this idea has been extended to elliptically contoured distributions, thus yielding various families of skew-elliptical distributions, the main topic of this book.

This book reviews the state-of-the-art advances in skew-elliptical distributions and provides many new developments in a single volume, collecting theoretical results and applications previously scattered throughout the literature. The main goal of this research area is to develop flexible parametric classes of distributions beyond the classical normal distribution. This is an exciting and fast-growing field of research that brings together both frequentist and Bayesian statisticians. Along with an explosion of interest for this new area of research has come a virtually unlimited potential for applications. The book is therefore divided into two parts, the first dealing with theory and the second dealing with applications.

The first part of the book covers theory and inference for skew-elliptical distributions. In Chapter 1, Dalla Valle introduces the classical skew-normal distribution in the univariate and multivariate setting. Then, in Chapter 2, González-Farías, Domínguez-Molina, and Gupta define an extension, the closed skew-normal distribution, which is closed under marginalization, conditioning, and certain additive operations. In Chapter 3, Liu and Dey present several extensions of the skew-normal distribution to skew-elliptical ones. In Chapter 4, Loperfido introduces a family of generalized skew-normal distributions, which allows for a wide range of possibilities to introduce skewness starting from the normal distribution. In Chapter 5, Genton introduces skew-symmetric and generalized skew-elliptical distributions, as well as flexible versions of them, which allow to model skewness, heavy tails, and multimodality. In Chapter 6, Arnold and Beaver describe several mechanisms that lead to the distributions described previously, arising from hidden truncations or selective sampling. In Chapter 7, Arellano-Valle and del Pino provide a unified approach to the construction of skewed distributions from symmetric ones. In Chapter 8, Chen presents various skewed link models for categorical response data. Finally, in Chapter 9, Liseo describes the use of skew-elliptical distributions in Bayesian inference.

The second part of the book covers applications and case studies, including areas such as economics, finance, oceanography, climatology, environmetrics, engineering, image processing, astronomy, and biomedical sciences. In Chapter 10, Ferreira and Steel discuss Bayesian multivariate regression models with skewed distributions for the errors, with an application to firm size in the UK. In Chapter 11, Adcock describes the use of the multivariate skew-normal distribution for capital asset pricing of UK stocks. In Chapter 12, De Luca and Loperfido define a skew-in-mean *GARCH* model with an application to returns of three European financial markets. In Chapter 13, Domínguez-Molina, González-Farías, and Ramos-Quiroga analyze various aspects of skew-normality in stochastic frontier analysis. In Chapter 14, Thompson and Shen use the skew-t distribution for the analysis of coastal flooding in Halifax, Canada. In Chapter 15, Naveau, Genton, and Ammann describe a skewed Kalman filter for the analysis of climatic impacts of large explosive volcanic eruptions on temperature time series over a few centuries. In Chapter 16, Kim, Ha, and Mallick define a skew-normal spatial process for the prediction of rainfall in Korea. In Chapter 17, Baloch, Krim, and Genton use flexible skew-symmetric distributions to analyze the shape of objects in images, including the human brain and heart. In Chapter 18, Eyer and Genton make use of regression with skewed errors to approach a challenging problem of astronomical distance determination. In Chapter 19, Sahu and Dey investigate the application of survival models with a skewed frailty. Finally, in Chapter 20, Ma, Genton, and Davidian relax the assumption of normality of random effects in linear mixed models by exploring the use of flexible generalized skew-elliptical distributions.

Such a book should be of great value for a special topic in a course on multivariate analysis. It should also be relevant to other fields that need flexible non-normal distributions for statistical analysis, as is illustrated in the second part of the book through applications and case studies. With a nice balance between theory and applications, this book is intended for a large audience of readers, including graduate students and researchers in statistics, statisticians, and practitioners from other fields.

Finally, I would like to thank all the contributors for providing thoughtful chapters in a timely fashion. Each chapter has been reviewed by two referees, and I would like to thank all the contributors for their help with this important task. In addition, I am grateful to Márcia Branco, Chris Cohen, Michael Dowd, and Jiuzhou Wang for serving as referees as well. In particular, I would like to thank Nicola Loperfido for refereeing many chapters of this book and for his constant support. The students in my "Multivariate Analysis" class at North Carolina State University during Spring 2004 also discovered many typos in the various chapters and provided useful comments. I would like to thank them all for their great help. Furthermore, I would like to thank Bob Stern and the staff at CRC Press, in particular, Chris Andreasen, Jessica Vakili, and Nishith Arora, for their encouragement and support during the production of this book.

Marc G. Genton

Contributors

Chris Adcock, University of Sheffield, UK

Caspar M. Ammann, National Center for Atmospheric Research, USA

Reinaldo B. Arellano-Valle, Pontificia Universidad Católica de Chile

Barry C. Arnold, University of California Riverside, USA

Sajjad Baloch, North Carolina State University, USA

Robert J. Beaver, University of California Riverside, USA

Ming-Hui Chen, University of Connecticut, USA

Alessandra Dalla Valle, Università di Padova, Italy

Marie Davidian, North Carolina State University, USA

Guido E. del Pino, Pontificia Universidad Católica de Chile

Giovanni De Luca, Università di Napoli "Parthenope," Italy

Dipak K. Dey, University of Connecticut, USA

J. Armando Domínguez-Molina, Universidad de Guanajuato, Mexico

Laurent Eyer, University of Geneva, Switzerland

José T. A. S. Ferreira, University of Warwick, UK

Marc G. Genton, North Carolina State University, USA

Graciela González-Farías, CIMAT, Mexico

Arjun K. Gupta, Bowling Green State University, USA

Eunho Ha, Yonsei University, Korea

Hyoung-Moon Kim, Texas A&M University, USA

Hamid Krim, North Carolina State University, USA

Brunero Liseo, Università di Roma "La Sapienza," Italy

Junfeng Liu, University of Connecticut, USA

Nicola M. R. Loperfido, Università di Urbino "Carlo Bo," Italy

Yanyuan Ma, North Carolina State University, USA

Bani K. Mallick, Texas A&M University, USA

Philippe Naveau, University of Colorado at Boulder, USA

Rogelio Ramos Quiroga, CIMAT, Mexico

Sujit K. Sahu, University of Southampton, UK

Yingshuo Shen, University of Hawaii, USA

Mark F. J. Steel, University of Warwick, UK

Keith R. Thompson, Dalhousie University, Canada

Contents

I Theory and Inference 1

1 The Skew-Normal Distribution **3**
Alessandra Dalla Valle

1.1 Introduction 3
1.2 The Univariate Skew-Normal Distribution 4
1.2.1 Moments 7
1.2.2 Cumulative Distribution Function 8
1.3 Estimation and Inference 9
1.4 Checking the Hypothesis of Skew-Normality 13
1.5 The Multivariate Skew-Normal Distribution 14
1.5.1 Genesis 15
1.5.2 The Cumulative Distribution Function 16
1.5.3 The Moment Generating Function 17
1.6 Extensions of Properties Holding in the Scalar Case 19
1.7 Quadratic Forms 20
1.8 The Conditional Distribution 22
1.9 Miscellanea 23
1.10 Future Research Problems 24

2 The Closed Skew-Normal Distribution **25**
Graciela González-Farías, J. Armando Domínguez-Molina, and Arjun K. Gupta

2.1 Introduction 25
2.2 Basic Results on the Multivariate CSN Distribution 26
2.3 Linear Transformations 29
2.4 Joint Distribution of Independent CSN Random Vectors 33
2.5 Sums of Independent CSN Random Vectors 35
2.6 Examples of Sums of Skew-Normal Random Vectors 37
2.6.1 Azzalini and Dalla Valle (1996) 37
2.6.2 Arnold and Beaver (2002) 38
2.6.3 Liseo and Loperfido (2003a) 39
2.7 A Multivariate Extended Skew-Elliptical Distribution 40
2.8 Concluding Remarks 41

3 Skew-Elliptical Distributions **43**
Junfeng Liu and Dipak K. Dey

3.1 Introduction 43
3.2 General Multivariate Skew-Elliptical Distributions 44
3.2.1 The Branco and Dey Approach 45
3.2.2 The Arnold and Beaver Approach 46
3.2.3 The Wang-Boyer-Genton Approach 47
3.2.4 The Dey and Liu Approach 48
Linear Constraint and Linear Combination of Type-1 (LCLC1) 49
Linear Constraint and Linear Combination of Type-2 (LCLC2) 49
3.3 Examples of Skew-Elliptical Distributions 50
3.3.1 Skew-Scale Mixture of Normal Distribution 50
Skew-Finite Mixture of Normal 51
Skew-Logistic Distribution 51
Skew-Stable Distribution 51
Skew-Exponential Power Distribution 52
Skew-t Distribution 52
Skew-Pearson Type II Distribution 53
3.3.2 Different Types of Multivariate Skew-Normal Distributions 53
Type A-$SN(\boldsymbol{\mu}, \Omega, \boldsymbol{\alpha})$ 53
Type B-$SN(\boldsymbol{\mu}, \Omega, D)$ 54
The Liseo and Loperfido Class of SN 54
The Domínguez-Molina-González-Farías-Gupta Class of SN 54
Skew-Elliptical Distribution $SE_k(\boldsymbol{\mu}, \Omega, \boldsymbol{\delta}, g^{(k+1)})$ 55
Skew-Elliptical Distribution $SE_m(\boldsymbol{\mu}, \Sigma, D, g^{(m)})$ 55
3.4 Some Properties of Skew-Elliptical Distributions 56
3.4.1 Moment Generating Functions 56
Skew-Scale Mixture of Normal Distributions 56
LCLC1 58
LCLC2 58
3.4.2 Marginal and Conditional Closure Property 58
3.4.3 Distribution of Quadratic Forms 59
LCLC1 59
LCLC2 59
Quadratic Forms from LCLC1/LCLC2 59
3.4.4 Other Distributional Properties 61
Properties of the Branco-Dey Skew-Elliptical Models 61
Density Shape of Univariate Skew-Elliptical Distributions 62
3.5 Concluding Remarks 62

4 Generalized Skew-Normal Distributions **65**
Nicola M. R. Loperfido

4.1 Introduction 65
4.2 Definition and Characterization 66
4.3 Transformations and Moments 68
4.4 Diagnostics 69
4.5 Parametric Inference 72
4.6 Australian Athletes Data 74
4.7 Concluding Remarks 76
4.8 Appendix: Proofs 77

5 Skew-Symmetric and Generalized Skew-Elliptical Distributions **81**
Marc G. Genton

5.1 Introduction 81
5.2 Skew-Symmetric Distributions 82
5.2.1 Invariance 82
5.2.2 Stochastic Representation and Simulations 85
5.2.3 Skew-Symmetric Representation of Multivariate Distributions 86
5.2.4 Example: Intensive Care Unit Data 86
5.3 Generalized Skew-Elliptical Distributions 88
5.4 Flexible Skew-Symmetric Distributions 90
5.4.1 Flexibility and Multimodality 91
5.4.2 Example: Australian Athletes Data 93
5.4.3 Locally Efficient Semiparametric Estimators 95
5.5 Concluding Remarks 100

6 Elliptical Models Subject to Hidden Truncation or Selective Sampling **101**
Barry C. Arnold and Robert J. Beaver

6.1 Introduction 101
6.2 Univariate Skew-Normal Models 101
6.3 Estimation for the Skew-Normal Distribution 103
6.4 Other Univariate Skewed Distributions 104
6.5 Multivariate Skewed Distributions 105
6.6 General Multivariate Skewed Distributions 107
6.6.1 Hidden Truncation Paradigm 107
6.6.2 Hidden Truncation, More General 107
6.6.3 Additive Component Paradigm 108
6.6.4 Additive Component, More General 108
6.6.5 The Jones Construction 108

6.7 Skew-Elliptical Distributions 109
6.8 Discussion 112

7 **From Symmetric to Asymmetric Distributions: A Unified Approach** **113**
Reinaldo B. Arellano-Valle and Guido E. del Pino

7.1 Introduction 113
7.2 Signs, Absolute Values, and Skewed Distributions 115
7.3 Latent Variables, Selection Models, Skewed Distributions 117
7.4 Symmetry, Invariance, and Skewness 118
7.4.1 Glossary and Basic Facts 118
7.4.2 Three Groups of Transformations 119
7.4.3 Conditional Representations 119
7.4.4 Density or Probability Functions for the Maximal Invariant 120
7.5 The $\mathcal{SI}$ Class of Sign Invariant Distributions 121
7.5.1 Examples and Main Results 121
7.5.2 Application to the Density Formula for a Skewed Distribution 123
7.6 A Stochastic Representation Associated with the $\mathcal{SI}$ Class 124
7.6.1 Moments of a Multivariate Skewed Distribution 125
7.6.2 Distribution of the Square of a Skewed Random Variable 126
7.7 Application to the Multivariate Skew-Normal Distribution 126
7.8 A Canonical Form for Skew-Elliptical Distributions 128

8 **Skewed Link Models for Categorical Response Data** **131**
Ming-Hui Chen

8.1 Introduction 131
8.2 Preliminaries 132
8.3 Importance of Links in Fitting Categorical Response Data 134
8.3.1 Relationship between Regression Coefficients under Different Links 134
8.3.2 Prediction under Different Links 135
8.4 General Skewed Link Models 137
8.4.1 Independent Binary and/or Ordinal Regression Models 137
8.4.2 Correlated Binary and/or Ordinal Regression Models 140
8.4.3 Discrete Choice Models 142
8.5 Bayesian Inference 143
8.6 Bayesian Model Assessment 144

8.6.1 Weighted L Measure 144
8.6.2 Conditional Predictive Ordinate 147
8.6.3 Deviance Information Criterion 148
8.7 Bayesian Model Diagnostics and Outlier Detection 148
8.7.1 Bayesian Latent Residuals 149
8.7.2 Bayesian CPO-Based Residuals 149
8.7.3 Observationwise Weighted L Measure 150
8.8 Concluding Remarks 151

9 Skew-Elliptical Distributions in Bayesian Inference 153
Brunero Liseo

9.1 Introduction 153
9.2 Skewed Prior Distributions for Location Parameters 154
9.2.1 Hierarchical Models with Linear Constraints 154
9.2.2 Efficiency of Linear Bayes Rules with Skewed Priors 156
9.2.3 Heavy Tail Priors 157
9.3 Skew-Elliptical Likelihood 158
9.3.1 Inferential Problems 159
9.3.2 Regression Models with SE Errors 160
9.3.3 Some Applications 162
9.4 Objective Bayesian Analysis of the Skew-Normal Model 163
9.4.1 The Scalar Case 164
9.4.2 Some Multivariate Results 166
9.5 Appendix 169

II Applications and Case Studies 173

10 Bayesian Multivariate Skewed Regression Modeling with an Application to Firm Size 175
José T. A. S. Ferreira and Mark F. J. Steel

10.1 Introduction 175
10.2 Multivariate Skewed Distributions 176
10.2.1 FS Skewed Distributions 177
10.2.2 SDB Skewed Distributions 178
10.3 Regression Models 179
10.3.1 Prior Distributions 179
10.3.2 Numerical Implementation 180
10.4 Application to Firm Size 181
10.4.1 Distribution of Firm Size 183
10.4.2 Analysis of Firm Growth 186
10.5 Discussion 189

11 Capital Asset Pricing for UK Stocks under the Multivariate Skew-Normal Distribution **191**
Chris Adcock

11.1 Introduction 191
11.2 The Multivariate Skew-Normal Model 194
11.3 The Market Model 196
11.4 Estimation Methodology and Data 198
11.5 Empirical Study 199
11.6 Summary and Conclusions 204

12 A Skew-in-Mean *GARCH* Model **205**
Giovanni De Luca and Nicola M. R. Loperfido

12.1 Introduction 205
12.2 Assumptions 207
12.3 News 208
12.4 Returns 210
12.5 Data 211
12.6 Estimation 213
12.7 Conclusions 215
12.8 Appendix 216

13 Skew-Normality in Stochastic Frontier Analysis **223**
J. Armando Domínguez-Molina, Graciela González-Farías, and Rogelio Ramos-Quiroga

13.1 Introduction 223
13.2 Estimation 225
13.2.1 Model Assumptions 226
Model I: Homoscedastic and Uncorrelated Errors 227
Model II: Heteroscedastic and Uncorrelated Errors 227
Model III: Correlated Errors 227
13.2.2 Likelihood 228
13.2.3 Estimation of Inefficiencies/Efficiencies 229
13.3 A Correlated Structure for the Compound Error 230
13.3.1 Simulated Example with Correlated Compound Errors 231
13.4 SFA with Skew-Elliptical Components 234
13.5 Conclusions 235
13.6 Appendix 236
13.6.1 Distributional Properties of Multivariate Compound Errors 236
13.6.2 Expectation of the Truncated Multivariate Normal Distribution 238
13.6.3 Efficiencies for Individual Errors of Model III 239

14 Coastal Flooding and the Multivariate Skew-*t* Distribution **243**
Keith R. Thompson and Yingshuo Shen

14.1 Introduction 243
14.2 A Seasonally Varying Skew-*t* Distribution 244
14.3 Observations of Coastal Sea Level 247
14.4 Fitting the Skew-*t* Distribution 250
14.5 Applications of the Skew-*t* Distribution 253
14.5.1 Quality Control of Sea Level Observations 253
14.5.2 Detecting Secular Changes in the Sea Level Distribution 254
14.5.3 Estimating Flooding Risk 255
14.6 Discussion 258

15 Time Series Analysis with a Skewed Kalman Filter **259**
Philippe Naveau, Marc G. Genton, and Caspar Ammann

15.1 Introduction 259
15.2 The Classical Kalman Filter 260
15.2.1 State-Space Model 260
15.2.2 The Kalman Filter Procedure in the Gaussian Case 261
15.3 A Skewed Kalman Filter 262
15.3.1 The Closed Skew-Normal Distribution 262
15.3.2 Extension of the Linear Gaussian State-Space Model 263
15.3.3 The Steps of our Skewed Kalman Filter 265
15.4 Applications to Paleoclimate Time Series 268
15.4.1 Multi-Process Linear Models 269
15.4.2 The Smoothing Spline Model for Trends 269
15.4.3 The Skewed Component 270
15.4.4 The State-Space Model 273
15.4.5 Simulations and Results 273
15.5 Conclusions 274
15.6 Appendix 275

16 Spatial Prediction of Rainfall Using Skew-Normal Processes **279**
Hyoung-Moon Kim, Eunho Ha, and Bani K. Mallick

16.1 Introduction 279
16.2 Data and Model 280
16.2.1 Automatic Weather Stations and Their Sensors 280
16.2.2 Model Description 280
16.2.3 Bayesian Analysis 282
16.3 Data Analyses 284
16.4 Discussion 288

17 Shape Representation with Flexible Skew-Symmetric Distributions **291**
Sajjad H. Baloch, Hamid Krim, and Marc G. Genton

17.1 Introduction 291
17.2 Problem Statement 292
17.3 Shape Analysis 294
17.3.1 Posterior Learning 296
17.3.2 Selection of a Distribution for the Angle 299
17.3.3 Overall Shape Distribution 300
17.3.4 Performance Assessment of the Learning Process 301
17.4 Experimental Results 302
17.4.1 Case Study 1 303
17.4.2 Case Study 2 304
17.4.3 Case Study 3 304
17.4.4 Case Study 4 305
17.4.5 Case Study 5 305
17.4.6 Sampling from Models 306
17.5 Conclusions 307
17.6 Appendix 308

18 An Astronomical Distance Determination Method Using Regression with Skew-Normal Errors **309**
Laurent Eyer and Marc G. Genton

18.1 Introduction 309
18.2 The Trigonometric Parallax 310
18.2.1 Astrometric Satellites 311
18.2.2 Some Statistical Aspects 311
Parallax Is a Positive Quantity 311
Parent Distribution 312
Formation of Samples 312
18.3 Famous Standard Candles: the Cepheids 312
18.4 Calibration of the Period-Luminosity Relation 314
18.4.1 Regression with Skew-Normal Errors 315
18.4.2 Discussion 316
Fixed Slope 316
Free Slope 317

19 On a Bayesian Multivariate Survival Model with a Skewed Frailty **321**
Sujit K. Sahu and Dipak K. Dey

19.1 Introduction 321
19.2 Frailty Models 323

19.2.1 Frailty Distributions 323
19.2.2 Comparison of Frailty Distributions 325
19.3 Baseline Hazard Function 326
19.4 Model Specification 327
19.4.1 Likelihood Specification 327
19.4.2 Prior Specification 328
19.4.3 Hyper-Parameter Values and Prior Sensitivity 329
19.5 Reversible Jump Steps 330
19.6 Examples 332
19.6.1 Kidney Infection Data Example 332
19.6.2 Litters Example 334
19.7 Conclusion 336

20 Linear Mixed Effects Models with Flexible Generalized Skew-Elliptical Random Effects 339
Yanyuan Ma, Marc G. Genton, and Marie Davidian

20.1 Introduction 339
20.2 FGSE Distributions and the Linear Mixed Effects Model 340
20.3 Implementation and Inference 343
20.3.1 Maximum Likelihood via the EM Algorithm 343
20.3.2 Bayesian Inference via Markov Chain Monte Carlo Simulation 346
20.4 Simulation Results 350
20.5 Application to Cholesterol Data 356
20.6 Discussion 358

References 359

Index 377

PART I

Theory and Inference

CHAPTER 1

The Skew-Normal Distribution

Alessandra Dalla Valle

1.1 Introduction

With the term skew-normal (SN) we refer to a parametric class of probability distributions that extends the normal distribution by an additional shape parameter that regulates the skewness, allowing for a continuous variation from normality to non-normality.

On the applied side, the skew-normal distribution as a generalization of the normal law is a natural choice in all practical situations in which there is some skewness present: a particularly valuable property is the continuity of the passage from the normal case to skewed distributions.

From the theoretical viewpoint, the SN class has the advantage of being mathematically tractable and of having a good number of properties in common with the standard normal distribution: for example, SN densities are unimodal, their support is the real line, and the square of an SN random variable is a χ^2 random variable with one degree of freedom.

The first systematic treatment of the SN class in the scalar case was given by Azzalini (1985, 1986); subsequently, Azzalini and Dalla Valle (1996) introduced a multivariate version of the skew-normal density, while Azzalini and Capitanio (1999) examined further probabilistic properties of the distribution, and also investigated the most relevant statistical aspects.

The development of the above-cited papers has encouraged a number of other developments and applications. Arnold, Beaver, Groeneveld, and Meeker (1993) used an SN distribution for an application to psychometric real data; Copas and Li (1997) showed the connections with the problem of the selection of a sample; Chen, Dey, and Shao (1999) started from a scalar SN distribution and defined a skewed link function in the generalized linear model for a binomial random variable. Among the computed generalizations of the SN class we cite the paper of Arnold and Beaver (2000b) in which there is also a generalization for some other distributions, such as, for example, multivariate skew-Cauchy. Recently, Azzalini and Capitanio (2003) proposed a generalization to skew-t distributions obtained through a perturbation of symmetry; Genton and Loperfido (2002) defined generalized skew-elliptical distributions, whereas Wang, Boyer, and Genton (2004a) studied skew-symmetric distributions.

From a Bayesian viewpoint, the papers of Liseo (1990) and Mukhopadhyay and Vidakovic (1995) followed that of O'Hagan and Leonard (1976), which highlighted that the SN distribution is a natural starting point when a one-sided constraint about the range of a parameter is inserted.

In econometrics, the pioneering paper of Aigner, Lovell, and Schmidt (1977) was fundamental in showing the strong connections between the SN class and stochastic frontier models for the analysis of the efficiency of firms. Some other applications of the SN class include the papers of Louis, Blenman, and Thatcher (1999), Caudill (1993), and Polachek and Robst (1998), about the analysis of financial market, property market, and labor market, respectively. Ball and Mankiw (1995) obtained the SN distribution as a natural choice for the distribution of relative-price changes that influence the rate of inflation. Adcock and Shutes (2001) presented a theoretical exposition of a model based upon the multivariate skew-normal distribution for portfolio selection.

This chapter introduces the skew-normal distribution in the univariate and multivariate case, studying in depth their most interesting and relevant features. In Section 1.2 we introduce the definition and the properties of the univariate skew-normal distribution, giving also the moment generating function and the cumulative distribution function. In Section 1.3 there are some considerations about the problems connected to the methods of estimation of the parameters and the solutions given in literature. Some statistical tests to check the hypothesis of skew-normality are presented in Section 1.4. In Section 1.5 we give the definition of the multivariate skew-normal distribution, its genesis and immediate properties. A number of properties that can be extended from the scalar case to the multivariate case are discussed in Section 1.6. Section 1.7 is devoted to quadratic forms, while in Section 1.8 the conditional skew-normal distribution is considered. Section 1.9 is concerned with results about the SN class in some particular contexts: the problem of reliability and the problem of finding regions of assigned probability and minimum volume. The chapter concludes with Section 1.10 in which some hints for future research are given.

1.2 The Univariate Skew-Normal Distribution

This section is devoted to the characterization of the univariate skew-normal distribution.

Definition 1.2.1 We say that Z is a skew-normal random variable with parameter $\alpha \in \mathbb{R}$, if it has the following probability density function (pdf)

$$g(z;\alpha) = 2\phi(z)\Phi(\alpha z), \qquad -\infty < z < \infty, \tag{1.1}$$

where $\phi(\cdot)$ and $\Phi(\cdot)$ are the standard normal pdf and the standard normal cumulative distribution function (cdf), respectively; then, for brevity, we shall say that $Z \sim SN(\alpha)$.

In practice, to fit real data, we work with a location and scale transformation $Y = \xi + \sigma Z$, with $\xi \in \mathbb{R}$ and $\sigma > 0$. The density of a random variable Y, written $Y \sim SN(\xi, \sigma, \alpha)$, is easily written as

$$g(y; \xi, \sigma, \alpha) = \frac{2}{\sigma}\phi\left(\frac{y-\xi}{\sigma}\right)\Phi\left(\alpha\frac{y-\xi}{\sigma}\right). \tag{1.2}$$

The inclusion of the location and scale parameters will be considered in great detail in Section 1.3. At present, we focus on the presentation of the characteristics and properties of the standard skew-normal, referring to density (1.1).

Figure 1.1 shows the shape of the pdf (1.1) for two values of α. As we see, if we increase the value of the skewness parameter we increase the skewness of the density.

The following properties are very immediate from the definition.

Property 1.2.1 The pdf of the $SN(0)$ is identical to the pdf of the $N(0, 1)$.

Property 1.2.2 As $\alpha \to \infty$, $g(z; \alpha)$ tends to $2\phi(z)I_{z>0}$, which is the half-normal pdf.

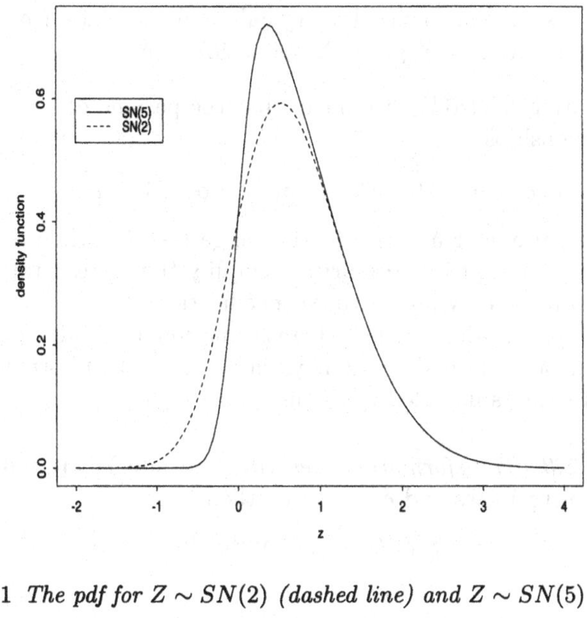

Figure 1.1 *The pdf for $Z \sim SN(2)$ (dashed line) and $Z \sim SN(5)$ (solid line).*

Property 1.2.3 If $Z \sim SN(\alpha)$, then $-Z \sim SN(-\alpha)$.

Property 1.2.4 The pdf of an $SN(\alpha)$ random variable is unimodal, because $\log(g(z;\alpha))$ is a concave function of z.

In the following we give some relevant results that help the reader in understanding the nature of the SN distribution and its relation to some other distributions.

Proposition 1.2.1 If Y and W are independent $N(0,1)$ random variables, and Z is set equal to Y conditionally on $\alpha Y > W$, then $Z \sim SN(\alpha)$.

Proposition 1.2.1 gives a quite efficient method to generate random numbers from a random variable with pdf given in Equation (1.1). It is in fact sufficient to generate Y and W iid $N(0,1)$, and put

$$Z = \begin{cases} Y & \text{if } \alpha Y > W, \\ -Y & \text{if } \alpha Y \leq W. \end{cases}$$

The relevance of the following two Propositions comes from the central role that they will have in Section 1.5 in obtaining the multivariate skew-normal distribution. The first one, also called "conditioning method," is a stochastic representation of the skew-normal random variable, or, equally, a method to obtain this variable, via an operation of truncation.

Proposition 1.2.2 (Conditioning method) If (X,Y) is a bivariate normal random vector with standardized marginals and correlation δ, then the conditional distribution of Y given $X > 0$ is $SN(\alpha(\delta))$.

Here, by writing $SN(\alpha(\delta))$, we mean that the parameter δ is related to α via the relationships

$$\alpha(\delta) = \delta/\sqrt{1-\delta^2}, \qquad \delta(\alpha) = \alpha/\sqrt{1+\alpha^2}. \tag{1.3}$$

Note that the parameter δ varies in the range $(-1,1)$, while α spans $\mathbb{R}$. From now on, because it is necessary to simplify the representation of the formulas, we will use only one of these specifications.

The second proposition, called "transformation method," performs a transformation on a normal random variable and a half-normal random variable, with weights included in the interval $(-1,1)$.

Proposition 1.2.3 (Transformation method) If Y_0 and Y_1 are independent $N(0,1)$ random variables and $\delta \in (-1,1)$, then

$$Z = \delta|Y_0| + (1-\delta^2)^{1/2}\, Y_1, \tag{1.4}$$

is $SN(\alpha(\delta))$.

Proposition 1.2.3 is well-known in the econometric literature on the stochastic frontier model, which starts from the work of Aigner *et al.* (1977). Later, Andel, Netuka, and Zvara (1984) obtain Equation (1.1) as the stationary distribution of a non-linear autoregressive process, with transition law given by Equation (1.4).

A probabilistic representation of the "transformation method" in terms of normal and truncated normal distributions is given by Henze (1986). This kind of representation helps us to understand how the SN class is structured and, in particular, the nature of the departure from normality.

It is also possible to present a third type of representation of the scalar skew-normal random variable, exhibited in the next proposition.

Proposition 1.2.4 If (X_0, X_1) is a bivariate normal random vector with standardized marginals and correlation ρ, then

$$\max(X_0, X_1) \sim SN(\alpha),$$

where $\alpha = \left(\frac{1-\rho}{1+\rho}\right)^{1/2}$.

This result has been given by Roberts (1966), in an early citation of the scalar SN distribution, and was later rediscovered by Loperfido (2002); the same conclusion follows as a special case of a result of H. N. Nagaraja, quoted by David (1981, Exercise 5.6.4).

The following proposition and the subsequent property highlight the strong connections between the skew-normal random variable and the normal random variable.

Proposition 1.2.5 If $Y \sim N(0,1)$ and $Z \sim SN(\alpha)$ then $|Z|$ and $|Y|$ have the same pdf.

As an immediate consequence of Proposition 1.2.5 we have the following relevant result.

Property 1.2.5 If $Z \sim SN(\alpha)$, then $Z^2 \sim \chi^2_1$.

1.2.1 Moments

As a consequence of the Property 1.2.5 we have that the even moments of the normal and of the skew-normal random variable are the same. To obtain the odd moments, we use the moment generating function (mgf) of Z, that is

$$M(t) = 2\exp(t^2)\Phi(\delta t), \tag{1.5}$$

where δ is related to α through the relationship (1.3).

From (1.5), we obtain the following:

$$\mathrm{E}(Z) = \sqrt{\frac{2}{\pi}}\delta, \tag{1.6}$$

$$\mathrm{Var}(Z) = 1 - \frac{2}{\pi}\delta^2, \tag{1.7}$$

$$\mathrm{S}(Z) = \frac{1}{2}(4-\pi)\mathrm{sign}(\alpha)\left(\frac{\alpha^2}{\frac{\pi}{2}+\left(\frac{\pi}{2}-1\right)\alpha^2}\right)^{3/2}, \tag{1.8}$$

$$\mathrm{K}(Z) = 2(\pi-3)\left(\frac{\alpha^2}{\frac{\pi}{2}+\left(\frac{\pi}{2}-1\right)\alpha^2}\right)^{2}, \tag{1.9}$$

where $\mathrm{S}(Z)$ and $\mathrm{K}(Z)$ are the measures of skewness and kurtosis, respectively. It can be shown that $\mathrm{S}(Z)$ varies in the interval $(-0.9953, 0.9953)$, while $\mathrm{K}(Z)$ takes values in the range $(-0.869, 0.869)$, see Azzalini (1985).

Henze (1986), as a consequence of Proposition 1.2.3, gives the following expression for the odds moments in closed form

$$\mathrm{E}(Z^{2k+1}) = \sqrt{\frac{2}{\pi}}\alpha(1+\alpha^2)^{-(k+1/2)}2^{-k}(2k+1)!\sum_{t=0}^{k}\frac{t!(2\alpha)^{2t}}{(2t+1)!(k-t)!}.$$

1.2.2 Cumulative Distribution Function

The cdf $G(z;\alpha)$ of a skew-normal random variable with pdf (1.1), given in Azzalini (1985), is

$$G(z;\alpha) = 2\int_{-\infty}^{z}\int_{-\infty}^{\alpha s}\phi(s)\phi(t)dtds. \tag{1.10}$$

The calculation of (1.10) can be obtained through the function $T(h;a)$ studied by Owen (1956), which gives the integral of the standard normal bivariate density over the region bounded by the lines $x = h$, $y = 0$, and $y = ax$ in the (x, y) plane. Then, we obtain

$$G(z;\alpha) = \Phi(z) - 2T(z;\alpha), \tag{1.11}$$

where

$$T(z;\alpha) = \int_{z}^{\infty}\int_{0}^{\alpha s}\phi(s)\phi(t)dtds.$$

It is easy to see that Equation (1.11) is also the cdf of a skew-normal random variable with negative values of z and α, if we recall that $T(h;\alpha)$ is a decreasing function of h and that it has the following properties

$$T(-h;\alpha) = T(h;\alpha), \quad -T(h;\alpha) = T(h;-\alpha), \quad 2T(h;1) = \Phi(h)\Phi(-h).$$

Owen (1956) gives an expression for $T(h;\alpha)$ in terms of an infinite series.

Property 1.2.3 and the properties of the function $T(h;a)$, lead to the following additional results. The first result shows that the SN cdf for negative α (and positive z) can be derived from the SN cdf for positive α (and negative z).

Property 1.2.6 $1 - G(-z;\alpha) = G(z;-\alpha)$.

The second result states that, for the particular case $\alpha = 1$, the SN cdf reduces to the square of the corresponding cdf of the normal class.

Property 1.2.7 $G(z;1) = \Phi(z)^2$.

Therefore, for every α, the distance between the normal cdf calculated for a positive value of z and the SN cdf for the same negative z has an exact representation, when we choose the supremum of this distance for varying z.

Property 1.2.8 $\sup_z |\Phi(z) - G(-z;\alpha)| = \pi^{-1} \arctan |\alpha|$.

1.3 Estimation and Inference

So far the treatment has been concerned with the most relevant theoretical features of the SN class of distributions. In particular, the properties of this class resemble those of the normal distribution and appear to justify its name, skew-normal. The purpose of this section is to consider, in an applied context, some of the issues connected with the analysis of data that have a unimodal empirical distribution with moderate skewness.

To estimate location, scale, and shape parameters of a random sample of values $y_1, \ldots, y_n$ from a $Y = \xi + \sigma Z$ random variable, with pdf (1.2), we can use the method of moments, or, more efficiently, the maximum likelihood method. Both of these methods present, however, several problems that are directly connected with the estimation of the shape parameter.

The method of moments cannot be applied when the sample skewness index goes out of the admissibility range, that is $(-0.9953, 0.9953)$, and this constitutes a heavy limitation in several practical situations.

The maximum likelihood method, especially when n is small, gives rise to estimates of the shape parameter reaching the frontier. This fact has a rigorous explanation if we suppose that all the observations in a sample coming from a random variable $Y \sim SN(0,1,\alpha)$, with $\alpha > 0$, are positive.

In fact, in this case, the log-likelihood (Liseo, 1990)

$$\log L(0,1,\alpha) = -\frac{1}{2}\sum_{i=1}^{n} y_i^2 + \sum_{i=1}^{n} \log \Phi(\alpha y_i)$$

is an increasing function of α when all observations are positive. Therefore, the maximum likelihood method produces $\hat{\alpha} = \infty$, estimating the data as

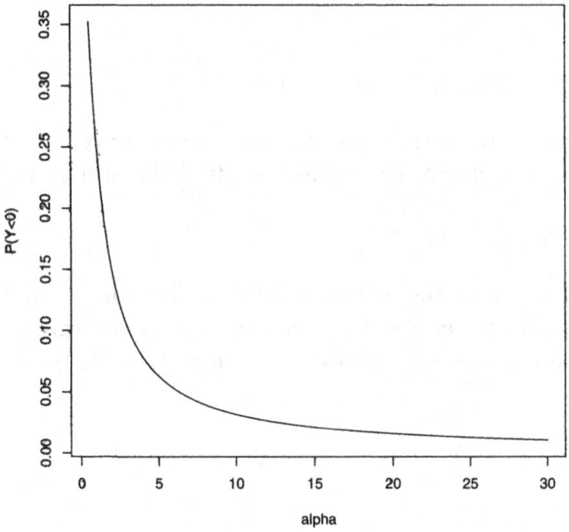

Figure 1.2 *Probability that $Y < 0$ for values of α ranging from 0 to 30.*

coming from a half-normal distribution. Figure 1.2 shows the probability that Y is less than 0 for values of α ranging from 0 to 30. As we see, when the shape parameter becomes greater, the probability that the signs of the data values in a sample are all positive is very high. Unfortunately, this problem arises also when the observations are not all positive.

Figure 1.3 gives the profile log-likelihood for a sample of size 50, called "frontier data" generated by an $SN(0,1,5)$ reported also in Azzalini and Capitanio (1999). In this case, the observations are not all positive, but the MLE estimation of the shape parameter produces $\hat{\alpha} = \infty$ and it is apparent from an inspection of the figure that the profile log-likelihood initially increases rapidly with α, then gradually becomes flatter. So, the likelihood is not able to estimate a single value of the shape parameter among all the possible ones. Let us try now to understand how the maximum likelihood method works. The maximization of the likelihood function

$$L(\xi,\sigma,\alpha) = 2^n\sigma^{-n}\prod_{i=1}^{n}\phi\left(\frac{y_i-\xi}{\sigma}\right)\Phi\left(\alpha\frac{y_i-\xi}{\sigma}\right),$$

with respect to (ξ,σ,α) leads to the following likelihood equations

$$\begin{cases} \sum_{i=1}^n z_i - \alpha\sum_{i=1}^n \beta_i = 0, \\ \sum_{i=1}^n z_i^2 - \alpha\sum_{i=1}^n \beta_i z_i - n = 0, \\ \sum_{i=1}^n \beta_i z_i = 0, \end{cases} \tag{1.12}$$

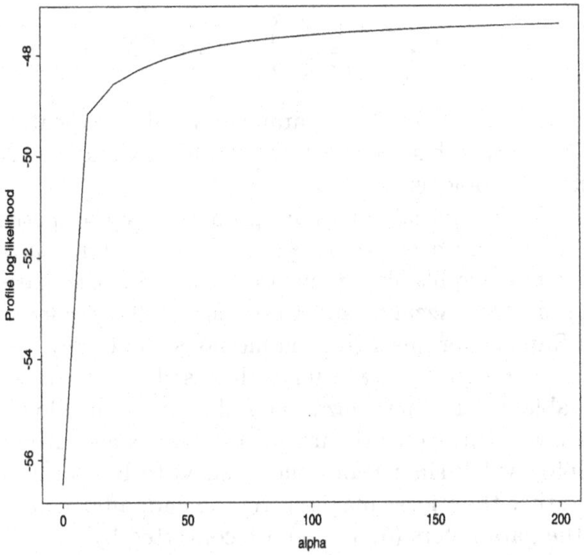

Figure 1.3 *Profile log-likelihood for the "frontier data."*

where $\beta_i = \frac{\phi(\alpha z_i)}{\Phi(\alpha z_i)}$ and $z_i = \frac{y_i - \xi}{\sigma}$. The system of equations (1.12) can be simplified in the following way (Chiogna, 1997)

$$\begin{cases} \hat{\xi} = \bar{y} - \hat{\alpha}\hat{\sigma}\bar{\beta}, \\ \hat{\sigma}^2 = s^2(1 - \hat{\alpha}^2\bar{\beta}^2)^{-1}, \\ \sum_{i=1}^n \beta_i z_i = 0, \end{cases}$$

where $\bar{y}$ and s^2 are the sample mean and the sample variance, respectively, and $\bar{\beta} = 1/n \sum_{i=1}^n \beta_i$. Since it is not possible to find a closed form expression for the maximum likelihood estimators as a function of the sample observations, a numerical maximization procedure has to be used.

The parameterization (ξ, σ, α) of the SN class is, however, inadequate for making inference because the Fisher information matrix I_α is given by $\sigma^2 I_\alpha$ equal to

$$\begin{pmatrix} (1+\alpha^2 c_0) & \left(\mathrm{E}(Z)\frac{1+2\alpha^2}{1+\alpha^2} + \alpha^2 c_1\right) & \left(\frac{(2/\pi)^{1/2}}{(1+\alpha^2)^{3/2}} - \alpha c_1\right)\sigma \\ \left(\mathrm{E}(Z)\frac{1+2\alpha^2}{1+\sigma^2} + \alpha^2 c_1\right) & (2+\alpha^2 c_2) & -\alpha c_2 \sigma \\ \left(\frac{(2/\pi)^{1/2}}{(1+\alpha^2)^{3/2}} - \alpha c_1\right)\sigma & -\alpha c_2 \sigma & c_2 \sigma^2 \end{pmatrix}$$

where $c_h = \mathrm{E}\left(Z^h \left(\phi(\alpha Z)/\Phi(\alpha Z)\right)^2\right)$, $h = 0, 1, 2, \ldots$, which is singular as $\alpha \to 0$. As a possible solution, Azzalini (1985) suggests considering the following reparameterization

$$Y = \theta_1 + \theta_2 \left(\frac{Z - \mathrm{E}(Z)}{\sqrt{\mathrm{Var}(Z)}} \right), \qquad Z \sim SN(\alpha), \tag{1.13}$$

where $(\theta_1, \theta_2, \gamma_1)$ is the new vector of parameters and γ_1 indicates the index of skewness. This approach solves both the singularity and the irregularity likelihood behavior problems.

To obtain the MLEs, initial values are produced by the method of moments. Then, a procedure of non-linear constrained iterative minimization of the negative log-likelihood function based on the parameterization $(\theta_1, \theta_2, \gamma_1)$ is used. Azzalini and Capitanio (1999) recommend using either Newton-Raphson or quasi-Newton methods and in few cases of non-convergence, the EM algorithm, which has the disadvantage of being rather slow, albeit reliable. For the maximization of the constrained log-likelihood, Pewsey (2000) instead proposes the use of the simplex algorithm of Nelder and Mead (1965), which simulation studies show to be suitable for identifying global rather than local maxima. At the end, after converging, the estimators of the parameters $(\theta_1, \theta_2, \gamma_1)$ are converted by

$$\begin{aligned}
\xi &= \theta_1 - \theta_2 \left(\frac{2\gamma_1}{4-\pi} \right)^{1/3}, \qquad (1.14)\\
\sigma &= \theta_2 \left[1 + \left(\frac{2\gamma_1}{4-\pi} \right)^{2/3} \right]^{1/2}, \\
\alpha &= \left(\frac{2\gamma_1}{4-\pi} \right)^{1/3} \left[\frac{2}{\pi} + \left(\frac{2\gamma_1}{4-\pi} \right)^{2/3} \left(\frac{2-\pi}{\pi} \right) \right],
\end{aligned}$$

into the estimators of (ξ, σ, α).

A very notable aspect in the estimation of parameters is that the achievement of the frontier affects only the shape parameter and not the location and scale parameters. In this respect, many works have been produced to date to solve the problem of the divergence of α.

A provisional solution proposed by Azzalini and Capitanio (1999) and followed also by Dalla Valle (2001) was to restart the maximization method and then stop it when the likelihood reaches a value that is not significantly lower than the maximum. This method is not completely satisfactory because it does not indicate how much below the maximum it is convenient to stay.

Sartori (2003) obtains a better result by using the method for bias prevention of the maximum likelihood estimator proposed by Firth (1993). This method consists of a modification of the score function such that the resulting estimator has lower bias than the maximum likelihood estimator. The appealing feature concerns the finiteness of this modified estimator.

Although this property is still unproved in the general case, Sartori (2003) showed that, in the SN case, the resulting estimator always exists and is finite. The evidence of simulation results is that this modified estimator underestimates the true value of the shape parameter when we have small samples. If we also have to estimate the location and scale parameters, we can use the modified estimator together with the maximum likelihood method.

1.4 Checking the Hypothesis of Skew-Normality

This section is devoted to statistical tests that have been developed to assess the hypothesis of skew-normality.

Salvan (1986) showed that if we test, in the SN case, the hypothesis of normality ($\alpha = 0$) for alternatives $\alpha > 0$, the most powerful invariant test against the alternative depends on the true value of α, therefore it is necessary to consider invariant tests that are only locally most powerful. So, by considering the ratio between invariant marginal likelihoods, it is demonstrated that the sample index of skewness is, for the SN class, the locally most powerful location and scale invariant statistic for testing normality.

This result is well considered by Pewsey (2000) in his attempt to find statistical tests to assess departures from an underlying normal or folded normal distribution. His idea is that, for data displaying a high degree of symmetry or skewness, it is more convenient (at least in terms of the number of parameters involved) to investigate the appropriateness of the two limiting cases of normality and folded normality, respectively. He suggests using the sample index of skewness, which is asymptotically normal with mean 0.9953 and variance $8.03572/n$ under the hypothesis of data coming from a folded normal (half-normal) distribution.

It is, however, necessary to be very prudent in its use because even for very large samples, its sampling distribution is positively skewed. In this case, it is possible to use some Monte Carlo-based approaches for significance testing: simulating data from the null model and then using, for example, a test based on the ranks of the sample index of skewness for the original and the simulated data.

Another kind of test is one that attempts to check if the skew-normal distribution is an appropriate model for a given set of data. Gupta and Chen (2001) present a table of the cdf given in Equation (1.10) for z ranging from 0.00 to 4.00 with increases of 0.01 in length, for different values of the shape parameter α. The table shows that the cdf (1.10) varies appreciably with α only in the neighborhood of zero, while, when α increases, it is almost constant.

This table makes the implementation of statistical tests easier: in particular, the testing methods discussed by these authors are the Kolmogorov-Smirnov test and the Pearson's χ^2 test. The test procedures are quite different in the two cases. Suppose that we have a random sample from a population with unknown cdf. For the Pearson's test, we need to partition the sample space into l disjoint intervals and then find the probability that an outcome falls inside a generic interval, under the assumption that the population from which data are coming has an SN distribution. Finally, the χ^2 test compares the real number of outcomes occurring in the intervals, with the expected number of outcomes falling in the intervals if we repeat the experiment n times. If we find a value of the χ^2 test that is too large when compared with that of a χ^2_{l-1}, for a given significance level, we reject the assumption of skew-normality. It is clear that we must have a large sample.

Alternatively, we can use the Kolmogorov-Smirnov test D which, in this case, is the maximum distance between the empirical cdf and the SN cdf at the sample points. If the value of D is too large when compared with that of D at a certain significance level, say β, we reject the hypothesis of skew-normality at the β level.

A different kind of test to check the hypothesis of skew-normality is given by Dalla Valle (2001). Given a random sample from a population with unknown cdf, we calculate the Anderson-Darling test under the hypothesis that the sample comes from the skew-normal distribution and then we transform this value according to a particular transformation calculated via a simulation procedure. This transformed value is then compared with one extracted from a table given in Dalla Valle (2001) that contains significance points of the transformed test at 0.05 significance level, for different sample sizes. If the value of the transformed test is larger than the corresponding significance point, we reject the hypothesis of skew-normality.

1.5 The Multivariate Skew-Normal Distribution

In this section we deal with the multivariate extension of the univariate skew-normal distribution introduced previously. In an applied context, this multivariate family appears to be very important, since in the multivariate case there are not many distributions available for dealing with non-normal data, primarily when the skewness of the marginals is quite moderate.

Azzalini (1985) extends the pdf given in Equation (1.1) into a multivariate density, but it is not satisfactory since the marginals are not skew-normals. On the contrary, the extension developed in Azzalini and Dalla Valle (1996) starts from the properties of the scalar skew-normal distribution to obtain the multivariate class with the marginals that are scalar skew-normal variates. The two different approaches based on two properties of the SN class lead to the same generalization, i.e., to the same class of distributions.

At first, we do not introduce location and scale parameters into the definitions, leaving them until later in order to simplify the notation.

Definition 1.5.1 A random vector $\mathbf{z} = (Z_1, \ldots, Z_p)^T$ is p-dimensional skew-normal, denoted by $\mathbf{z} \sim SN_p(\Omega, \boldsymbol{\alpha})$, if it is continuous with pdf

$$g(\mathbf{z}) = 2\phi_p(\mathbf{z}; \Omega)\Phi(\boldsymbol{\alpha}^T \mathbf{z}), \quad \mathbf{z} \in \mathbb{R}^p, \tag{1.15}$$

where $\phi_p(\mathbf{z}; \Omega)$ denotes the pdf of the p-dimensional multivariate normal distribution with standardized marginals and correlation matrix Ω.

In the same way, as in the univariate case, we can refer to $\boldsymbol{\alpha}$ as the vector of shape parameters, even if it is clear from Equation (1.19) below that its derivation is more complex. If $p = 2$, the pdf given in Equation (1.15) becomes

$$g(z_1, z_2) = 2\,\phi_2(z_1, z_2; \omega)\Phi\left(\alpha_1 z_1 + \alpha_2 z_2\right), \tag{1.16}$$

where ω is the off-diagonal element of Ω.

1.5.1 *Genesis*

To construct the Definition 1.5.1, the starting point is to consider a p-dimensional normal random vector $\mathbf{y} = (Y_1, \ldots, Y_p)^T$ with standardized marginals and correlation matrix Ψ, independent of $Y_0 \sim N(0, 1)$ such that

$$\begin{pmatrix} Y_0 \\ \mathbf{y} \end{pmatrix} \sim N_{p+1}\left(\mathbf{0}, \begin{pmatrix} 1 & \mathbf{0}^T \\ \mathbf{0} & \Psi \end{pmatrix}\right). \tag{1.17}$$

If $\delta_1, \ldots, \delta_p$ are each in the interval $(-1, 1)$, we obtain

$$Z_j = \delta_j |Y_0| + (1 - \delta_j^2)^{1/2}\, Y_j, \qquad j = 1, \ldots, p, \tag{1.18}$$

where, by Proposition 1.2.3, $Z_j \sim SN(\alpha(\delta_j))$. The pdf of $\mathbf{z} = (Z_1, \ldots, Z_p)^T$ is, after some algebra, equal to Equation (1.15), where

$$\boldsymbol{\alpha}^T = \frac{\boldsymbol{\lambda}^T \Psi^{-1} \Delta^{-1}}{(1 + \boldsymbol{\lambda}^T \Psi^{-1} \boldsymbol{\lambda})^{1/2}}, \tag{1.19}$$

$$\Delta = \operatorname{diag}((1 - \delta_1^2)^{1/2}, \ldots, (1 - \delta_p^2)^{1/2}),$$
$$\Omega = \Delta(\Psi + \boldsymbol{\lambda}\boldsymbol{\lambda}^T)\Delta, \tag{1.20}$$

Ψ a $p \times p$ correlation matrix, and $\boldsymbol{\lambda} = (\alpha(\delta_1), \ldots, \alpha(\delta_p))^T$. If $p = 2$, then α_1 and α_2 in Equation (1.16) are defined in the following way:

$$\begin{aligned} \alpha_1 &= \frac{\delta_1 - \delta_2 \omega}{\{(1 - \omega^2)(1 - \omega^2 - \delta_1^2 - \delta_2^2 + 2\delta_1\delta_2\omega)\}^{1/2}}, \\ \alpha_2 &= \frac{\delta_2 - \delta_1 \omega}{\{(1 - \omega^2)(1 - \omega^2 - \delta_1^2 - \delta_2^2 + 2\delta_1\delta_2\omega)\}^{1/2}}. \end{aligned}$$

It is, however, possible to show that, if we consider another property of the SN class, we finally arrive at the definition of a random vector having (1.15) as pdf. In fact, if $\mathbf{x} = (X_0, X_1, \ldots, X_p)^T$ is a $(p + 1)$-dimensional multivariate normal random vector with standardized marginals, such that $\mathbf{x} \sim N_{p+1}(\mathbf{0}, \Omega^*)$, and the correlation matrix is

$$\Omega^* = \begin{pmatrix} 1 & \delta_1 & \dots & \delta_p \\ \delta_1 & & & \\ \vdots & & \Omega & \\ \delta_p & & & \end{pmatrix}, \tag{1.21}$$

we can consider the distribution of $(X_1, \dots, X_p)$ given $X_0 > 0$. Each of these conditional distributions is, by Proposition 1.2.2, $SN(\alpha(\delta_j))$, for $j = 1, \dots, p$, so, we can say that $(X_1, \dots, X_p)$ conditionally on $X_0 > 0$ is a multivariate skew-normal random vector and, furthermore, show that its pdf is equivalent to Equation (1.15) with another parameterization.

So, we assert that the pairs $(\boldsymbol{\alpha}, \Psi)$ and $(\boldsymbol{\alpha}, \Omega)$ are equivalent tools for the identification of a member of the multivariate class of distributions introduced with Definition 1.5.1. Comparing the two generating methods, we can in fact realize that the transformation operated in Equation (1.18) is equivalent to the conditioning of Proposition 1.2.2 because the $N(0,1)$ density is symmetric. For an algebraic proof of this result, see the appendix of Azzalini and Capitanio (1999).

Therefore, we come to the same class of densities, provided that the generic element ω_{ij} of Ω and the corresponding element ψ_{ij} of Ψ are related as follows

$$\omega_{ij} = \delta_i \delta_j + \psi_{ij}\{(1-\delta_i^2)(1-\delta_j^2)\}^{1/2}, \qquad i > 0, j > 0. \tag{1.22}$$

Note that the positive definiteness of the correlation matrix Ω^* in Equation (1.21) requires that some conditions on its elements are satisfied. In particular, fixing the δ_j's, the restrictions are moved to the elements of the sub-matrix Ω in Equation (1.21).

It is clear that, in the multivariate case, we also need to introduce location and scale parameters. So, if $\boldsymbol{\xi} = (\xi_1, \dots, \xi_p)^T$ and $S = \text{diag}(\sigma_1, \dots, \sigma_p)$ are location and scale parameters, the density function of $\mathbf{y} = \boldsymbol{\xi} + S\mathbf{z}$, where $\mathbf{z} \sim SN_p(\Omega, \boldsymbol{\alpha})$ is

$$g(\mathbf{y}) = 2\phi_p(\mathbf{y}; \boldsymbol{\xi}, S\Omega S)\Phi(\boldsymbol{\alpha}^T S^{-1}(\mathbf{y} - \boldsymbol{\xi})),$$

and we write $\mathbf{y} \sim SN_p(\boldsymbol{\xi}, S\Omega S, \boldsymbol{\alpha})$.

1.5.2 The Cumulative Distribution Function

The following result also highlights that in the multivariate case there is a strong relation between the skew-normal and the normal distribution. In fact, the cdf of $\mathbf{z} \sim SN_p(\Omega, \boldsymbol{\alpha})$ is

$$\begin{aligned} G(\mathbf{z}) &= \mathrm{P}(Z_1 \le z_1, \dots, Z_p \le z_p) \\ &= 2\int_{-\infty}^{z_1} \cdots \int_{-\infty}^{z_p} \phi(\mathbf{z};\Omega)\Phi(\boldsymbol{\alpha}^T\mathbf{z})\, dz_1 \cdots dz_p \\ &= 2\int_{-\infty}^{\boldsymbol{\alpha}^T\mathbf{z}} \int_{-\infty}^{z_1} \cdots \int_{-\infty}^{z_p} \phi_p(\mathbf{z};\Omega)\phi(u)\, du\, dz_1 \cdots dz_p, \end{aligned}$$

for $\mathbf{z} \in \mathbb{R}^p$, with $\boldsymbol{\alpha}$ defined by Equation (1.19). This last expression can also be written in the following way

$$G(\mathbf{z}) = 2\,\mathrm{P}(\widetilde{Y}_0 \le 0, Y_1 \le z_1, \dots, Y_p \le z_p), \tag{1.23}$$

i.e., as a pdf of the $(p+1)$-dimensional normal vector $(\widetilde{Y}_0, Y_1, \dots, Y_p)^T$, where $(Y_0, Y_1, \dots, Y_p)^T$ has distribution (1.17), and $\widetilde{Y}_0 = Y_0 - \boldsymbol{\alpha}^T\mathbf{y}$. Equation (1.23) demonstrates the equivalence between the cdf of the multivariate skew-normal random vector and the cdf of a random vector distributed as

$$\begin{pmatrix} \widetilde{Y}_0 \\ \mathbf{y} \end{pmatrix} \sim N_{p+1}\left(\mathbf{0}, \begin{pmatrix} 1+\boldsymbol{\alpha}^T\Psi\boldsymbol{\alpha} & -\boldsymbol{\alpha}^T\Psi \\ -\Psi\boldsymbol{\alpha} & \Psi \end{pmatrix}\right). \tag{1.24}$$

1.5.3 The Moment Generating Function

The calculation of the mgf of $\mathbf{z}$ with pdf given in Equation (1.15) leads to

$$\begin{aligned} M(\boldsymbol{t}) &= 2\int_{\mathbb{R}^p} \exp(\boldsymbol{t}^T\mathbf{z})\, \phi_p(\mathbf{z};\Omega)\, \Phi\left(\boldsymbol{\alpha}^T\mathbf{z}\right)\, d\mathbf{z} \\ &= 2\exp\left\{\frac{1}{2}\boldsymbol{t}^T\Omega\,\boldsymbol{t}\right\}\Phi\left\{\frac{\boldsymbol{\alpha}^T\Omega\,\boldsymbol{t}}{\{1+\boldsymbol{\alpha}^T\Omega\,\boldsymbol{\alpha}\}^{1/2}}\right\}, \end{aligned} \tag{1.25}$$

which, in the bivariate case, assumes the following final form

$$M(t_1, t_2) = 2\exp\left\{\frac{1}{2}\left(t_1^2 + 2\omega\, t_1 t_2 + t_2^2\right)\right\}\Phi(t_1\delta_1 + t_2\delta_2).$$

Through very simple algebra, we find that the cumulant generating function (cgf) of $\mathbf{z}$ is

$$K(\boldsymbol{t}) = \log M(\boldsymbol{t}) = \frac{1}{2}\boldsymbol{t}^T\Omega\boldsymbol{t} + \log\{2\Phi(\boldsymbol{\delta}^T\boldsymbol{t})\}, \tag{1.26}$$

where

$$\boldsymbol{\delta} = \frac{1}{(1+\boldsymbol{\alpha}^T\Omega\boldsymbol{\alpha})^{1/2}}\Omega\boldsymbol{\alpha}.$$

A direct consequence is that the mean vector is

$$\mathrm{E}(\mathbf{z}) = (2/\pi)^{1/2}\boldsymbol{\delta},$$

and the variance matrix is

$$\mathrm{Var}(\mathbf{z}) = \Omega - \frac{2}{\pi}\boldsymbol{\delta}\boldsymbol{\delta}^T.$$

The calculation of the correlations between two generic components of the skew-normal random vector $\mathbf{z}$ is quite simple and leads, in the case of the transformation method, to the expression

$$\begin{aligned} \rho_{ij} &= \text{Corr}(Z_i, Z_j) \quad i,j = 1,\ldots,p, \\ &= \frac{\psi_{ij}\{(1-\delta_i^2)(1-\delta_j^2)\}^{1/2} + \delta_i\delta_j(1-\frac{2}{\pi})}{\{(1-\frac{2}{\pi}\delta_i^2)(1-\frac{2}{\pi}\delta_j^2)\}^{1/2}}, \end{aligned} \tag{1.27}$$

which, in the case of the conditioning method, and taking into account Equation (1.22), becomes

$$\rho_{ij} = \frac{\omega_{ij} - \frac{2}{\pi}\delta_i\delta_j}{\{(1-\frac{2}{\pi}\delta_i^2)(1-\frac{2}{\pi}\delta_j^2)\}^{1/2}}.$$

It is now quite interesting to analyze more carefully the nature of the restrictions on the elements of the matrix Ω^* in the bivariate case. In fact, for being positive definite, we require that all the determinants of the principal sub-matrices of Ω^* are positive. So, we have that ω must satisfy

$$\delta_1\delta_2 - \{(1-\delta_1^2)(1-\delta_2^2)\}^{1/2} < \omega < \delta_1\delta_2 + \{(1-\delta_1^2)(1-\delta_2^2)\}^{1/2}. \tag{1.28}$$

From another point of view, ω has to be included into the two admissibility surfaces, that is, the extremes of the inequality in Equation (1.28).

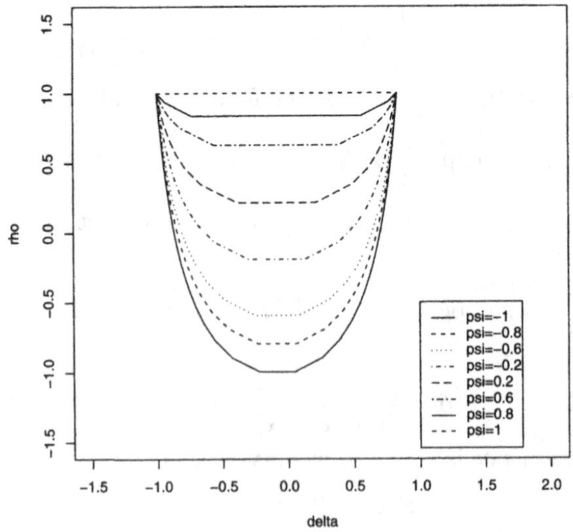

Figure 1.4 *Relation between ρ and δ for varying ψ.*

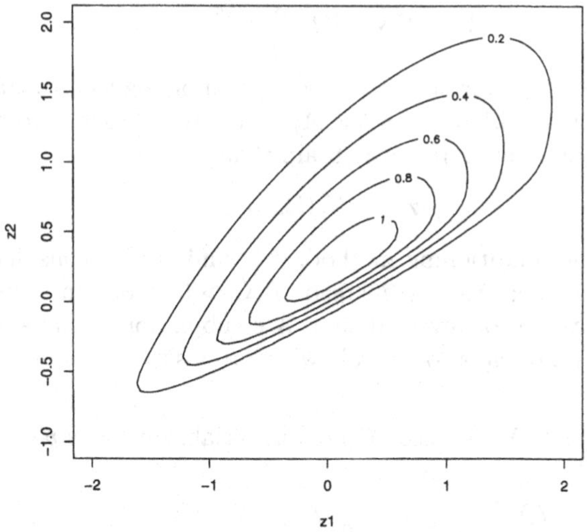

Figure 1.5 *Contour plot of the* SN_2 *for* $\delta_1 = 0.4$, $\delta_2 = 0.9$, *and* $\omega = 0.75$.

A three-dimensional graphical representation of this surfaces is not very informative, and it is better to refer to a classical two-dimensional representation. In particular, imposing $\delta_1 = \delta_2 = \delta$, it is of some interest to investigate the correlation ρ between Z_1 and Z_2 in the bivariate case, as a function of δ and ψ or δ and ω according to the considered parameterization. Figure 1.4 shows the relation between ρ and δ when ψ assumes values $-1, -0.8, -0.6, -0.2, 0.2, 0.6, 0.8, 1$. It is interesting to note that in the general case in which $\delta_1 \neq \delta_2$, the upper bound for ρ is less than 1. To represent the graphical form of the pdf in the bivariate case we choose the contour plots instead of three-dimensional graphs. The evidence is that, in general, the shape of the bivariate density is far from being elliptical and is strongly influenced by the value of ω: little variations in the value of ω induce a radical transformation of the form of the distribution. In Figure 1.5, there is an example of contour plot when the shape parameters are different, that is $\delta_1 = 0.4$, $\delta_2 = 0.9$, and ω is equal to 0.75. The presence of skewness is apparent from inspection of the figure.

1.6 Extensions of Properties Holding in the Scalar Case

There is a number of properties of the scalar SN distribution that also hold in the multivariate case. A generalization of the Property 1.2.3 is stated in the next proposition.

Proposition 1.6.1 If $\mathbf{z} \sim SN_p(\Omega, \boldsymbol{\alpha})$, and A is a $p \times p$ non-singular matrix, then

$$A\mathbf{z} \sim SN_p(A\Omega A^T, A^{-T}\boldsymbol{\alpha}).$$

In the case of the transformation method, by applying the $A\mathbf{z}$ transformation of Proposition 1.6.1 with all the A_{jj} equal to -1, and zero otherwise, to the Y_j's in Equation (1.18), we obtain that

$$-\mathbf{z} \sim SN_p(\Omega, -\boldsymbol{\alpha}). \tag{1.29}$$

In the case of the conditioning method, it is sufficient to consider the censoring operation when $X_0 < 0$, instead of $X_0 > 0$ to obtain a distribution with the shape vector of reversed sign. A combination of the above result and Equation (1.29) leads to the following proposition.

Proposition 1.6.2 If X_0 is a scalar random variable and $\mathbf{x}$ is p-dimensional, such that

$$\begin{pmatrix} X_0 \\ \mathbf{x} \end{pmatrix} \sim N_{p+1}\left(\mathbf{0}, \begin{pmatrix} 1 & \boldsymbol{\delta}^T \\ \boldsymbol{\delta} & \Omega \end{pmatrix}\right),$$

and $\mathbf{z}$ is defined by

$$\mathbf{z} = \begin{cases} \mathbf{x} & \text{if } X_0 > 0, \\ -\mathbf{x} & \text{otherwise,} \end{cases}$$

then $\mathbf{z} \sim SN_p(\Omega, \boldsymbol{\alpha})$ where $\boldsymbol{\alpha} = (1 - \boldsymbol{\delta}^T\Omega^{-1}\boldsymbol{\delta})^{-1/2}\Omega^{-1}\boldsymbol{\delta}$.

1.7 Quadratic Forms

We have seen the strong relations between the normal and the skew-normal distribution. We also know that, in the multivariate case, an appealing feature that characterizes the multivariate normal distribution is the good tractability of linear and quadratic forms. This is also true for the multivariate skew-normal distribution and was studied by Azzalini and Dalla Valle (1996), Azzalini and Capitanio (1999), Genton, He, and Liu (2001), and Loperfido (2001), among others. This section presents the most relevant outcomes.

The first result is a generalization of the appealing Property 1.2.5, that we simply obtain by observing that $\mathbf{z}^T\Omega^{-1}\mathbf{z} \stackrel{d}{=} \mathbf{x}^T\Omega^{-1}\mathbf{x}$ in distribution, where $\mathbf{x} \sim N_p(\mathbf{0}, \Omega)$.

Proposition 1.7.1 If $\mathbf{z} \sim SN_p(\Omega, \boldsymbol{\alpha})$ and Ω is given by the matrix in Equation (1.20), then

$$\mathbf{z}^T\Omega^{-1}\mathbf{z} \sim \chi^2_p. \tag{1.30}$$

It is important to note that this fact holds irrespective of the values of the shape parameter vector $\boldsymbol{\alpha}$. This property is a particular case of a set of more general results reported by Azzalini and Capitanio (1999), whose final result was the extension of the Fisher-Cochran theorem to the SN case.

Under the same assumptions of Proposition 1.7.1, a χ^2 distribution with p degrees of freedom can also be obtained if we substitute instead of Ω, in Equation (1.30), a more general matrix which has to satisfy some requirements.

Proposition 1.7.2 If $\mathbf{z} \sim SN_p(\Omega, \boldsymbol{\alpha})$ and B is a symmetric positive semi-definite $p \times p$ matrix of rank k such that $B\Omega B = B$, then $\mathbf{z}^T B\mathbf{z} \sim \chi^2_k$.

See the proof in Azzalini and Capitanio (1999). As an immediate consequence of this fact, we have the following result.

Proposition 1.7.3 If $\mathbf{z} \sim SN_p(\Omega, \boldsymbol{\alpha})$ and C is a full rank $p \times k$ matrix $(k \leq p)$, then $\mathbf{z}^T C(C^T \Omega C)^{-1} C^T \mathbf{z} \sim \chi^2_k$.

The following notable statement highlights the conditions about the independence of quadratic forms.

Proposition 1.7.4 If $\mathbf{z} \sim SN_p(\Omega, \boldsymbol{\alpha})$ and B_i is a symmetric positive semi-definite $p \times p$ matrix of rank $k_i, (i = 1, \dots, h)$ such that

1. $B_i \Omega B_j = O$, the zero matrix, for $i \neq j$,
2. $\boldsymbol{\alpha}^T \Omega B_i \Omega \boldsymbol{\alpha} \neq 0$ for at most one i,

then the quadratic forms $\mathbf{z}^T B_i \mathbf{z}, (i = 1, \dots, h)$ are mutually independent.

As final result, in the following proposition, we have the generalization of the Fisher-Cochran theorem for the SN case.

Proposition 1.7.5 If $\mathbf{z} \sim SN_p(I_p, \boldsymbol{\alpha})$ and $(B_1, \dots, B_p)$ are symmetric $p \times p$ matrices of rank $(k_1, \dots, k_h)$, respectively, such that $\sum_{i=1}^p B_i = I_p$ and $B_i \boldsymbol{\alpha} \neq \mathbf{0}$ for at most one choice of i, then the quadratic forms $\mathbf{z}^T B_i \mathbf{z}$ are independent $\chi^2_{k_i}$ if and only if $\sum_{i=1}^p k_i = p$.

Loperfido (2001) has generalized these results about quadratic forms, through the following two propositions.

Proposition 1.7.6 Let $\mathbf{x} \sim SN_p(\mathbf{0}, \Omega, \boldsymbol{\alpha})$. Then the distribution of $\mathbf{x}\mathbf{x}^T$ is Wishart with scale parameter Ω and 1 degree of freedom:

$$\mathbf{x}\mathbf{x}^T \sim W(\Omega, 1).$$

A direct consequence of Proposition 1.7.6 is the following, which generalizes the above result.

Proposition 1.7.7 Let $\mathbf{x} \sim SN_p(\mathbf{0}, \Omega, \boldsymbol{\alpha})$ and let $B_1, \ldots, B_m$ be square matrices of order p. Then the joint distribution of $(\mathbf{x}^T B_1 \mathbf{x}, \ldots, \mathbf{x}^T B_m \mathbf{x})$ does not depend on $\boldsymbol{\alpha}$.

Genton *et al.* (2001) derive the first four moments of a random vector $\mathbf{z}$ with a skew-normal distribution $SN_p(\boldsymbol{\xi}, \Omega, \boldsymbol{\alpha})$. The mean and the variance of the quadratic form $\mathbf{z}^T A\mathbf{z}$ and the covariance between this latter and the quadratic form $\mathbf{z}^T B\mathbf{z}$, where A and B are symmetric $p \times p$ matrices, derived by Genton *et al.* (2001), are

$$\begin{aligned}
\mathrm{E}(\mathbf{z}^T A\mathbf{z}) &= \mathrm{tr}[A\Omega] + \boldsymbol{\xi}^T A\boldsymbol{\xi} + 2\sqrt{\frac{2}{\pi}}\boldsymbol{\xi}^T A\boldsymbol{\delta}, \\
\mathrm{Var}(\mathbf{z}^T A\mathbf{z}) &= 2\mathrm{tr}[(A\Omega)^2] + 4\boldsymbol{\xi}^T(A\Omega A)(\boldsymbol{\xi} + 2\sqrt{\frac{2}{\pi}}\boldsymbol{\delta}) \\
&\quad -2\sqrt{\frac{2}{\pi}}\left[2(\boldsymbol{\delta}^T A\boldsymbol{\delta})(\boldsymbol{\xi}^T A\boldsymbol{\delta}) + 2\sqrt{\frac{2}{\pi}}\left(\boldsymbol{\xi}^T A\boldsymbol{\delta}\right)^2\right], \\
\mathrm{Cov}(\mathbf{z}^T A\mathbf{z}, \mathbf{z}^T B\mathbf{z}) &= 2\mathrm{tr}[A\Omega B\Omega] + 2\boldsymbol{\xi}^T(A\Omega B + B\Omega A)\left(\boldsymbol{\xi} + 2\sqrt{\frac{2}{\pi}}\boldsymbol{\delta}\right) \\
&\quad -2\sqrt{\frac{2}{\pi}}\Big[\left(\boldsymbol{\delta}^T A\boldsymbol{\delta}\right)\left(\boldsymbol{\xi}^T B\boldsymbol{\delta}\right) + \left(\boldsymbol{\xi}^T A\boldsymbol{\delta}\right)\left(\boldsymbol{\delta}^T B\boldsymbol{\delta}\right) \\
&\quad +2\sqrt{\frac{2}{\pi}}\left(\boldsymbol{\xi}^T A\boldsymbol{\delta}\right)\left(\boldsymbol{\xi}^T B\boldsymbol{\delta}\right)\Big].
\end{aligned}$$

Another issue is concerned with the applications of quadratic forms of skew-normal vectors to time series and spatial statistics. In particular, a relevant result given by Genton *et al.* (2001) is that the first four moments of the sample autocorrelation function and the sample variogram do not depend on the shape parameter if the location parameter is proportional to a vector of ones. Moreover, Loperfido (2001) also shows that their distributions do not depend on the shape parameter. Further discussions on this topic can be found in Genton and Loperfido (2002), and Wang *et al.* (2004a).

1.8 The Conditional Distribution

Let $(Z_1, Z_2)^T \sim SN_2$. The conditional pdf of Z_2 given $Z_1 = z_1$ is

$$\phi_c(z_2|z_1;\omega)\frac{\Phi(\alpha_1 z_1 + \alpha_2 z_2)}{\Phi(\alpha_1 z_1)}, \tag{1.31}$$

where $\phi_c(z_2|z_1;\omega)$ denotes the conditional pdf associated with a bivariate normal variable with standardized marginals and correlation ω. Equation

(1.31) is a member of the extended skew-normal class of densities, introduced by Azzalini (1985).

The mgf of Z_2 given Z_1 is

$$M(t_2) = \exp\left[\frac{1}{2}\left\{(1-\omega^2)t_2^2 + 2\omega t_2 z_1\right\}\right] \Phi\left\{\frac{\delta_1 z_1 + t_2(\delta_2 - \omega\delta_1)}{(1-\delta_1^2)^{1/2}}\right\} / \Phi(\alpha_1 z_1). \tag{1.32}$$

By differentiating Equation (1.32), we obtain, after some algebra, the expressions for the conditional mean and variance:

$$\begin{aligned} \mathrm{E}(Z_2|Z_1 = z_1) &= \omega z_1 + \left\{\frac{\delta_2 - \omega\delta_1}{(1-\delta_1^2)^{1/2}}\right\} H(-\alpha_1 z_1), \\ \mathrm{Var}(Z_2|Z_1 = z_1) &= 1 - \omega^2 - \frac{(\delta_2 - \omega\delta_1)^2}{1-\delta_1^2} H(-\alpha_1 z_1)\left\{\alpha_1 z_1 + H(-\alpha_1 z_1)\right\}, \end{aligned}$$

where $H(x) = \phi(x)/\Phi(-x)$ denotes the hazard function of the standard normal density.

1.9 Miscellanea

This section presents some topics about the SN class that appear in some particular frameworks.

In the context of the strength-stress models, originating from the problem of the reliability R of a component of strength Y subjected to a stress X, where we write $R = \mathrm{P}(X < Y)$, it could be interesting to study the behavior of the skew-normal distribution, and to compare it with that of the normal distribution. The estimation of the probability R has a practical relevance in many areas, like, for example, clinical trials or genetics, when we think of R as the measure of the effect of a treatment and X, Y as the responses for, respectively, a treatment group and a control group.

Gupta and Brown (2001) obtain the failure rate, the mean residual life function, and the reliability function of the SN class. They show that these functions are equivalent in the sense that, given one of them, the other two are univocally determined. In particular, the failure rate of the skew-normal distribution is increasing just like that of the normal distribution. They also give an interesting expression for the calculation of the measure of the difference between two populations, in terms of probability, $\mathrm{P}(X < Y)$,

$$\mathrm{P}(X < Y) = \frac{1}{\pi}\arctan\left(\frac{\delta_2 - \delta_1}{\sqrt{2 - \delta_1^2 - \delta_2^2}}\right) + \frac{1}{2},$$

where X and Y are skew-normal random variables with shape parameters α_1 and α_2, respectively, and δ_i is related to α_i, $i = 1, 2$, according to Equations (1.3).

Azzalini (2001) tackles the problem of the construction of regions with assigned probability p and minimum volume, for the SN class. He considers a region of the form

$$R = \{\mathbf{x} : f(\mathbf{x}) \geq q_0\},$$

for a suitable value of q_0, such that $\mathrm{P}(R) = p$, and gives an approximate but highly accurate solution. He remarks, in fact, that a region like

$$R = \{\mathbf{x} : \mathbf{x}^T \Omega^{-1} \mathbf{x} \leq q_p\}, \tag{1.33}$$

where q_p is the p-th quantile of the χ^2_p distribution, in the case of $\mathbf{x} \sim N_p(\mathbf{0}, \Omega)$ is of exact probability and minimum volume, but in the SN case lacks the latter property.

1.10 Future Research Problems

Attention is now turned to some open problems connected with theoretical unexplored features and with applications of the SN class.

The first issue concerns the extension of Proposition 1.2.4 to the multivariate case. Loperfido (2002) points out a considerable consequence of this result in the univariate case, that is, the availability of statistical tests, having skew-normal sampling distribution, for the hypothesis that the means of two normal populations are greater than a given value.

The following issues concern the inferential context. In the first place, it could be very useful to know what type of mechanism produces, with non-negligible frequency, samples having a likelihood function for which the maximization gives estimates of the shape parameter that are infinite. Apart from the considerations of Section 1.3, the problem is still open.

Another interesting issue is confidence intervals for the parameters of the SN class, in the univariate and multivariate case, which clearly would represent an important step in the inferential field.

As a final issue, it would be very useful to understand why the Newton-Raphson algorithm for numerical maximization of the likelihood function does not work well when the location parameter is known and to propose some alternative methods.

CHAPTER 2

The Closed Skew-Normal Distribution

Graciela González-Farías, J. Armando Domínguez-Molina, and Arjun K. Gupta

2.1 Introduction

The multivariate skew-normal distributions have been introduced as a generalization of the normal distribution to model, in a natural way, skewness features in the distribution. These families also have properties similar to the normal distribution. However, two important properties have been absent: the closure for the joint distribution of independent members of the multivariate skew-normal family and the closure under linear combinations other than those given by non-singular matrices.

The multivariate closed skew-normal (CSN) distribution, as will be defined in Section 2.2, has more properties similar to the normal distribution than any other. González-Farías, Domínguez-Molina, and Gupta (2003), show that for a random vector with a CSN distribution, all column (row) full rank linear transformations are in the family of CSN distributions. They also prove closure under sums of independent CSN random vectors and the closure for the joint distribution of independent CSN distributions, thus providing a result that characterizes the CSN distributions.

Earlier, Roberts (1966) introduced the idea of a univariate skewed distribution as a way to model the minimum between two bivariate normal random variables. Later, O'Hagan and Leonard (1976) gave a univariate version of a skew-normal in the context of prior distributions. The first formal version of the skew-normal distribution was given by Azzalini (1985, 1986). For a multivariate skew-normal distribution reference can be made to Azzalini and Dalla Valle (1996), Azzalini and Capitanio (1999), Branco and Dey (2001), just to mention a few. An important review for the univariate and multivariate skew-normal is Arnold and Beaver (2002). Also Arnold, Beaver, Groeneveld, and Meeker (1993) present an interesting field of applications for skew-normal distributions, in particular for screening procedures. Liseo and Loperfido (2003a) obtained a multivariate skew-normal density as a direct extension to the multivariate case of the O'Hagan and Leonard (1976) procedure; Domínguez-Molina, González-Farías, and Ramos-Quiroga (2003) provide an example of the use and construction of the multivariate CSN distribution in stochastic frontier analysis.

An approach that comprises some recent proposals in the literature of non-symmetric densities is given in Azzalini and Capitanio (2003). They propose a general procedure to perturb a multivariate density satisfying a weak form of multivariate symmetry, and to generate a whole set of non-symmetric densities.

We present a motivation for the CSN distribution, as well as some of the basic properties, such as its moment generating function (mgf), that allows us to establish other important properties of interest in Section 2.2. Section 2.3 provides results for linear transformations and shows a characterization for the CSN. In Section 2.4 we present the joint distribution of independent CSN random vectors. Section 2.5 discusses the distribution of the sum of independent CSN random vectors. Section 2.6 presents some of the proposals for the multivariate skew-normal as particular cases of the CSN distributions. In Section 2.7 we present a generalization to extended skew-elliptical (ESE) distributions.

2.2 Basic Results on the Multivariate CSN Distribution

We start with a definition of the closed skew-normal distribution.

Definition 2.2.1 Consider $p \geq 1, q \geq 1$, $\boldsymbol{\mu} \in \mathbb{R}^p$, $\boldsymbol{\nu} \in \mathbb{R}^q$, D an arbitrary $q \times p$ matrix, Σ and Δ positive definite matrices of dimensions $p \times p$ and $q \times q$, respectively. Then the probability density function (pdf) of the CSN distribution is given by:

$$g_{p,q}(\mathbf{y}) = C\phi_p(\mathbf{y}; \boldsymbol{\mu}, \Sigma)\, \Phi_q(D(\mathbf{y} - \boldsymbol{\mu}); \boldsymbol{\nu}, \Delta), \quad \mathbf{y} \in \mathbb{R}^p, \tag{2.1}$$

with:

$$C^{-1} = \Phi_q\left(\mathbf{0}; \boldsymbol{\nu}, \Delta + D\Sigma D^T\right), \tag{2.2}$$

where $\phi_p(\cdot; \boldsymbol{\eta}, \Psi)$, $\Phi_p(\cdot; \boldsymbol{\eta}, \Psi)$ are the pdf and cumulative distribution function (cdf) of a p-dimensional normal distribution. Here $\boldsymbol{\eta} \in \mathbb{R}^p$ denotes the mean and Ψ is a $p \times p$ covariance matrix. We denote a p-dimensional random vector distributed according to a CSN distribution with parameters $q, \boldsymbol{\mu}, \Sigma, D, \boldsymbol{\nu}, \Delta$ by $\mathbf{y} \sim CSN_{p,q}(\boldsymbol{\mu}, \Sigma, D, \boldsymbol{\nu}, \Delta)$.

The proof that $g_{p,q}$ is indeed a density function is given in Domínguez-Molina *et al.* (2003).

One interesting idea is that using a simple extension of the multivariate case of the Copas and Li (1997) model, we can generate a CSN random vector in the following way. Copas and Li (1997) present a model for missing data (Y is observed if $Z > 0$) or comparative trials (a subject is allocated to treatment A if $Z > 0$ and to treatment B if $Z \leq 0$) given by:

$$Y = \boldsymbol{\beta}^T \mathbf{x} + \sigma\varepsilon_1, \quad \varepsilon_1 \sim N(0,1),$$

$$Z = \boldsymbol{\gamma}^T \mathbf{x} + \varepsilon_2, \quad \varepsilon_2 \sim N(0,1),$$

where $(\varepsilon_1, \varepsilon_2)^T \sim N_2(\mathbf{0}, \Sigma)$, $\Sigma = \begin{pmatrix} 1 & \rho \\ \rho & 1 \end{pmatrix}$, and $\boldsymbol{\beta}, \boldsymbol{\gamma}, \mathbf{x}$ are p-vectors.

Note that:

$$\begin{pmatrix} Y \\ Z \end{pmatrix} \sim N_2\left[\begin{pmatrix} \boldsymbol{\beta}^T\mathbf{x} \\ \boldsymbol{\gamma}^T\mathbf{x} \end{pmatrix}, \begin{pmatrix} \sigma^2 & \sigma\rho \\ \sigma\rho & 1 \end{pmatrix}\right].$$

It is not difficult to see that $g(y|\mathbf{x}, z > 0)$ is given by:

$$g(y|\mathbf{x}, z > 0) = \frac{\sigma^{-1}\phi\left(\frac{y-\boldsymbol{\beta}^T\mathbf{x}}{\sigma}\right)\Phi\left(\frac{\boldsymbol{\gamma}^T\mathbf{x}+\rho\sigma^{-1}(y-\boldsymbol{\beta}^T\mathbf{x})}{\sqrt{1-\rho^2}}\right)}{\Phi(\boldsymbol{\gamma}^T\mathbf{x})}, \tag{2.3}$$

where $\phi(\cdot)$ and $\Phi(\cdot)$ denote the pdf and the cdf of the standard normal distribution.

Let us consider a multivariate version of the Copas and Li (1997) model, given as:

$$\begin{aligned} \mathbf{w}_0 &= \boldsymbol{\mu} + \mathbf{u}_1 \\ \mathbf{z} &= -\boldsymbol{\nu} + D\mathbf{u}_1 + \mathbf{u}_2, \end{aligned}$$

where $\mathbf{u}_1 \sim N_p(\mathbf{0}, \Sigma)$ and $\mathbf{u}_2 \sim N_q(\mathbf{0}, \Delta)$ are independent random vectors, and $D(q \times p)$ is an arbitrary matrix, $\boldsymbol{\mu} \in \mathbb{R}^p, \boldsymbol{\nu} \in \mathbb{R}^q$, and $\Delta(q \times q) > 0$. The joint distribution of $\mathbf{w}_0$ and $\mathbf{z}$ is:

$$\begin{pmatrix} \mathbf{w}_0 \\ \mathbf{z} \end{pmatrix} \sim N_{p+q}\left[\begin{pmatrix} \boldsymbol{\mu} \\ -\boldsymbol{\nu} \end{pmatrix}, \begin{pmatrix} \Sigma & \Sigma D^T \\ D\Sigma & \Delta + D\Sigma D^T \end{pmatrix}\right].$$

From Arellano-Valle, del Pino, and San Martín (2002, Equation (5.1.)) we have:

$$g_{\mathbf{w}_0|\{\mathbf{z}\geq \mathbf{0}\}}(\mathbf{w}|\mathbf{z} \geq \mathbf{0}) = \frac{g_{\mathbf{w}}(\mathbf{w})}{\mathrm{P}(\mathbf{z} \geq \mathbf{0})}\mathrm{P}(\mathbf{z} \geq \mathbf{0}|\mathbf{w}_0 = \mathbf{w}). \tag{2.4}$$

Thus the conditional density of $\mathbf{w}_0$ given $\mathbf{z} \geq 0$ is:

$$g_{\mathbf{w}_0|\{\mathbf{z}\geq \mathbf{0}\}}(\mathbf{w}|\mathbf{z} \geq \mathbf{0}) = C\phi_p(\mathbf{w}; \boldsymbol{\mu}, \Sigma)\,\Phi_q(D(\mathbf{w} - \boldsymbol{\mu}); \boldsymbol{\nu}, \Delta), \tag{2.5}$$

where C is given in (2.2). Note that the density (2.5) is the same as (2.1). Thus, we see that a natural generalization of the Copas and Li (1997) model is, indeed, a multivariate *CSN* distribution.

By rewriting (2.3), and after some algebra, we have:

$$g(y|\mathbf{x}, z > 0) = \frac{\phi\left(y; \boldsymbol{\beta}^T\mathbf{x}, \sigma^2\right)\Phi\left(\frac{\rho}{\sigma}\left(y - \boldsymbol{\beta}^T\mathbf{x}\right); -\boldsymbol{\gamma}^T\mathbf{x}, 1 - \rho^2\right)}{\Phi(0; -\boldsymbol{\gamma}^T\mathbf{x}, 1)}. \tag{2.6}$$

As we shall see, (2.6) corresponds to the multivariate closed skew-normal with parameters $p = q = 1$, $\mu = \boldsymbol{\beta}^T\mathbf{x}$, $\sigma^2 = \sigma^2$, $D = \rho/\sigma$, $\nu = -\boldsymbol{\gamma}^T\mathbf{x}$, and $\Delta = 1 - \rho^2$, that is:

$$y|\{\mathbf{x}, z > 0\} \sim CSN_{1,1}\left(\boldsymbol{\beta}^T\mathbf{x}, \sigma^2, \frac{\rho}{\sigma}, -\boldsymbol{\gamma}^T\mathbf{x}, 1 - \rho^2\right).$$

Now, it is easy to see that the density defined in (2.1) includes some other skewed distribution like the density given by Azzalini and Dalla Valle (1996, Equation (2.3)) with parameters $\boldsymbol{\mu} = \mathbf{0}$, $\Omega = \Sigma$, $\boldsymbol{\alpha} = \Delta^{-1/2} D^T$, which can be obtained by letting $q = 1, \boldsymbol{\nu} = \mathbf{0}$ in (2.1), i.e.,

$$\begin{aligned} g_{p,1}(\mathbf{y}) &= 2\phi_p(\mathbf{y}; \boldsymbol{\mu}, \Sigma)\, \Phi(D(\mathbf{y} - \boldsymbol{\mu}); \mathbf{0}, \Delta) \\ &\equiv 2\phi_p(\mathbf{y}; \boldsymbol{\mu}, \Sigma)\, \Phi\left(\Delta^{-1/2} D(\mathbf{y} - \boldsymbol{\mu})\right). \end{aligned} \tag{2.7}$$

Also, if we take $\boldsymbol{\nu} = \mathbf{0}$ and $\Delta = I_p$ then the density (2.1) reduces to:

$$\begin{aligned} g_{p,p}(\mathbf{y}) &= \frac{1}{\Phi_q(\mathbf{0}; \mathbf{0}, I_p + D\Sigma D^T)} \phi_p(\mathbf{y}; \boldsymbol{\mu}, \Sigma)\, \Phi_p(D(\mathbf{y} - \boldsymbol{\mu}); \mathbf{0}, I_p) \\ &\equiv \frac{1}{\Phi_q(\mathbf{0}; \mathbf{0}, I_p + D\Sigma D^T)} \phi_p(\mathbf{y}; \boldsymbol{\mu}, \Sigma)\, \Phi_p(D(\mathbf{y} - \boldsymbol{\mu})), \end{aligned}$$

which is the density of the distribution $SN_p(\boldsymbol{\mu}, \Sigma, D)$ defined by Gupta, González-Farías, and Domínguez-Molina (2004).

Remark 1 Let $\mathbf{z} \sim N_q(\boldsymbol{\nu}, \Delta)$, and $\mathbf{y} \sim N_p(\boldsymbol{\mu}, \Sigma)$, $\mathbf{z}$ and $\mathbf{y}$ independent. Taking $\mathbf{w} = \mathbf{z} - D(\mathbf{y} - \boldsymbol{\mu})$, the distribution of $\mathbf{y}|\{\mathbf{w} \leq \mathbf{0}\}$ leads to (2.1). If we focus on the joint multivariate normal distribution of $(\mathbf{y}^T, \mathbf{w}^T)^T$, it could help to understand some other analytical result. A relevant property here is that the conditional distributions are also normal. For example, the joint behavior of two independent CSN distributions follows by choosing $(\mathbf{y}_1^T, \mathbf{w}_1^T)^T$ and $(\mathbf{y}_2^T, \mathbf{w}_2^T)^T$ to be independent. Then $(\mathbf{y}_1^T, \mathbf{y}_2^T)^T$ and $(\mathbf{w}_1^T, \mathbf{w}_2^T)^T$ are jointly normal and the conditional distribution given $(\mathbf{w}_1^T, \mathbf{w}_2^T)^T \leq \mathbf{0}$ is CSN. Also, if the dimension of $\mathbf{y}_1$ and $\mathbf{y}_2$ are the same, then $\mathbf{y}_1 + \mathbf{y}_2$ and $(\mathbf{w}_1^T, \mathbf{w}_2^T)^T$ are jointly normal and the conditional distribution given $(\mathbf{w}_1^T, \mathbf{w}_2^T)^T \leq \mathbf{0}$ is again CSN. A similar argument is presented in Domínguez-Molina *et al.* (2003) to study the properties of the CSN distribution.

The main differences of the CSN distribution with respect to other definitions are: the constant C (which can make computations quite difficult sometimes); the inclusion of an extra parameter $\boldsymbol{\nu}$, which allows the possibility of closure properties for the conditional densities; the inclusion of an extra parameter Δ, which allows the possibility of closure properties for the marginal densities; the inclusion of $\Phi_q(\cdot)$ for $q \geq 1$, which allows the possibility of closure properties for the sum, and the joint distribution of independent CSN random vectors. The parameters $\boldsymbol{\mu}$, Σ, and D have the same meaning as in the other skew-normal families, that is, location, scale, and skewness parameters.

The distribution associated to the pdf (2.1) can be found directly as it is shown in the following lemma and the proof is given in Domínguez-Molina *et al.* (2003).

Lemma 2.2.1 The distribution function of $\mathbf{y} \sim CSN_{p,q}(\boldsymbol{\mu}, \Sigma, D, \boldsymbol{\nu}, \Delta)$ is given by:

$$G_{p,q}(\mathbf{y}) = C\Phi_{p+q}\left[\begin{pmatrix}\mathbf{y}\\ \mathbf{0}\end{pmatrix}; \begin{pmatrix}\boldsymbol{\mu}\\ \boldsymbol{\nu}\end{pmatrix}, \begin{pmatrix}\Sigma & -\Sigma D^T\\ -D\Sigma & \Delta + D\Sigma D^T\end{pmatrix}\right],$$

where C is given in (2.2).

In order to derive many of the most important results for the CSN distribution we need the moment generating function that we present here without a proof.

Lemma 2.2.2 If $\mathbf{y} \sim CSN_{p,q}(\boldsymbol{\mu}, \Sigma, D, \boldsymbol{\nu}, \Delta)$, then the moment generating function of $\mathbf{y}$ is:

$$M_{\mathbf{y}}(\mathbf{t}) = \frac{\Phi_q\left(D\Sigma\mathbf{t}; \boldsymbol{\nu}, \Delta + D\Sigma D^T\right)}{\Phi_q\left(\mathbf{0}; \boldsymbol{\nu}, \Delta + D\Sigma D^T\right)} e^{\mathbf{t}^T\boldsymbol{\mu} + \frac{1}{2}\mathbf{t}^T\Sigma\mathbf{t}}, \quad \mathbf{t} \in \mathbb{R}^p. \tag{2.8}$$

Domínguez-Molina *et al.* (2003) establish the closure under marginalization and conditioning using a conditional argument that, we think, would help later with numerical problems that are present in the estimation of parameters in this family of distributions.

Remark 2 In Section 2.7, we present a possible extension to skew-elliptical distributions that are closed under conditional, marginal, and linear transformations (full row rank), for which the closed skew-normal is a particular case. It seems like a possible characterization because the CSN is given by the properties of closure under sums and joint distribution of independent and identically distributed (iid) CSN random vectors.

In the next section, we give a direct proof for the closure under marginalization and the conditional distributions of the CSN distribution.

2.3 Linear Transformations

In this section we present three results for CSN distributions:

- The CSN family is closed under full row rank linear transformations.
- The CSN family admits a characterization similar to the Cramér-Wold result for the normal distribution.
- The CSN family is closed under full column rank linear transformations (defining the singular skew-normal distribution).

The CSN distribution is closed under translations and scalar multiplications, i.e., for an arbitrary constant vector $\mathbf{b} \in \mathbb{R}^p$:

$$\mathbf{y} \sim CSN_{p,q}(\boldsymbol{\mu}, \Sigma, D, \boldsymbol{\nu}, \Delta) \Rightarrow \mathbf{y} + \mathbf{b} \sim CSN_{p,q}(\boldsymbol{\mu} + \mathbf{b}, \Sigma, D, \boldsymbol{\nu}, \Delta),$$

and if c is a real number

$$c\mathbf{y} \sim CSN_{p,q}\left(c\boldsymbol{\mu}, \Sigma c^2, Dc^{-1}, \boldsymbol{\nu}, \Delta\right). \tag{2.9}$$

Thus, without loss of generality, we may study closure under full row rank linear transformations, by considering only the case

$$\mathbf{x} = A\mathbf{y},$$

where $\mathbf{y}$ is $p \times 1$ and A is $n \times p$, $n \geq 1$.

Note that when $n = 1$ the scalar random variable $\mathbf{x} = A\mathbf{y}$ is the linear combination of the elements of the p-vector $\mathbf{y}$.

Proposition 2.3.1 Let $\mathbf{y} \sim CSN_{p,q}(\boldsymbol{\mu}, \Sigma, D, \boldsymbol{\nu}, \Delta)$ and let A be an $n \times p$ $(n \leq p)$ matrix of rank n. Then

$$A\mathbf{y} \sim CSN_{n,q}(\boldsymbol{\mu}_A, \Sigma_A, D_A, \boldsymbol{\nu}, \Delta_A),$$

where

$$\boldsymbol{\mu}_A = A\boldsymbol{\mu}, \quad \Sigma_A = A\Sigma A^T, \quad D_A = D\Sigma A^T \Sigma_A^{-1},$$

and

$$\Delta_A = \Delta + D\Sigma D^T - D\Sigma A^T \Sigma_A^{-1} A\Sigma D^T.$$

Proof. For $\mathbf{t} \in \mathbb{R}^n$ the mgf of $A\mathbf{y}$ is given by:

$$\begin{aligned} M_{A\mathbf{y}}(\mathbf{t}) &= M_{\mathbf{y}}\left(A^T\mathbf{t}\right) \\ &= \frac{\Phi_q\left(D\Sigma A^T\mathbf{t}; \boldsymbol{\nu}, \Delta + D\Sigma D^T\right)}{\Phi_q\left(\mathbf{0}; \boldsymbol{\nu}, \Delta + D\Sigma D^T\right)} e^{\mathbf{t}^T A\boldsymbol{\mu} + \frac{1}{2}\mathbf{t}^T A\Sigma A^T \mathbf{t}}. \end{aligned}$$

Given that $\Sigma > 0$ and $\operatorname{rank}(A) = n$, we have that $\Sigma_A = A\Sigma A^T$ is a non-singular matrix. By noting that:

$$\Phi_q\left(D\Sigma A^T\mathbf{t}; \boldsymbol{\nu}, \Delta + D\Sigma D^T\right) = \Phi_q\left(D\Sigma A^T\Sigma_A^{-1}\Sigma_A\mathbf{t}; \boldsymbol{\nu}, \Delta + D\Sigma D^T\right),$$

and using $\boldsymbol{\mu}_A$, Σ_A, D_A, and Δ_A as defined above, we get:

$$M_{A\mathbf{y}}(t) = \frac{\Phi_q\left(D_A\Sigma_A\mathbf{t}; \boldsymbol{\nu}, \Delta_A + D_A\Sigma_A D_A^T\right)}{\Phi_q\left(\mathbf{0}; \boldsymbol{\nu}, \Delta_A + D_A\Sigma_A D_A^T\right)} e^{\mathbf{t}^T\boldsymbol{\mu}_A + \frac{1}{2}\mathbf{t}^T\Sigma_A\mathbf{t}}.$$

□

Remark 3 If $n = 1$ in Proposition 2.3.1 and if $\mathbf{a} \neq \mathbf{0}$ is an arbitrary p-vector in $\mathbb{R}^p$, then

$$\mathbf{a}^T\mathbf{y} \sim CSN_{1,q}(\mu_{\mathbf{a}}, \Sigma_{\mathbf{a}}, D_{\mathbf{a}}, \boldsymbol{\nu}, \Delta_{\mathbf{a}}), \tag{2.10}$$

where

$$\mu_{\mathbf{a}} = \mathbf{a}^T \boldsymbol{\mu},\ \Sigma_{\mathbf{a}} = \mathbf{a}^T \Sigma \mathbf{a},\ D_{\mathbf{a}} = D\Sigma\mathbf{a}\Sigma_{\mathbf{a}}^{-1}$$
$$\Delta_{\mathbf{a}} = \Delta + D\Sigma D^T - D\Sigma\mathbf{a}\mathbf{a}^T\Sigma D^T\Sigma_{\mathbf{a}}^{-1}.$$

Remark 4 Letting $\mathbf{a} = \mathbf{e}_i$ in (2.10), where $\mathbf{e}_i$ is the i-th unit vector in $\mathbb{R}^p$, it is straightforward to evaluate the marginal distribution of $\mathbf{y}_i$, $i = 1, \ldots, p$, as it is given in the following Lemma.

Lemma 2.3.1 (Marginal density) Let $\mathbf{y}$ be partitioned as $\mathbf{y} = \begin{pmatrix} \mathbf{y}_1 \\ \mathbf{y}_2 \end{pmatrix} \begin{matrix} k \\ p-k \end{matrix}$ where $\mathbf{y}$ is a random vector distributed as $CSN_{p,q}(\boldsymbol{\mu}, \Sigma, D, \boldsymbol{\nu}, \Delta)$. If we take $A = [I_k\ O]$, with O a $k \times (p-k)$ zero matrix, then $\mathbf{y}_1 = A\mathbf{y}$, and as a consequence of Proposition 2.3.1 we get that

$$\mathbf{y}_1 \sim CSN_{k,q}(\boldsymbol{\mu}_1,\ \Sigma_{11},\ D^\star,\ \boldsymbol{\nu},\ \Delta^\star),$$

where

$$D^\star = D_1 + D_2\Sigma_{21}\Sigma_{11}^{-1}, \quad \Delta^\star = \Delta + D_2\Sigma_{22\cdot 1}D_2^T,$$

$\Sigma_{22\cdot 1} = \Sigma_{22} - \Sigma_{21}\Sigma_{11}^{-1}\Sigma_{12}$, and $\boldsymbol{\mu}_1$, Σ_{11}, Σ_{22}, Σ_{12}, Σ_{21} came from the partition

$$\mu = \begin{pmatrix} \mu_1 \\ \mu_2 \end{pmatrix} \begin{matrix} k \\ p-k \end{matrix}, \quad \Sigma = \begin{matrix} & \begin{matrix} k & p-k \end{matrix} \\ & \begin{pmatrix} \Sigma_{11} & \Sigma_{12} \\ \Sigma_{21} & \Sigma_{22} \end{pmatrix} \end{matrix} \begin{matrix} k \\ p-k \end{matrix}$$

and D_1, D_2 from

$$D = \begin{matrix} \begin{matrix} k & p-k \end{matrix} \\ \begin{pmatrix} D_1 & D_2 \end{pmatrix} \end{matrix}\ q.$$

Proposition 2.3.2 If $\mathbf{y} \sim CSN_{p,q}(\boldsymbol{\mu}, \Sigma, D, \boldsymbol{\nu}, \Delta)$ then for two subvectors $\mathbf{y}_1$ and $\mathbf{y}_2$,where $\mathbf{y}^T = (\mathbf{y}_1^T, \mathbf{y}_2^T)$, $\mathbf{y}_1$ is k-dimensional, $1 \leq k \leq p$, and $\boldsymbol{\mu}, \Sigma, D$ are partitioned as above, then the conditional distribution of $\mathbf{y}_2$ given $\mathbf{y}_1 = \mathbf{y}_{10}$ is

$$CSN_{p-k,q}\left(\boldsymbol{\mu}_2 + \Sigma_{21}\Sigma_{11}^{-1}(\mathbf{y}_{10} - \boldsymbol{\mu}_1), \Sigma_{22\cdot 1}, D_2, \boldsymbol{\nu} - D^\star(\mathbf{y}_{10} - \boldsymbol{\mu}_1), \Delta\right).$$

Note that the dimension q is not affected by linear transformations. The following result is a direct consequence of Proposition 2.3.1; however, its importance becomes clear because it provides a characterization for the closed multivariate skew-normal random vector in the same way as we have it for the multivariate normal distribution.

Corollary 2.3.1 (Characterization of the CSN distribution) The vector $\mathbf{y} \sim CSN_{p,q}(\mu, \Sigma, D, \boldsymbol{\nu}, \Delta)$ if, and only if, $\mathbf{a}^T\mathbf{y} \sim CSN_{1,q}(\boldsymbol{\mu}_{\mathbf{a}}, \Sigma_{\mathbf{a}}, D_{\mathbf{a}}, \boldsymbol{\nu}_{\mathbf{a}}, \Delta_{\mathbf{a}})$, for every $\mathbf{a} \neq \mathbf{0}$, p-vector in $\mathbb{R}^p$, where $\mu_{\mathbf{a}}, \Sigma_{\mathbf{a}}, D_{\mathbf{a}}, \boldsymbol{\nu}_{\mathbf{a}}, \Delta_{\mathbf{a}}$ are given in Remark 3.

Proof. We shall only prove sufficiency, since necessity follows in a straightforward manner. Observe that if $\mathbf{a}^T\mathbf{y} \sim CSN_{1,q}(\mu_\mathbf{a}, \Sigma_\mathbf{a}, D_\mathbf{a}, \boldsymbol{\nu}_\mathbf{a}, \Delta_\mathbf{a})$ for every $\mathbf{a} \neq \mathbf{0}$ then from (2.8) we get:

$$M_{\mathbf{a}^T\mathbf{y}}(s) = \frac{\Phi_q\left(D_\mathbf{a}\Sigma_\mathbf{a}s; \boldsymbol{\nu}, \Delta_\mathbf{a} + D_\mathbf{a}\Sigma_\mathbf{a}D_\mathbf{a}^T\right)}{\Phi_q\left(\mathbf{0}; \boldsymbol{\nu}, \Delta_\mathbf{a} + D_\mathbf{a}\Sigma_\mathbf{a}D_\mathbf{a}^T\right)} e^{s\mu_\mathbf{a} + \frac{1}{2}s^2\Sigma_\mathbf{a}}, \quad s \in \mathbb{R}.$$

Now, given that $M_{\mathbf{a}^T\mathbf{y}}(s) = M_\mathbf{y}(\mathbf{a}s)$, taking $s = 1$ we get:

$$M_\mathbf{y}(\mathbf{a}) = \frac{\Phi_q\left(D_\mathbf{a}\Sigma_\mathbf{a}; \boldsymbol{\nu}, \Delta_\mathbf{a} + D_\mathbf{a}\Sigma_\mathbf{a}D_\mathbf{a}^T\right)}{\Phi_q\left(\mathbf{0}; \boldsymbol{\nu}, \Delta_\mathbf{a} + D_\mathbf{a}\Sigma_\mathbf{a}D_\mathbf{a}^T\right)} e^{\mu_\mathbf{a} + \frac{1}{2}\Sigma_\mathbf{a}},$$

and noting that $D_\mathbf{a}\Sigma_\mathbf{a} = D\Sigma\mathbf{a}$ and $\Delta_\mathbf{a} + D_\mathbf{a}\Sigma_\mathbf{a}D_\mathbf{a}^T = \Delta + D\Sigma D^T$ $M_\mathbf{y}(\mathbf{a})$ the expression for $M_\mathbf{y}(\mathbf{a})$ reduces to:

$$M_\mathbf{y}(\mathbf{a}) = \frac{\Phi_q\left(D\Sigma\mathbf{a}; \boldsymbol{\nu}, \Delta + D\Sigma D^T\right)}{\Phi_q\left(\mathbf{0}; \boldsymbol{\nu}, \Delta + D\Sigma D^T\right)} e^{\mathbf{a}^T\boldsymbol{\mu} + \frac{1}{2}\mathbf{a}^T\Sigma\mathbf{a}},$$

and, since $\mathbf{a}$ is arbitrary, by (2.8) we get that $\mathbf{y} \sim CSN_{p,q}(\boldsymbol{\mu}, \Sigma, D, \boldsymbol{\nu}, \Delta)$. □

We consider now a linear transformation $A\mathbf{y}$ where A is an $n \times p$ matrix, with $n > p$, i.e., we consider the case of linear transformations that are full column rank instead of full row rank. In this case the scale parameter matrix will not be a full rank matrix.

Proposition 2.3.3 (The singular skew-normal (SSN) distribution) Let $\mathbf{y} \sim CSN_{p,q}(\boldsymbol{\mu}, \Sigma, D, \boldsymbol{\nu}, \Delta)$ and let A be an $n \times p$ $(n > p)$ matrix of rank p. Then

$$A\mathbf{y} \sim SSN_{n,q}\left(A\boldsymbol{\mu},\ A\Sigma A^T,\ D\left(A^TA\right)^{-1}A^T,\ \boldsymbol{\nu},\ \Delta\right), \tag{2.11}$$

where $SSN_{n,q}$ will be used to denote a singular skew-normal distribution.

Proof. For $\mathbf{t} \in \mathbb{R}^n$ the mgf of $A\mathbf{y}$ is given by:

$$\begin{aligned}
M_{A\mathbf{y}}(\mathbf{t}) &= \mathrm{E}\left(e^{\mathbf{t}^TA\mathbf{y}}\right) \\
&= M_\mathbf{y}\left(A^T\mathbf{t}\right) \\
&= \frac{\Phi_q\left(D\Sigma A^T\mathbf{t}; \boldsymbol{\nu}, \Delta + D\Sigma D^T\right)}{\Phi_q\left(\mathbf{0}; \boldsymbol{\nu}, \Delta + D\Sigma D^T\right)} e^{\mathbf{t}^TA\boldsymbol{\mu} + \frac{1}{2}\mathbf{t}^TA\Sigma A^T\mathbf{t}} \\
&= \frac{\Phi_q\left(DB\left(A\Sigma A^T\right)\mathbf{t}; \boldsymbol{\nu}, \Delta + (DB)\left(A\Sigma A^T\right)(DB)^T\right)}{\Phi_q\left(\mathbf{0}; \boldsymbol{\nu}, \Delta + (DB)\left(A\Sigma A^T\right)(DB)^T\right)} \\
&\quad \times e^{\mathbf{t}^TA\boldsymbol{\mu} + \frac{1}{2}\mathbf{t}^TA\Sigma A^T\mathbf{t}},
\end{aligned}$$

where $B = \left(A^TA\right)^{-1}A^T$. □

Note that $A\Sigma A^T$ is singular, therefore the distribution of $A\mathbf{y}$ cannot have a density; in this case we will call it singular skew-normal distribution, and we will use the notation given in (2.11).

2.4 Joint Distribution of Independent CSN Random Vectors

In this section we show that the joint distribution of independent CSN random vectors is again CSN distributed. This property, together with the closure under linear transformations, allows us to obtain an important result: the closure under sum of independent CSN random vectors (not necessarily identically distributed).

In what follows we will use the symbols " $\otimes$ " and " $\oplus$ " to indicate the Kronecker matrix product and the matrix direct sum operator (see Horn and Johnson, 1991, pp. 12 and 242): for any two matrices A and B, $A \otimes B = (a_{ij}B)$ and $A \oplus B$ gives as a result a block diagonal matrix, i.e.,

$$A \oplus B = \begin{bmatrix} A & O \\ O & B \end{bmatrix}.$$

Read $\bigoplus_{i=1}^{m} A_i$ as $A_1 \oplus \cdots \oplus A_m$, for any matrices $A_1, \ldots, A_m$. It is easy to see that $\bigoplus_{i=1}^{m} A = I_m \otimes A$, where I_m is the $m \times m$ identity matrix.

Proposition 2.4.1 If $\mathbf{y}_1, \ldots, \mathbf{y}_n$ are independent random vectors with $\mathbf{y}_i \sim CSN_{p_i,q_i}(\boldsymbol{\mu}_i, \Sigma_i, D_i, \boldsymbol{\nu}_i, \Delta_i)$ the joint distribution of $\mathbf{y}_1, \ldots, \mathbf{y}_n$ is

$$\mathbf{y} = \left(\mathbf{y}_1^T, \ldots, \mathbf{y}_n^T\right)^T \sim CSN_{p^\dagger,q^\dagger}\left(\boldsymbol{\mu}^\dagger, \Sigma^\dagger, D^\dagger, \boldsymbol{\nu}^\dagger, \Delta^\dagger\right),$$

where

$$p^\dagger = \sum_{i=1}^{n} p_i, \; q^\dagger = \sum_{i=1}^{n} q_i, \; \boldsymbol{\mu}^\dagger = \left(\boldsymbol{\mu}_1^T, \ldots, \boldsymbol{\mu}_n^T\right)^T, \; \Sigma^\dagger = \bigoplus_{i=1}^{n} \Sigma_i,$$

and

$$D^\dagger = \bigoplus_{i=1}^{n} D_i, \quad \boldsymbol{\nu}^\dagger = \left(\boldsymbol{\nu}_1^T, \ldots, \boldsymbol{\nu}_n^T\right)^T, \quad \Delta^\dagger = \bigoplus_{i=1}^{n} \Delta_i.$$

Proof. For $\mathbf{y} = \left(\mathbf{y}_1^T, \ldots, \mathbf{y}_n^T\right)^T$, $\mathbf{y}_i \in \mathbb{R}^{p_i}$, the density function of $\mathbf{y}$ is given by:

$$\begin{aligned} g(\mathbf{y}) &= \prod_{i=1}^{n} f_{p_i,q_i}\left(\mathbf{y}_i; \boldsymbol{\mu}_i, \Sigma_i, D_i, \boldsymbol{\nu}_i, \Delta_i\right) \\ &= \prod_{i=1}^{n} \frac{\phi_{p_i}\left(\mathbf{y}_i; \boldsymbol{\mu}_i, \Sigma_i\right) \Phi_{q_i}\left(D_i\left(\mathbf{y}_i - \boldsymbol{\mu}_i\right); \boldsymbol{\nu}_i, \Delta_i\right)}{\Phi_{q_i}\left(\mathbf{0}; \boldsymbol{\nu}_i, \Delta_i + D_i \Sigma_i D_i^T\right)} \\ &= \frac{\prod_{i=1}^{n} \phi_{p_i}\left(\mathbf{y}_i; \boldsymbol{\mu}_i, \Sigma_i\right) \prod_{i=1}^{n} \Phi_{q_i}\left(D_i\left(\mathbf{y}_i - \boldsymbol{\mu}_i\right); \boldsymbol{\nu}_i, \Delta_i\right)}{\prod_{i=1}^{n} \Phi_{q_i}\left(\mathbf{0}; \boldsymbol{\nu}_i, \Delta_i + D_i \Sigma_i D_i^T\right)} \end{aligned}$$

$$= \frac{\phi_{p^\dagger}\left(\mathbf{y};\boldsymbol{\mu}^\dagger,\Sigma^\dagger\right)\Phi_{q^\dagger}\left(D^\dagger\left(\mathbf{y}-\boldsymbol{\mu}^\dagger\right);\boldsymbol{\nu}^\dagger,\Delta^\dagger\right)}{\prod_{i=1}^{n}\Phi_{q_i}\left(\mathbf{0};\boldsymbol{\nu}_i,\Delta_i+D_i\Sigma_iD_i^T\right)},$$

and by noting that:

$$\begin{aligned}\bigoplus_{i=1}^{n}\left(\Delta_i+D_i\Sigma_iD_i^T\right) &= \bigoplus_{i=1}^{n}\Delta_i+\bigoplus_{i=1}^{n}D_i\Sigma_iD_i^T\\ &= \bigoplus_{i=1}^{n}\Delta_i+\left(\bigoplus_{i=1}^{n}D_i\right)\left(\bigoplus_{i=1}^{n}\Sigma_i\right)\left(\bigoplus_{i=1}^{n}D_i^T\right)\\ &= \Delta^\dagger+D^\dagger\Sigma^\dagger D^{\dagger T},\end{aligned}$$

we get:

$$g\left(\mathbf{y}\right) = \frac{\phi_{p^\dagger}\left(\mathbf{y};\boldsymbol{\mu}^\dagger,\Sigma^\dagger\right)\Phi_{q^\dagger}\left(D^\dagger\left(\mathbf{y}-\boldsymbol{\mu}^\dagger\right);\boldsymbol{\nu}^\dagger,\Delta^\dagger\right)}{\Phi_{q^\dagger}\left(\mathbf{0};\boldsymbol{\nu}^\dagger,\Delta^\dagger+D^\dagger\Sigma^\dagger D^{\dagger T}\right)}.$$

□

Also note that $\bigoplus_{i=1}^{n}M = I_n \otimes M$ in the iid case will simplify the expression for the joint distribution as follows.

Corollary 2.4.1 If $\mathbf{y}_1,\ldots,\mathbf{y}_n$ are independent and identically distributed random vectors from the $CSN_{p,q}\left(\boldsymbol{\mu},\Sigma,D,\boldsymbol{\nu},\Delta\right)$ distribution, then the parameters of the joint distribution of $\mathbf{y}_1,\ldots,\mathbf{y}_n$ are

$$p^\dagger = np, \quad q^\dagger = nq, \quad \boldsymbol{\mu}^\dagger = \mathbf{1}_n\otimes\boldsymbol{\mu}, \quad \Sigma^\dagger = I_n\otimes\Sigma,$$

and

$$D^\dagger = I_n\otimes D, \quad \boldsymbol{\nu}^\dagger = \mathbf{1}_n\otimes\boldsymbol{\nu}, \quad \Delta^\dagger = I_n\otimes\Delta.$$

The following result establishes conditions for the independence of linear forms of CSN random vectors.

Corollary 2.4.2 Let $\mathbf{y} \sim CSN_{p,q}\left(\boldsymbol{\mu},\Sigma,D,\boldsymbol{\nu},\Delta\right)$ and let A be an $(n\times p)$ matrix of rank $n \le p$. Let $k \le \min\left(n,q\right)$ be a positive integer. Moreover, suppose that D and A are partitioned as $D = \left(D_1^T,\ldots,D_k^T\right)^T$, $D_i\ (r_i\times p)$, $\sum_{i=1}^{k} r_i = q$ and $A = \left(A_1^T,\ldots,A_k^T\right)^T$, $A_i\left(m_i\times p\right)$ of rank m_i, $\sum_{i=1}^{k} m_i = n$. If

(i) $A_i\Sigma A_j^T = O \quad i\neq j$,

(ii) $D_i\Sigma A_j^T = O \quad i\neq j$,

(iii) $D_i\Sigma D_j^T = O$ (or, in a more general case, if $\Delta + D\Sigma D^T$ is a block diagonal matrix),

(iv) $\Delta = \bigoplus_{i=1}^{k}\Delta_i$, $\Delta_i\left(r_i\times r_i\right) > 0$, then $A_i\mathbf{y}$ is independent of $A_j\mathbf{y}$, $i\neq j$.

Proof. From Proposition 2.3.1 we have that the mgf of $A_i\mathbf{y}$ is given by:

$$M_{A_i\mathbf{y}}(\mathbf{t}_i) = \frac{\Phi_q\left(D\Sigma A_i^T\mathbf{t}_i;\boldsymbol{\nu},\Delta+D\Sigma D^T\right)}{\Phi_q\left(\mathbf{0};\boldsymbol{\nu},\Delta+D\Sigma D^T\right)}e^{\mathbf{t}_i^T A_i\boldsymbol{\mu}+\frac{1}{2}\mathbf{t}_i^T A_i\Sigma A_i^T\mathbf{t}_i}. \tag{2.12}$$

From (i) through (iv), we obtain, respectively:

$$A\Sigma A^T = \bigoplus_{i=1}^k A_i\Sigma A_i^T,\quad D\Sigma A^T = \bigoplus_{i=1}^k D_i\Sigma A_i^T,\quad D\Sigma D = \bigoplus_{i=1}^k D_i\Sigma D_i^T,$$

and:

$$\Delta + D\Sigma D^T = \bigoplus_{i=1}^k\left(\Delta_i + D_i\Sigma D_i^T\right),$$

where $\mathbf{t}_i,\boldsymbol{\mu}_i\in\mathbb{R}^{m_i}$ and $\boldsymbol{\nu}_i\in\mathbb{R}^{r_i}$.

With the help of the former identities we conclude that:

$$\frac{\Phi_q\left(D\Sigma A_i^T\mathbf{t}_i;\boldsymbol{\nu},\Delta+D\Sigma D^T\right)}{\Phi_q\left(\mathbf{0};\boldsymbol{\nu},\Delta+D\Sigma D^T\right)} = \frac{\Phi_{r_i}\left(D_i\Sigma_i A_i^T\mathbf{t}_i;\boldsymbol{\nu}_i,\Delta_i+D_i\Sigma_i D_i^T\right)}{\Phi_{r_i}\left(\mathbf{0};\boldsymbol{\nu}_i,\Delta_i+D_i\Sigma_i D_i^T\right)}. \tag{2.13}$$

Now, using (i) through (iv) the mgf of $A\mathbf{y}$ is:

$$\begin{aligned} M_{A\mathbf{y}}(\mathbf{t}) &= \frac{\Phi_q\left(D\Sigma A^T\mathbf{t};\boldsymbol{\nu},\Delta+D\Sigma D^T\right)}{\Phi_q\left(\mathbf{0};\boldsymbol{\nu},\Delta+D\Sigma D^T\right)}e^{\mathbf{t}^T A\boldsymbol{\mu}+\frac{1}{2}\mathbf{t}^T A\Sigma A^T\mathbf{t}} \\ &= \prod_{i=1}^k\frac{\Phi_{r_i}\left(D_i\Sigma_i A_i^T t_i;\boldsymbol{\nu}_i,\Delta_i+D_i\Sigma_i D_i^T\right)}{\Phi_{r_i}\left(\mathbf{0};\boldsymbol{\nu}_i,\Delta_i+D_i\Sigma_i D_i^T\right)}e^{\mathbf{t}_i^T A_i\boldsymbol{\mu}+\frac{1}{2}\mathbf{t}_i^T A_i\Sigma A_i^T\mathbf{t}_i}, \end{aligned}$$

and from (2.12) and (2.13) we get:

$$\begin{aligned} M_{A\mathbf{y}}(\mathbf{t}) &= \prod_{i=1}^k\frac{\Phi_q\left(D\Sigma A_i^T\mathbf{t}_i;\boldsymbol{\nu},\Delta+D\Sigma D^T\right)}{\Phi_q\left(\mathbf{0};\boldsymbol{\nu},\Delta+D\Sigma D^T\right)}e^{\mathbf{t}_i^T A_i\boldsymbol{\mu}+\frac{1}{2}\mathbf{t}_i^T A_i\Sigma A_i^T\mathbf{t}_i} \\ &= \prod_{i=1}^k M_{A_i\mathbf{y}}(\mathbf{t}_i), \end{aligned}$$

where $\mathbf{t} = (\mathbf{t}_1^T,\dots,\mathbf{t}_k^T)^T$, $\boldsymbol{\mu} = (\boldsymbol{\mu}_1^T,\dots,\boldsymbol{\mu}_k^T)^T$, $\mathbf{t}_i$, $\boldsymbol{\mu}_i\in\mathbb{R}^{m_i}$, $\boldsymbol{\nu} = \left(\boldsymbol{\nu}_1^T,\dots,\boldsymbol{\nu}_k^T\right)^T$, $\boldsymbol{\nu}_i\in\mathbb{R}^{r_i}$. □

2.5 Sums of Independent CSN Random Vectors

In this section we present the main result corresponding to the sum of independent random vectors in the family of the CSN distributions. As we can see from Proposition 2.4.1, the role of the parameter q will be relevant since summation will have an effect only on this dimension parameter adding up the dimensions of all summands.

Proposition 2.5.1 If $\mathbf{y}_1, \ldots, \mathbf{y}_n$ are independent random vectors with $\mathbf{y}_i \sim CSN_{p,q_i}(\boldsymbol{\mu}_i, \Sigma_i, D_i, \boldsymbol{\nu}_i, \Delta_i)$, $i = 1, \ldots, n$, then

$$\sum_{i=1}^{n} \mathbf{y}_i \sim CSN_{p,q^\star}\left(\boldsymbol{\mu}^\star, \Sigma^\star, D^\star, \boldsymbol{\nu}^\star, \Delta^\star\right),$$

where:

$$q^\star = \sum_{i=1}^{n} q_i, \quad \boldsymbol{\mu}^\star = \sum_{i=1}^{n} \boldsymbol{\mu}_i, \quad \Sigma^\star = \sum_{i=1}^{n} \Sigma_i,$$

$$D^\star = \left(\Sigma_1 D_1^T, \ldots, \Sigma_n D_n^T\right)^T \left(\sum_{i=1}^{n} \Sigma_i\right)^{-1}, \quad \boldsymbol{\nu}^\star = \left(\boldsymbol{\nu}_1^T, \ldots, \boldsymbol{\nu}_n^T\right)^T,$$

and:

$$\Delta^\star = \Delta^\dagger + D^\dagger \Sigma^\dagger D^{\dagger T} - \left[\bigoplus_{i=1}^{n} (D_i \Sigma_i)\right] \left(\sum_{i=1}^{n} \Sigma_i\right)^{-1} \left[\bigoplus_{i=1}^{n} (\Sigma_i D_i^T)\right],$$

where $\Delta^\dagger, D^\dagger$, and $\Sigma^\dagger$ are as in Proposition 2.4.1.

Proof. Let $\mathbf{y} = \left(\mathbf{y}_1^T, \ldots, \mathbf{y}_n^T\right)^T$ and $A = \mathbf{1}_n^T \otimes I_p$. Observe that $\sum_{i=1}^{n} \mathbf{y}_i = A\mathbf{y}$, where $\mathbf{y}$ is an (np)-vector and A is a $p \times (np)$ matrix of rank p. Thus, by Propositions 2.3.1 and 2.4.1: $A\mathbf{y} \sim CSN_{p,nq}\left(\boldsymbol{\mu}_A^\dagger,\ \Sigma_A^\dagger,\ D_A^\dagger,\ \boldsymbol{\nu}_A^\dagger,\ \Delta_A^\dagger\right)$, where:

$$\boldsymbol{\mu}_A^\dagger = A\boldsymbol{\mu}^\dagger, \quad \Sigma_A^\dagger = A\Sigma^\dagger A^T, \quad D_A^\dagger = D^\dagger \Sigma^\dagger A^T \left(A\Sigma^\dagger A^T\right)^{-1},$$

and:

$$\boldsymbol{\nu}_A^\dagger = \boldsymbol{\nu}^\dagger,\ \Delta_A^\dagger = \Delta^\dagger + D^\dagger \Sigma^\dagger D^{\dagger T} - D^\dagger \Sigma^\dagger A^T \left(A\Sigma^\dagger A^T\right)^{-1} A\Sigma^\dagger D^{\dagger T},$$

where $\boldsymbol{\mu}^\dagger, \Sigma^\dagger, D^\dagger, \boldsymbol{\nu}^\dagger$, and $\Delta^\dagger$ are given in Proposition 2.4.1.

Now, it is easy to see that:

$$A\boldsymbol{\mu}^\dagger = \sum_{i=1}^{n} \boldsymbol{\mu}_i, \quad A\Sigma^\dagger A^T = \sum_{i=1}^{n} \Sigma_i, \quad D^\dagger \Sigma^\dagger A^T = \bigoplus_{i=1}^{n} (D_i \Sigma_i),$$

$$A\Sigma^\dagger D^{\dagger T} = \left(D^\dagger \Sigma^\dagger A^T\right)^T = \bigoplus_{i=1}^{n} \left(\Sigma_i D_i^T\right), \quad \text{and} \quad D^\dagger \boldsymbol{\mu}^\dagger = \bigoplus_{i=1}^{n} (D_i \boldsymbol{\mu}_i).$$

□

For the iid case we have the following corollary.

Corollary 2.5.1 If $\mathbf{y}_1, \ldots, \mathbf{y}_n$ are iid random vectors following the distribution $\mathbf{y}_i \sim CSN_{p,q}(\boldsymbol{\mu}, \Sigma, D, \boldsymbol{\nu}, \Delta)$, $i = 1, \ldots, n$, then:

$$\sum_{i=1}^{n} \mathbf{y}_i \sim CSN_{p,q^\star}\left(\boldsymbol{\mu}^\star, \Sigma^\star, D^\star, \boldsymbol{\nu}^\star, \Delta^\star\right),$$

where:

$$q^\star = nq, \quad \boldsymbol{\mu}^\star = n\boldsymbol{\mu}, \quad \Sigma^\star = n\Sigma, \quad D^\star = \frac{1}{n}\mathbf{1}_n \otimes D,$$

$$\boldsymbol{\nu}^\star = \mathbf{1}_n \otimes \boldsymbol{\nu}, \quad \Delta^\star = I_n \otimes \left(\Delta + D\Sigma D^T\right) - \frac{1}{n}\mathbf{1}_n\mathbf{1}_n^T \otimes \left(D\Sigma D^T\right).$$

Proof. It is only necessary to illustrate the calculations for $\Delta^\star$. By Proposition 2.5.1 we get:

$$\begin{aligned}
\Delta^\star &= \bigoplus_{i=1}^{n}\Delta + \left(\bigoplus_{i=1}^{n} D\right)\left(\bigoplus_{i=1}^{n}\Sigma\right)\left(\bigoplus_{i=1}^{n} D\right)^T - \frac{1}{n}\left(\mathbf{1}_n \otimes D\Sigma\right)\Sigma^{-1}\left(\mathbf{1}_n \otimes D\Sigma\right)^T \\
&= I_n \otimes \Delta + \left(I_n \otimes D\right)\left(I_n \otimes \Sigma\right)\left(I_n \otimes D^T\right) - \frac{1}{n}\left(\mathbf{1}_n \otimes D\right)\left(\mathbf{1}_n \otimes D\Sigma\right)^T \\
&= I_n \otimes \Delta + I_n \otimes D\Sigma D^T - \frac{1}{n}\mathbf{1}_n\mathbf{1}_n^T \otimes D\Sigma D^T \\
&= I_n \otimes \left(\Delta + D\Sigma D^T\right) - \frac{1}{n}\mathbf{1}_n\mathbf{1}_n^T \otimes D\Sigma D^T.
\end{aligned}$$

□

Domínguez-Molina *et al.* (2003) give an alternative proof of the distributional closure of sums of independent CSN random vectors. Their proof is based on a conditional argument.
Example. If $n = 2$ and $\mathbf{y}_i \sim CSN_{p,q_i}\left(\boldsymbol{\mu}_i, \Sigma_i, D_i, \boldsymbol{\nu}_i, \Delta_i\right)$, $i = 1, 2$, we get

$$\mathbf{y}_1 + \mathbf{y}_2 \sim CSN_{p,q_1+q_2}\left(\boldsymbol{\mu}_1 + \boldsymbol{\mu}_2, \Sigma_1 + \Sigma_2, D^\star, \boldsymbol{\nu}^\star, \Delta^\star\right),$$

where:

$$D^\star = \begin{pmatrix} D_1\Sigma_1\left(\Sigma_1 + \Sigma_2\right)^{-1} \\ D_2\Sigma_2\left(\Sigma_1 + \Sigma_2\right)^{-1} \end{pmatrix}, \quad \Delta^\star = \begin{pmatrix} A_{11} & A_{12} \\ A_{21} & A_{22} \end{pmatrix},$$

and

$$\begin{aligned}
A_{11} &= \Delta_1 + D_1\Sigma_1 D_1^T - D_1\Sigma_1\left(\Sigma_1 + \Sigma_2\right)^{-1}\Sigma_1 D_1^T, \\
A_{22} &= \Delta_2 + D_2\Sigma_2 D_2^T - D_2\Sigma_2\left(\Sigma_1 + \Sigma_2\right)^{-1}\Sigma_2 D_2^T, \\
A_{12} &= -D_1\Sigma_1\left(\Sigma_1 + \Sigma_2\right)^{-1}\Sigma_2 D_2^T.
\end{aligned}$$

2.6 Examples of Sums of Skew-Normal Random Vectors

From the results of Sections 2.3, 2.4, and 2.5, we are able to obtain the distribution of linear transformations (column or row full rank), joint distribution, and sums of independent random vectors of any of the following densities, when we write them as particular cases of the CSN distribution.

2.6.1 Azzalini and Dalla Valle (1996)

Let $\boldsymbol{\mu}, \boldsymbol{\delta}, \mathbf{y}$ be vectors in $\mathbb{R}^p$, Σ a $p \times p$ positive definite matrix. If $\mathbf{y}$ has a multivariate skew-normal distribution according to Azzalini and Dalla

Valle (1996), its density function is:

$$\begin{aligned} g(\mathbf{y}) &= 2\phi_p(\mathbf{y};\boldsymbol{\mu},\Sigma)\,\Phi\left(\boldsymbol{\delta}^T(\mathbf{y}-\boldsymbol{\mu})\right) \\ &= 2\phi_p(\mathbf{y};\boldsymbol{\mu},\Sigma)\,\Phi\left(\boldsymbol{\delta}^T(\mathbf{y}-\boldsymbol{\mu});0,1\right), \end{aligned}$$

that is,

$$\mathbf{y} \sim CSN_{p,1}\left(\boldsymbol{\mu},\Sigma,\boldsymbol{\delta}^T,0,1\right).$$

If $\mathbf{y}_1,\ldots,\mathbf{y}_n$ are iid random vectors each from a $CSN_{p,1}\left(\boldsymbol{\mu},\Sigma,\boldsymbol{\delta}^T,0,1\right)$ distribution, then $\mathbf{s}=\sum_{i=1}^{n}\mathbf{y}_i$ is distributed as:

$$CSN_{p,n}\left(n\boldsymbol{\mu},n\Sigma,\frac{1}{n}\mathbf{1}_n\otimes\boldsymbol{\delta}^T,\mathbf{0},I_n\otimes\left(1+\boldsymbol{\delta}^T\Sigma\boldsymbol{\delta}\right)-\frac{1}{n}\mathbf{1}_n\mathbf{1}_n^T\otimes\left(\boldsymbol{\delta}^T\Sigma\boldsymbol{\delta}\right)\right). \tag{2.14}$$

As a particular case, if $Y_1,\ldots,Y_n$ are iid random variables, each one with the Azzalini (1985) distribution, i.e., $Y_i \sim CSN_{1,1}(0,1,\delta,0,1)$, $i=1,\ldots,n$, then

$$\sum_{i=1}^{n} Y_i \sim CSN_{1,n}\left(0,n,\tfrac{1}{n}\boldsymbol{\delta},\mathbf{0},\Delta^\star\right),$$

where:

$$\boldsymbol{\delta}=\delta\mathbf{1}_n=(\delta,\ldots,\delta)^T,\ \text{ and }\Delta^\star=\left(1+\delta^2\right)I_n-\frac{1}{n}\boldsymbol{\delta}\boldsymbol{\delta}^T,$$

and by (2.14) and (2.9) we obtain that the distribution of the mean is:

$$\bar{Y}=\frac{1}{n}\sum_{i=1}^{n}Y_i \sim CSN_{1,n}\left(0,\tfrac{1}{n},\boldsymbol{\delta},\mathbf{0},\Delta^\star\right).$$

2.6.2 Arnold and Beaver (2002)

Let $\lambda_0\in\mathbb{R}$, $\boldsymbol{\mu},\boldsymbol{\lambda}_1,\mathbf{y}\in\mathbb{R}^p$, Σ a $(p\times p)$ positive definite matrix, and consider the density function:

$$g(\mathbf{y}) \propto \exp\left(-\frac{1}{2}(\mathbf{y}-\boldsymbol{\mu})^T\Sigma^{-1}(\mathbf{y}-\boldsymbol{\mu})\right)\Phi\left(\frac{\delta_0+\boldsymbol{\delta}_1^T\Sigma^{-1/2}(\mathbf{y}-\boldsymbol{\mu})}{\sqrt{1-\boldsymbol{\delta}_1^T\boldsymbol{\delta}_1}}\right), \tag{2.15}$$

with $\delta_0 = \lambda_0/\sqrt{1+\boldsymbol{\lambda}_1^T\boldsymbol{\lambda}_1}$ and $\boldsymbol{\delta}_1 = \boldsymbol{\lambda}_1/\sqrt{1+\boldsymbol{\lambda}_1^T\boldsymbol{\lambda}_1}$. Then:

$$
\begin{aligned}
&g(\mathbf{y}) \\
&= \frac{1}{(2\pi)^{p/2}|\Sigma|^{1/2}\Phi(\delta_0)}\exp\left(-\frac{1}{2}(\mathbf{y}-\boldsymbol{\mu})^T\Sigma^{-1}(\mathbf{y}-\boldsymbol{\mu})\right) \\
&\quad\times\Phi\left(\frac{\delta_0+\boldsymbol{\delta}_1^T\Sigma^{-1/2}(\mathbf{y}-\boldsymbol{\mu})}{\sqrt{1-\boldsymbol{\delta}_1^T\boldsymbol{\delta}_1}}\right) \\
&= \frac{1}{\Phi(\delta_0)}\phi_p(\mathbf{y};\boldsymbol{\mu},\Sigma)\,\Phi\left(\frac{\boldsymbol{\delta}_1^T}{\sqrt{1-\boldsymbol{\delta}_1^T\boldsymbol{\delta}_1}}\Sigma^{-1/2}(\mathbf{y}-\boldsymbol{\mu});-\frac{\delta_0}{\sqrt{1-\boldsymbol{\delta}_1^T\boldsymbol{\delta}_1}},1\right).
\end{aligned}
$$

Thus, if $\mathbf{y}$ is a random vector with density function given by (2.15), we have that:

$$
\begin{aligned}
\mathbf{y} &\sim CSN_{p,1}\left(\boldsymbol{\mu},\Sigma,\frac{\boldsymbol{\delta}_1^T}{\sqrt{1-\boldsymbol{\delta}_1^T\boldsymbol{\delta}_1}}\Sigma^{-1/2},\frac{-\delta_0}{\sqrt{1-\boldsymbol{\delta}_1^T\boldsymbol{\delta}_1}},1\right) \\
&= CSN_{p,1}\left(\boldsymbol{\mu},\Sigma,\boldsymbol{\lambda}_1^T\Sigma^{-1/2},-\lambda_0,1\right),
\end{aligned}
$$

and hence, the distribution of the sum of n iid random vectors with density (2.15) is

$$
\begin{aligned}
\mathbf{s} \sim CSN_{p,n}\Big(&n\boldsymbol{\mu},n\Sigma,\frac{1}{n}\mathbf{1}_n\otimes\left(\boldsymbol{\lambda}_1^T\Sigma^{-1/2}\right), \\
&-\mathbf{1}_n\otimes\lambda_0, I_n\otimes\left(1+\boldsymbol{\lambda}_1^T\boldsymbol{\lambda}_1\right)-\frac{1}{n}\mathbf{1}_n\mathbf{1}_n^T\otimes\left(\boldsymbol{\lambda}_1^T\boldsymbol{\lambda}_1\right)\Big).
\end{aligned}
$$

2.6.3 Liseo and Loperfido (2003a)

Let $\mathbf{y}$ and $\boldsymbol{\mu}$ be vectors in $\mathbb{R}^p$ and C be a full row rank $k \times p$ matrix and $\mathbf{d} \in \mathbb{R}^k$. Consider the positive definite $p \times p$ matrices, Σ and Ω. Liseo and Loperfido (2003a) proposed the following density function:

$$
\begin{aligned}
g(\mathbf{y}) &= K\phi_p(\mathbf{y},\boldsymbol{\mu},\Sigma+\Omega)\,\Phi_k\left(\mathbf{0};C\Delta\left(\Sigma^{-1}\mathbf{y}+\Omega^{-1}\boldsymbol{\mu}\right)+\mathbf{d},C\Delta C^T\right) \qquad (2.16)\\
&= K\phi_p(\mathbf{y},\boldsymbol{\mu},\Sigma+\Omega)\,\Phi_k\left(-C\Delta\Sigma^{-1}(\mathbf{y}-\boldsymbol{\mu});C\boldsymbol{\mu}+\mathbf{d},C\Delta C^T\right),
\end{aligned}
$$

where $K^{-1} = \Phi_k\left(\mathbf{0};C\boldsymbol{\mu}+\mathbf{d},C\Omega C^T\right)$ and $\Delta^{-1} = \Sigma^{-1}+\Omega^{-1}$.
Then if $\mathbf{y}$ is a random vector with density given by (2.16) we get that

$$
\mathbf{y} \sim CSN_{p,k}\left(\boldsymbol{\mu},\Sigma+\Omega,-C\Delta\Sigma^{-1},C\boldsymbol{\mu}+\mathbf{d},C\Delta C^T\right). \qquad (2.17)
$$

If $\mathbf{y}_1,\ldots,\mathbf{y}_n$ are iid random vectors, each one following the distribution

(2.17), then the distribution of the sum or the mean can be easily obtained by Corollary 2.5.1 or Proposition 2.5.1.

For all the above cases of multivariate skew-normal distributions, it is a straightforward matter to compute the distribution of the sum in the non-identically distributed case.

2.7 A Multivariate Extended Skew-Elliptical Distribution

In this section we consider elliptical random vectors whose density function exists and $P(\mathbf{y} = \mathbf{0}) = 0$. Suppose that a p-dimensional random vector $\mathbf{y}$ has a density of the form:

$$f_p(\mathbf{y}; \boldsymbol{\eta}, \Theta, h) = |\Theta|^{-1/2} h\left((\mathbf{y} - \boldsymbol{\eta})^T \Theta^{-1} (\mathbf{y} - \boldsymbol{\eta})\right), \tag{2.18}$$

where $\boldsymbol{\eta} \in \mathbb{R}^p$, Θ is a $p \times p$ positive definite matrix and $h(\cdot)$ is a non-negative function of a scalar variable such that:

$$\pi^{\frac{p}{2}} \int_0^\infty t^{\frac{p}{2}-1} h(t)\, dt = \Gamma\left(\tfrac{p}{2}\right).$$

If $\mathbf{y}$ has density function given by (2.18) we say that $\mathbf{y}$ has an elliptically contoured distribution (or elliptically symmetric distribution) and it will be denoted by $\mathbf{y} \sim EC_p(\boldsymbol{\eta}, \Theta, h)$, where $h(\cdot)$ is called the pdf generator of the elliptically contoured distribution. We recommend Fang, Kotz, and Ng (1990) as a comprehensive treatise concerning symmetric and elliptically symmetric distributions.

It is possible to obtain a multivariate extended skew-elliptical (ESE) in a way similar to the one we used to construct the CSN distribution given at the beginning of this chapter.

Let:

$$\begin{pmatrix} \mathbf{w}_0 \\ \mathbf{z} \end{pmatrix} \sim EC_{q+p}\left(\begin{pmatrix} \boldsymbol{\mu} \\ -\boldsymbol{\nu} \end{pmatrix}, \begin{pmatrix} \Sigma & \Sigma D^T \\ D\Sigma & \Delta + D\Sigma D^T \end{pmatrix}, h\right), \tag{2.19}$$

where $\mathbf{w}_0$ is a random p-vector, $\mathbf{z}$ is a random q-vector, $\boldsymbol{\mu} \in \mathbb{R}^p, \boldsymbol{\nu} \in \mathbb{R}^q$, D is an arbitrary $q \times p$ matrix, and Σ and Δ are $p \times p$ and $q \times q$ positive definite matrices, respectively.

From Theorem 2.18 of Fang *et al.* (1990) we deduce that:

$$\mathbf{z}|\{\mathbf{w}_0 = \mathbf{w}\} \sim EC_q\left(-\boldsymbol{\nu} + D(\mathbf{w} - \boldsymbol{\mu}), \Delta, h_{s(\mathbf{w})}\right),$$

where

$$h_a(u) = \frac{h(a+u)}{h_2(a)}, \quad a, u > 0, \tag{2.20}$$

with h_2 being the pdf generator of the marginal distribution of $\mathbf{z}$.

Then using Equation (5.1) of Arellano-Valle *et al.* (2002) it follows that the conditional density of $\mathbf{w}_0 | \{\mathbf{z} \geq \mathbf{0}\}$ is given by:

$$f_{\mathbf{w}_0|\{\mathbf{z}\geq 0\}}\left(\mathbf{w}|\{\mathbf{z}\geq \mathbf{0}\}\right) = \frac{f_p\left(\mathbf{w};\boldsymbol{\mu},\Sigma,h_1\right)F_q\left[D\left(\mathbf{w}-\boldsymbol{\mu}\right);\boldsymbol{\nu},\Delta,h_{s(\mathbf{w})}\right]}{F_q\left(\mathbf{0};\boldsymbol{\nu},\Delta+D\Sigma D^T,h_2\right)}, \tag{2.21}$$

where $f_p(\cdot;\cdot,\cdot,h_1)$ is as in (2.18), $h_{s(\mathbf{w})}$ is given in (2.20), and $F_q\left(\cdot;\cdot,\cdot,h^{\dagger}\right)$ is the cdf of a random vector with pdf generator $h^{\dagger}$. We will denote a random p-vector, $\mathbf{y}$, distributed according to an ESE distribution with parameters $\boldsymbol{\mu},\Sigma,D,\boldsymbol{\nu},\Delta$ and pdf generator h by $\mathbf{y}\sim ESE_{p,q}\left(\boldsymbol{\mu},\Sigma,D,\boldsymbol{\nu},\Delta,h\right)$. The CSN distribution is obtained by taking $h(u)=(2\pi)^{-\left(\frac{p+q}{2}\right)}e^{-u/2}$.

Using the closure properties of elliptically contoured distributions for linear transformations in (2.19) it is easy to prove that the ESE distributions are closed under full row rank linear transformations. This and other properties are being presently developed.

2.8 Concluding Remarks

The CSN is a generalization of the multivariate skew-normal distribution, such that it is possible to preserve some important properties of the normal distribution, for instance, being closed under:

- Marginalization
- Conditional distributions
- Linear transformations (full column, or row, rank)
- Sums of independent random variables from this family
- Joint distribution of independent random variables in this family.

On the other hand, there are still several other problems to be solved that range from the numerical calculation of some maximum likelihood estimators to the definition of singular densities for this type of distributions, extensions to matrix representations, and so on.

Note that when we apply a linear transformation to a CSN random vector, the value of q is not affected, which is natural since the effect of a linear transformation will be, at the most, to reduce the dimension of $\mathbf{y}$, i.e., p. On the other hand, if we are adding independent CSN random vectors, the dimension p has to be fixed in order to sum them, and what Proposition 2.4 tells us is that the dimension q is the one that changes. If we think about the conditional argument given in Domínguez-Molina *et al.* (2003), when we add those random variables the effect would be that the number of conditions are adding up. In many practical cases $q=1$, and it will not be necessary to estimate this extra parameter, and when we sum, we simply have $q=k$, k being the number of summands. As a consequence of Proposition 2.4, it is a straightforward matter to see that a CSN distribution is not a stable distribution.

In terms of estimation, even for the univariate case, it is known that we have problems of identifiability depending on the parameterization used, see Pewsey (2000). Similar problems are expected for the multivariate case besides the usual computational problems, as given in Azzalini and Capitanio (1999). These issues have not been completely resolved and are being developed at present for CSN distributions.

Acknowledgments

This study was partially supported by CONACYT research grant 39017E and CONCYTEG research grant 03-16-K118-027, Anexo 01.

CHAPTER 3

Skew-Elliptical Distributions

Junfeng Liu and Dipak K. Dey

3.1 Introduction

It is well-known that symmetric distributions are not suitable for modeling all types of data. Therefore, there is a strong motivation to construct skewed distribution with nice properties. Azzalini (1985, 1986) formalized the univariate skew-normal distribution having the probability density function (pdf) of the form $2\phi(x)\Phi(\alpha x)$, $x, \alpha \in \mathbb{R}$. Azzalini and Dalla Valle (1996) extended the results to the multivariate setting with the pdf of the form $2\phi_p(\mathbf{x}; \boldsymbol{\mu}, \Omega)\Phi(\boldsymbol{\alpha}^T(\mathbf{x} - \boldsymbol{\mu}))$, $\mathbf{x}, \boldsymbol{\alpha} \in \mathbb{R}^p$, which has been generalized later by many authors. To incorporate departures from normality, the first approach taken is to consider elliptical distributions. Consider a p-dimensional random vector $\mathbf{x}$ having a pdf of the form

$$f(\mathbf{x}|\boldsymbol{\mu}, \Omega, g^{(p)}) = |\Omega|^{-\frac{1}{2}} g^{(p)}((\mathbf{x} - \boldsymbol{\mu})^T \Omega^{-1}(\mathbf{x} - \boldsymbol{\mu})), \ \mathbf{x} \in \mathbb{R}^p, \tag{3.1}$$

where $g^{(p)}(u)$ is a non-increasing function from $\mathbb{R}_+$ to $\mathbb{R}_+$ defined by

$$g^{(p)}(u) = \frac{\Gamma(p/2)}{\pi^{p/2}} \frac{g(u;p)}{\int_0^\infty r^{p/2-1} g(r;p)dr}, \tag{3.2}$$

and $g(u;p)$ is a non-increasing function from $\mathbb{R}_+$ to $\mathbb{R}_+$ such that the integral $\int_0^\infty r^{p/2-1} g(r;p)dr$ exists. We will always assume the existence of the pdf $f(\mathbf{x}|\boldsymbol{\mu}, \Omega, g^{(p)})$. The function $g^{(p)}$ is often called the *density generator* of the random vector $\mathbf{x}$. Note that the function $g(u;p)$ provides the kernel of $\mathbf{x}$ and other terms in $g^{(p)}$ constitute the normalizing constant for the pdf f. In addition, the function g, hence $g^{(p)}$, may depend on other parameters that would be clear from the context. For example, in the case of the t distribution, the additional parameter will be the degrees of freedom. The density f defined above represents a broad class of distributions called the *elliptically contoured* distributions, and we will use the notation $\mathbf{x} \sim EC_p(\boldsymbol{\mu}, \Omega, g^{(p)})$ from now on in this chapter. Let $F\left(\mathbf{x}|\boldsymbol{\mu}, \Omega, g^{(p)}\right)$ denote the cumulative distribution function (cdf) of $\mathbf{x}$.

In the fields of biology, economics, psychology, sociology, and so on, often the application backgrounds exhibit non-normal error structures. The skewness is one of the significant aspects for such departures from normality. It is suggested that the error structures in this case should be handled in a quite flexible way, i.e., beyond the prevalent normality or multivariate normality framework. Arnold and Beaver (2002) gave a comprehensive review of the literature on multivariate skewness construction, interpretation, and property research. Their explanation for the skewness mechanism is related to hidden truncation and/or selective reporting. A very convenient, flexible, and well-behaved multivariate distribution class is the multivariate skew-elliptical distribution (see Branco and Dey, 2001) which can be used for modeling skewed distributions with heavy tails. Branco and Dey (2001) presented a class of multivariate skew-elliptical distributions that includes several unimodal elliptical and spherical distributions. Their method is developed by introducing a skewness parameter. The new distribution brings additional flexibility of modeling skewed data that can be used in regression and calibration in the presence of skewness. They studied the properties of such skew-elliptical distributions, and we discuss some of them in Section 3.4. Arnold and Beaver (2002) interpreted univariate skew-normal, non-normal univariate skewed distributions, multivariate skew-normal, and non-normal skewed multivariate distributions. They also discussed their construction of skew-elliptical distributions, which included the method of Branco and Dey (2001) as a special case.

A recent paper by Dey and Liu (2003) utilized linear constraint and linear combination (LCLC1 and LCLC2) on a general random vector from an elliptically contoured distribution to construct a very general class of multivariate skew-elliptical distributions. They further showed the equivalence of the LCLC1 and LCLC2 constructions under very mild conditions and consequently many previous constructions became special cases of their results. The important property of a multivariate elliptical distribution is conditional closure, which they used to obtain the closed form density of multivariate skew-elliptical distributions. They used random vector transformations to obtain the closed form of LCLC2 distributions and simultaneously proved the equivalence of LCLC1 and LCLC2 under a mild condition.

The organization of this chapter is as follows. Section 3.2 describes general multivariate skew-elliptical distribution. Section 3.3 gives examples of skew-elliptical distributions. Section 3.4 shows some properties of skew-elliptical distributions, and Section 3.5 concludes with remarks.

3.2 General Multivariate Skew-Elliptical Distributions

The following two lemmas are crucial for the skewness construction. Their proof follows from Fang, Kotz, and Ng (1990).

Lemma 3.2.1 Suppose $\mathbf{y} \sim EC_p(\boldsymbol{\mu}, \Omega, g^{(p)})$, and R is a matrix of size $k \times p$, then $R\mathbf{y} \sim EC_k(R\boldsymbol{\mu}, R\Omega R^T, g^{(k)})$, where the marginal generator $g^{(k)}$ is defined as

$$g^{(k)}(u) = \frac{2\pi^{\frac{k}{2}}}{\Gamma(\frac{k}{2})} \int_0^\infty g^{(k+1)}(r^2+u)r^{k-1}dr, \quad u \geq 0.$$

Lemma 3.2.2 If $\mathbf{y} \sim EC_p(\boldsymbol{\mu}, \Omega, g^{(p)})$, and $\mathbf{y}$ is partitioned as $\mathbf{y} = (\mathbf{y}_1, \mathbf{y}_2)^T$, where $\mathbf{y}_1$ is $p_1 \times 1$ and $\mathbf{y}_2$ is $p_2 \times 1$ with

$$\begin{pmatrix} \mathbf{y}_1 \\ \mathbf{y}_2 \end{pmatrix} \sim EC_p\left(\begin{pmatrix} \boldsymbol{\mu}_1 \\ \boldsymbol{\mu}_2 \end{pmatrix}, \begin{pmatrix} \Omega_{11} & \Omega_{12} \\ \Omega_{21} & \Omega_{22} \end{pmatrix}, g^{(p)} \right), \quad p = p_1 + p_2,$$

then

$$\mathbf{y}_1|\mathbf{y}_2 = \mathbf{y}_2^* \sim EC_{p_1}(\boldsymbol{\mu}_{1.2}, \Omega_{11.2}, g^{(p_1)}_{q(\mathbf{y}_2^*)}),$$

where

$$\begin{aligned} \boldsymbol{\mu}_{1.2} &= \boldsymbol{\mu}_1 + \Omega_{12}\Omega_{22}^{-1}(\mathbf{y}_2^* - \boldsymbol{\mu}_2), \\ \Omega_{11.2} &= \Omega_{11} - \Omega_{12}\Omega_{22}^{-1}\Omega_{21}, \\ q(\mathbf{y}_2^*) &= (\mathbf{y}_2^* - \boldsymbol{\mu}_2)^T\Omega_{22}^{-1}(\mathbf{y}_2^* - \boldsymbol{\mu}_2), \\ g_{q(\mathbf{y})}(u) &= g^{(k+1)}(u + q(\mathbf{y}))/g^{(k)}(q(\mathbf{y})), \\ q(\mathbf{y}) &= (\mathbf{y} - \boldsymbol{\mu})^T\Omega^{-1}(\mathbf{y} - \boldsymbol{\mu}). \end{aligned}$$

3.2.1 The Branco and Dey Approach

By a conditioning method, Branco and Dey (2001) extended Azzalini and Dalla Valle's (1996) results to multivariate skew-elliptical distributions. Consider a random vector $\mathbf{x} = (X_1, \ldots, X_p)^T$. Let $\mathbf{x}^* = (X_0, \mathbf{x}^T)^T$ be a $(p+1)$-dimensional random vector, such that $\mathbf{x}^* \sim EC_{p+1}(\boldsymbol{\mu}^*, \Sigma, \varphi)$, where $\boldsymbol{\mu}^* = (0, \boldsymbol{\mu}^T)^T$, $\boldsymbol{\mu} = (\mu_1, \ldots, \mu_p)^T$, φ is the characteristic function, and the scale parameter matrix Σ has the form

$$\Sigma = \begin{pmatrix} 1 & \boldsymbol{\delta}^T \\ \boldsymbol{\delta} & \Omega \end{pmatrix},$$

with $\boldsymbol{\delta} = (\delta_1, \ldots, \delta_p)^T$. Here Ω is the scale matrix associated with the vector $\mathbf{x}$. We say that the random vector $\mathbf{y} = [\mathbf{x}|X_0 > 0]$ has a skew-elliptical distribution and denote by $\mathbf{y} \sim SE_p(\boldsymbol{\mu}, \Omega, \boldsymbol{\delta}, \varphi)$, where $\boldsymbol{\delta}$ is the skewness parameter. If the density of the random vector $\mathbf{x}^*$ exists and $\mathrm{P}(\mathbf{x}^* = \mathbf{0}) = 0$, then the pdf of $\mathbf{y}$ will be of the form

$$f_{\mathbf{y}}(\mathbf{y}) = 2f_{g^{(p)}}(\mathbf{y})F_{g(q(\mathbf{y}))}(\boldsymbol{\alpha}^T(\mathbf{y} - \boldsymbol{\mu})), \tag{3.3}$$

where $f_{g^{(p)}}(\cdot)$ is the pdf of an $EC_p(\boldsymbol{\mu}, \Omega, g^{(p)})$ and $F_{g(q(\mathbf{y}))}$ is the cdf of an $EC_1(0, 1, g_{q(\mathbf{y})})$, with

$$\begin{aligned} \boldsymbol{\alpha}^T &= \frac{\boldsymbol{\delta}^T\Omega^{-1}}{(1-\boldsymbol{\delta}^T\Omega^{-1}\boldsymbol{\delta})^{1/2}}, \\ g^{(p)}(u) &= \frac{2\pi^{\frac{p}{2}}}{\Gamma(\frac{p}{2})}\int_0^\infty g^{(p+1)}(r^2+u)r^{p-1}dr, \quad u \geq 0, \\ g_{q(\mathbf{y})}(u) &= g^{(p+1)}(u+q(\mathbf{y}))/g^{(p)}(q(\mathbf{y})), \\ q(\mathbf{y}) &= (\mathbf{y}-\boldsymbol{\mu})^T\Omega^{-1}(\mathbf{y}-\boldsymbol{\mu}). \end{aligned} \tag{3.4}$$

It is denoted by $\mathbf{y} \sim SE_p(\boldsymbol{\mu}, \Omega, \boldsymbol{\delta}, g^{(p+1)})$, where $g^{(p+1)}$ is the generator function as given in (3.4) with p replaced by $p+1$. From (3.4) and the positive definiteness of the Σ matrix, it follows that $\boldsymbol{\delta}$ and Ω must satisfy the condition $\boldsymbol{\delta}^T\Omega^{-1}\boldsymbol{\delta} < 1$. Using the relation between the elliptical generator functions they obtained an alternative and convenient expression for the pdf of the skew-elliptical distribution as

$$f_{\mathbf{y}}(\mathbf{y}) = 2|\Omega|^{-1/2}\int_{-\infty}^{\alpha^T(\mathbf{y}-\boldsymbol{\mu})} g^{(p+1)}(r^2 + (\mathbf{y}-\boldsymbol{\mu})^T\Omega^{-1}(\mathbf{y}-\boldsymbol{\mu}))dr. \tag{3.5}$$

3.2.2 The Arnold and Beaver Approach

Arnold and Beaver (2002) considered a p-dimensional random vector $\mathbf{w}$ which has an elliptically contoured distribution, e.g.,

$$\mathbf{w} = \boldsymbol{\mu} + \Sigma^{1/2}\mathbf{z}, \tag{3.6}$$

where $\mathbf{z}$ is a spherically symmetric random vector. Such a random vector $\mathbf{z}$ may be represented as

$$\mathbf{z} = R\mathbf{u},$$

where the random variable R and the random vector $\mathbf{u} = (U_1, \ldots, U_p)^T$ are independent, with $\mathrm{P}(R > 0) = 1$, $R \sim F_R$, and $\mathbf{u}$ is uniformly distributed over the unit p-sphere. The distribution of $\mathbf{z}$ determines the distribution of $\mathbf{w}$. The components of $\mathbf{w}$ are dependent, and so are the components of $\mathbf{z}$. It is known that the components of $\mathbf{z}$ will be independent only if R has a particular chi-distribution, i.e., R is a constant multiple of the square root of a χ^2 distribution with p degrees of freedom. Then $\mathbf{z}$ will be $N_p(\mathbf{0}, I_p)$ and $\mathbf{w}$ will also have independent components if $\Sigma^{1/2}$ is orthogonal.

Beginning with a $(p+1)$-dimensional elliptically contoured random vector of the form (3.6), say $(W_0, W_1, \ldots, W_p)^T$, Arnold and Beaver (2002) considered the conditional distribution of $(W_1, \ldots, W_p)^T$ given that $W_0 > c$, which they called a p-dimensional skew-elliptical density, where c is a constant. If we consider a special case in which R has an appropriate chi-distribution, this skewed model will reduce to the model obtained by applying transformations to the density

$$f(\mathbf{w};\mathbf{0},\boldsymbol{\alpha}_1) = 2[\prod_{i=1}^{p}\psi(\mathbf{w}_i)]\Psi(\boldsymbol{\alpha}_1^T\mathbf{w}), \tag{3.7}$$

where ψ is the pdf, Ψ is the cdf. They chose $\mathbf{z}$ such that $\mathbf{u}$ has Cauchy marginals to get a skew-Cauchy distribution. It is well-known that many spherically symmetric random vectors can be represented as scale mixtures of $N_p(\mathbf{0},\sigma^2 I_p)$ random vectors. For such distributions and corresponding skewed counterparts, Branco and Dey (2001) obtained expressions for moments and moment generating functions by suitable integration of the corresponding expressions for their normal and skew-normal counterparts; see Section 3.4 for details.

Suppose $\mathbf{w} = (W_1,\ldots,W_p)^T$ and U are independent with densities given by ψ_1, …, ψ_p and cdf Ψ_0, respectively. The conditional distribution of $\mathbf{w}$ given that $\alpha_0 + \boldsymbol{\alpha}_1^T\mathbf{w} > U$, where $\alpha_0 \in \mathbb{R}$ and $\boldsymbol{\alpha}_1 \in \mathbb{R}^p$ is

$$f_{\mathbf{w}|A}(\mathbf{w}) = \frac{\prod_{i=1}^{p}\psi_i(\mathbf{w}_i)\Psi_0(\alpha_0+\boldsymbol{\alpha}_1^T\mathbf{w})}{\mathrm{P}(A)}, \tag{3.8}$$

where $A = \{\alpha_0 + \boldsymbol{\alpha}_1^T\mathbf{w} > U\}$. If we assume a joint density of $\mathbf{w}$ as $\psi(\mathbf{w})$, the above density will be

$$f_{\mathbf{w}|A}(\mathbf{w}) = \frac{\psi(\mathbf{w})\Psi_0(\alpha_0+\boldsymbol{\alpha}_1^T\mathbf{w})}{\mathrm{P}(A)}. \tag{3.9}$$

Arnold and Beaver (2002) pointed out that, if we begin with $\psi(\mathbf{w})$, an elliptically contoured density, then formula (3.9) included the Branco-Dey approaches as special cases. A different type of skew-elliptical distribution by Azzalini and Capitanio (1999) is also a special case of it.

Arnold, Castillo, and Sarabia (2002) used a conditional specification paradigm to construct a skew-normal density of the form

$$f(\mathbf{x}) \propto \phi(\mathbf{x})\Phi(P_{K^*}(\mathbf{x})), \tag{3.10}$$

where P_{K^*} is a polynomial in $\mathbf{x}$ of degree K^*. For example, for $p=1$ and $K^*=2$, we obtain the quadratically skew-normal (QSN) density:

$$f(x;\alpha_0,\alpha_1,\alpha_2) \propto \phi(x)\Phi(\alpha_0+\alpha_1 x+\alpha_2 x^2).$$

3.2.3 The Wang-Boyer-Genton Approach

In this section we will review a skewness construction given by Wang, Boyer, and Genton (2004a). They constructed a random vector with pdf

$$2f(\mathbf{x}-\boldsymbol{\xi})\pi(\mathbf{x}-\boldsymbol{\xi}), \tag{3.11}$$

where $f:\mathbb{R}^p \to \mathbb{R}_+$ is a pdf symmetric around $\mathbf{0}$, $\pi:\mathbb{R}^p \to [0,1]$ is a *skewing* function with $0 \le \pi(\mathbf{x}) \le 1$, $\pi(-\mathbf{x}) = 1-\pi(\mathbf{x})$, and $\boldsymbol{\xi}$ is any point in $\mathbb{R}^p$, which is defined as *skew-symmetric distribution* with respect to $\boldsymbol{\xi}$ with *symmetric component* f and *skewed component* π. The construction procedure of Wang *et al.* (2004a) is quite straightforward. Compared to the LCLC1/LCLC2 skewness mechanism of Dey and Liu (2003), their skewness argument is motivated by a Bayesian simulation. Let $\mathbf{y} \sim f(\mathbf{y} - \boldsymbol{\xi})$ and $U \sim U[0,1]$, then the skewed random vector is produced by a conditional distribution

$$\mathbf{y} | U \leq \pi(\mathbf{y} - \boldsymbol{\xi}). \tag{3.12}$$

Using mathematical tools, they decomposed any pdf $g : \mathbb{R}^p \to \mathbb{R}_+$, with $\boldsymbol{\xi}$ being any point in $\mathbb{R}^p$, into $g(\mathbf{x}) = 2f_{\boldsymbol{\xi}}(\mathbf{x} - \boldsymbol{\xi})\pi_{\boldsymbol{\xi}}(\mathbf{x} - \boldsymbol{\xi})$, where

$$f_{\boldsymbol{\xi}}(\mathbf{s}) = \frac{1}{2}\{g(\boldsymbol{\xi} + \mathbf{s}) + g(\boldsymbol{\xi} - \mathbf{s})\}, \tag{3.13}$$

$$\pi_{\boldsymbol{\xi}}(\mathbf{s}) = \frac{g(\boldsymbol{\xi} + \mathbf{s})}{g(\boldsymbol{\xi} + \mathbf{s}) + g(\boldsymbol{\xi} - \mathbf{s})}. \tag{3.14}$$

Within the setup given by (3.11), Ma and Genton (2004) constructed a class of flexible skew-symmetric (FSS) distributions through approximating the skewing function π on a compact set, e.g., $\pi_K(\mathbf{x}) = H(P_K(\mathbf{x}))$, where H is any cdf of a continuous random variable symmetric around 0 and P_K is an odd polynomial of order K. When the pdf f is elliptically contoured, they defined it as flexible generalized skew-elliptical ($FGSE$) distributions, whereas the flexible generalized skew-normal ($FGSN$) distributions are defined by

$$2\phi_p(\mathbf{x}; \boldsymbol{\xi}, \Omega)\Phi(P_K(A(\mathbf{x} - \boldsymbol{\xi}))), \tag{3.15}$$

where $\Omega^{-1} = A^T A$. This skewed class incorporates skewness, heavy tails, and multimodality.

3.2.4 The Dey and Liu Approach

Suppose the random vector $\mathbf{y} \sim EC_p(\boldsymbol{\mu}, \Omega, g^{(p)})$ and satisfies the following linear constraint:

$$R\mathbf{y} + \mathbf{d} \leq \mathbf{0}, \tag{3.16}$$

where R is a given matrix of dimension $k \times p$, $k \leq p$, and $\mathbf{d}$ is a vector of dimension k. Then, defining p_c as the probability of the constraint set, we have

$$\begin{aligned} p_c &= \mathrm{P}(R\mathbf{y} + \mathbf{d} \leq \mathbf{0}) = \mathrm{P}(R\mathbf{y} - R\boldsymbol{\mu} \leq -\mathbf{d} - R\boldsymbol{\mu}) \\ &= F(-\mathbf{d} - R\boldsymbol{\mu} | \mathbf{0}, R\Omega R^T, g^{(k)}), \end{aligned}$$

where F is the cdf of an elliptically contoured distribution with location $\mathbf{0}$, scale $R\Omega R^T$, and density generator $g^{(k)}$. Further partitioning R as $R = (R_1, R_2)$, we can express the constraint (3.16) as

$$R_1\mathbf{y}_1 + R_2\mathbf{y}_2 + \mathbf{d} \leq \mathbf{0}, \tag{3.17}$$

where the dimensions of the vectors of $\mathbf{y}_1$ and $\mathbf{y}_2$ are respectively p_1 and p_2. The dimension of R_1 is $k \times p_1$, the dimension of R_2 is $k \times p_2$. The matrix R_1 is of full row rank, and R_2 is an arbitrary matrix such that the matrix multiplication is valid.

Linear Constraint and Linear Combination of Type-1 (LCLC1)

Suppose the dimension of $\mathbf{y}_1$ is not equal to the dimension of $\mathbf{y}_2$, i.e., $p_1 \neq p_2$. Consider the distribution of $C_2\mathbf{y}_2$, where C_2 is a non-singular square matrix with dimension $p_2 \times p_2$. Here $\mathbf{y}_1$ is used only as an auxiliary variable for producing skewness for $\mathbf{y}_2$, it does not show up in the final expression for the pdf of the skewed distribution. Under the conditions in LCLC1, if C_2 is a non-singular matrix of dimension $p_2 \times p_2$, then the density function for $\mathbf{x} = C_2\mathbf{y}_2$ under the constraint (3.17) is

$$\frac{F(R_1\Omega_{12}\Omega_{22}^{-1}\boldsymbol{\mu}_2 - R_1\boldsymbol{\mu}_1 - \mathbf{d} - (R_2 + R_1\Omega_{12}\Omega_{22}^{-1})C_2^{-1}\mathbf{x}|\mathbf{0}, \Omega_{\mathbf{x}}, g_{\mathbf{x}}^{(k)})}{F(-R\boldsymbol{\mu} - \mathbf{d}|0, R\Omega R^T, g^{(k)})}$$
$$\times f(\mathbf{x}|C_2\boldsymbol{\mu}_2, C_2\Omega_{22}C_2^T, g^{(p_2)}),$$

where k is the number of rows in R, and

$$\begin{aligned}\Omega_{\mathbf{x}} &= R_1(\Omega_{11} - \Omega_{12}\Omega_{22}^{-1}\Omega_{21})R_1^T,\\ g_{\mathbf{x}}^{(k)} &= g^{(k)}_{(C_2^{-1}\mathbf{x}-\boldsymbol{\mu}_2)^T\Omega_{22}^{-1}(C_2^{-1}\mathbf{x}-\boldsymbol{\mu}_2)}.\end{aligned}$$

Linear Constraint and Linear Combination of Type-2 (LCLC2)

If the dimension of $\mathbf{y}_1$ is the same as that of $\mathbf{y}_2$, i.e., $p_1 = p_2$, we consider the distribution of $C_1\mathbf{y}_1 + C_2\mathbf{y}_2$ under the constraint (3.17), with $R_1C_1^{-1} - R_2C_2^{-1}$ full row rank, where C_1, C_2 are non-singular square matrices. Under the conditions in LCLC2, the pdf for $\mathbf{x} = C_1\mathbf{y}_1 + C_2\mathbf{y}_2$ is

$$\frac{F(-C_3(\boldsymbol{\mu}_1^{wx} - \Omega_{12}^{wx}\Omega_{22}^{wx\,-1}\boldsymbol{\mu}_2^{wx}) - \mathbf{d} - \Omega_4\mathbf{x}|\mathbf{0}, C_3\Omega_{11}^{*}C_3^T, g_{a^*}^{(k)})}{F(-\mathbf{d} - R\boldsymbol{\mu}|\mathbf{0}, R\Omega R^T, g^{(k)})}$$
$$\times f(\mathbf{x}|\boldsymbol{\mu}_2^{wx}, \Omega_{22}^{wx}; g^{(p_1)}),$$

where

$$\Omega_4 = R_2C_2^{-1} + C_3\Omega_{12}^{wx}\Omega_{22}^{wx\,-1},$$

and

$$\boldsymbol{\mu}^{wx} = \begin{pmatrix} \boldsymbol{\mu}_1^{wx} \\ \boldsymbol{\mu}_2^{wx} \end{pmatrix} = \begin{pmatrix} C_1\boldsymbol{\mu}_1 \\ C_1\boldsymbol{\mu}_1 + C_2\boldsymbol{\mu}_2 \end{pmatrix},$$

$$
\begin{aligned}
\Omega^{wx} &= \begin{pmatrix} C_1 & O \\ C_1 & C_2 \end{pmatrix} \begin{pmatrix} \Omega_{11} & \Omega_{12} \\ \Omega_{21} & \Omega_{22} \end{pmatrix} \begin{pmatrix} C_1^T & C_1^T \\ O^T & C_2^T \end{pmatrix} = \begin{pmatrix} \Omega_{11}^{wx} & \Omega_{12}^{wx} \\ \Omega_{21}^{wx} & \Omega_{22}^{wx} \end{pmatrix}, \\
\Omega_{11}^{wx} &= C_1\Omega_{11}C_1^T, \\
\Omega_{12}^{wx} &= C_1\Omega_{11}C_1^T + C_1\Omega_{12}C_2^T, \\
\Omega_{21}^{wx} &= C_1\Omega_{11}C_1^T + C_2\Omega_{21}C_1^T, \\
\Omega_{22}^{wx} &= C_1\Omega_{11}C_1^T + C_2\Omega_{21}C_1^T + C_1\Omega_{12}C_2^T + C_2\Omega_{22}C_2^T, \\
C_3 &= R_1C_1^{-1} - R_2C_2^{-1}, \\
\Omega_{11}^{*} &= \Omega_{11}^{wx} - \Omega_{12}^{wx}\Omega_{22}^{wx\,-1}\Omega_{21}^{wx}, \\
a^* &= (\mathbf{x} - \boldsymbol{\mu}_2^{wx})^T\Omega_{22}^{wx\,-1}(\mathbf{x} - \boldsymbol{\mu}_2^{wx}).
\end{aligned} \tag{3.18}
$$

Dey and Liu (2003) showed that LCLC2 and LCLC1 are equivalent in a reparameterization sense. The necessary condition to be satisfied is only that $R_1C_1^{-1} - R_2C_2^{-1}$ is of full rank, which is very easy to check. Thus within the LCLC1 framework it is very easy to construct "skewness." Sahu, Dey, and Branco (2003) use a special case of LCLC2 to analyze a real data set and obtained explicit forms of the Gibbs sampling layout. Based on their work, we point out that the LCLC2 setup can easily be applied to Gibbs' sampler in many modelings, such as the WinBugs software, by using the form of parameter summation as a "composite" parameter. The theoretical foundation is the equivalence property between LCLC1 and LCLC2. A recent example is given in Liu and Dey (2003) on multilevel binomial regression modeling.

3.3 Examples of Skew-Elliptical Distributions

In this section, we review several examples of skew-elliptical distributions.

3.3.1 Skew-Scale Mixture of Normal Distribution

The density function is

$$
\begin{aligned}
f_{\mathbf{y}}(\mathbf{y}) = {} & 2|\Omega|^{-1/2} \int_{-\infty}^{\boldsymbol{\alpha}^T(\mathbf{y}-\boldsymbol{\mu})} \int_0^\infty (2\pi K(\eta))^{-(k+1)/2} \\
& \times \exp\left\{-\tfrac{r^2+(\mathbf{y}-\boldsymbol{\mu})^T\Omega^{-1}(\mathbf{y}-\boldsymbol{\mu})}{2K(\eta)}\right\} dH(\eta)dr \\
= {} & 2\int_0^\infty |\Omega|^{-1/2}(2\pi K(\eta))^{-k/2} \exp\left\{-\tfrac{(\mathbf{y}-\boldsymbol{\mu})^T\Omega^{-1}(\mathbf{y}-\boldsymbol{\mu})}{2K(\eta)}\right\} \\
& \times\Phi\left\{\tfrac{\boldsymbol{\alpha}^T(\mathbf{y}-\boldsymbol{\mu})}{K(\eta)^{1/2}}\right\} dH(\eta).
\end{aligned}
$$

Thus the density function reduces to

$$
f_{\mathbf{y}}(\mathbf{y}) = 2\int_0^\infty \phi_k(\mathbf{y};\boldsymbol{\mu},K(\eta)\Omega)\Phi\left\{\frac{\boldsymbol{\alpha}^T(\mathbf{y}-\boldsymbol{\mu})}{K(\eta)^{1/2}}\right\} dH(\eta). \tag{3.19}
$$

One particular case of this distribution is the skew-normal distribution, for which H is degenerate, with $K(\eta) = 1$. In this case the density is given by

$$2\phi_k(\mathbf{y}; \boldsymbol{\mu}, \Omega)\Phi\left\{\boldsymbol{\alpha}^T(\mathbf{y} - \boldsymbol{\mu})\right\}.$$

Therefore, if $g^{(k+1)}(\cdot)$ is the generator function of a scale mixture of normal, it follows that $f_{\mathbf{y}}(\cdot)$ is again a scale mixture of skew-normal distribution. The following are special cases of the skew-scale mixture of normals.

Skew-Finite Mixture of Normal

If the generator function is

$$g^{(k+1)}(u) = \sum_{i=1}^{n} p_i(2\pi K(\eta_i))^{-(k+1)/2} \exp\{-u/2K(\eta_i)\},$$

with $0 \le p_i \le 1$ and $\sum_{i=1}^{n} p_i = 1$, then the distribution H is a discrete measure on $\{\eta_1, \ldots, \eta_n\}$ with probabilities $p_1, \ldots, p_n$, respectively. The density of the skew-finite mixture of normal is given by

$$f_{\mathbf{y}}(\mathbf{y}) = 2\sum_{i=1}^{n} p_i\phi_k(\mathbf{y}; \boldsymbol{\mu}, K(\eta_i)\Omega)\Phi\left\{\frac{\boldsymbol{\alpha}^T(\mathbf{y} - \boldsymbol{\mu})}{K(\eta_i)^{1/2}}\right\},$$

which is again a finite mixture of skew-normal distributions. In this case, for simplicity, we often take $K(\eta_i) = 1$, $i = 1, \ldots, n$.

Skew-Logistic Distribution

The generator function is

$$g^{(k+1)}(u) = \frac{\exp(-u)}{1 + \exp(-u)}, \quad u > 0.$$

Choy (1995) pointed out that the logistic distribution is a special case of the scale mixture of normals, when $K(\eta) = 4\eta^2$ and η follows an asymptotic Kolmogorov distribution with density

$$f(\eta) = 8\textstyle\sum_{k=1}^{\infty}(-1)^{k+1}k^2\eta\exp\{-2k^2\eta^2\}.$$

Skew-Stable Distribution

This is obtained by choosing $K(\eta) = 2\eta$ and the mixture distribution $dH(\eta) = S^P(\tilde{\alpha}, 1)$, where the pdf of the positive stable distribution $S^P(\tilde{\alpha}, 1)$ in the polar form is given by

$$h_{SP}(\eta|\tilde{\alpha}, 1) = \frac{\tilde{\alpha}}{1 - \tilde{\alpha}}\eta^{-(\tilde{\alpha}/(1-\tilde{\alpha})+1)} \int_0^1 s(u)\exp\left\{-\frac{s(u)}{\eta^{\frac{\tilde{\alpha}}{1-\tilde{\alpha}}}}\right\} du, \qquad (3.20)$$

for $0 < \tilde{\alpha} < 1$ with

$$s(u) = \left\{ \frac{\sin(\tilde{\alpha}\pi u)}{\sin(\pi u)} \right\}^{\frac{\tilde{\alpha}}{1-\tilde{\alpha}}} \left\{ \frac{\sin[(1-\tilde{\alpha})\pi u]}{\sin(\pi u)} \right\}.$$

Note that when $\tilde{\alpha} = 1$, we get a skew-Cauchy distribution. The skew-normal distribution can also be obtained from the skew-stable by taking $\tilde{\alpha} \to 1$.

Skew-Exponential Power Distribution

This is obtained with $K(\eta) = 1/(2c_0\eta)$ and $h(\eta) = (1/\eta)^{\frac{k+1}{2}} h_{SP}(\eta|\tilde{\alpha}, 1)$, where $h_{SP}(\eta|\tilde{\alpha}, 1)$ is given by (3.20) and c_0 is defined by

$$c_0 = \frac{\Gamma[3/2\tilde{\alpha}]}{\Gamma[1/2\tilde{\alpha}]}, \quad \text{and} \quad \frac{1}{2} < \tilde{\alpha} < 1.$$

The parameter c_0 is called the kurtosis parameter. Further references on the symmetric exponential power family of distributions are given in Andrews and Mallows (1974), and Choy (1995). Again the skew-normal and the skew-double exponential distributions can be obtained by taking $\tilde{\alpha} = 1$ and $\tilde{\alpha} = 1/2$, respectively.

Skew-t Distribution

The generator function of a skew-t distribution is given by

$$g_{\nu,\tau}(u) = C(\nu, \tau)[\tau + u]^{-\frac{\nu+k+1}{2}},$$

where

$$C(\nu, \tau) = \frac{\Gamma[\frac{\nu+k+1}{2}]\tau^{\nu/2}}{\Gamma[\nu/2]\pi^{\frac{k+1}{2}}}.$$

It follows that

$$\begin{aligned} f_{\mathbf{y}}(\mathbf{y}) = \; & 2|\Omega|^{-1/2} \int_{-\infty}^{\boldsymbol{\alpha}^T(\mathbf{y}-\boldsymbol{\mu})} C(\nu, \tau)[\tau + r^2 \\ & +(\mathbf{y}-\boldsymbol{\mu})^T\Omega^{-1}(\mathbf{y}-\boldsymbol{\mu})]^{-(\nu+k+1)/2} dr. \end{aligned}$$

Considering $\tau^* = \tau + (\mathbf{y}-\boldsymbol{\mu})^T\Omega^{-1}(\mathbf{y}-\boldsymbol{\mu})$, $\nu^* = \nu + k$ and

$$\begin{aligned} C^*(\nu, \tau) &= \frac{\Gamma[\frac{\nu+k+1}{2}](\tau+(\mathbf{y}-\boldsymbol{\mu})^T\Omega^{-1}(\mathbf{y}-\boldsymbol{\mu}))^{(\nu+k)/2}}{\Gamma[(\nu+k)/2]\pi^{1/2}} \\ &= \frac{\Gamma[\frac{\nu^*+1}{2}](\tau^*)^{\nu^*/2}}{\Gamma[\nu^*/2]\pi^{1/2}}, \end{aligned}$$

then the density $f_{\mathbf{y}}(\mathbf{y})$ equals to

$$\begin{aligned} f_{\mathbf{y}}(\mathbf{y}) = \; & 2|\Omega|^{-1/2} \frac{\tau^{\nu/2}[\tau+(\mathbf{y}-\boldsymbol{\mu})^T\Omega^{-1}(\mathbf{y}-\boldsymbol{\mu})]^{-(\nu+k)/2}\Gamma[(\nu+k)/2]}{\Gamma[\frac{\nu}{2}]\pi^{k/2}} \\ & \times \int_{-\infty}^{\boldsymbol{\alpha}^T(\mathbf{y}-\boldsymbol{\mu})} C^*(\nu, \tau)[\tau^* + r^2]^{-(\nu^*+1)/2} dr. \end{aligned}$$

So the pdf of the multivariate skew-t is given by

$$f_{\mathbf{y}}(\mathbf{y}) = 2 f_{\nu,\tau}(\mathbf{y}; \boldsymbol{\mu}, \Omega) F_{\nu^*,\tau^*}(\boldsymbol{\alpha}^T(\mathbf{y}-\boldsymbol{\mu})),$$

where $f_{\nu,\tau}(\cdot;\boldsymbol{\mu},\Omega)$ is the pdf of a k-variate generalized Student t distribution with location and scale parameters $\boldsymbol{\mu}$ and Ω, respectively, and $F_{\nu^*,\tau^*}(\cdot)$ is the cdf of a univariate standard generalized t distribution.

Skew-Pearson Type II Distribution

The generator function of a skew-Pearson type II distribution is given by

$$g^{(k+1)}(u) = \frac{\Gamma[m+1+(k+1)/2]}{\Gamma[m+1]\pi^{(k+1)/2}}(1-u)^m,\ 0 < u < 1,\ m > -1.$$

Considering a generalized version of the Pearson type II distribution with an additional parameter $s > 0$, then the generator function for this is reduced to

$$g^{(k+1)}(u) = \frac{\Gamma[m+1+(k+1)/2]}{\Gamma[m+1]\pi^{(k+1)/2}}s^{-(m+(k+1)/2)}(s-u)^m,$$

where $0 < u < s$, $m > -1$, and $s > 0$. Then the density $f_{\mathbf{y}}(\mathbf{y})$ equals to

$$\begin{aligned} & 2|\Omega|^{-1/2}\int_{-\sqrt{s}}^{\boldsymbol{\alpha}^T(\mathbf{y}-\boldsymbol{\mu})}\tfrac{\Gamma[m+1+(k+1)/2]}{\Gamma[m+1]\pi^{(k+1)/2}}s^{-(m+(k+1)/2)} \\ & \times[s-(r^2+(\mathbf{y}-\boldsymbol{\mu})^T\Omega^{-1}(\mathbf{y}-\boldsymbol{\mu}))]^m dr \\ = \ & 2|\Omega|^{-1/2}\tfrac{\Gamma[m+1+(k+1)/2]s^{-(m+(k+1)/2)}}{\Gamma[m+1+1/2]\pi^{k/2}(s-(\mathbf{y}-\boldsymbol{\mu})^T\Omega^{-1}(\mathbf{y}-\boldsymbol{\mu}))^{-(m+1/2)}} \\ & \times F_{m,s^*}(\boldsymbol{\alpha}^T(\mathbf{y}-\boldsymbol{\mu})), \end{aligned}$$

where $s^* = s-(\mathbf{y}-\boldsymbol{\mu})^T\Omega^{-1}(\mathbf{y}-\boldsymbol{\mu})$. Considering $m^* = m+1/2$, it follows that

$$f_{\mathbf{y}}(\mathbf{y}) = 2f_{m^*,s}(\mathbf{y};\boldsymbol{\mu},\Omega)F_{m,s^*}(\boldsymbol{\alpha}^T(\mathbf{y}-\boldsymbol{\mu})),$$

where $f_{m^*,s}(;\boldsymbol{\mu},\Omega)$ is the pdf of a k-variate generalized Pearson type II distribution with location and scale parameters $\boldsymbol{\mu}$ and Ω, respectively, and $F_{m,s^*}(\cdot)$ is the cdf of a univariate standard generalized Pearson type II distribution.

3.3.2 Different Types of Multivariate Skew-Normal Distributions

In this section we consider two types of multivariate skew-normal distributions from the literature and show how they can be constructed as a special case from the LCLC1/LCLC2 approaches.

Type A-SN$(\boldsymbol{\mu},\Omega,\boldsymbol{\alpha})$

Azzalini and Dalla Valle (1996) constructed a multivariate skew-normal density of the form $2\phi_p(\mathbf{x};\boldsymbol{\mu},\Omega)\Phi(\boldsymbol{\alpha}^T(\mathbf{x}-\boldsymbol{\mu}))$, $\mathbf{x}\in\mathbb{R}^p$. Their method is included in the LCLC1 setup by choosing

$$R_1 = -1,\ R_2 = \mathbf{0}^T,\ R = (R_1, R_2),\ \mathbf{d} = \mathbf{0},\ C_2 = I_p,$$

$$\boldsymbol{\mu} = (0,\boldsymbol{\mu}^T)^T,\ g = \phi,\ \Omega_{11} = 1,\ \Omega_{21} = \boldsymbol{\delta},\ \Omega_{22} = \Omega.$$

Type B-$SN(\boldsymbol{\mu}, \Omega, D)$

Gupta, González-Farías, and Domínguez-Molina (2004) constructed a multivariate skew-normal density of the form $\frac{\Phi_p(D(\mathbf{x}-\boldsymbol{\mu});\mathbf{0},I_p)}{\Phi_p(\mathbf{0};\mathbf{0},I_p+D\Omega D^T)}\phi_p(\mathbf{x};\boldsymbol{\mu},\Omega)$, $\mathbf{x} \in \mathbb{R}^p$. Their method is included in the LCLC1 setup by choosing

$$R_1 = I_q,\ R_2 = -D \in \mathbb{R}^{q\times p},\ R = (I_q, -D),\ \mathbf{d} = \mathbf{0} \in \mathbb{R}^q,\ C_2 = I_p,$$

$$\boldsymbol{\mu} = ((D\boldsymbol{\mu})^T, \boldsymbol{\mu}^T)^T,\ g = \phi,\ \Omega_{11} = I_q,\ \Omega_{21} = O \in \mathbb{R}^{p\times q},\ \Omega_{22} = \Sigma \in \mathbb{R}^{p\times p}.$$

The Liseo and Loperfido Class of SN

Liseo and Loperfido (2003a) constructed a broader class of multivariate skew-normal distributions with the density function

$$\frac{\Phi_k(\mathbf{0}; C\Delta(\Sigma^{-1}\mathbf{x} + \Omega^{-1}\boldsymbol{\mu}) + \mathbf{d}, C\Delta C^T)}{\Phi_k(\mathbf{0}; C\boldsymbol{\mu} + \mathbf{d}, C\Omega C^T)}\phi_p(\mathbf{x}; \boldsymbol{\mu}, \Sigma + \Omega),$$

where

$$\Delta^{-1} = \Sigma^{-1} + \Omega^{-1},\ \ \Delta = \Sigma(\Sigma + \Omega)^{-1}\Omega.$$

It is a special case of LCLC2 with

$$\boldsymbol{\mu} = \begin{pmatrix} \boldsymbol{\mu} \\ \mathbf{0} \end{pmatrix},\ \Omega = \begin{pmatrix} \Omega & O \\ O & \Sigma \end{pmatrix},\ R_1 = C,\ R_2 = O,\ C_1 = I_p,\ C_2 = I_p.$$

Another way to illustrate the Liseo and Loperfido (2003a) method is to consider $\mathbf{x}|\mathbf{x}_0 \sim N_p(\mathbf{x}_0, \Sigma)$ and $\mathbf{x}_0 \sim N_p(\boldsymbol{\mu}, \Omega)$ with the linear constraints $K^T\mathbf{x}_0 + \mathbf{d} \leq \mathbf{0}$, where the dimension of K is $p \times m$, $m \leq p$, such that K is full rank. This can be regarded as

$$\mathbf{x} = (N_p(\boldsymbol{\mu}, \Omega) \text{ with the linear constraints } \ K^T\mathbf{x} + \mathbf{d} \leq \mathbf{0}) + N_p(\mathbf{0}, \Sigma),$$

where $N_p(\boldsymbol{\mu}, \Omega)$, the normal with linear constraints, and $N_p(\mathbf{0}, \Sigma)$ are independent (a special case of linear combinations).

Here we observe that the Type A-SN and Type B-SN are both derived from the LCLC1, whereas the Liseo and Loperfido (2003a) construction is derived from the LCLC2. The reparameterization equivalence was shown by Dey and Liu (2003).

The Domínguez-Molina-González-Farías-Gupta Class of SN

Domínguez-Molina, González-Farías, and Gupta (2003) constructed closed skew-normal (CSN) distributions that include Type B-SN as a special case. The density is of the form

$$f_{p,q}(\mathbf{x}; \boldsymbol{\mu}, \Sigma, D, \boldsymbol{\nu}, \Delta) = \frac{\Phi_q(D\mathbf{x}; \boldsymbol{\nu}, \Delta)}{\Phi_q(D\boldsymbol{\mu}; \boldsymbol{\nu}, \Delta + D\Sigma D^T)}\phi_p(\mathbf{x}; \boldsymbol{\mu}, \Sigma).$$

They used the notation $\mathbf{x} \sim CSN_{p,q}(\boldsymbol{\mu}, \Sigma, D, \boldsymbol{\nu}, \Delta)$. If we take $\boldsymbol{\nu} = D\boldsymbol{\mu}$ and $\Delta = I_p$, then the density reduces to

$$\begin{aligned} f_{p,q}(\mathbf{x}; \boldsymbol{\mu}, \Sigma, D, D\boldsymbol{\mu}, I_p) &= \tfrac{\Phi_p(D(\mathbf{x}-\boldsymbol{\mu});\mathbf{0},I_p)}{\Phi_p(\mathbf{0};\mathbf{0},I_p+D\Sigma D^T)} \phi_p(\mathbf{x}; \boldsymbol{\mu}, \Sigma) \\ &\sim \text{Type B-}SN(\boldsymbol{\mu}, \Omega, D). \end{aligned}$$

The construction of Domínguez-Molina *et al.* (2003) is a special case of the LCLC1. To obtain their results, we only need to take the multivariate normal distribution in the elliptically contoured distribution framework and define

$$\begin{cases} \begin{pmatrix} \mathbf{z} \\ \mathbf{x} \end{pmatrix} \sim N_{q+p}\left(\begin{pmatrix} \boldsymbol{\nu} \\ \boldsymbol{\mu} \end{pmatrix}, \begin{pmatrix} \Delta & O \\ O & \Sigma \end{pmatrix} \right), \\ \begin{pmatrix} I_q & -D \end{pmatrix} \begin{pmatrix} \mathbf{z} \\ \mathbf{x} \end{pmatrix} \leq \mathbf{0}, \end{cases}$$

and calculate the marginal density of $\mathbf{x}$, i.e., $R_1 = I_q$, $R_2 = -D \in \mathbb{R}^{q \times p}$, $\Omega = \begin{pmatrix} \Delta & O \\ O & \Sigma \end{pmatrix}$. The calculation of the density of $\mathbf{x}$ given $\mathbf{z} - D\mathbf{x} \leq \mathbf{0}$ produces the results from Domínguez-Molina *et al.* (2003).

Skew-Elliptical Distribution $SE_k(\boldsymbol{\mu}, \Omega, \boldsymbol{\delta}, g^{(k+1)})$

The construction by Branco and Dey (2001) is given in Section 3.2.1.

Skew-Elliptical Distribution $SE_m(\boldsymbol{\mu}, \Sigma, D, g^{(m)})$

Sahu *et al.* (2003) used the following construction. Suppose $\boldsymbol{\epsilon}$ and $\mathbf{z}$ are two m-dimensional random vectors. Let $\boldsymbol{\mu}$ be an m-dimensional vector and Σ be an $m \times m$ positive definite matrix. Assume that

$$\begin{pmatrix} \boldsymbol{\epsilon} \\ \mathbf{z} \end{pmatrix} \sim EC_{2m}\left(\begin{pmatrix} \boldsymbol{\mu} \\ \mathbf{0} \end{pmatrix}, \begin{pmatrix} \Sigma & O \\ O & I_m \end{pmatrix}, g^{(2m)} \right).$$

They considered a skew-elliptical class of distributions by using the transformation

$$\mathbf{x} = D\mathbf{z} + \boldsymbol{\epsilon},$$

where D is a diagonal matrix with elements $\delta_1, \ldots, \delta_m$. Define the vector $\boldsymbol{\delta} = (\delta_1, \ldots, \delta_m)^T$. The class is developed by considering the random variable $[\mathbf{x}|\mathbf{z} > \mathbf{0}]$ where $\mathbf{z} > \mathbf{0}$ means $Z_i > 0$ for $i = 1, \ldots, m$. Then

$$\begin{aligned} &f_{\mathbf{x}}(\mathbf{x}; \boldsymbol{\mu}, \Sigma, D, g^{(m)}) = 2^m f_{\mathbf{x}}(\mathbf{x}|\boldsymbol{\mu}, \Sigma + D^2, g^{(m)}) \\ &\times F\left(\left[I_m - D(\Sigma + D^2)^{-1} D \right]^{-1/2} D(\Sigma + D^2)^{-1}(\mathbf{x} - \boldsymbol{\mu}) | \mathbf{0}, I_m, g^{(m)}_{q(\mathbf{x}-\boldsymbol{\mu})} \right), \end{aligned}$$

where

$$g_a^{(m)}(u) = \frac{\Gamma(m/2)}{\pi^{m/2}} \frac{g(a+u; 2m)}{\int_0^\infty r^{m/2-1} g(a+r; 2m) dr}, \quad a > 0,$$

and

$$q(\mathbf{x}-\boldsymbol{\mu}) = (\mathbf{x}-\boldsymbol{\mu})^T(\Sigma + D^2)^{-1}(\mathbf{x}-\boldsymbol{\mu}).$$

This density matches with the one obtained by Branco and Dey (2001) only in the univariate case. The derived skew-normal distribution is

$$\begin{aligned} f(\mathbf{x}|\boldsymbol{\mu},\Sigma,D) &= 2^m|\Sigma + D^2|^{-1/2}\phi_m\Big((\Sigma + D^2)^{-1/2}(\mathbf{x}-\boldsymbol{\mu})\Big) \\ &\quad\times \Phi_m\Big(\left(I_m - D(\Sigma + D^2)^{-1}D\right)^{-1/2} D(\Sigma + D^2)^{-1}(\mathbf{x}-\boldsymbol{\mu})\Big), \end{aligned}$$

where ϕ_m and Φ_m denote the density and cdf of the m-dimensional normal distribution with mean $\mathbf{0}$ and covariance matrix identity. It is different from the Type A-SN distribution. It is a special case of LCLC2 with

$$\boldsymbol{\mu} = \begin{pmatrix} \mathbf{0} \\ \boldsymbol{\mu} \end{pmatrix}, \Omega = \begin{pmatrix} I_m & O \\ O & \Sigma \end{pmatrix}, R_1 = -I_m, R_2 = O,$$
$$\mathbf{d} = \mathbf{0}, C_1 = D \in \mathbb{R}^{m\times m}, C_2 = I_m.$$

3.4 Some Properties of Skew-Elliptical Distributions

3.4.1 Moment Generating Functions

In this section, we state the moment generating functions of different classes of skew-elliptical distributions.

Skew-Scale Mixture of Normal Distributions

Branco and Dey (2001) derived a general expression for the moment generating function of a skew-scale mixture of normal distributions; see Section 3.3. Referring to their notation, let $\mathbf{z} \sim SE_k(\boldsymbol{\mu}, \Omega, g^{(k+1)})$, where $g^{(k+1)}$ is the generator function for a $(k+1)$-variate scale mixture of normal distributions.

Considering Equation (3.19) and $\mathbf{t} \in \mathbb{R}^k$, we have

$$\begin{aligned} M(\mathbf{t}) &= \mathrm{E}[\exp(\mathbf{t}^T\mathbf{z})] \\ &= 2|\Omega|^{-1/2}\int_{\mathbb{R}^k}\int_0^\infty (2\pi K(\eta))^{-k/2}\exp\left\{\mathbf{t}^T\mathbf{z} - \frac{\mathbf{z}^T\Omega^{-1}\mathbf{z}}{2K(\eta)}\right\} \\ &\quad\times\Phi\left(\frac{\boldsymbol{\alpha}^T\mathbf{z}}{K(\eta)^{1/2}}\right)dH(\eta)d\mathbf{z}. \end{aligned}$$

Using Proposition 4 in Azzalini and Dalla Valle (1996), it follows that

$$\begin{aligned} & M_{SMN}(\mathbf{t}) \\ = \; & 2\int_0^\infty \exp\left\{\frac{\mathbf{t}^T K(\eta)^{1/2}\Omega K(\eta)^{1/2}\mathbf{t}}{2}\right\}\Phi\left\{\frac{\boldsymbol{\alpha}^T\Omega K(\eta)\mathbf{t}}{(1+\boldsymbol{\alpha}^T\Omega\boldsymbol{\alpha}/K(\eta))^{1/2}}\right\}dH(\eta) \\ = \; & \int_0^\infty M_{SN}(K(\eta)^{1/2}\mathbf{t})dH(\eta), \end{aligned}$$

where $M_{SMN}(\cdot)$ is the moment generating function of the skew-scale mixture of normal distributions and $M_{SN}(\cdot)$ is the moment generating function for skew-normal distributions. They further used the moment generating function to obtain the mean vector and covariance matrix of a univariate skew-normal distribution that are given by

$$\mathrm{E}_{SN}[Z] = \left(\frac{2}{\pi}\right)^{1/2}\delta,$$

$$\mathrm{Var}_{SN}[Z] = 1 - \frac{2\delta^2}{\pi},$$

$$\mathrm{Cov}_{SN}[Z_i, Z_j] = w_{ij} - \frac{2}{\pi}\delta_i\delta_j,$$

where w_{ij} is the (i,j)-th element of the matrix Ω.

Subsequently, for the univariate skew-scale mixture of normal distributions, we have the following results:

$$\mathrm{E}_{SMN}[Z] = \left(\frac{2}{\pi}\right)^{1/2}\delta\mathrm{E}[K(\eta)^{1/2}], \quad \text{if} \quad \mathrm{E}[K(\eta)^{1/2}] < \infty,$$

$$\mathrm{Var}_{SMN}[Z] = \mathrm{E}[K(\eta)] - \frac{2\delta^2}{\pi}\mathrm{E}[K(\eta)^{1/2}], \quad \text{if} \quad \mathrm{E}[K(\eta)] < \infty,$$

$$\mathrm{Cov}_{SMN}[Z_i, Z_j] = w_{ij}\mathrm{E}[K(\eta)] - \frac{2}{\pi}\delta_i\delta_j\mathrm{E}^2[K(\eta)^{1/2}].$$

In general, for the mean vector $\mathbf{m}$ and the covariance matrix M of the skew-scale mixture of normal distributions, we have

$$\mathbf{m} = \left(\frac{2}{\pi}\right)^{1/2}\mathrm{E}[K(\eta)^{1/2}]\boldsymbol{\delta},$$

and

$$M = \mathrm{E}[K(\eta)]\Omega - \frac{2\mathrm{E}^2[K(\eta)^{1/2}]}{\pi}\boldsymbol{\delta}\boldsymbol{\delta}^T.$$

An example from their paper is the skew-t distribution with $K(\eta) = 1/\eta$ and $\eta \sim Gamma(\nu/2, \nu/2)$. If $\nu > 1$ then $\mathrm{E}[K(\eta)^{1/2}] < \infty$ and

$$\mathrm{E}[K(\eta)^{1/2}] = \frac{\Gamma[(\nu-1)/2]\sqrt{\nu}}{\Gamma[\nu/2]\sqrt{2}}.$$

Thus

$$\mathrm{E}_{ST}[Z] = \frac{\delta\Gamma[(\nu-1)/2]}{\Gamma[\nu/2]}\sqrt{\frac{\nu}{\pi}}.$$

If $\nu > 2$ then $\mathrm{E}[1/\eta] < \infty$ and $\mathrm{E}[1/\eta] = \frac{\nu}{\nu-2}$, so

$$\mathrm{Var}_{ST}[Z] = \frac{\nu}{\nu-2} - \frac{\delta^2\nu}{\pi}\left\{\frac{\Gamma\left(\frac{\nu-1}{2}\right)}{\Gamma\left(\frac{\nu}{2}\right)}\right\}^2.$$

LCLC1

Dey and Liu (2003) showed that, under the LCLC1 condition, the moment generating function of $\mathbf{x}$ is

$$\begin{aligned} M_{\mathbf{x}}(\mathbf{t}) &= \frac{1}{\Phi_p(-R\boldsymbol{\mu}-\mathbf{d}|\mathbf{0},R\Omega R^T)}\exp(\boldsymbol{\mu}_2^T C_2^T\mathbf{t} + \tfrac{1}{2}\mathbf{t}^T C_2\Omega_{22}C_2^T\mathbf{t}) \\ &\times\Phi_p\Bigg(-\mathbf{d} - R_1(\boldsymbol{\mu}_1 - \Omega_{12}\Omega_{22}^{-1}\boldsymbol{\mu}_2) \\ &-(R_2 - R_1\Omega_{12}\Omega_{22}^{-1})(\boldsymbol{\mu}_2 + \Omega_{22}C_2^T\mathbf{t}), \\ &\mathbf{0}, (R_2 - R_1\Omega_{12}\Omega_{22}^{-1})\Omega_{22}(R_2 - R_1\Omega_{12}\Omega_{22}^{-1})^T \\ &+R_1(\Omega_{11} - \Omega_{12}\Omega_{22}^{-1}\Omega_{21})R_1^T\Bigg). \end{aligned}$$

LCLC2

Dey and Liu (2003) also pointed out that, under the LCLC2 condition, the moment generating function of $\mathbf{x}$ is

$$\begin{aligned} M_{\mathbf{x}}(\mathbf{t}) &= \frac{1}{\Phi_p(-R\boldsymbol{\mu}-\mathbf{d}|\mathbf{d},R\Omega R^T)}\exp(\boldsymbol{\mu}_2^{wx\,T}\mathbf{t} + \tfrac{1}{2}\mathbf{t}^T\Omega_{22}^{wx}\mathbf{t}) \\ &\times\Phi_p\Bigg(-\mathbf{d} - C_3(\boldsymbol{\mu}_1^{wx} - \Omega_{12}^{wx}\Omega_{22}^{wx\,-1}\boldsymbol{\mu}_2^{wx}) \\ &-(R_2C_2^{-1} + C_3\Omega_{12}^{wx}\Omega_{22}^{wx\,-1})(\boldsymbol{\mu}_2^{wx} + \Omega_{22}^{wx}\mathbf{t}), \\ &\mathbf{0}, (R_2C_2^{-1} + C_3\Omega_{12}^{wx}\Omega_{22}^{wx\,-1})\Omega_{22}^{wx}(R_2C_2^{-1} + C_3\Omega_{12}^{wx}\Omega_{22}^{wx\,-1})^T \\ &+C_3(\Omega_{11}^{wx} - \Omega_{12}^{wx}\Omega_{22}^{wx\,-1}\Omega_{21}^{wx})C_3^T\Bigg), \end{aligned}$$

where the notations are defined in (3.18).

3.4.2 Marginal and Conditional Closure Property

The marginal and conditional distributions are closed under the LCLC1 setup; the proof is given in Dey and Liu (2003). The marginal and conditional closures are carried over to the LCLC2 setup by a reparameterization equivalence under a mild condition (Dey and Liu, 2003).

3.4.3 Distribution of Quadratic Forms

LCLC1

In the LCLC1 setup, suppose A is a symmetric idempotent matrix, then the mgf of $\mathbf{x}^T A\mathbf{x}$ is given by

$$
\begin{aligned}
&M_{\mathbf{x}^T A\mathbf{x}}(t) = \tfrac{1}{\Phi_p(-R\boldsymbol{\mu}-\mathbf{d};\mathbf{0},R\Omega R^T)} \\
&\times|I-2tAC_2\Omega_{22}C_2^T|^{-1/2}\exp[-\tfrac{1}{2}\boldsymbol{\mu}_2^T\Omega_{22}^{-1}C_2^{-1} \\
&(I_p-(I_p-2tAC_2\Omega_{22}C_2^T)^{-1})\,C_2\boldsymbol{\mu}_2] \\
&\times\Phi_p\Bigg(R_1\Omega_{12}\Omega_{22}^{-1}\boldsymbol{\mu}_2-R_1\boldsymbol{\mu}_1-\mathbf{d}-(R_2+R_1\Omega_{12}\Omega_{22}^{-1})C_2^{-1}\boldsymbol{\mu}_2^T C_2^T \\
&(I-2tAC_2\Omega_{22}C_2^T)^{-1},\mathbf{0},(R_2+R_1\Omega_{12}\Omega_{22}^{-1})\Omega_{22}C_2^T \\
&(I-2tAC_2\Omega_{22}C_2^T)^{-1}C_2^{T^{-1}}(R_2+R_1\Omega_{12}\Omega_{22}^{-1})+\Omega_x\Bigg).
\end{aligned}
$$

The proof is given in Dey and Liu (2003).

LCLC2

In the LCLC2 setup, suppose A is a symmetric idempotent matrix, then the mgf of $\mathbf{x}^T A\mathbf{x}$ is

$$
\begin{aligned}
&M_{\mathbf{x}^T A\mathbf{x}}(t) = \tfrac{1}{\Phi_p(-R\boldsymbol{\mu}-\mathbf{d};\mathbf{0},R\Omega R^T)} \\
&\times|I_p-2tA\Omega_{22}^{wx}|^{-1/2}\exp\Bigg(-\tfrac{1}{2}\boldsymbol{\mu}_2^{wx\,T}\Omega_{22}^{wx^{-1}}\left(I_p-(I_p-2tA\Omega_{22}^{wx})^{-1}\right)\Bigg) \\
&\times\Phi_p\Bigg(-C_3(\boldsymbol{\mu}_1^{wx}-\Omega_{12}^{wx}\Omega_{22}^{wx^{-1}}\boldsymbol{\mu}_2^{wx})-\mathbf{d} \\
&-(R_2C_2^{-1}+C_3\Omega_{12}^{wx}\Omega_{22}^{wx^{-1}})(I_p-2tA\Omega_{22}^{wx})^{T^{-1}}\boldsymbol{\mu}_2^{wx};\mathbf{0}, \\
&(R_2C_2^{-1}+C_3\Omega_{12}^{wx}\Omega_{22}^{wx^{-1}})\Omega_{22}^{wx}(I_p-2tA\Omega_{22}^{wx})^{-1} \\
&(R_2C_2^{-1}+C_3\Omega_{12}^{wx}\Omega_{22}^{wx^{-1}})^T+C_3\Omega_{11}^{*}C_3^T\Bigg).
\end{aligned}
$$

Again, the notations are defined in (3.18). The proof can be implemented directly from the density function or using the equivalence with the LCLC1 and plugging in the LCLC1 results after reparameterization.

Based on the moment generating functions for LCLC1 and LCLC2, we can get the properties of the distribution of general multivariate skew-normal quadratic forms.

Quadratic Forms from LCLC1/LCLC2

Under the LCLC1 setup (Section 3.2), if $R_1\boldsymbol{\mu}_1 + R_2\boldsymbol{\mu}_2 + \mathbf{d} = \mathbf{0}$, then $(\mathbf{y}_2 - \boldsymbol{\mu}_2)^T\Omega_{22}^{-1}(\mathbf{y}_2 - \boldsymbol{\mu}_2)$ has the same distribution as R^2, where R is the radial variable in the stochastic representation of the original $\mathbf{y}_2$. This is

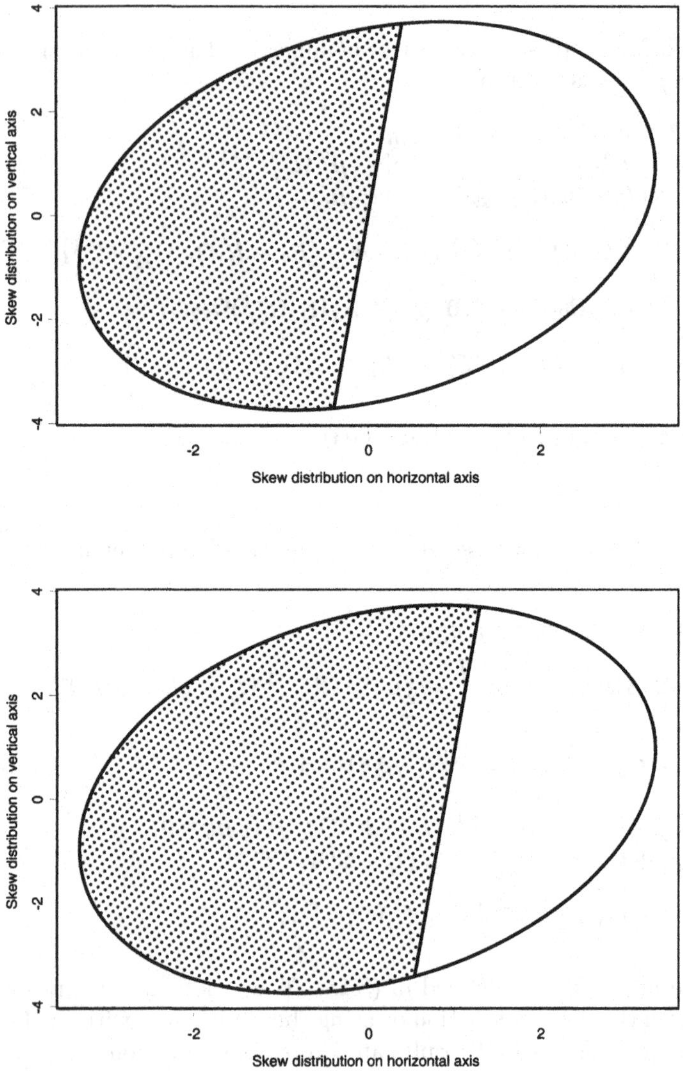

Figure 3.1 *Q-Symmetric (top) and Q-non-symmetric (bottom) skew-elliptical distributions.*

because of the symmetric partition and the linear constraint. A straightforward graphical illustration is given in Figure 3.1, where the linear constraint plane passes through the original mean vector $(\boldsymbol{\mu}_1, \boldsymbol{\mu}_2)$ of their joint distribution. A special case (see Proposition 3.4.1 below) is proved in Branco and Dey (2001).

Unfortunately, sometimes the multivariate skew-elliptical random vector $\mathbf{x}$ from LCLC1/LCLC2 does not have a straightforward quadratic form $\mathbf{x}^T A\mathbf{x}$, see Figure 3.1, (right panel). For the normal case, it turns out that we can obtain distribution properties by the analytical formula of $M_{\mathbf{y}^T A\mathbf{y}}(t)$ using a transformation, i.e., by considering $(A^{1/2}\mathbf{x})^T I_p(A^{1/2}\mathbf{x})$. Thus, for the broad class of skew-scale mixtures of normal distributions, we have an analytical representation for the quadratic form using an integral formula.

If we use this symmetric/non-symmetric partition as a criterion to distinguish the influence on the centered quadratic form, we end up with two definitions: "*Q-symmetric*" skew-elliptical distributions (Figure 3.1, top panel) and "*Q-non-symmetric*" skew-elliptical distributions (Figure 3.1, bottom panel). They are different from the "skew-symmetric" definition in Wang *et al.* (2004a).

3.4.4 Other Distributional Properties

Properties of the Branco-Dey Skew-Elliptical Models

We start with two conclusions from Fang and Zhang (1990, p. 65). First, the random vector $\mathbf{x} \sim EC_k(\boldsymbol{\mu}, \Sigma, g)$ if and only if $\mathbf{x} \stackrel{d}{=} \boldsymbol{\mu} + RA^T\boldsymbol{\mu}^{(k)}$, where $R \geq 0$ is a random variable independent of $\boldsymbol{\mu}^{(k)}$, R is one to one with the generator function g, A is a $k \times k$ matrix such that $A^T A = \Sigma$, and $\boldsymbol{\mu}^{(k)}$ has a uniform distribution on the unit sphere in $\mathbb{R}^k$. This is called a stochastic representation for $\mathbf{x}$, where R is the radial variable. Second, if $\mathbf{x} \stackrel{d}{=} \boldsymbol{\mu} + RA^T\boldsymbol{\mu}^{(k)} \sim EC_k(\boldsymbol{\mu}, \Sigma, g)$, then $(\mathbf{x} - \boldsymbol{\mu})^T\Sigma^{-1}(\mathbf{x} - \boldsymbol{\mu}) \stackrel{d}{=} R^2$.

Branco and Dey (2001) derived similar results for the skew-elliptical distributions.

Proposition 3.4.1 If $\mathbf{y} \sim SE_k(\boldsymbol{\mu}, \Omega, \boldsymbol{\delta}, g)$ then $(\mathbf{y} - \boldsymbol{\mu})^T\Omega^{-1}(\mathbf{y} - \boldsymbol{\mu})$ has the same distribution as R^2, where R is the radial variable in the stochastic representation of $\mathbf{y}$.

Proposition 3.4.2 Let C be a non-singular matrix of dimension $k \times k$, $\mathbf{b}$ a $k \times 1$ vector, and $\mathbf{y}^* = \mathbf{b} + C\mathbf{y}$, where $\mathbf{y} \sim SE_k(\boldsymbol{\mu}, \Omega, \boldsymbol{\delta}, g)$. Then

$$\mathbf{y}^* \sim SE_k(\boldsymbol{\mu}^*, \Omega^*, \boldsymbol{\delta}^*, g),$$

with $\boldsymbol{\mu}^* = \mathbf{b} + C\boldsymbol{\mu}$, $\Omega^* = C^T\Omega C$, and $\boldsymbol{\delta}^* = C\boldsymbol{\delta}$.

Wang *et al.* (2004a) also gave some invariance properties of their construction.

Density Shape of Univariate Skew-Elliptical Distributions

The univariate skew-elliptical distribution produced from the $(\mathbf{y}, \mathbf{z})$ pair is unimodal. The proof for the normal case was given by Liu and Dey (2003). The conclusion also holds for general univariate skew-elliptical distributions. Gupta and Brown (2001) explored the application of skew-normal distributions to reliability analysis. They showed it is log-concave. An $SN(\boldsymbol{\alpha})$ skew-normal variable has an increasing failure rate, and therefore a decreasing mean residual life. Wang *et al.* (2004a) studied the multimodality of the symmetric component and the monotonicity of the skewing component in their construction framework; see Section 3.2.

3.5 Concluding Remarks

When we look back at the development of skewed multivariate distributions, there have been many construction ideas. Theoretically the construction strategies would be many. Several authors tried to "skew" a multivariate distribution in such a way that they should be convenient, flexible, reasonable, straightforward, and statistically tractable. The skewness mechanism has been developed in a more and more comprehensive way. If we focus on statistical constructions, the vast literature can be summarized by three tracks. The first method is by order statistics, such as $X_{(n)} \propto F_X^n(\cdot) f_X(\cdot)$ derived from an iid univariate symmetric random variable group with $F_X^n(\cdot)$ to produce very smooth skewness. As a matter of fact, the first appearance of the skew-normal distribution dates back to Birnbaum (1950), where a $(k + l)$-dimensional random variable $(X_1, \ldots, X_k, Y_1, \ldots, Y_l) = (\mathbf{x}^T, \mathbf{y}^T)$ was considered with a joint probability distribution assumed to be non-singular multi-normal. The absolute distribution of $\mathbf{x}$ under linear truncation $\sum_{j=1}^{l} a_j Y_j \geq \tau$ was studied. The determination of τ for given expectation or quantile of $\mathbf{x}$ after truncation and minimizing the rejected part of the population was also described, which is a special case of LCLC1. Roberts (1966) considered $Z = \min(X, Y)$, where (X, Y) was a correlated bivariate normal random vector. It can be verified that it is a mixture of two univariate skew-normal densities with forms $2\phi(z)\Phi(\alpha z)$ in Azzalini (1985, 1986) based on two regions $X = \min(X, Y), X \leq Y$, and $Y = \min(X, Y), Y \leq X$. Azzalini and Dalla Valle (1996) initiated their ideas on the "conditional" approach, which is a special case of LCLC. Since then, many authors have made contributions to the LCLC track, which includes Branco and Dey's (2001) work on a general class of multivariate skew-elliptical distributions. In the review paper by Arnold and Beaver (2002), they summarized the "skewness" developments of this type. Starting from the well-known form of $2\phi(z)\Phi(\alpha z)$, $z \in \mathbb{R}$ by Azzalini (1985, 1986), they expanded the initial simple skewness mechanism. The linear constraint idea actually appeared in their construction, although they called it "conditional/hidden truncation." The starting point is an independent normal random variable pair (Y, Z). In this case they pointed out the reparameterization equivalence between $Z|Y > 0$ and $Z = \delta|Y_1| + \sqrt{1-\delta^2}Y_2$, where (Y_1, Y_2) is also an independent normal random variable pair. Note that these two cases are expanded by Dey and Liu (2003) by considering a whole correlated general elliptical random vector undergoing linear manipulations.

Under the independence assumption and by the conditioning method, Arnold and Beaver (2002) constructed non-normal or arbitrary skewed multivariate distributions. They also considered the multiple constraint models, which were similar to the Dey and Liu (2003) approach from a mathematical and implementational point of view.

The third track is to introduce a "skewing function." The Wang-Boyer-Genton approach (2004a) has the advantage that every distribution can be represented as a skew-symmetric distribution. The expression of the pdf is analytical.

If we look at each approach to date, it is obvious that there are tradeoffs among all of them. The approach of Dey and Liu (2003) is very comprehensive from a mathematical and parametric perspective. It did not consider some mechanism imposed subjectively, e.g., their skewness foundation is the inherent mathematical and statistical structures within the original multivariate elliptical distributions. On the other hand, the Wang-Boyer-Genton approach (2004a) expanded the skewness by introducing a skewing function π, thus it becomes more subjective in order to model multiple modes and flat peaks. Although the arbitrary choice of π causes much more flexibility, the symmetric component $f(\mathbf{x}-\boldsymbol{\xi})$ and skewed component $\pi(\mathbf{x}-\boldsymbol{\xi})$ have the same arbitrary location parameter $\boldsymbol{\xi}$. It is difficult to classify the skew-symmetric representations caused by π. The production mechanism of Bayesian sampling (3.12) can be interpreted via "hidden truncation/selective reporting" as in Arnold and Beaver (2002).

Let $\mathbf{x}_1$ $(k_1 \times 1)$ and $\mathbf{x}_2$ $(k_2 \times 1)$ be random vectors with a joint distribution such that $\mathbf{x}_2$ has density $f_{\mathbf{x}_2}$. Also, let f_A be the conditional density of $\mathbf{x}_2$ given $\mathbf{x}_1 \in A$, for some set A of $\mathbb{R}^{k_1}$. Arellano-Valle, del Pino, and San Martín (2002) showed that

$$f_A(\mathbf{x}_2^*) = f_{\mathbf{x}_2}(\mathbf{x}_2^*)\frac{\mathrm{P}(\mathbf{x}_1 \in A|\mathbf{x}_2 = \mathbf{x}_2^*)}{\mathrm{P}(\mathbf{x}_1 \in A)}.$$

Suppose $\mathbf{x}_1 = R\mathbf{y}+\mathbf{d}$, $\mathbf{x}_2 = C\mathbf{y}+\mathbf{e}$, with $\mathbf{y} \sim EC_p(\boldsymbol{\mu}, \Omega, g^{(p)})$. Both LCLC1 and LCLC2 could be unified in the above setup. But the LCLC setup enjoys more intriguing applicable properties. The capability of dealing with arbitrary correlation structure between two parts, e.g., truncated and symmetric, can be taken as original quantities and ensuing accumulated part (with flexible correlation). A simple example is, original normally distributed infant body weights plus accumulated additional weights (something like a truncated normal distribution).

More crucial issues are related to inference for the general multivariate skew-elliptical distributions. Arnold and Beaver (2002) showed some calculation results such as MLE based on the likelihood surface. For the LCLC1/LCLC2, some cautions must be kept in mind for the degrees of freedom of the estimated parameters. This is because some parameters may become redundant after proper reparameterization. The goodness-of-fit test during modeling is also a problem. The Bayesian skewness modeling approaches in the literature are in Sahu *et al.* (2003) working on a linear regression with a skewed error structure, and in Liu and Dey (2003), who used Bayesian model diagnostic criteria to evaluate the different types of random effects modeling based on LCLC2, e.g., in multilevel binomial regression.

CHAPTER 4

Generalized Skew-Normal Distributions

Nicola M. R. Loperfido

4.1 Introduction

Formal properties of the normal distribution belong to the core of statistical theory. Real data, however, are often skewed, multimodal, or censored (Hill and Dixon, 1982). Generalizations of the normal distribution should retain its relevant properties while dealing with the above non-normal features.

Skewness can be modeled through the multivariate skew-normal distribution introduced by Azzalini and Dalla Valle (1996), a reasonable compromise between mathematical tractability and shape flexibility. Its density is

$$2\phi_p(\mathbf{z};\boldsymbol{\xi},\Omega)\,\Phi\left(\boldsymbol{\alpha}^T(\mathbf{z}-\boldsymbol{\xi})\right), \qquad \mathbf{z},\boldsymbol{\alpha}\in\mathbb{R}^p, \tag{4.1}$$

where $\phi_p(\mathbf{z};\boldsymbol{\xi},\Omega)$ denotes the probability density function (pdf) of a p-dimensional normal distribution centered at $\boldsymbol{\xi}$ with scale matrix Ω and Φ denotes the cumulative distribution function (cdf) of a standard normal distribution. When a random vector $\mathbf{z}$ has the pdf (4.1), we write $\mathbf{z}\sim SN_p(\boldsymbol{\xi},\Omega,\boldsymbol{\alpha})$. The vector $\boldsymbol{\alpha}$ controls shape and the special case $\boldsymbol{\alpha}=\mathbf{0}$ corresponds to the normal distribution. Despite the presence of an additional parameter, skew-normal distributions resemble the normal ones in several ways: they are unimodal, their support is the p-dimensional real space, and $(\mathbf{z}-\boldsymbol{\xi})^T\Omega^{-1}(\mathbf{z}-\boldsymbol{\xi})\sim\chi^2_p$.

Unfortunately, skewness and kurtosis coefficients of the skew-normal distribution are bounded (Azzalini, 1985). Moreover, skew-normal distributions do not model multimodality. Hence the need for further generalizations of the normal law. Genton and Loperfido (2002) introduced the generalized skew-elliptical (GSE) distributions, whose density is

$$2f\left(\Omega^{-1/2}(\mathbf{z}-\boldsymbol{\xi})\right)\pi\left(\Omega^{-1/2}(\mathbf{z}-\boldsymbol{\xi})\right), \tag{4.2}$$

where f is the pdf of a spherical distribution and the skewing function π satisfies $0\le\pi(\mathbf{z})=1-\pi(-\mathbf{z})\le 1$. GSE distributions also generalize other skewed extensions of normal distributions: skew-elliptical (Azzalini and Capitanio, 1999; Branco and Dey, 2001), multivariate skew-t (Branco and Dey, 2001), and multivariate skew-Cauchy (Arnold and Beaver, 2000b). GSE distributions show a remarkable invariance property: distributions of even functions of GSE random vectors do not depend on the skewing function, when the location parameter is the zero vector. The same property holds for skew-symmetric distributions, a more general class introduced by Wang, Boyer, and Genton (2004a), whose pdf is

$$2f(\mathbf{z}-\boldsymbol{\xi})\,\pi(\mathbf{z}-\boldsymbol{\xi}), \tag{4.3}$$

where f is a pdf such that $f(\mathbf{z}) = f(-\mathbf{z})$ and $0 \le \pi(\mathbf{z}) = 1 - \pi(-\mathbf{z}) \le 1$ as above. Wang *et al.* (2004a) show that skew-symmetric distributions are selection models (Bayarri and De Groot, 1992). Selection models arise in inference from non-random samples (Copas and Li, 1997) where different observations have different probabilities of being included in the sample. This situation is rather common in biomedical research, as showed by Ludbrook and Dudley (1998).

Generalized skew-normal (GSN) distributions are the most interesting examples in the GSE class and hence in the skew-symmetric one. In the first place, functions of GSN random vectors can have Wishart, Hotelling, or Wilks distributions. In the second place, GSN distributions admit sufficient statistics, uniformly minimum variance unbiased (UMVU) estimators and uniformly most powerful (UMP) tests. In the third place, GSN distributions provide motivating counterexamples for a critical look at commonly used diagnostic methods for normality.

The chapter is structured as follows. Section 4.2 recalls the definition and a characterization of GSN distributions. Section 4.3 deals with transformations and moments. Section 4.4 examines some problems with diagnostic methods for normality when the data have a GSN distribution. Section 4.5 deals with inference on the parameters of GSN distributions. Section 4.6 applies results in the previous sections to a statistical analysis of the Australian Athletes data set. Section 4.7 hints at some research problems. The appendix contains proofs of new results.

4.2 Definition and Characterization

GSN densities are basically twice the product of a normal density and a non-negative skewing function shifting density from vector $\mathbf{z}$ to vector $-\mathbf{z}$. More formally:

Definition 4.2.1 A p-dimensional random vector $\mathbf{z}$ is said to be generalized skew-normal with location parameter $\boldsymbol{\xi}$, scale parameter Ω, and skewing function π, that is, $\mathbf{z} \sim GSN_p(\boldsymbol{\xi}, \Omega, \pi)$, if its density function is

$$2\phi_p(\mathbf{z};\boldsymbol{\xi},\Omega)\,\pi(\mathbf{z}-\boldsymbol{\xi}), \tag{4.4}$$

where $0 \leq \pi(-\mathbf{z}) = 1 - \pi(\mathbf{z}) \leq 1$ and $\phi_p(\mathbf{z}; \boldsymbol{\xi}, \Omega)$ is the pdf of a p-dimensional normal density centered at $\boldsymbol{\xi}$ with scale matrix Ω.

In order to show that GSN distributions are GSE distributions, it suffices to let $\pi(\mathbf{z} - \boldsymbol{\xi}) = \tilde{\pi}(\Omega^{-1/2}(\mathbf{z} - \boldsymbol{\xi}))$, where $0 \leq \tilde{\pi}(-\mathbf{z}) = 1 - \tilde{\pi}(\mathbf{z}) \leq 1$. Notice the following:

- A multivariate normal distribution is a GSN distribution with $\pi(\mathbf{z}) \equiv 1/2$.
- The skewing function π is a parameter of infinite dimension.
- The parameters $\boldsymbol{\xi}$ and Ω do not necessarily equal $\mathrm{E}(\mathbf{z})$ and $\mathrm{Var}(\mathbf{z})$.

Nadarajah and Kotz (2003) show several examples of GSN distributions. They can also model skewness, multimodality, and truncation, through an appropriate choice of the skewing function (Ma and Genton, 2004). As an example, consider the following univariate GSN distribution:

$$2\phi(z)\,\Phi\left(\sum_{i=0}^{\tilde{K}} \beta_i z^{2i+1}\right), \qquad \tilde{K} \in \mathbb{N},\ \beta_i \in \mathbb{R}. \tag{4.5}$$

The standard normal distribution, the skew-normal distribution, bimodal distributions, and the half-normal distribution are special cases of the above model:

- $\beta_i = 0,\ i = 0, \ldots, \tilde{K}$: standard normal distribution,
- $\beta_i = 0,\ i = 1, \ldots, \tilde{K}$: skew-normal distribution,
- $\beta_0 = -\beta_1 = 1,\ \beta_i = 0,\ i = 2, \ldots, \tilde{K}$: bimodal distribution,
- $\beta_0 \to +\infty; \ \ldots\ ; \beta_{\tilde{K}} \to +\infty$: half-normal distribution.

Azzalini and Capitanio (1999) proposed random censoring as a method for generating certain skew-symmetric distributions. Wang *et al.* (2004a) showed that the same method generates all skew-symmetric distributions. The following proposition is a direct implication of these results and shows that all GSN distributions are selection models with respect to the normal distribution. This makes GSN distributions eligible for modeling nonrandom samples from normal distributions (an application is shown in Section 4.6).

Proposition 4.2.1 The distribution of a random vector $\mathbf{z}$ is $GSN_p(\boldsymbol{\xi}, \Omega, \pi)$ if and only if there exists a skewing function π, a uniform random variable Y on the interval $[0, 1]$, and a random vector $\mathbf{x} \sim N_p(\mathbf{0}, \Omega)$ independent of Y, such that

$$\mathbf{z} = \begin{cases} \mathbf{x} - \boldsymbol{\xi}, & \text{if } Y \leq \pi(\mathbf{x} - \boldsymbol{\xi}), \\ \boldsymbol{\xi} - \mathbf{x}, & \text{if } Y > \pi(\mathbf{x} - \boldsymbol{\xi}). \end{cases}$$

4.3 Transformations and Moments

Linear and quadratic functions in normal random vectors play a fundamental role in statistical inference. There are many known results regarding their moments and distributions. Similar results also hold for linear and quadratic functions in GSN random vectors.

It is well-known that linear functions of normal random vectors are normal, too (see, for example, Mardia, Kent, and Bibby, 1979). Azzalini and Dalla Valle (1996) showed that linear functions of skew-normal random vectors are skew-normal. A similar property holds for GSN distributions: linear functions in GSN random vectors have a GSN distribution. More formally:

Proposition 4.3.1 Let $\mathbf{z} \sim GSN_p(\boldsymbol{\xi},\Omega,\pi)$ and let A be a $k \times p$ matrix of rank k. Then $A\mathbf{z}+\mathbf{b} \sim GSN_p\left(A\boldsymbol{\xi}+\mathbf{b},A\Omega A^T,\pi^*\right)$, where π^* is a skewing function.

It easily follows that marginals of GSN distributions are GSN distributions, too. In fact marginal distributions are associated to vector projections, which in turn can be represented as linear functions of the vector itself. For example, Z_1 is a linear function $\mathbf{v}^T\mathbf{z}$ of $\mathbf{z} = (Z_1,\ldots,Z_p)^T$, where $\mathbf{v} = (1,0,\ldots,0)^T$.

Despite the variety of shapes they can achieve, all GSN distributions have something in common: the density of $(\mathbf{z}-\boldsymbol{\xi})^T \Omega^{-1} (\mathbf{z}-\boldsymbol{\xi})$ is always chi-squared. Moreover, the skewing function controls the shape, but not the Euclidean distance of the GSN random vector from its location parameter. Both results follow from the invariance property of the GSN class (Genton and Loperfido, 2002), that is:

Proposition 4.3.2 Let $\mathbf{z} \sim GSN_p(\boldsymbol{\xi},\Omega,\pi)$. Then the following hold:

1. The distribution of $(\mathbf{z}-\boldsymbol{\xi})(\mathbf{z}-\boldsymbol{\xi})^T$ is Wishart with scale parameter Ω and one degree of freedom.
2. Distributions of even functions of $(\mathbf{z}-\boldsymbol{\xi})$ do not depend on the skewing function.
3. The distribution of $(\mathbf{z}-\boldsymbol{\xi})^T \Omega^{-1} (\mathbf{z}-\boldsymbol{\xi})$ is chi-squared with p degrees of freedom.

Part 2 of the above proposition was first proved for GSE distributions (Genton and Loperfido, 2002) and later for skew-symmetric distributions by Wang *et al.* (2004a,b), see also Azzalini and Capitanio (2003). Even functions include quadratic forms. Hence all results on quadratic forms of normal random vectors also hold for quadratic forms in $(\mathbf{z}-\boldsymbol{\xi})$. The following are some examples of even functions in the real vector $\mathbf{z} = (z_1,z_2)^T$:

$$|z_1|, \quad z_1 \sin z_2, \quad |z_2| \cos z_1, \quad z_1/z_2, \quad \mathbf{z}^T A\mathbf{z}.$$

The expectation and variance of a GSN random vector might not be equal to the location parameter $\boldsymbol{\xi}$ and the scale parameter Ω, respectively. However, the Euclidean distance of the expectation from $\boldsymbol{\xi}$ cannot be greater than the square root of the largest eigenvalue of Ω. The difference between the i-th diagonal element of Ω and the variance of the i-th component of the GSN random vector is non-negative, but the sum of these differences is not greater than the largest eigenvalue of Ω. More formally:

Proposition 4.3.3 Let $\mathbf{z} \sim GSN_p(\boldsymbol{\xi}, \Omega, \pi)$ and let λ_1 be the largest eigenvalue of Ω. Then:

$$\|\mathrm{E}(\mathbf{z}) - \boldsymbol{\xi}\| \leq \sqrt{\lambda_1},$$
$$0 \leq \mathrm{tr}(\Omega) - \mathrm{tr}[\mathrm{Var}(\mathbf{z})] \leq \lambda_1.$$

Azzalini and Dalla Valle (1996) show that the moment generating function (mgf) of $SN_p(\mathbf{0}, \Omega, \boldsymbol{\alpha})$ is $M(\mathbf{t}) = 2\exp\left(-\mathbf{t}^T\Omega\mathbf{t}/2\right)\Phi\left(\Omega\mathbf{t}/\sqrt{1+\mathbf{t}^T\Omega\mathbf{t}}\right)$. Notice that it is twice the mgf of $N_p(\mathbf{0}, \Omega)$ multiplied by a skewing function. The same holds under the more general assumption of generalized skew-normality:

Proposition 4.3.4 Let $\mathbf{z} \sim GSN_p(\mathbf{0}, \Omega, \pi)$ and let $M(\mathbf{t})$ be the mgf of $\mathbf{z}$. Then $M(\mathbf{t}) = 2\exp\left(-\mathbf{t}^T\Omega\mathbf{t}/2\right)\pi^*(\mathbf{t})$, where $0 \leq \pi^*(\mathbf{t}) = 1 - \pi^*(-\mathbf{t}) \leq 1$.

The above proposition implies that the sum of a normal random vector and a GSN one is again a GSN random vector, when the vectors are independent:

Corollary 4.3.1 Let the vectors $\mathbf{z} \sim GSN_p(\mathbf{0}, \Omega, \pi)$ and $\mathbf{x} \sim N_p(\mathbf{0}, \Psi)$ be independent. Then the mgf of $\mathbf{z}+\mathbf{x}$ is $M(\mathbf{t}) = 2\exp\left(-\mathbf{t}^T(\Omega + \Psi)\mathbf{t}/2\right)\pi^*(\mathbf{t})$ where $0 \leq \pi^*(\mathbf{t}) = 1 - \pi^*(-\mathbf{t}) \leq 1$.

4.4 Diagnostics

Model diagnostics is a fundamental step in statistical analysis and it often means checking whether the data are normal. Examples in this section show that commonly used diagnostic methods fail to tell normality from generalized skew-normality.

A diagnostic method can be graphical in nature, as happens for projection onto principal axes. It represents data in the two-dimensional space that retains as much information as possible about the data themselves. This does not mean that generalized skew-normality of data is retained as well, since the first principal components of a generalized skew-normal random vector can be normal, as is shown in the following proposition.

Proposition 4.4.1 Let $\mathbf{a}_1, \ldots, \mathbf{a}_p$ be the eigenvectors associated with the eigenvalues $\lambda_1 > \ldots > \lambda_p$ of the matrix Ω. Moreover, let $\mathbf{z} \sim SN_p(\boldsymbol{\xi}, \Omega, c\mathbf{a}_i)$, $i = 1, \ldots, p$, and $c \in \mathbb{R}$. Then the following hold:

1. The matrices Ω and $\mathrm{Var}(\mathbf{z})$ have the same eigenvectors.
2. The eigenvalues of $\mathrm{Var}(\mathbf{z})$ associated with $\mathbf{a}_1, \ldots, \mathbf{a}_{i-1}, \mathbf{a}_{i+1}, \ldots, \mathbf{a}_p$ are $\lambda_1, \ldots, \lambda_{i-1}, \lambda_{i+1}, \ldots, \lambda_p$.
3. The eigenvalue of $\mathrm{Var}(\mathbf{z})$ associated with $\mathbf{a}_i$ is given by the expression $\left[\pi\lambda_i + (\pi - 2)\, c^2\lambda_i^2\right] / \left[\pi\left(1 + c^2\lambda_i\right)\right]$.
4. When $i \neq j, h$, $\mathbf{a}_j^T\mathbf{z}$ and $\mathbf{a}_h^T\mathbf{z}$ are independent and normally distributed.

Assume now that $i = p > 2$ and consider projections on principal axes for checking the normality of a sample $\mathbf{z}_1, \ldots, \mathbf{z}_n$ from $\mathbf{z} \sim SN_p(\boldsymbol{\xi}, \Omega, c\mathbf{a}_i)$. It is necessary to plot the points $\left(\widehat{\mathbf{a}}_1^T\mathbf{z}_1, \widehat{\mathbf{a}}_2^T\mathbf{z}_1\right), \ldots, \left(\widehat{\mathbf{a}}_1^T\mathbf{z}_n, \widehat{\mathbf{a}}_2^T\mathbf{z}_n\right)$, where $\widehat{\mathbf{a}}_1$ and $\widehat{\mathbf{a}}_2$ are the eigenvectors associated with the largest eigenvalues of the sample covariance matrix S. When the sample size is large enough, then the quantities $\left(\widehat{\mathbf{a}}_1^T\mathbf{z}_1, \widehat{\mathbf{a}}_2^T\mathbf{z}_1\right), \ldots, \left(\widehat{\mathbf{a}}_1^T\mathbf{z}_n, \widehat{\mathbf{a}}_2^T\mathbf{z}_n\right)$ approximate $\left(\mathbf{a}_1^T\mathbf{z}_1, \mathbf{a}_2^T\mathbf{z}_1\right)$, $\ldots$, $\left(\mathbf{a}_1^T\mathbf{z}_n, \mathbf{a}_2^T\mathbf{z}_n\right)$, respectively. Large sample sizes, together with the above proposition, imply that the scatterplot of $\left(\widehat{\mathbf{a}}_1^T\mathbf{z}_1, \widehat{\mathbf{a}}_2^T\mathbf{z}_1\right), \ldots, \left(\widehat{\mathbf{a}}_1^T\mathbf{z}_n, \widehat{\mathbf{a}}_2^T\mathbf{z}_n\right)$ looks exactly as if the data come from a multivariate normal distribution. The same argument also applies to projections on other subspaces.

Healy's plot (1968) is another graphical method for checking normality. It is operationally defined as follows. For a sample $\mathbf{z}_1, \ldots, \mathbf{z}_n$ of p-dimensional vectors whose mean and variance are $\overline{\mathbf{z}}$ and S respectively:

1. Compute the Mahalanobis distances $D_i = (\mathbf{z}_i - \overline{\mathbf{z}})^T S^{-1}(\mathbf{z}_i - \overline{\mathbf{z}})$.
2. Rearrange $D_1, \ldots, D_n$ in increasing order to obtain $D_{(1)} \leq \ldots \leq D_{(n)}$.
3. Plot $D_{(i)}$ against q_i, where $\mathrm{P}(Y \leq q_i) = i/n$ and $Y \sim \chi_p^2$.

When the sample size is large enough and the data are independent and identically distributed (iid) according to a normal law, the points $\left(D_{(i)}, q_i\right)$ are close to a straight line, since $\mathrm{E}\left(D_{(i)}\right) \approx q_i$. The same holds when the data have GSN densities (not necessarily identical) whose expectations are equal to the common location parameter. More formally:

Proposition 4.4.2 Let $\mathbf{z}_1, \ldots, \mathbf{z}_n$ be iid observations from $GSN_p(\boldsymbol{\xi}, \Omega, \pi)$, where $\mathrm{E}(\mathbf{z}_i) = \boldsymbol{\xi}$, $i = 1, \ldots, n$. Then the asymptotic distribution of the Mahalanobis distances $D_1, \ldots, D_n$ does not depend on the skewing function π.

GSN distributions whose expectations equal their location parameters can be easily obtained through mixtures of SN distributions. As an example, consider the random vector $\mathbf{z} \sim GSN_p(\boldsymbol{\xi}, \Omega, \pi)$, where

$$\pi(\mathbf{z}) = \frac{\sum_{i=1}^{3} \Phi\left(\boldsymbol{\alpha}_i^T \mathbf{z}\right) \sqrt{1 + \boldsymbol{\alpha}_i^T \Omega \boldsymbol{\alpha}_i}}{\sum_{j=1}^{3} \sqrt{1 + \boldsymbol{\alpha}_j^T \Omega \boldsymbol{\alpha}_j}}, \qquad \boldsymbol{\alpha}_1 + \boldsymbol{\alpha}_2 + \boldsymbol{\alpha}_3 = \mathbf{0}.$$

A straightforward application of results on moments of SN distributions (Azzalini and Dalla Valle, 1996; Genton, He, and Liu, 2001) shows that $\mathrm{E}(\mathbf{z}) = \boldsymbol{\xi}$.

Model diagnostic is also based on formal testing. The sample skewness $b_{1,p}$ and the sample kurtosis $b_{2,p}$ are often used to test the adequacy of the normal law. They are defined as follows (Mardia, 1970):

$$b_{1,p} = \frac{1}{n^2} \sum_{i=1}^{n} \sum_{j=1}^{n} \left[(\mathbf{z}_i - \overline{\mathbf{z}})^T S^{-1} (\mathbf{z}_j - \overline{\mathbf{z}}) \right]^3,$$

$$b_{2,p} = \frac{1}{n} \sum_{i=1}^{n} \left[(\mathbf{z}_i - \overline{\mathbf{z}})^T S^{-1} (\mathbf{z}_i - \overline{\mathbf{z}}) \right]^2.$$

When the data are iid normal, the following asymptotic result holds (Mardia, 1970):

$$\frac{b_{2,p} - p(p+2)}{\sqrt{8p(p+2)/n}} \overset{a}{\sim} N(0,1).$$

It also holds under more general assumptions: data can have a GSN distribution and they do not need to be identically distributed. More formally:

Proposition 4.4.3 Let $b_{2,p}$ be the sample kurtosis of the independent p-dimensional random vectors $\mathbf{z}_1, \ldots, \mathbf{z}_n$ where $\mathbf{z}_i \sim GSN_p(\boldsymbol{\xi}, \Omega, \pi_i)$, $\mathrm{E}(\mathbf{z}_i) = \boldsymbol{\xi}$. Then

$$\frac{b_{2,p} - p(p+2)}{\sqrt{8p(p+2)/n}} \overset{a}{\sim} N(0,1). \tag{4.6}$$

Proof of the above proposition easily follows from $b_{2,p}$ being a continuous function of $(\mathbf{z}_1 - \overline{\mathbf{z}})^T S^{-1} (\mathbf{z}_1 - \overline{\mathbf{z}}), \ldots, (\mathbf{z}_n - \overline{\mathbf{z}})^T S^{-1} (\mathbf{z}_n - \overline{\mathbf{z}})$. A similar result holds for $b_{1,p}$. From the inferential point of view, the above proposition implies that Mardia's tests for normality cannot discriminate between normality and generalized skew-normality, when the sample size is large enough.

Results in this section highlight limitations of diagnostic methods when the data are a random sample from a GSN distribution because they can easily mislead us into believing that the parent population is normal. The same results motivate the use of the same methods when the sampled population is normal but the data have a GSN distribution because of non-random sampling. In this situation they retain their inferential properties despite non-random sampling.

4.5 Parametric Inference

This section deals with inference from GSN data that are non-random samples from a normal distribution. Under these circumstances, interest focuses on location and scale parameters, corresponding to the population's mean and variance, respectively. Skewing functions become nuisance parameters; they are not relevant by themselves but they influence the data distribution. Invariance properties of GSN densities lead to statistics whose sampling distributions do not depend on the skewing functions.

First consider testing the hypothesis $H_0 : \boldsymbol{\xi} = \boldsymbol{\xi}_0$ versus $H_1 : \boldsymbol{\xi} \neq \boldsymbol{\xi}_0$ on the mean $\boldsymbol{\xi}$ of a normal distribution $N_p(\boldsymbol{\xi}, \Omega)$ using a non-random sample $\mathbf{z}_1, \ldots, \mathbf{z}_n$ of independent GSN data: $\mathbf{z}_i \sim GSN_p(\boldsymbol{\xi}, \Omega, \pi_i)$. The sampling distribution of the Hotelling statistic depends on skewing functions, that is, on nuisance parameters. Hence the need for another test statistic. A possible choice is a trivial modification of Mardia's index of kurtosis $b_{2,p}$, whose asymptotic properties are shown below.

Proposition 4.5.1 Let $\widetilde{S}$ and $\widetilde{b}_{2,p}$ be defined as follows:

$$\widetilde{S} = \frac{1}{n}\sum_{i=1}^{n}(\mathbf{z}_i - \boldsymbol{\xi})(\mathbf{z}_i - \boldsymbol{\xi})^T, \qquad \widetilde{b}_{2,p} = \frac{1}{n}\sum_{i=1}^{n}\left[(\mathbf{z}_i - \boldsymbol{\xi})^T \widetilde{S}^{-1}(\mathbf{z}_i - \boldsymbol{\xi})\right]^2,$$

where $\mathbf{z}_1, \ldots, \mathbf{z}_n$ are iid p-dimensional random vectors whose distribution is $\mathbf{z}_i \sim GSN_p(\boldsymbol{\xi}, \Omega, \pi_i)$. Then

$$\frac{\widetilde{b}_{2,p} - p(p+2)}{\sqrt{8p(p+2)/n}} \overset{a}{\sim} N(0,1). \tag{4.7}$$

When $\boldsymbol{\xi}$ is known, $\widetilde{b}_{2,p}$ is a statistic whose asymptotic distribution is completely specified. Hence $\widetilde{b}_{2,p}$ can be used for testing $H_0 : \boldsymbol{\xi} = \boldsymbol{\xi}_0$ versus $H_1 : \boldsymbol{\xi} \neq \boldsymbol{\xi}_0$.

The skewing function π can be interpreted as a parameter with infinite dimension, so that sufficient and ancillary statistics are defined for GSN distributions. For example, consider a random vector $\mathbf{z}$ whose density $f(\mathbf{z}; \theta, \gamma)$ depends on the parameters θ and γ. A statistic $\mathbf{t}$ is said to be partially sufficient for θ if its marginal distribution depends on θ only and it is sufficient for θ for any given value of γ (Basu and Pereira, 1983; Reid, 1995). Loperfido (2001) shows that the sum of squares and products (SSP) matrix is partially sufficient for the scale parameter Ω when the rows of the data matrix are independent skew-normal $SN_p(\mathbf{0}, \Omega, \boldsymbol{\alpha}_i)$ random vectors. Similar results hold for GSN random vectors.

Under the same assumptions, the SSP matrix divided by the number of observations is the maximum likelihood estimator and the UMVU estimator for the scale matrix. Moreover, its distribution is Wishart. More formally (Genton and Loperfido, 2002):

Proposition 4.5.2 Let $\widehat{\Omega} = Z^T Z/n$, where Z is an $n \times p$ matrix whose rows $\mathbf{z}_1, \ldots, \mathbf{z}_n$ are independent and $\mathbf{z}_i \sim GSN_p(\mathbf{0}, \Omega, \pi_i)$. Then the following hold:

- $\widehat{\Omega}$ is the maximum likelihood estimator of Ω.
- $\widehat{\Omega}$ is the UMVU estimator of Ω.
- $\widehat{\Omega}$ is partially sufficient for Ω.
- The distribution of $\widehat{\Omega}$ is Wishart: $\widehat{\Omega} \sim W(\Omega/n, n)$.

Invariance properties of GSN distributions can also be used for testing hypotheses on the scale matrix. Many likelihood-based tests for scale matrices with normal data $\mathbf{z}_i \sim N_p(\mathbf{0}, \Omega)$ maintain their properties when the data are $\mathbf{z}_i \sim GSN_p(\mathbf{0}, \Omega, \pi_i)$. The following proposition focuses on likelihood ratio tests (LRT) for the equality of two scale matrices. It shows that the functional form of the LRT statistic and the power function of the LRT test do not depend on the skewing functions. Hence the skewing functions do not need to be specified, allowing a high degree of robustness with respect to departures from normality and maintaining many optimality properties of the LRT. More formally (Genton and Loperfido, 2002):

Proposition 4.5.3 Let $\mathbf{z}_1, \ldots, \mathbf{z}_n, \mathbf{y}_1, \ldots, \mathbf{y}_m$ be independent p-dimensional random vectors such that $\mathbf{z}_i \sim GSN_p(\mathbf{0}, \Omega, \pi_i)$ and $\mathbf{y}_j \sim GSN_p(\mathbf{0}, \Psi, \omega_j)$. Moreover, let L be the likelihood ratio test statistic for the null hypothesis $H_0 : \Omega = \Psi$ against the alternative $H_1 : \Omega \neq \Psi$. Then the following hold:

- L is invariant with respect to the skewing functions.
- The power function of the LRT does not depend on the skewing functions.
- When the null hypothesis is true and the sample size is large, the distribution of $-2\log L$ is chi-squared with $p(p+1)/2$ degrees of freedom.
- When the null hypothesis is true, the following statistic has a Wilks' distribution with parameters $p, m, n-m$:

$$W = \frac{\left|\sum_{j=1}^{m} \mathbf{y}_j \mathbf{y}_j^T\right|}{\left|\sum_{j=1}^{m} \mathbf{y}_j \mathbf{y}_j^T + \sum_{i=1}^{n} \mathbf{z}_i \mathbf{z}_i^T\right|}.$$

The above result holds even when the observations are not identically distributed, since the skewing functions might not be equal. Moreover, the assumption of the location parameter $\boldsymbol{\xi}$ being zero is not restrictive when $\boldsymbol{\xi}$ is known or the sample is large and a consistent estimator of $\boldsymbol{\xi}$ is available.

4.6 Australian Athletes Data

This section applies results from previous sections to inference from non-random samples. The Australian Institute of Sport (AIS) collected heights and weights of 102 male athletes, competing in different events; see Cook and Weisberg (1994). AIS data are significantly skewed and Azzalini and Dalla Valle (1996) showed that the bivariate skew-normal distribution gave a good fit. Arnold and Beaver (2000b) modeled the same data through the skew-Cauchy distribution, which also accounts for heavier tails. Genton and Loperfido (2002) modeled skewness, sampling bias, and individual differences through the GSN distribution. We shall first outline their motivation for the GSN distribution.

The average height (in centimeters) and weight (in kilograms) of adult Australian males is 174.8 and 82.0, respectively (Australian Bureau of Statistics, 1995). Hence the standard assumption that the joint distribution of heights (H) and weights (W) for adult Australian males is bivariate normal implies that

$$\begin{pmatrix} H \\ W \end{pmatrix} \sim N_2\left(\begin{pmatrix} 174.8 \\ 82.0 \end{pmatrix}, \Omega\right). \tag{4.8}$$

The joint distribution of height and weight of an adult Australian male can be represented as follows:

$$\begin{aligned} H &= 174.8 + \lambda_H U + \gamma_H \xi_H, \\ W &= 82.0 + \lambda_W U + \gamma_W \xi_W, \end{aligned} \tag{4.9}$$

where U, ξ_H and ξ_W are independent standard normal variables and

$$\begin{pmatrix} \lambda_H^2 + \gamma_H^2 & \lambda_H \lambda_W \\ \lambda_H \lambda_W & \lambda_W^2 + \gamma_W^2 \end{pmatrix} = \Omega.$$

It easily follows that the above model is a single factor model, where U is the common factor and ξ_H, ξ_W are the specific factors for height and weight respectively. The former can be interpreted as a proxy for physical fitness, which is clearly above average for all individuals in the AIS data set: the sample only includes gifted athletes. More formally:

$$\left.\begin{pmatrix} H \\ W \end{pmatrix}\right| \text{individual in the data set} \Leftrightarrow \left.\begin{pmatrix} H \\ W \end{pmatrix}\right| U > 0.$$

Azzalini and Dalla Valle (1996) show that the distribution of $H, W | U > 0$ (and hence the joint distribution of heights and weights for the athletes in the sample) is bivariate skew-normal under the above assumptions. More precisely:

$$\left.\begin{pmatrix} H \\ W \end{pmatrix}\right| \text{individual in the data set} \sim SN_2\left(\begin{pmatrix} 174.8 \\ 82.0 \end{pmatrix}, \Omega, \boldsymbol{\alpha}\right), \tag{4.10}$$

where $\boldsymbol{\alpha}$ is a function of λ_H, γ_H, λ_W, and γ_W. Let $\mathbf{z}_1, \ldots, \mathbf{z}_{102}$ be the vectors of differences between the observed heights, weights and the corresponding populations' averages:

$$\mathbf{z}_i = \begin{pmatrix} \text{Height of the } i\text{-th male athlete}-174.8 \\ \text{Weight of the } i\text{-th male athlete}-82.0 \end{pmatrix}.$$

Sample bias can be modeled in a more general way, by assuming that $\mathbf{z}_1, \ldots, \mathbf{z}_{102}$ are GSN random vectors:

$$\mathbf{z}_i \sim GSN_2(\mathbf{0}, \Omega, \pi_i), \qquad i = 1, \ldots, 102. \tag{4.11}$$

This model admits UMVU and maximum likelihood estimates of the scale matrix Ω (Genton and Loperfido, 2002):

$$\widetilde{S} = \frac{1}{102}\sum_{i=1}^{102} \mathbf{z}_i\mathbf{z}_i^T = \begin{pmatrix} 176.47 & 70.31 \\ 70.31 & 152.68 \end{pmatrix}. \tag{4.12}$$

Despite the above arguments, however, the GSN distribution might not be the appropriate model for the AIS data. The sampled population might not be normal, nor can the sampling bias be modeled through a skewing function $\pi(\mathbf{z}) = 1 - \pi(-\mathbf{z})$. The p-value of the statistical test based on $\widetilde{b}_{2,p}$ is close to 0.1, so that the null hypothesis of generalized skew-normality is not rejected at the 0.05 level.

Genton and Loperfido (2002) assume that the scale matrices of all observations are equal to each other, even for athletes competing in different events. This assumption is rather restrictive and needs to be tested. As an example, consider the subsample of male swimmers. Let $\mathbf{y}_1, \ldots, \mathbf{y}_{13}$ be defined as follows:

$$\mathbf{y}_i = \begin{pmatrix} \text{Height of the } i\text{-th male swimmer}-174.8 \\ \text{Weight of the } i\text{-th male swimmer}-82.0 \end{pmatrix}, \quad i = 1, \ldots, 13.$$

Let Y be the matrix whose i-th row is $\mathbf{y}_i^T$, that is,

$$Y = \begin{pmatrix} -2.1 & -15.0 \\ 1.7 & -7.6 \\ 8.2 & -2.7 \\ 19.6 & 5.5 \\ 18.6 & 1.5 \\ 5.4 & -4.0 \\ 8.2 & -4.0 \\ 9.2 & 3.0 \\ 17.9 & 2.7 \\ 12.4 & 10.0 \\ 9.1 & -9.7 \\ 17.2 & 1.0 \\ 15.6 & 14.9 \end{pmatrix}.$$

The corresponding UMVU estimate of their scale matrix is

$$\frac{1}{13}Y^TY = \begin{pmatrix} 160.14 & 33.80 \\ 33.80 & 60.61 \end{pmatrix}.$$

It is not intuitively clear whether differences between the above matrix and $\widetilde{S}$ are due to random errors only. Hence the need for a formal test of the hypothesis $H_0 : \Omega_S = \Omega$ versus $H_1 : \Omega_S \neq \Omega$, where Ω_S is the scale matrix of the swimmers' heights and weights. We shall use the test statistic

$$W = \frac{\left|\sum_{j=1}^{13} \mathbf{y}_j\mathbf{y}_j{}^T\right|}{\left|\sum_{i=1}^{102} \mathbf{z}_i\mathbf{z}_i{}^T\right|}.$$

From the previous section, we know that the distribution of the test statistic, under the null hypothesis, is Wilks' distribution:

$$W|H_0 \sim \Lambda(2, 13, 102).$$

The observed value of the test statistic is 0.006325, and the corresponding p-value is 0.187. Hence we do not reject at the 0.05 level the hypothesis of Ω_S and Ω being equal.

4.7 Concluding Remarks

Generalized skew-normal distributions include the normal ones and often arise in inference from non-random samples. Well-known goodness-of-fit methods might easily lead to the erroneous belief that GSN data are normal. Inferential properties of statistical methods commonly used in the normal case also hold for GSN data. However, their application is limited by the following problems, which should guide future research:

Characterization. When do GSN distributions arise? Normality is often motivated through the central limit theorem or through an entropy-based argument. Skew-normality naturally arises in threshold models (Aigner, Lovell, and Schmidt, 1977) and order statistics (Loperfido, 2002). No similar results are available at the moment for GSN distributions.

Diagnostic. Goodness-of-fit methods might fail to discriminate between normal and GSN distributions. This is not a problem when further inference does not depend on the skewing function, but it is a problem otherwise. Hence the need to define tests for normality versus other members of the GSN class and tests for generalized skew-normality versus other distributions.

Estimation. Inference on the scale parameter is very easy when location is known. The same holds when the sample size is large and a consistent estimator for location is available. Unfortunately, no such estimator is known at present time. A similar problem exists for interval estimation (and hence hypothesis testing): there is no interval that is known to contain the location parameter with given probability.

4.8 Appendix: Proofs

Proof of Proposition 4.3.1: In order to keep the notation simple we shall assume that $\boldsymbol{\xi} = \mathbf{b} = \mathbf{0}$. The case $\boldsymbol{\xi}, \mathbf{b} \neq \mathbf{0}$ is a trivial generalization. Let g be the density of the random vector $A\mathbf{z}$ and

$$\mathbf{x} \sim N_p\left(\mathbf{0},\Omega\right), \qquad \mathbf{u} = A\mathbf{x}, \qquad \pi^*\left(\mathbf{a}\right) = \frac{g\left(\mathbf{a}\right)}{g\left(\mathbf{a}\right) + g\left(-\mathbf{a}\right)}.$$

From Proposition 4.2.1 we know that

$$\mathbf{z} \sim GSN_p\left(\mathbf{0},\Omega,\pi\right) \Leftrightarrow \mathbf{z} = \begin{cases} \mathbf{x}, & \text{if } Y \leq \pi\left(\mathbf{x}\right), \\ -\mathbf{x}, & \text{if } Y > \pi\left(\mathbf{x}\right), \end{cases}$$

where π is a skewing function and $Y \sim U(0,1)$ is independent of $\mathbf{x}$. Equivalently,

$$\mathbf{z} \sim \mathbf{x}|Y \leq \pi\left(\mathbf{x}\right) \Rightarrow A\mathbf{z} \sim A\mathbf{x}|Y \leq \pi\left(\mathbf{x}\right) \Rightarrow -A\mathbf{z} \sim -A\mathbf{x}|Y \leq \pi\left(\mathbf{x}\right).$$

Hence

$$-A\mathbf{z} \sim A\mathbf{x}|Y > \pi\left(\mathbf{x}\right).$$

The density of $\mathbf{u}$ can be represented as the mixture of conditional densities:

$$\begin{aligned}&\phi_k\left(\mathbf{u}; A\boldsymbol{\xi}, A\Omega A^T\right)\\ &= f\left(\mathbf{u}|Y \leq \pi\left(\mathbf{x}\right)\right) \mathrm{P}\left(Y \leq \pi\left(\mathbf{x}\right)\right) + f\left(\mathbf{u}|Y \geq \pi\left(\mathbf{x}\right)\right) \mathrm{P}\left(Y > \pi\left(\mathbf{x}\right)\right).\end{aligned}$$

By symmetry, $\mathrm{P}(Y \leq \pi\left(\mathbf{x}\right)) = \mathrm{P}(Y > \pi\left(\mathbf{x}\right))$. Then the above equation is equivalent to the following one:

$$2\phi_k\left(\mathbf{u}; A\boldsymbol{\xi}, A\Omega A^T\right) = f\left(\mathbf{u}|Y \leq \pi\left(\mathbf{x}\right)\right) + f\left(\mathbf{u}|Y > \pi\left(\mathbf{x}\right)\right).$$

Recall that $-A\mathbf{x}|Y \leq \pi\left(\mathbf{x}\right) \sim A\mathbf{x}|Y > \pi\left(\mathbf{x}\right)$. Hence

$$2\phi_k\left(\mathbf{u}; A\boldsymbol{\xi}, A\Omega A^T\right) = f\left(\mathbf{u}|Y \leq \pi\left(\mathbf{x}\right)\right) + f\left(-\mathbf{u}|Y \leq \pi\left(\mathbf{x}\right)\right).$$

By definition $f\left(\mathbf{u}|Y \leq \pi\left(\mathbf{x}\right)\right) = g\left(\mathbf{u}\right)$. Then

$$2\phi_k\left(\mathbf{u}; A\boldsymbol{\xi}, A\Omega A^T\right) = g\left(\mathbf{u}\right) + g\left(-\mathbf{u}\right).$$

Recall now the definition of $\pi^*\left(\mathbf{u}\right)$:

$$g\left(\mathbf{u}\right) = 2\phi_k\left(\mathbf{u}; A\boldsymbol{\xi}, A\Omega A^T\right) \pi^*\left(\mathbf{u}\right),$$

and the proof is complete. □

Proof of Proposition 4.3.3: We shall first prove that $\|\mathrm{E}(\mathbf{z}) - \boldsymbol{\xi}\| \leq \sqrt{\lambda_1}$. Let $\mathbf{x} = \mathbf{z} - \boldsymbol{\xi}$ and recall the standard decomposition of the variance matrix:

$$\mathrm{Var}\left(\mathbf{x}\right) = \mathrm{E}\left(\mathbf{x}\mathbf{x}^T\right) - \mathrm{E}\left(\mathbf{x}\right)\mathrm{E}\left(\mathbf{x}^T\right).$$

By Proposition 4.3.2, $\mathbf{x}\mathbf{x}^T \sim W(\Omega, 1)$, so that $\mathrm{E}(\mathbf{x}\mathbf{x}^T) = \Omega$. Hence:

$$\mathrm{Var}(\mathbf{x}) = \Omega - \mathrm{E}(\mathbf{x})\,\mathrm{E}(\mathbf{x}^T).$$

Consider now the variance of the linear combination $\mathbf{x}^T\mathrm{E}(\mathbf{x})$:

$$\mathrm{Var}\left[\mathbf{x}^T\mathrm{E}(\mathbf{x})\right] = \mathrm{E}(\mathbf{x}^T)\,\Omega\mathrm{E}(\mathbf{x}) - \left(\mathrm{E}(\mathbf{x}^T)\,\mathrm{E}(\mathbf{x})\right)^2.$$

By definition, λ_1 is the largest eigenvalue of the positive semidefinite matrix Ω. Hence a quadratic form $\mathbf{v}^T\Omega\mathbf{v}$ cannot be greater than $\lambda\mathbf{v}^T\mathbf{v}$ (Mardia *et al.*, 1979, p. 479). It follows that

$$\mathrm{Var}\left[\mathbf{x}^T\mathrm{E}(\mathbf{x})\right] \leq \lambda_1\mathrm{E}(\mathbf{x}^T)\,\mathrm{E}(\mathbf{x}) - \left(\mathrm{E}(\mathbf{x}^T)\,\mathrm{E}(\mathbf{x})\right)^2.$$

The variance of $\mathbf{x}^T\mathrm{E}(\mathbf{x})$ is a non-negative quantity. A little algebra shows that

$$0 \leq \mathrm{E}(\mathbf{x}^T)\,\mathrm{E}(\mathbf{x}) \leq \lambda_1.$$

By definition $||\mathbf{v}|| = \sqrt{\mathbf{v}^T\mathbf{v}}$ and $\mathbf{x} = \mathbf{z} - \boldsymbol{\xi}$. Hence:

$$||\mathrm{E}(\mathbf{z}) - \boldsymbol{\xi}|| \leq \sqrt{\lambda_1},$$

and this part of the proof is complete.

We shall now prove that $0 \leq \mathrm{tr}(\Omega) - \mathrm{tr}(\mathrm{Var}(\mathbf{z})) \leq \lambda_1$. The previous part of the proof implies that $\mathrm{Var}(\mathbf{z}) = \mathrm{Var}(\mathbf{x})$ and that trace of $\mathrm{Var}(\mathbf{x})$ can be written as follows:

$$\mathrm{tr}(\mathrm{Var}(\mathbf{x})) = \mathrm{tr}(\Omega) - \mathrm{tr}\left(\mathrm{E}(\mathbf{x})\,\mathrm{E}(\mathbf{x}^T)\right).$$

Standard properties of the trace of a matrix lead to the following equation:

$$\mathrm{tr}(\mathrm{Var}(\mathbf{x})) = \mathrm{tr}(\Omega) - \mathrm{E}(\mathbf{x}^T)\,\mathrm{E}(\mathbf{x}).$$

Since $0 \leq \mathrm{E}(\mathbf{x}^T)\mathrm{E}(\mathbf{x}) \leq \lambda_1$, a little algebra shows that

$$0 \leq \mathrm{tr}(\Omega) - \mathrm{tr}(\mathrm{Var}(\mathbf{x})) \leq \lambda_1,$$

and the proof is complete. □

Proof of Proposition 4.3.4: Let

$$M(\mathbf{t}) = \mathrm{E}\left(\exp\left(\mathbf{t}^T\mathbf{z}\right)\right), \qquad \pi^*(\mathbf{t}) = \frac{M(\mathbf{t})}{M(\mathbf{t}) + M(-\mathbf{t})}.$$

From Proposition 4.2.1 we know that

$$\mathbf{z} \sim GSN_p(\mathbf{0}, \Omega, \pi) \Leftrightarrow \mathbf{z} = \begin{cases} \mathbf{x}, & \text{if } Y \leq \pi(\mathbf{x}), \\ -\mathbf{x}, & \text{if } Y > \pi(\mathbf{x}), \end{cases}$$

where $\mathbf{x} \sim N_p(\mathbf{0}, \Omega)$, π is a skewing function, and $Y \sim U(0,1)$ is independent of $\mathbf{x}$. Hence

$$-\mathbf{z} \sim \mathbf{x}|Y > \pi(\mathbf{x}).$$

Now apply the associative property of expected values to $\mathrm{E}(e^{\mathbf{t}^T\mathbf{z}})$:

$$\mathrm{E}\left(e^{\mathbf{t}^T\mathbf{z}}|Y \leq \pi(\mathbf{x})\right)\mathrm{P}(Y \leq \pi(\mathbf{x})) + \mathrm{E}\left(e^{\mathbf{t}^T\mathbf{z}}|Y \geq \pi(\mathbf{x})\right)\mathrm{P}(Y > \pi(\mathbf{x})).$$

The mgf of $\mathbf{x} \sim N_p(\mathbf{0}, \Omega)$ is $\exp\left(-\mathbf{t}^T\Omega\mathbf{t}/2\right)$ (Mardia *et al.*, 1979). Moreover $\mathrm{P}(Y \leq \pi(\mathbf{x})) = \mathrm{P}(Y > \pi(\mathbf{x}))$. Then the above equation is equivalent to the following:

$$2\exp\left(-\mathbf{t}^T\Omega\mathbf{t}/2\right) = \mathrm{E}\left(e^{\mathbf{t}^T\mathbf{z}}|Y \leq \pi(\mathbf{x})\right) + \mathrm{E}\left(e^{\mathbf{t}^T\mathbf{z}}|Y > \pi(\mathbf{x})\right).$$

We already know that $-\mathbf{z}|Y \leq \pi(\mathbf{x}) \sim \mathbf{z}|Y > \pi(\mathbf{x})$. Hence

$$2\exp\left(-\mathbf{t}^T\Omega\mathbf{t}/2\right) = \mathrm{E}\left(e^{\mathbf{t}^T\mathbf{z}}|Y \leq \pi(\mathbf{x})\right) + \mathrm{E}\left(e^{-\mathbf{t}^T\mathbf{z}}|Y \leq \pi(\mathbf{x})\right).$$

By definition $\mathrm{E}\left(e^{\mathbf{t}^T\mathbf{z}}|Y \leq \pi(\mathbf{x})\right) = M(\mathbf{t})$. Then

$$M(\mathbf{t}) = \frac{2\exp\left(-\mathbf{t}^T\Omega\mathbf{t}/2\right)M(\mathbf{t})}{M(\mathbf{t}) + M(-\mathbf{t})}.$$

Recall now the definition of $\pi^*(\mathbf{t})$:

$$M(\mathbf{t}) = 2\exp\left(-\mathbf{t}^T\Omega\mathbf{t}/2\right)\pi^*(\mathbf{t}),$$

and the proof is complete. □

Proof of Proposition 4.4.1: Without loss of generality we can assume that $i = p$. By assumption the distribution of $\mathbf{z}$ is multivariate skew-normal and hence $\mathrm{Var}(\mathbf{z})$ can be represented as follows (Azzalini and Dalla Valle, 1996):

$$\mathrm{Var}(\mathbf{z}) = \Omega - \frac{2}{\pi}\frac{c^2\Omega\mathbf{a}_p\mathbf{a}_p^T\Omega}{1 + c^2\mathbf{a}_p^T\Omega\mathbf{a}_p}.$$

By assumption, $\mathbf{a}_1, \ldots, \mathbf{a}_p$ are the eigenvectors associated with the eigenvalues $\lambda_1 > \ldots > \lambda_p$ of the matrix Ω. Now apply ordinary properties of eigenvectors:

$$\mathrm{Var}(\mathbf{z}) = \sum_{i=1}^{p}\lambda_i\mathbf{a}_i\mathbf{a}_i^T - \frac{2}{\pi}\frac{c^2\lambda_p^2}{1 + c^2\lambda_p}\mathbf{a}_p\mathbf{a}_p^T.$$

A little algebra shows that

$$\mathrm{Var}(\mathbf{z}) = \sum_{i=1}^{p-1}\lambda_i\mathbf{a}_i\mathbf{a}_i^T + \frac{\pi\lambda_p + (\pi - 2)c^2\lambda_p^2}{\pi(1 + c^2\lambda_p)}\mathbf{a}_p\mathbf{a}_p^T. \tag{4.13}$$

The above equation implies that $\mathrm{Var}(\mathbf{z})$ and Ω have the same eigenvectors. The i-th eigenvalue of $\mathrm{Var}(\mathbf{z})$ equals the i-th eigenvalue of Ω, for $i = 1, \ldots, p - 1$. The p-th eigenvalue of $\mathrm{Var}(\mathbf{z})$ is given by the expression $\left[\pi\lambda_p + (\pi - 2)c^2\lambda_p^2\right] / \left[\pi\left(1 + c^2\lambda_p\right)\right]$. Hence the proof of parts 1, 2, and 3 is complete.

We shall now prove part 4. Let $\mathbf{y} = (Y_1, \ldots, Y_p)^T = A\mathbf{z}$, where A is a matrix whose rows are the eigenvectors $\mathbf{a}_1, \ldots, \mathbf{a}_p$. Hence the density of $\mathbf{y}$ is

$$f(y_1, \ldots, y_p) = 2\Phi(cy_p) \prod_{i=1}^{p} \frac{\exp(-y_i^2/2\lambda_i)}{\sqrt{2\pi\lambda_i}}.$$

Integration with respect to y_p leads to the joint density of $Y_1, \ldots, Y_{p-1}$, that is

$$f(y_1, \ldots, y_{p-1}) = \prod_{i=1}^{p-1} \frac{\exp(-y_i^2/2\lambda_i)}{\sqrt{2\pi\lambda_i}}.$$

The above density characterizes a normal random vector whose components are independent. The proof is then complete. □

Proof of Proposition 4.4.2: We shall first introduce the following notation:

$$\widetilde{S} = \frac{1}{n}\sum_{i=1}^{n} (\mathbf{z}_i - \boldsymbol{\xi})(\mathbf{z}_i - \boldsymbol{\xi})^T, \qquad \widetilde{D}_i = (\mathbf{z}_i - \boldsymbol{\xi})^T \widetilde{S}^{-1}(\mathbf{z}_i - \boldsymbol{\xi}),$$

$$\mathbf{d} = (D_1, \ldots, D_n), \qquad \widetilde{\mathbf{d}} = \left(\widetilde{D}_1, \ldots, \widetilde{D}_n\right).$$

The vector $\mathbf{d}$ is an even function of $(\mathbf{z}_1 - \boldsymbol{\xi}, \ldots, \mathbf{z}_n - \boldsymbol{\xi})$. Hence its distribution does not depend on the weight function (Proposition 4.3.2). We shall also write $\mathbf{x}_n \to \mathbf{x}$ to denote that the random sequence $\{\mathbf{x}_n\}$, $n \in \mathrm{N}$, converges in law to the random vector $\mathbf{x}$. By assumption $\mathrm{E}(\mathbf{z}_i) = \boldsymbol{\xi}$. Now apply some well-known asymptotic results:

$$\overline{\mathbf{z}} \to \boldsymbol{\xi}, \quad \widetilde{S} - S \to O,$$

and

$$\left((\mathbf{z}_1 - \overline{\mathbf{z}})^T S^{-1}(\mathbf{z}_1 - \overline{\mathbf{z}}), \ldots, (\mathbf{z}_n - \overline{\mathbf{z}})^T S^{-1}(\mathbf{z}_n - \overline{\mathbf{z}})\right) - \widetilde{\mathbf{d}} \to \mathbf{0}.$$

Hence $D_1, \ldots, D_n$ has the same asymptotic distribution as $\widetilde{\mathbf{d}}$, which does not depend on the weight function. The proof is then complete. □

CHAPTER 5

Skew-Symmetric and Generalized Skew-Elliptical Distributions

Marc G. Genton

5.1 Introduction

A popular approach to model departures from normality consists of modifying a symmetric probability density function (pdf) of a random variable, or of a random vector in the multivariate setting, in a multiplicative fashion, thereby introducing skewness. This idea has been in the literature for a long time, but it has been thoroughly implemented for the univariate normal distribution by Azzalini (1985, 1986), yielding the so-called skew-normal distribution. An extension to the multivariate case was then introduced by Azzalini and Dalla Valle (1996). Statistical applications of the multivariate skew-normal distribution were presented by Azzalini and Capitanio (1999), who also briefly discussed an extension to elliptical densities. Since then, several authors have tried to generalize these results to skewing arbitrary symmetric pdf's with very general forms of multiplicative functions.

This chapter is devoted to some of these generalizations and is organized as follows. In Section 5.2, we present skew-symmetric distributions, a very general class of skewed distributions introduced by Wang, Boyer, and Genton (2004a). In particular, we discuss various properties of these distributions, their stochastic representation, a simple procedure for simulations, skew-symmetric representations, and an illustrative example. In Section 5.3, we present generalized skew-elliptical distributions, a particular class of skew-symmetric distributions introduced by Genton and Loperfido (2002) where the symmetric pdf is also required to be elliptically contoured. We discuss various properties of these distributions and show that this class is of great interest for applications. In Section 5.4, we present a flexible class of skew-symmetric and generalized skew-elliptical distributions as introduced by Ma and Genton (2004). This approach is based on an approximation of the skewing function involving odd polynomials and allows for multimodality in the resulting distributions. We discuss also semiparametric estimators and efficiency bounds recently presented by Ma, Genton, and Tsiatis (2003). Concluding remarks are presented in Section 5.5.

5.2 Skew-Symmetric Distributions

Skew-symmetric distributions have been introduced by Wang *et al.* (2004a). This family of distributions is constructed with a skewing function that we define next.

Definition 5.2.1 (Skewing function) A function $\pi : \mathbb{R}^p \to [0,1]$ is a skewing function if it satisfies $\pi(\mathbf{z}) + \pi(-\mathbf{z}) = 1$ for all $\mathbf{z} \in \mathbb{R}^p$.

Notice that the cdf of a symmetric univariate distribution is a skewing function, but a skewing function is not necessarily a cdf, in particular, a skewing function is not required to be increasing. We now define skew-symmetric distributions.

Definition 5.2.2 (Skew-symmetric distribution) Let $f : \mathbb{R}^p \to \mathbb{R}_+$ be a continuous pdf symmetric around $\mathbf{0}$, i.e., $f(-\mathbf{z}) = f(\mathbf{z})$ for all $\mathbf{z} \in \mathbb{R}^p$, and $\pi : \mathbb{R}^p \to [0,1]$ a skewing function. A random vector $\mathbf{z} \in \mathbb{R}^p$ has a skew-symmetric distribution with location parameter $\boldsymbol{\xi} \in \mathbb{R}^p$, denoted by $\mathbf{z} \sim SS_p(\boldsymbol{\xi}, f, \pi)$, if its pdf is

$$2f(\mathbf{z} - \boldsymbol{\xi})\pi(\mathbf{z} - \boldsymbol{\xi}). \tag{5.1}$$

It is straightforward to show that (5.1) integrates to one and thus is a valid pdf. When π is identically a constant $(1/2)$, the resulting pdf is simply f and therefore symmetric. For non-constant skewing functions π, the resulting pdf is skewed and can even be multimodal; see Section 5.4. Note also that $\mathbf{z} - \boldsymbol{\xi}$ in (5.1) can be replaced by the standardized quantity $\Omega^{-1/2}(\mathbf{z} - \boldsymbol{\xi})$, where Ω is a positive definite scale matrix. This form is sometimes more convenient, for instance when the pdf f is elliptically contoured.

5.2.1 Invariance

Skew-symmetric distributions possess interesting invariance properties for even functions, that is for functions τ satisfying $\tau(-\mathbf{z}) = \tau(\mathbf{z})$ for all $\mathbf{z} \in \mathbb{R}^p$.

Proposition 5.2.1 If $\mathbf{z} \sim SS_p(\boldsymbol{\xi}, f, \pi)$, then the distribution of $\tau(\mathbf{z} - \boldsymbol{\xi})$, where τ is an even function, does not depend on the skewing function π.

The proof is straightforward and given by Wang *et al.* (2004a). Proposition 5.2.1 holds in particular for quadratic forms in the vector $\mathbf{z}$. Quadratic forms play an important role in multivariate analysis, for example when defining Mahalanobis distances. As an illustration, consider $\mathbf{z} \sim SS_p(\boldsymbol{\xi}, \phi_p(\cdot; \mathbf{0}, \Omega), \pi)$, where ϕ_p is the p-dimensional normal pdf and Ω is a non-singular covariance matrix of rank p. Then $(\mathbf{z} - \boldsymbol{\xi})^T\Omega^{-1}(\mathbf{z} - \boldsymbol{\xi}) \sim \chi^2_p$. When $p = 1$, Wang, Boyer, and Genton (2004b) show that $Z^2 \sim \chi^2_1$ if and only if there exists a skewing function π such that the pdf of Z is $2\phi(z)\pi(z)$.

Quadratic forms of random vectors naturally arise in time series and spatial statistics, where they measure temporal and spatial dependence, respectively. Quadratic forms possess many useful properties if the data distribution is elliptical. However, the same quadratic forms might mislead statistical inference when the sampled process has a skew-symmetric distribution.

We first examine in detail the above statement with respect to the second-order stationary time-series $\{Z_t : t \in \mathbb{Z}\}$. We denote by $m = \mathrm{E}(Z_t)$ the mean of the process and by $\nu(h) = \mathrm{Cov}(Z_{t+h}, Z_t)$, $\forall t, h \in \mathbb{Z}$, the autocovariance function of Z_t at temporal lag h (e.g., Brockwell and Davis, 1991). The classical method of moments estimator for the autocovariance function is

$$\hat{\nu}(h) = \frac{1}{n}\sum_{i=1}^{n-h}(Z_{i+h}-\bar{Z})(Z_i-\bar{Z}), \quad 0 \le h \le n-1, \quad \bar{Z} = \frac{1}{n}\sum_{i=1}^{n} Z_i. \quad (5.2)$$

Note that $\hat{\nu}(h)$ is a quadratic form in the data vector $\mathbf{z} = (Z_1, \ldots, Z_n)^T$:

$$\hat{\nu}(h) = \frac{1}{n}\mathbf{z}^T M D(h) M \mathbf{z}, \quad (5.3)$$

where $M = I_n - (1/n)\mathbf{1}_n\mathbf{1}_n^T$, $D(h) = (P(h) + P(h)^T)/2$, $0 \le h \le n-1$, and $P(h)$ is an $n \times n$ matrix with ones on the h-th upper diagonal and zeros elsewhere, $1 \le h \le n-1$, with $P(0) = I_n$. The following proposition shows that the distribution of the sample autocovariance function may not be affected by the skewness in the distribution of the time series.

Proposition 5.2.2 If $\mathbf{z} \sim SS_n(m\mathbf{1}_n, f, \pi)$, then the distribution of the sample autocovariance function $\hat{\nu}(h)$ does not depend on the skewing function π.

Proof. By the stationarity assumption, the location parameter is proportional to $\mathbf{1}_n$. Hence, the sample autocovariance function $\hat{\nu}(h)$ does not depend on the location parameter itself, which can therefore be set equal to zero. Since $\hat{\nu}(0), \ldots, \hat{\nu}(n-1)$ are quadratic forms in $\mathbf{z}$, and thus even functions of $\mathbf{z}$, we can apply Proposition 5.2.1 and prove the result. □

Proposition 5.2.2 implies that autocovariance functions computed from the vectors $\mathbf{x} \sim N_n(m\mathbf{1}_n, \Omega)$ and $\mathbf{z} \sim SN_n(m\mathbf{1}_n, \Omega, \boldsymbol{\alpha})$ are identically distributed. When Ω is a Toeplitz matrix, however, $\mathbf{x}$ is derived from a second-order stationary process, while $\mathbf{z}$, in general, is not. More precisely $\mathrm{E}(\mathbf{x}) = m\mathbf{1}_n$ and $\mathrm{Var}(\mathbf{x}) = \Omega$, but Genton, He, and Liu (2001) showed that:

$$\begin{aligned}\mathrm{E}(\mathbf{z}) &= m\mathbf{1}_n + \sqrt{\frac{2}{\pi}}\frac{\Omega\boldsymbol{\alpha}}{\sqrt{1+\boldsymbol{\alpha}^T\Omega\boldsymbol{\alpha}}},\\ \mathrm{Var}(\mathbf{z}) &= \Omega - \frac{2}{\pi}\frac{\Omega\boldsymbol{\alpha}\boldsymbol{\alpha}^T\Omega}{1+\boldsymbol{\alpha}^T\Omega\boldsymbol{\alpha}}.\end{aligned}$$

The sample autocovariance function is not affected by skewness even when $\mathbf{z}$ is derived from a second-order stationary process, i.e., when $\boldsymbol{\alpha} = c\Omega^{-1}\mathbf{1}_n$, where c is a constant. Similar invariance properties also arise under the more general assumption that $\mathbf{x}$ is symmetrically (or elliptically) distributed according to a pdf f and $\mathbf{z} \sim SS_n(m\mathbf{1}_n, f, \pi)$.

As a second example, consider the estimation of the variogram in spatial statistics (e.g., Cressie, 1993). Let $\{Z(\mathbf{x}) : \mathbf{x} \in D \subset \mathbb{R}^d\}$, $d \geq 1$, be a spatial stochastic process, intrinsically stationary, from which one realization $Z(\mathbf{x}_1), \ldots, Z(\mathbf{x}_n)$ has been obtained. The variogram $2\gamma(\mathbf{h}) = \mathrm{Var}(Z(\mathbf{x}+\mathbf{h}) - Z(\mathbf{x}))$ is the cornerstone to the spatial interpolation method named kriging. It has been recognized in the geostatistics literature that skewness is often present in data sets, for instance, in mining applications. This issue has also been raised by Kim and Mallick (2001) when analyzing rainfall data. Here, we investigate the effect of skewness on the sample variogram estimator:

$$2\hat{\gamma}(\mathbf{h}) = \frac{1}{N_{\mathbf{h}}} \sum_{N(\mathbf{h})} (Z(\mathbf{x}_i) - Z(\mathbf{x}_j))^2, \quad \mathbf{h} \in \mathbb{R}^d, \tag{5.4}$$

where $N(\mathbf{h}) = \{(\mathbf{x}_i, \mathbf{x}_j) : \mathbf{x}_i - \mathbf{x}_j = \mathbf{h}\}$ and $N_{\mathbf{h}}$ is the cardinality of $N(\mathbf{h})$. Again, we can rewrite the estimator $2\hat{\gamma}(\mathbf{h})$ as a quadratic form in the data vector $\mathbf{z} = (Z(\mathbf{x}_1), \ldots, Z(\mathbf{x}_n))^T$ with the spatial design matrix $A(\mathbf{h})$; see, e.g., Genton (1998), Gorsich, Genton, and Strang (2002), as:

$$2\hat{\gamma}(\mathbf{h}) = \frac{1}{N_{\mathbf{h}}} \mathbf{z}^T A(\mathbf{h}) \mathbf{z}. \tag{5.5}$$

The following proposition shows that the sample variogram estimator may not be affected by the possible skewness in the distribution of the spatial data. This will also be true for other measures of spatial dependence based on quadratic forms of the data, such as the Geary and the Moran contiguity indices (see, e.g., Cliff and Ord, 1981), which are used for modeling spatial relationships in regression residuals.

Proposition 5.2.3 If $\mathbf{z} \sim SS_n(m\mathbf{1}_n, f, \pi)$, then the distribution of the sample variogram estimator $2\hat{\gamma}(\mathbf{h})$ does not depend on the skewing function π.

The proof is omitted since it is similar to the proof of Proposition 5.2.2. As an application of Proposition 5.2.3, consider the spatial skew-normal process defined by Kim and Mallick (2001) and assume it is stationary. It follows that a realization $Z(\mathbf{x}_1), \ldots, Z(\mathbf{x}_n)$ from that process has a multivariate skew-normal distribution $SN_n(m\mathbf{1}_n, \Omega, \boldsymbol{\alpha})$. In general, the scale matrix Ω and the shape parameter $\boldsymbol{\alpha}$ depend on the spatial locations $\mathbf{x}_1, \ldots, \mathbf{x}_n$. By Proposition 5.2.3, the distribution of the sample variogram estimator (5.4) does not depend on the shape parameter $\boldsymbol{\alpha}$, although the variance-covariance matrix of the spatial process Z does; see Genton *et al.* (2001) and references therein. Here again, these invariance properties also arise under the more general assumption of a skew-symmetric spatial process.

5.2.2 Stochastic Representation and Simulations

The skew-symmetric distribution (5.1) has a simple stochastic representation that provides a constructive description of a random vector having a skew-symmetric distribution. This allows for straightforward and quick simulation procedures.

Consider a p-dimensional random vector $\mathbf{x}$ with symmetric pdf $f(\mathbf{x}-\boldsymbol{\xi})$ and Y a random variable symmetrically distributed around zero, independent of $\mathbf{x}$, with univariate cdf H. Define a random vector $\mathbf{z}$ as being equal to $\mathbf{x}$ conditionally on the event $\{w(\mathbf{x}-\boldsymbol{\xi}) > Y\}$, where w is an odd function. Then $\mathbf{z}$ has a skew-symmetric distribution given by the pdf (5.1) with a skewing function $\pi(\mathbf{z}) = H(w(\mathbf{z}))$. Effectively, setting $\boldsymbol{\xi} = \mathbf{0}$ without loss of generality, the cdf of $\mathbf{z}$ is easily seen to be

$$\mathrm{P}(\mathbf{z} \leq \mathbf{z}^*) = 2\int_{-\infty}^{\mathbf{z}^*} f(\mathbf{x})H(w(\mathbf{x}))d\mathbf{x}, \tag{5.6}$$

and thus yields the desired pdf (5.1) after differentiation with respect to $\mathbf{z}^*$. Note that $H(w(\mathbf{z}))$ is a valid skewing function since $H(w(-\mathbf{z})) = H(-w(\mathbf{z})) = 1 - H(w(\mathbf{z}))$. Further details can be found in Wang *et al.* (2004a) and Ma and Genton (2004). As a particular case of the above stochastic representation, the multivariate skew-normal distribution of Azzalini and Dalla Valle (1996) uses $f(\mathbf{x}-\boldsymbol{\xi}) = \phi_p(\mathbf{x}-\boldsymbol{\xi};\mathbf{0},\Omega)$, $H = \Phi$, and $w(\mathbf{z}) = \boldsymbol{\alpha}^T\mathbf{z}$.

We have now the following simple and quick procedure to simulate pseudo-realizations from the skew-symmetric distribution (5.1). First, simulate a p-dimensional random vector $\mathbf{x}$ with symmetric pdf $f(\mathbf{x}-\boldsymbol{\xi})$ and a uniform random variable U on the interval $[0,1]$. Then, define $\mathbf{z}$ by

$$\mathbf{z} = \begin{cases} \mathbf{x} & \text{if } U \leq \pi(\mathbf{x}-\boldsymbol{\xi}), \\ 2\boldsymbol{\xi}-\mathbf{x} & \text{if } U > \pi(\mathbf{x}-\boldsymbol{\xi}). \end{cases} \tag{5.7}$$

Then, $\mathbf{z} \sim SS_p(\boldsymbol{\xi}, f, \pi)$. This procedure is, in fact, an extension of a similar one proposed by Azzalini and Capitanio (1999, 2003). If $f(\mathbf{x}-\boldsymbol{\xi}) = \phi_p(\mathbf{x}-\boldsymbol{\xi};\mathbf{0},\Omega)$, $H = \Phi$, and $\boldsymbol{\xi} = \mathbf{0}$, then the simulation procedure simplifies to:

$$\mathbf{z} = \begin{cases} \mathbf{x} & \text{if } U \leq \Phi(\boldsymbol{\alpha}^T\mathbf{x}), \\ -\mathbf{x} & \text{if } U > \Phi(\boldsymbol{\alpha}^T\mathbf{x}). \end{cases} \tag{5.8}$$

Then, $\mathbf{z} \sim SN_p(\boldsymbol{\xi}, \Omega, \boldsymbol{\alpha})$, and this method reduces to the one proposed by Azzalini and Dalla Valle (1996) to simulate from the skew-normal distribution, by noticing that $\Phi^{-1}(U)$ is a standard normal random variable.

5.2.3 Skew-Symmetric Representation of Multivariate Distributions

Wang *et al.* (2004a) show that skew-symmetric distributions can be used to represent any continuous multivariate distribution. Effectively, let the function $g : \mathbb{R}^p \to \mathbb{R}_+$ be a multivariate pdf and let $\boldsymbol{\xi}$ be any point in $\mathbb{R}^p$. Then g can be written as

$$g(\mathbf{x}) = 2f_{\boldsymbol{\xi}}(\mathbf{x} - \boldsymbol{\xi})\pi_{\boldsymbol{\xi}}(\mathbf{x} - \boldsymbol{\xi}), \tag{5.9}$$

where $f_{\boldsymbol{\xi}}$ is a pdf, symmetric around $\mathbf{0}$, and $\pi_{\boldsymbol{\xi}}$ is a skewing function. This representation is unique for any $\boldsymbol{\xi}$, and it is determined by

$$f_{\boldsymbol{\xi}}(\mathbf{s}) = \frac{g(\boldsymbol{\xi} + \mathbf{s}) + g(\boldsymbol{\xi} - \mathbf{s})}{2}, \tag{5.10}$$

$$\pi_{\boldsymbol{\xi}}(\mathbf{s}) = \frac{g(\boldsymbol{\xi} + \mathbf{s})}{g(\boldsymbol{\xi} + \mathbf{s}) + g(\boldsymbol{\xi} - \mathbf{s})}. \tag{5.11}$$

The representation (5.9) is called the *skew-symmetric representation* of the pdf g with respect to $\boldsymbol{\xi}$. Note that we can always write $f_{\boldsymbol{\xi}}(\mathbf{0}) = g(\boldsymbol{\xi})$ and $\pi_{\boldsymbol{\xi}}(\mathbf{0}) = 1/2$.

Several properties about the continuity and differentiability of $f_{\boldsymbol{\xi}}$ and $\pi_{\boldsymbol{\xi}}$ related to the continuity of g and to its gradient are derived by Wang *et al.* (2004a). In addition, multimodality and concavity of the symmetric pdf $f_{\boldsymbol{\xi}}$ and monotonicity of the skewing function $\pi_{\boldsymbol{\xi}}$ are related to the properties of the pdf g. For instance, it is shown that if g has a global maximum at $\boldsymbol{\xi}$, then $f_{\boldsymbol{\xi}}$ has a global maximum at $\mathbf{0}$. Another interesting property is that $\pi_{\boldsymbol{\xi}}(\mathbf{s})$ has the same monotonicity as $g(\boldsymbol{\xi} + \mathbf{s})/g(\boldsymbol{\xi} - \mathbf{s})$.

The choice of the vector $\boldsymbol{\xi}$ is also briefly discussed by Wang *et al.* (2004a). It would seem natural to choose $\boldsymbol{\xi}$ as the mean, the median, or the mode of g. However, many other choices of $\boldsymbol{\xi}$ are possible and each of them will imply, along with the characteristics of g, different shapes and properties of the symmetric pdf $f_{\boldsymbol{\xi}}$ and the skewing function $\pi_{\boldsymbol{\xi}}$. The skew-symmetric representation (5.9) is of theoretical and conceptual interest because it generalizes the ideas of skew-normal and other skew-elliptical distributions. In practice however, a parametric family of symmetric (or elliptically contoured) pdf's f and of skewing functions π are chosen, and the skew-symmetric construction (5.1) is used. The vector $\boldsymbol{\xi}$ is then a location parameter that can be estimated from data. For example, this is the case when the skew-normal distribution is considered.

5.2.4 Example: Intensive Care Unit Data

An obvious advantage of the skew-symmetric distributions is their great flexibility. From (5.1), any combination of a symmetric pdf and a skewing function results in a valid pdf. Candidates for the symmetric pdf include unimodal, multimodal, spherical, elliptical, and any non-elliptical symmetric distribution, e.g., a product of univariate symmetric pdf's. Typical candidates for the skewing function are of the form $\pi(\boldsymbol{\alpha}^T\mathbf{x})$, where π could be either a cdf or not, either monotone or not, or any convex combination of such π functions:

$$\sum_{i=1}^{k} l_i\pi_i(\boldsymbol{\alpha}_i^T\mathbf{x}), \quad l_1 + \cdots + l_k = 1, \ l_i \geq 0.$$

Intuitively, $\pi(\boldsymbol{\alpha}^T\mathbf{x})$ models skewness in the $\boldsymbol{\alpha}$ direction, because $\|\boldsymbol{\alpha}^T\mathbf{x}\|$ is maximized for $\mathbf{x} = c\boldsymbol{\alpha}$, where c is a constant, and minimized ($= 0$) for $\mathbf{x}$ orthogonal to $\boldsymbol{\alpha}$. If apparent skewness is present in different directions, it makes sense to use the above convex combination to model the skewness. Indeed, it yields a skew-symmetric pdf with a mixture of skewing functions, each of which models skewness in one direction.

As a simple illustrative example, we consider a data set of size $n = 200$ from a study on the survival of patients following admission to an adult intensive care unit. We are interested in modeling the joint distribution of systolic blood pressure (SYS) at admission (in mm Hg) and heart rate (HRA) at admission (beats/min). Mardia's tests of multivariate normality, based on the multivariate measures of skewness $b_{1,2} = 0.3$ and kurtosis $b_{2,2} = 9.5$ (Mardia, Kent, and Bibby, 1979, p. 148), yield respectively the p-values 0.04 and 0.01. Thus, both tests reject the null hypothesis that the data are bivariate normal at the 0.05 level, suggesting the need for a distribution that exhibits both skewness and larger-than-normal kurtosis.

We fit a bivariate skew-symmetric pdf with a t-pdf with ν degrees of freedom as the symmetric part and a $N(0,1)$ cdf as the skewing function:

$$2|\Omega|^{-1/2}t\left(\Omega^{-1/2}(\mathbf{x}-\boldsymbol{\xi});\nu\right)\Phi\left(\boldsymbol{\alpha}^T\Omega^{-1/2}(\mathbf{x}-\boldsymbol{\xi})\right), \tag{5.12}$$

to the data. This illustrates the fact that the skewing function does not need to be a cdf from the same family of distributions as the symmetric pdf, and can therefore be chosen to favor simplicity. In addition, note that the kurtosis parameter ν is in the symmetric component, whereas the skewness parameter $\boldsymbol{\alpha}$ is in the skewed component. The contour plot of the fitted skew-symmetric distribution by maximum likelihood is shown in Figure 5.1. The fitted marginal distributions of SYS and HRA are depicted in

Table 5.1 *Fitted values of the skew-symmetric model (5.12) to the SYS/HRA data.*

$\hat{\xi}_1$	$\hat{\xi}_2$	$\hat{\Omega}_{11}$	$\hat{\Omega}_{22}$	$\hat{\Omega}_{12}$	$\hat{\alpha}_1$	$\hat{\alpha}_2$	$\hat{\nu}$
139.4	73.5	936.1	1185.7	−217.8	−0.02	1.8	12.6

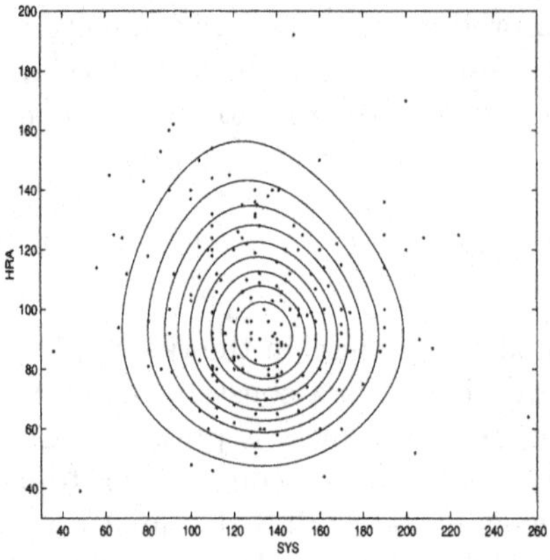

Figure 5.1 *Fit of the bivariate skew-symmetric pdf (5.12) to the SYS/HRA data.*

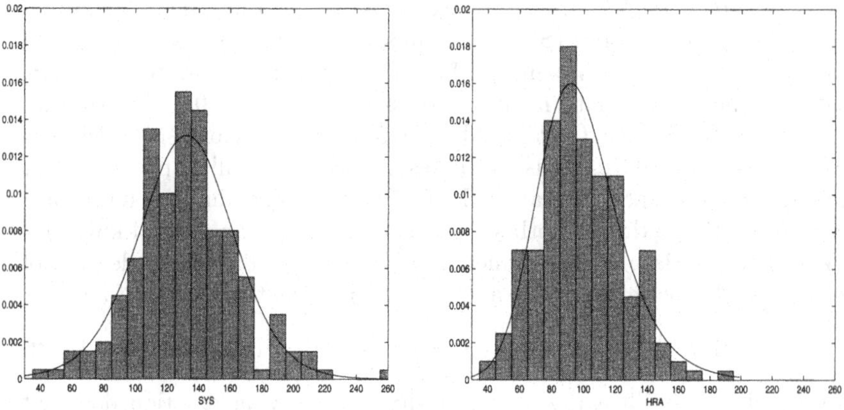

Figure 5.2 *Fit of the marginals of the pdf (5.12) to the SYS/HRA data.*

Figure 5.2. The fitted parameters are listed in Table 5.1, indicating skewness for the HRA variable and moderate heavy-tail behavior. A likelihood ratio test for the null hypothesis $H_0 : \alpha_1 = \alpha_2 = 0 \ \& \ \nu = \infty$ yields a p-value of 0.012 and therefore we reject a bivariate normal model at the 0.05 level.

5.3 Generalized Skew-Elliptical Distributions

Generalized skew-elliptical distributions were introduced by Genton and Loperfido (2002) and are special cases of the skew-symmetric distributions described in Definition 5.2.2 for which the symmetric pdf is now also required to be elliptically contoured; see, e.g., Fang, Kotz, and Ng (1990). We now give a formal definition.

Definition 5.3.1 (Generalized skew-elliptical distribution) Let $f : \mathbb{R}^p \to \mathbb{R}_+$ be a continuous elliptically contoured pdf symmetric around $\mathbf{0}$, i.e., $f(-\mathbf{z}) = f(\mathbf{z})$ for all $\mathbf{z} \in \mathbb{R}^p$, and $\pi : \mathbb{R}^p \to [0,1]$ a skewing function. A random vector $\mathbf{z} \in \mathbb{R}^p$ has a generalized skew-elliptical distribution with location parameter $\boldsymbol{\xi} \in \mathbb{R}^p$, denoted by $\mathbf{z} \sim GSE_p(\boldsymbol{\xi}, f, \pi)$, if its pdf is

$$2f(\mathbf{z} - \boldsymbol{\xi})\pi(\mathbf{z} - \boldsymbol{\xi}). \tag{5.13}$$

Important special types of generalized skew-elliptical distributions are the skew-normal (Azzalini, 1985; Azzalini and Dalla Valle, 1996), skew-t (Branco and Dey, 2001; Azzalini and Capitanio, 2003; Sahu, Dey, and Branco, 2003), skew-logistic (Wahed and Ali, 2001), skew-Cauchy (Arnold and Beaver, 2000b), skew-slash (Wang and Genton, 2004), and other skew-elliptical ones (Azzalini and Capitanio, 1999; Branco and Dey, 2001; Sahu *et al.*, 2003). All these distributions are skew-symmetric and therefore they possess the same properties as those described in the previous sections.

For instance, the invariance property implies that if $\mathbf{z} \sim GSE_p(\boldsymbol{\xi}, f, \pi)$, then the distribution of $\tau(\mathbf{z} - \boldsymbol{\xi})$, where τ is an even function, does not depend on the skewing function π. As a consequence, Genton and Loperfido (2002) show that if $\mathbf{z}_i \sim GSE_p(\boldsymbol{\xi}, f, \pi_i)$, $i = 1, \ldots, n$, are independent random vectors and $S = \sum_{i=1}^n \mathbf{z}_i \mathbf{z}_i^T / n$, then the joint distribution of $\mathbf{z}_1^T S^{-1} \mathbf{z}_1, \ldots, \mathbf{z}_n^T S^{-1} \mathbf{z}_n$ does not depend on $\pi_1, \ldots, \pi_n$. This is in particular true when $f(\mathbf{z} - \boldsymbol{\xi}) = \phi_p(\mathbf{z} - \boldsymbol{\xi}; \mathbf{0}, \Omega)$, the multivariate normal pdf, i.e., when the distribution is generalized skew-normal. In that case, the result is an extension of the one in multivariate normal theory. Genton and Loperfido (2002) use the invariance property to show that certain likelihood ratio tests and associated power do not depend on the skewing functions $\pi_1, \ldots, \pi_n$.

The stochastic representation of GSE distributions is the same as for SS distributions, and thus their simulation can be carried out with the simple procedure associated with Equation (5.7). The example provided in the previous section about SYS/HRA data actually uses a GSE distribution since the t-pdf is elliptically contoured; see Equation (5.12). The GSE family of distributions is useful to model skewness departures from elliptically contoured models. However, they also arise naturally in prospective studies and in inference from non-random samples (selection models), i.e., collections of observations that are not independent and identically distributed (iid) according to the pdf of the population; see, e.g., Copas and Li (1997). Such an example is discussed by Genton and Loperfido (2002) when analyzing a data set of heights and weights of Australian athletes. Specifically, they make the standard assumption that the joint distribution of adult Australians is bivariate normal, but the available sample comes from a selection model since only gifted athletes are included. This naturally gives rise to a GSN distribution and the invariance properties of this model are used to perform likelihood ratio tests.

5.4 Flexible Skew-Symmetric Distributions

A flexible class of skew-symmetric, and in particular of generalized skew-elliptical, distributions has been introduced by Ma and Genton (2004). The main idea is to notice that any skewing function π can be written as

$$\pi(\mathbf{z}) = H(w(\mathbf{z})), \tag{5.14}$$

where $H : \mathbb{R} \to [0,1]$ is the cdf of a continuous random variable symmetric around zero, and $w : \mathbb{R}^p \to \mathbb{R}$ is an odd continuous function, that is, $w(-\mathbf{z}) = -w(\mathbf{z})$. In fact, for a chosen H such that H^{-1} exists, $w(\mathbf{z}) = H^{-1}(\pi(\mathbf{z}))$ is a continuous odd function. Azzalini and Capitanio (2003) have used this representation to define certain skewed distributions by perturbation of symmetry. Note that the representation (5.14) is however not unique because many choices of H are possible. By using an odd polynomial P_K of order K to approximate the function w, we obtain a flexible family of skew-symmetric distributions.

Definition 5.4.1 (Flexible skew-symmetric distribution) Let $f : \mathbb{R}^p \to \mathbb{R}_+$ be a continuous pdf symmetric around $\mathbf{0}$, i.e., $f(-\mathbf{z}) = f(\mathbf{z})$ for all $\mathbf{z} \in \mathbb{R}^p$, $H : \mathbb{R} \to [0,1]$ the cdf of a continuous random variable symmetric around 0, P_K an odd polynomial of order K, and $\pi_K : \mathbb{R}^p \to [0,1]$ the skewing function defined by $\pi_K(\mathbf{z}) = H(P_K(\mathbf{z}))$. A random vector $\mathbf{z} \in \mathbb{R}^p$ has a flexible skew-symmetric distribution with location parameter $\boldsymbol{\xi} \in \mathbb{R}^p$, denoted by $\mathbf{z} \sim FSS_p(\boldsymbol{\xi}, f, \pi_K)$, if its pdf is

$$2f(\mathbf{z} - \boldsymbol{\xi})\pi_K(\mathbf{z} - \boldsymbol{\xi}). \tag{5.15}$$

Because FSS distributions are a particular type of SS distributions, they enjoy the same properties discussed in the previous sections, such as invariance, stochastic representation, and simulation procedure. If the pdf f is such that $f(\mathbf{x}) \to 0$ when $\|\mathbf{x}\|_2 \to \infty$ in the L^2 norm, then Ma and Genton (2004) have proved that the class of FSS distributions is dense in the class of SS distributions under the L^∞ norm. This means that skew-symmetric distributions can be approximated arbitrarily well by their flexible versions. Moreover, the choice of the pdf H is not essential, but it should be chosen to facilitate computations. A natural choice is $H = \Phi$ or the cdf corresponding to the symmetric pdf f. When the pdf f is elliptically contoured, then (5.15) defines flexible generalized skew-elliptical ($FGSE$) distributions. Of particular interest for applications are the flexible generalized skew-normal ($FGSN$) distributions defined by

$$2\phi_p(\mathbf{z}; \boldsymbol{\xi}, \Omega)\Phi(P_K(A(\mathbf{z} - \boldsymbol{\xi}))), \tag{5.16}$$

and flexible generalized skew-t ($FGST$) distributions defined by

$$2t_p(\mathbf{z}; \boldsymbol{\xi}, \Omega, \nu)T(P_K(A(\mathbf{z} - \boldsymbol{\xi})); \nu), \tag{5.17}$$

where we use the Choleski decomposition $\Omega^{-1} = A^T A$, t_p denotes a p-dimensional multivariate t pdf, and T denotes a univariate t cdf, both with degrees of freedom ν.

5.4.1 *Flexibility and Multimodality*

The use of an odd polynomial P_K in the skewing function π_K results in a flexible class of distributions that can capture skewness, heavy tails, and even multimodality. Effectively, when the degree K of the odd polynomial

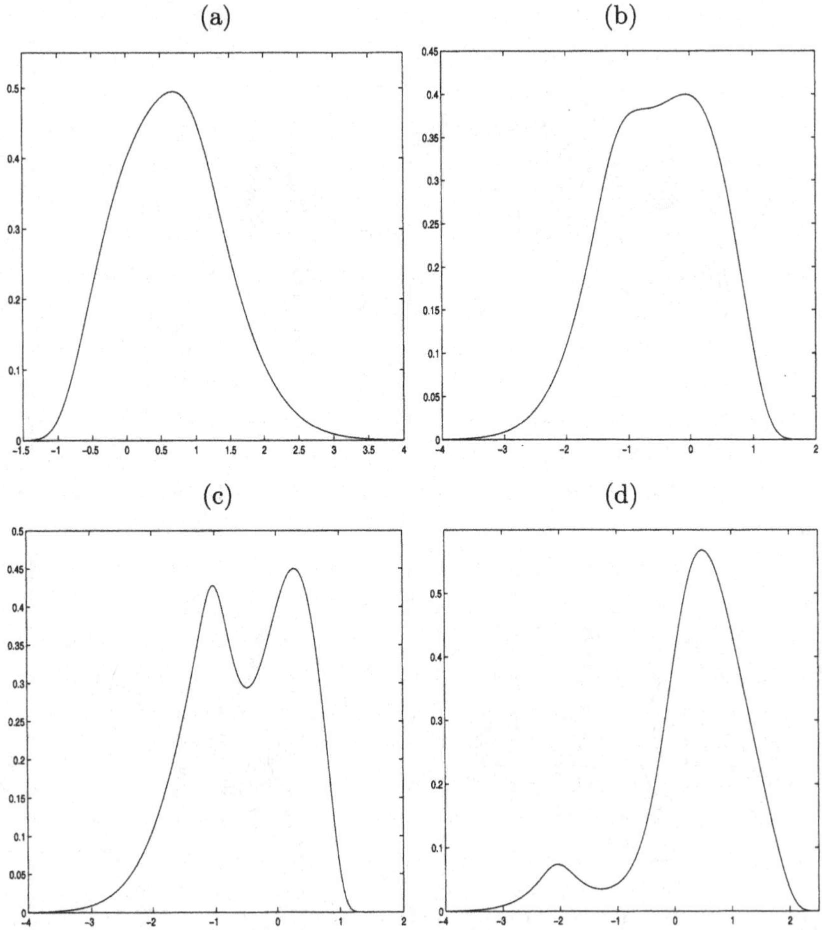

Figure 5.3 *Four univariate FGSN pdf's (5.18) with* $K = 3$, $\xi = 0$, *and* $\sigma^2 = 1$*: (a)* $\alpha_1 = 0.816$, $\alpha_2 = 0.712$*; (b)* $\alpha_1 = -0.079$, $\alpha_2 = -0.682$*; (c)* $\alpha_1 = 0.944$, $\alpha_2 = -2.120$*; (d)* $\alpha_1 = 1.863$, $\alpha_2 = -0.523$.

becomes large, the maximum number of modes in the pdf (5.15) increases, whereas only one mode is possible when $K = 1$. Ma and Genton (2004) proved that in the univariate case with $K = 3$, the $FGSN$ pdf

$$2\phi_1(z;\xi,\sigma^2)\Phi(\alpha_1(z-\xi)/\sigma+\alpha_2(z-\xi)^3/\sigma^3) \tag{5.18}$$

has at most two modes. Figure 5.3 depicts the pdf (5.18) for $\xi = 0$, $\sigma^2 = 1$, and various values of the coefficients α_1 and α_2. Figure 5.4 depicts the bivariate contours of four $FGSN$ pdf's (5.16) with $K = 3$, $\boldsymbol{\xi} = \mathbf{0}$, $\Omega = I_2$,

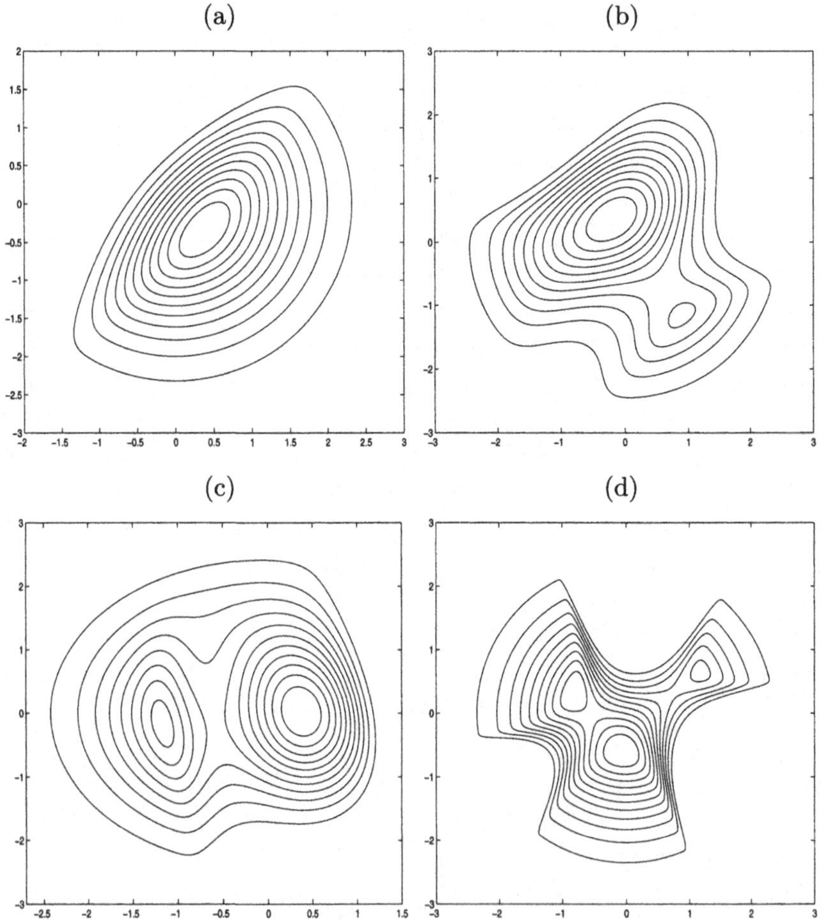

Figure 5.4 *Bivariate contours of four FGSN pdf's (5.16) with* $K = 3$, $\boldsymbol{\xi} = \mathbf{0}$, $\Omega = I_2$, *and various choices of coefficients in the polynomial* P_K*: (a) unimodal; (b) bimodal; (c) bimodal; (d) trimodal.*

and various choices of coefficients in the polynomial P_K, yielding unimodal, bimodal, and trimodal distributions.

5.4.2 Example: Australian Athletes Data

Fitting an FSS distribution to data can be carried out by maximizing the likelihood function based on (5.15) for a given order K using standard optimization techniques. The order K can be identified with likelihood ratio tests since the models (5.15) are nested when K decreases, for a given symmetric pdf f and skewing function π_K. An alternative is to use model selection criteria such as AIC (twice the difference between the log-likelihood and the number of parameters) or BIC (twice the log-likelihood minus the number of parameters multiplied by the logarithm of the sample size). However, the degree $K = 3$ seems to provide enough flexibility for practical applications involving unimodal and bimodal distributions. Ma and Genton (2004) describe three examples of fitting FSS distributions: fiber-glass data, old Swiss 1,000 franc bills data, and simulated data from a mixture of two normal distributions. In this section, we discuss the fitting of an $FGSN$ distribution to the Australian athletes data set discussed by Cook and Weisberg (1994).

The Australian athletes data set consists of several variables measured on 202 athletes and we focus on height (Ht) and body mass index (BMI). We fit a bivariate flexible generalized skew-normal ($FGSN$) distribution (5.16) with $K = 1$ and $K = 3$, i.e., the polynomial P_K takes the form:

$$P_3(z_1, z_2) = \alpha_1 z_1 + \alpha_2 z_2 + \alpha_3 z_1^3 + \alpha_4 z_2^3 + \alpha_5 z_1^2 z_2 + \alpha_6 z_1 z_2^2. \qquad (5.19)$$

The parameters $\boldsymbol{\xi} = (\xi_1, \xi_2)^T$, $A = (a_{ij})$, and $\alpha_1, \ldots, \alpha_6$ are estimated by maximizing the corresponding likelihood function, and are listed in Table

Table 5.2 *Fitted values of the bivariate FGSN model (5.16) for $K = 1$ and $K = 3$ on the Australian athletes Ht/BMI data.*

	$\hat{\xi}_1$	$\hat{\xi}_2$	$\hat{a}_{11}$	$\hat{a}_{12}$	$\hat{a}_{22}$
$K = 1$	180.51	19.98	0.011	−0.005	0.06
$K = 3$	180.87	24.40	0.012	−0.012	0.11

	$\hat{\alpha}_1$	$\hat{\alpha}_2$	$\hat{\alpha}_3$	$\hat{\alpha}_4$	$\hat{\alpha}_5$	$\hat{\alpha}_6$
$K = 1$	−0.83	3.11	—	—	—	—
$K = 3$	0.21	−1.26	0.18	0.16	−0.18	−0.76

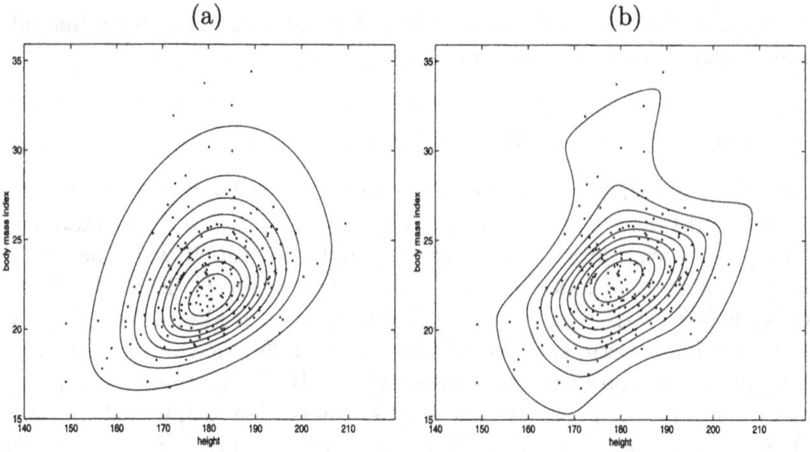

Figure 5.5 *Bivariate contours of the fitted FGSN model (5.16) to the Australian athletes Ht/BMI data by maximum likelihood: (a) $K = 1$; (b) $K = 3$.*

5.2. The fit for $K = 1$ corresponds to a skew-normal distribution advocated by Azzalini and Dalla Valle (1996).

The contours of the fitted bivariate pdf's are depicted in Figure 5.5(a) for $K = 1$ and in Figure 5.5(b) for $K = 3$. A likelihood ratio test (LRT) for the null hypothesis $H_0 : \alpha_3 = \ldots = \alpha_6 = 0$, using the approximate asymptotic distribution χ^2_4 is used as well as the AIC and BIC criteria. The results are tabulated in Table 5.3. All three methods suggest that $K = 3$ is a better model than $K = 1$. We further test for $K = 5$. The result of the likelihood ratio test relative to $K = 3$, as well as the AIC and BIC criteria, are shown in Table 5.3. Although the p-value of the LRT and the AIC score indicate that $K = 5$ is a better fit for the data, BIC suggests that $K = 5$ imposes too much model complexity for the gain. We decide to adopt a more complex model only when all three methods indicate so, hence we keep $K = 3$ as our final model. Notice that the final model is unimodal and reveals a complex

Table 5.3 *Likelihood ratio test and model selection criteria for $K = 1, 3, 5$ on the Ht/BMI data.*

	LRT (p-value)	AIC	BIC
$K = 1$	—	−1103.6	−1126.7
$K = 3$	0.0002	−1089.0	−1125.4
$K = 5$	0.0119	−1084.7	−1140.9

shape that cannot be captured by a skew-normal distribution. In addition, it is interesting to remark that the 10 observations with largest BMI seem to be modeled as "outliers" by the distribution with $K = 3$. We further discuss this point of view in the next section.

5.4.3 Locally Efficient Semiparametric Estimators

The generalized skew-elliptical distributions (5.13) whose pdf can be written as

$$2f(\mathbf{z}; \boldsymbol{\xi}, \Omega)\pi(\mathbf{z} - \boldsymbol{\xi}), \tag{5.20}$$

where f is an elliptically contoured pdf with location parameter $\boldsymbol{\xi}$ and scale matrix parameter Ω, fit naturally into a semiparametric model framework. Here we consider $\boldsymbol{\beta} = (\boldsymbol{\xi}^T, \text{vec}(\Omega)^T)^T$ as the parameters of interest (vec denotes the operator of vectorization of a matrix) and the function π as an infinite dimensional nuisance parameter. Such models arise from selection models, case control studies, and also from a robust statistics point of view, as suggested in the previous section.

For instance, consider the case of a p-dimensional random vector $\mathbf{z}$ distributed according to a pdf $f(\mathbf{z}; \boldsymbol{\beta})$, where $\boldsymbol{\beta}$ is a q-dimensional vector of unknown parameters. In order to make inference about $\boldsymbol{\beta}$, the usual statistical analysis assumes that a random sample $\mathbf{z}_1, \ldots, \mathbf{z}_n$ from $f(\mathbf{z}; \boldsymbol{\beta})$ can be observed. However, there are many situations where such a random sample might not be available, for instance if it is too difficult or too costly to obtain. If the probability density function is distorted by some multiplicative non-negative weight function $w(\mathbf{z}; \boldsymbol{\beta}, \boldsymbol{\alpha})$, where $\boldsymbol{\alpha}$ denotes some r-dimensional vector of additional unknown parameters, then the observed data is a random sample from a distribution with pdf

$$g(\mathbf{z}; \boldsymbol{\beta}, \boldsymbol{\alpha}) = f(\mathbf{z}; \boldsymbol{\beta}) \frac{w(\mathbf{z}; \boldsymbol{\beta}, \boldsymbol{\alpha})}{\mathrm{E}\left(w(\mathbf{z}; \boldsymbol{\beta}, \boldsymbol{\alpha})\right)}, \tag{5.21}$$

where g is said to be the pdf of a weighted distribution; see Rao (1985) and references therein. In particular, if the observed data are obtained only from a selected portion of the population of interest, then (5.21) is called a selection model. For example, this can happen if the observation vector $\mathbf{z}$ of characteristics of a certain population is measured only for individuals who manifest a certain disease due to cost or ethical reasons; see the survey article by Bayarri and DeGroot (1992) and references therein. For such problems, the goal is to find consistent and asymptotically normal estimators of $\boldsymbol{\beta}$ in the presence of the nuisance weight function w. We can immediately see that (5.20) is a particular type of selection model where the weight function w is a skewing function, i.e., $w(-\mathbf{z}; \boldsymbol{\beta}, \boldsymbol{\alpha}) = 1 - w(\mathbf{z}; \boldsymbol{\beta}, \boldsymbol{\alpha})$ and thus $\mathrm{E}\left(w(\mathbf{z}; \boldsymbol{\beta}, \boldsymbol{\alpha})\right) = 1/2$.

Another setting where *GSE* distributions arise is the modeling of case-control data in prospective studies (see, e.g., Wacholder and Weinberg,

1994; Zhang, 2000). Let $\mathbf{z}_1, \ldots, \mathbf{z}_n$ be a random sample from an elliptically contoured distribution with pdf f, and let $d_i \in \{0,1\}$, $i = 1, \ldots, n$ be the observed value of a dichotomous random variable D_i associated with the i-th observation. Suppose we model the probability $\mathrm{P}(D_i = 0|\mathbf{z}_i = \mathbf{z}_i^*) = \pi(\mathbf{z}_i^*)$. Prospective studies focus on the conditional distribution of $\mathbf{z}_i$ given $D_i = d_i$. From Bayes' theorem, we obtain

$$g(\mathbf{z}_i|D_i = 0) = \frac{f(\mathbf{z}_i)\,\pi(\mathbf{z}_i)}{\mathrm{E}(\pi(\mathbf{z}))}, \quad g(\mathbf{z}_i|D_i = 1) = \frac{f(\mathbf{z}_i)(1 - \pi(\mathbf{z}_i))}{1 - \mathrm{E}(\pi(\mathbf{z}))}. \tag{5.22}$$

If we impose $\pi(\mathbf{z}_i) + \pi(-\mathbf{z}_i) = 1$, then it follows that $f(\mathbf{z}_i|D_i = d_i)$ is the pdf of a GSE distribution. This condition on π naturally arises in models such as the logistic regression model where the skewing function $\pi(\mathbf{z}_i) = \exp(\boldsymbol{\beta}^T\mathbf{z}_i)/\{1+\exp(\boldsymbol{\beta}^T\mathbf{z}_i)\}$, and the probit regression model where $\pi(\mathbf{z}_i) = \Phi(\boldsymbol{\beta}^T\mathbf{z}_i)$, when the intercept is assumed to be equal to zero.

The above situations can also be viewed from the setting of robust statistics. Indeed, if $f(\mathbf{z};\boldsymbol{\beta})$ is the central model of interest, then the weight function w in (5.21) can be seen as a contaminating function. For instance, if f is an elliptically contoured pdf, then w generates asymmetric outliers in the observed sample from g. This has been noticed in the previous section from Figure 5.5(b). The goal is then to derive robust estimators of $\boldsymbol{\beta}$, that is again to provide consistent and asymptotically normal estimators of $\boldsymbol{\beta}$ in the presence of a certain class of the nuisance weight function w.

As mentioned previously, we are interested in inference on the parameters $\boldsymbol{\xi}$ and Ω in (5.20), which represent the mean and the covariance matrix of the population of which only samples from a particular subpopulation are available. The skewing function is assumed to be differentiable and to satisfy $0 \leq \pi(\mathbf{z}) \leq 1$ and $\pi(-\mathbf{z}) = 1 - \pi(\mathbf{z})$, but we make no additional assumptions on π. Therefore, we are considering a semiparametric model and can use the semiparametric theory developed by Bickel, Klaassen, Ritov, and Wellner (1993), and Newey (1990). In this setting, regular asymptotically linear (RAL) estimators are represented through their influence function in the Hilbert space consisting of mean zero square integrable random functions with the covariance inner product. For the univariate generalized skew-normal (GSN) distributions

$$g(z) = \frac{2}{\sigma}\phi\left(\frac{z-\xi}{\sigma}\right)\pi\left(\frac{z-\xi}{\sigma}\right), \tag{5.23}$$

locally efficient semiparametric estimators were thoroughly studied by Ma *et al.* (2003). We summarize some of their results here.

Consider a random sample $Z_1, \ldots, Z_n$ from (5.23). For any even function $v(z)$ such that $\int v(z)\phi(z)d\mu(z) = 0$, where $\mu(z)$ is the Lebesgue measure for which densities are defined, the equation

$$\sum_{i=1}^{n} v((Z_i - \xi)/\sigma) = 0, \tag{5.24}$$

yields a regular asymptotically linear (RAL) estimator for $\boldsymbol{\beta} = (\xi, \sigma)^T$, among which, the semiparametric efficient estimator is given by the equations

$$\begin{cases} \sum_{i=1}^{n} \left[\frac{Z_i - \xi}{\sigma} \left(2\pi_0 \left(\frac{Z_i - \xi}{\sigma} \right) - 1 \right) - 2\pi_{01} \left(\frac{Z_i - \xi}{\sigma} \right) \right] = 0, \\ \sum_{i=1}^{n} \left[(Z_i - \xi)^2 - \sigma^2 \right] = 0, \end{cases} \tag{5.25}$$

where $\pi_{01}(z) = d\pi_0(z)/dz$ and π_0 represents the true skewing function. This provides us a way of constructing RAL estimators as long as we can find a suitable function v. For instance, we can take any even function $h(z)$ and construct $v(z) = h(x) - \int h(z)\phi(z)d\mu(z)$. If we take h to be z^{2k}, then the corresponding v functions are $v(z) = z^2 - 1, v(z) = z^4 - 3, v(z) = z^6 - 15$, and so on.

Assume the solution to the efficient estimator given in Equation (5.25) is $\hat{\boldsymbol{\beta}}$, then $n^{1/2}(\hat{\boldsymbol{\beta}} - \boldsymbol{\beta}_0) \rightarrow N_2(\mathbf{0}, \{\mathrm{E}(\mathbf{S}_{eff}\mathbf{S}_{eff}^T)\}^{-1})$ in distribution, where the efficient score function $\mathbf{S}_{eff}$ is given by

$$\left[\frac{z - \xi_0}{\sigma_0^2} \left\{ 2\pi_0 \left(\frac{z - \xi_0}{\sigma_0} \right) - 1 \right\} - \frac{2}{\sigma_0} \pi_{01} \left(\frac{z - \xi_0}{\sigma_0} \right), \frac{(z - \xi_0)^2}{\sigma_0^3} - \frac{1}{\sigma_0} \right]^T.$$

Here, the smallest variance of the estimate is given by $A = \{\mathrm{E}(\mathbf{S}_{eff}\mathbf{S}_{eff}^T)\}^{-1}$ and has the form

$$\sigma_0^2 \begin{pmatrix} \int [\{2\pi_0(z) - 1\}^2 + 4\pi_{01}(z)^2] \phi(z) d\mu(z) & 4 \int \pi_{01}(z)\phi(z)d\mu(z) \\ 4 \int \pi_{01}(z)\phi(z)d\mu(z) & 2 \end{pmatrix}^{-1}.$$

Notice that when $\pi(z) \equiv 1/2$, the first component of the efficient score vector is 0, in which case an efficient semiparametric estimator does not exist. Similar phenomena have been observed by Lee and Berger (2001) in Bayesian analysis of selection models, where a constant weight function (corresponding to $\pi(z) \equiv 1/2$ in our case) has to be ruled out *a priori* to any analysis.

The efficient estimator defined by Equation (5.25) depends on using the true skewing function π_0, which is unknown to us. Any choice of a skewing function in Equation (5.25) will however lead to a consistent asymptotically normal estimator for $\boldsymbol{\beta}$. This can be shown by noticing that $[z(2\pi(z) - 1) - 2\pi_1(z), z^2 - 1]^T$ is a valid candidate for the function $v(z)$, where $\pi_1(z) = d\pi(z)/dz$. In practice, we generally posit a model for π in terms of a finite set of parameters $\boldsymbol{\alpha}$, say, $\pi((z-\xi)/\sigma;\boldsymbol{\alpha})$, and then estimate $\boldsymbol{\alpha}$ using an estimator $\hat{\boldsymbol{\alpha}}$. We use

$$\sum_{i=1}^{n} F(Z_i;\xi,\sigma,\hat{\boldsymbol{\alpha}}) = \mathbf{0} \tag{5.26}$$

to denote estimators of the form in Equation (5.25) with $\pi_0((z-\xi)/\sigma)$ replaced by $\pi((z-\xi)/\sigma;\hat{\boldsymbol{\alpha}})$. Notice that $\mathrm{E}\{F(Z_i;\xi,\sigma,\boldsymbol{\alpha})\} = \mathbf{0}$ for all values $\boldsymbol{\alpha}$, hence $\mathrm{E}\{\partial F(Z_i;\xi,\sigma,\boldsymbol{\alpha})/\partial\boldsymbol{\alpha}\} = \mathbf{0}$ assuming sufficiently smooth conditions on F to interchange the expectation and the partial derivative. If the true skewing function belongs to this parametric model then $\pi(\cdot,\hat{\boldsymbol{\alpha}})$ will converge to $\pi_0(\cdot)$. But even if the parametric model does not contain the true $\pi_0(\cdot)$, the estimate $\hat{\boldsymbol{\alpha}}$ will generally converge to a constant $\boldsymbol{\alpha}^*$ and $\pi(\cdot,\hat{\boldsymbol{\alpha}})$ will converge to some skewing function $\pi(\cdot,\boldsymbol{\alpha}^*)$. As long as $n^{1/2}(\hat{\boldsymbol{\alpha}}-\boldsymbol{\alpha}^*)$ is bounded in probability, the asymptotic distribution of $\hat{\boldsymbol{\beta}}$ obtained by using $\pi(\cdot,\hat{\boldsymbol{\alpha}})$ is asymptotically the same as that which uses $\pi(\cdot,\boldsymbol{\alpha}^*)$, which is consistent and asymptotically normal. Specifically, assume $2\phi((z-\xi)/\sigma)\pi((z-\xi)/\sigma;\boldsymbol{\alpha})/\sigma$ is a parametric model and $n^{1/2}(\hat{\boldsymbol{\alpha}}-\boldsymbol{\alpha}^*)$ is bounded in probability, then the two RAL estimators $\sum_{i=1}^n F(Z_i;\xi,\sigma,\boldsymbol{\alpha}^*) = \mathbf{0}$ and $\sum_{i=1}^n F(Z_i;\xi,\sigma,\hat{\boldsymbol{\alpha}}) = \mathbf{0}$ are asymptotically equivalent, i.e., if $(\hat{\xi}_1,\hat{\sigma}_1)$ is the solution to the first equation, and $(\hat{\xi}_2,\hat{\sigma}_2)$ is the solution to the second equation, then $n^{1/2}(\hat{\xi}_1-\hat{\xi}_2)\to 0$ and $n^{1/2}(\hat{\sigma}_1-\hat{\sigma}_2)\to 0$ in probability. However, if the parametric model does contain the truth, then the estimator for $\boldsymbol{\beta}$ in (5.26) will be semiparametric efficient. Such estimators are referred to as locally efficient.

The efficiency of an estimator depends on how close the true π_0 is to the parametric family $\pi(z;\boldsymbol{\alpha})$. Ma and Genton (2004) proposed to construct the parametric submodel by using $\Phi(P_K(z))$ to approximate $\pi(z;\boldsymbol{\alpha})$, where P_K is an odd polynomial of order K, see the previous section on FSS distributions. In general, the relation between the efficiency loss and the "distance" between π_0 and $\{\pi(z;\boldsymbol{\alpha})\}$ is given as follows. Let $\nu(z) = \pi(z;\boldsymbol{\alpha}) - \pi_0(z)$, $\theta = \int 4[\partial\{\nu(z)\phi(z)\}/\partial z]^2/\phi(z)d\mu(z)$. The most efficient semiparametric estimator of the form in Equation (5.26) has efficiency $A + \min_{\boldsymbol{\alpha}}(\theta)B$, where A is given above, and

$$B = \frac{\sigma_0^2}{\left[\mathrm{E}(2\pi_0(Z)-1+2Z\pi_{01}(Z)-2\pi_{02}(Z))-2\mathrm{E}(Z)^2\right]^2}\begin{pmatrix} 1 & -\mathrm{E}(Z) \\ -\mathrm{E}(Z) & \mathrm{E}(Z)^2 \end{pmatrix},$$

which does not depend on the estimator. Here π_{02} denotes $d^2\pi_0(z)/dz^2$, the expectations are taken with respect to $2\phi(z)\pi_0(z)$.

The $\boldsymbol{\alpha}$ that minimizes θ cannot be given explicitly since this depends on the true π_0 that is unknown to us. Typically in these problems a parametric model is assumed in terms of both $\boldsymbol{\beta}$ and $\boldsymbol{\alpha}$ and maximum likelihood is used to estimate both sets of parameters. However, if the model for the skewing function is not correct then the MLE for $\boldsymbol{\beta}$ will be biased. What is proposed by Ma *et al.* (2003) is to use the MLE estimate for $\hat{\boldsymbol{\alpha}}$ in the

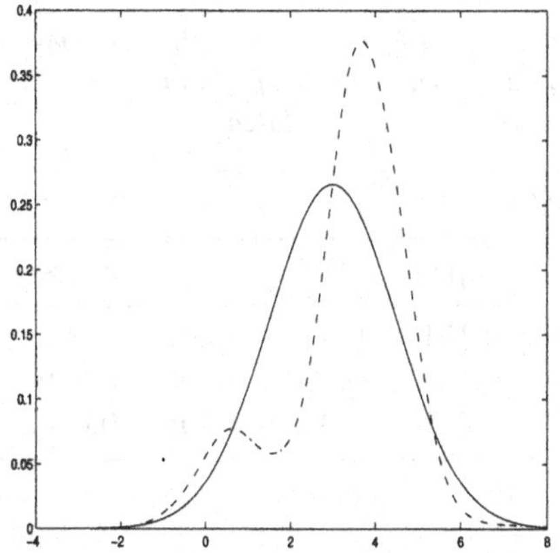

Figure 5.6 *Population distribution (solid line) and sample distribution (dashed line).*

π function and proceed to estimate ξ and σ using the semiparametric estimating equation with π_0 replaced by π. This estimator will be consistent and asymptotically normal even if the model for π was incorrectly specified and will be semiparametric efficient if it is correctly specified. We illustrate this result by means of a small Monte Carlo experiment.

We generate 2,000 data sets of size 200 each from the distribution $\frac{2}{\sigma}\phi\{(z-\xi)/\sigma\}\Phi[(\sin\{2(z-\xi)/\sigma\}]$ with $\sigma = 1.5$ and $\xi = 3$. We approximate the true $\pi_0(z) = \Phi[(\sin\{2(z-\xi)/\sigma\}]$ with $\pi_K(z) = H[P_K\{(z-\xi)/\sigma\}]$ where H is the logistic cdf, i.e., $H(z) = 1/\{1+\exp(-z)\}$, and P_K is an odd polynomial of order K. Figure 5.6 depicts the true $N(3, 1.5^2)$ distribution of the population (solid line) and the distribution from which we get a sample (dashed line). The goal is to estimate the distribution of the population based on the biased sample. Because we don't know the true skewing function π_0, we use π_K, $K = 1, 3$. We calculate the empirical standard errors (emp.se) of the estimates and also the average of the estimated standard errors (est.se). The estimated standard error is calculated via the usual sandwich matrix of M-estimators; see, e.g., Stefanski and Boos (2002). In addition, we give the coverage of 95% confidence intervals. The simulation results are shown in Table 5.4. We can verify that all three estimators are unbiased, while the estimator with the true posited π_0 has the smallest standard error. The standard error for π_3 is smaller than for π_1 because π_3 approximates π_0 better. In fact, as shown in Ma and Genton (2004), π_0 can be approximated arbitrarily well if we allow the order of the odd polynomial to increase sufficiently, hence the estimator will approach the most efficient one.

Finally, the derivation of the semiparametric estimators and their prop-

Table 5.4 *Simulation results on ξ and σ with different posited skewing function π_1, π_3 and the true π_0. The true values of ξ and σ are 3 and 1.5. The sample size is 200, and 2,000 data sets are simulated.*

	$\xi(3)$				$\sigma(1.5)$			
	mean	est.se	emp.se	95%	mean	est.se	emp.se	95%
π_1	3.015	0.158	0.151	0.922	1.502	0.089	0.083	0.948
π_3	2.984	0.144	0.134	0.960	1.505	0.089	0.083	0.949
π_0	3.004	0.100	0.097	0.949	1.498	0.078	0.077	0.945

erties is not specific to univariate GSN distributions and hence can be generalized to other GSE and SS distributions.

5.5 Concluding Remarks

This chapter has presented skew-symmetric (SS) and generalized skew-elliptical (GSE) distributions, as well as flexible classes that can model skewness, heavy tails, and multimodality. Several properties of these distributions have been discussed along with illustrative examples. Locally efficient semiparametric estimators have also been presented for univariate generalized skew-normal (GSN) distributions. Those estimators play an important role in selection models and robustness analysis.

The flexibility of the above distributions will prove very useful in applications. For example, this book includes applications of $FGSE$ distributions in modeling random effects in linear mixed models, as well as the use of FSS distributions in shape analysis.

CHAPTER 6

Elliptical Models Subject to Hidden Truncation or Selective Sampling

Barry C. Arnold and Robert J. Beaver

6.1 Introduction

Hidden truncation (or selective sampling) models have received considerable attention in the last 20 years. The seminal paper for reigniting interest in such distributions was Azzalini (1985). He introduced the skew-normal density from which a cornucopia of variants have evolved. Earlier discussion of related distributions (e.g., Birnbaum, 1950) did not lead to a flurry of developments. The time for such developments was ripe in the late 20th century, partially because of the ready availability of powerful computers for any simulation and numerical integration that might be required. We will survey the spectrum of hidden truncation models with special attention being paid to those involving elliptically contoured components in their development. The resulting models provide flexible alternatives to the somewhat restrictive elliptically contoured distributions for modeling multivariate data. Specifically, in Section 6.2 we present various univariate skew-normal models and discuss estimation issues in Section 6.3. Other univariate skewed distributions are presented in Section 6.4. We describe multivariate skewed distributions in Section 6.5 and further generalizations in Section 6.6. Finally, we discuss skew-elliptical distributions in Section 6.7 and conclude in Section 6.8.

6.2 Univariate Skew-Normal Models

Definition 6.2.1 We will say that a random variable Z has a skew-normal distribution with parameters α_0 and α_1 if its density function is of the form:

$$f(z; \alpha_0, \alpha_1) = \phi(z)\Phi(\alpha_0 + \alpha_1 z)/\Phi\left(\frac{\alpha_0}{\sqrt{1+\alpha_1^2}}\right). \tag{6.1}$$

Here $\alpha_0 \in \mathbb{R}$, $\alpha_1 \in \mathbb{R}$, and $\phi(z)$ and $\Phi(z)$ denote the standard normal density and distribution functions, respectively.

If a random variable Z has its density of the form (6.1) we will write $Z \sim SN(\alpha_0, \alpha_1)$. It is not difficult to verify that (6.1) is a valid density function, i.e., integrates to 1. Skew-normal densities are asymmetric unless $\alpha_1 = 0$, in which case the density reduces to that of a standard normal random variable. Azzalini (1985) introduced the model (6.1) with the restriction that $\alpha_0 = 0$. This leads to a slightly simpler form of the density,

$$f(z; 0, \alpha_1) = 2\phi(z)\Phi(\alpha_1 z). \tag{6.2}$$

There is an intimate relationship between such skewed models and certain weighted distributions arising as a consequence of hidden truncation.

Consider a two dimensional random vector $(X_1, X_2)^T$ with a classical bivariate normal density with mean vector $(\mu_1, \mu_2)^T$ and variance covariance matrix $\Sigma = (\sigma_{ij})$. Suppose that only X_1 is observed and moreover it is only observed if $X_2 > c$, a constant. Thus we have an instance of hidden truncation with respect to the unseen variable X_2, which must exceed the threshold c in order for X_1 to be observable. The conditional distribution of X_1 given $X_2 > c$ will, up to a location and scale change, be of the form (6.1). Specifically one finds:

$$f_{X_1|X_2>c}(x_1) = \frac{1}{\sqrt{\sigma_{11}}}\phi\left(\frac{x_1 - \mu_1}{\sqrt{\sigma_{11}}}\right)\Phi\left(\frac{\rho(\frac{x_1-\mu_1}{\sqrt{\sigma_{11}}}) - \theta_c}{\sqrt{1-\rho^2}}\right)/\Phi(\theta_c), \tag{6.3}$$

for $\theta_c = -(c - \mu_2)/\sqrt{\sigma_{22}}$.

An alternative derivation of the model (6.1) was suggested by Arnold and Beaver (2003). It involves an additive component construction. We begin with two independent and identically distributed (iid) $N(0,1)$ random variables X_0 and X_1 and define $X_0(d)$ to be X_0 truncated below at d. Now define

$$Y = \nu_0 + \nu_1 X_0(d) + \nu_2 X_1. \tag{6.4}$$

By considering the moment generating function of such a random variable it is possible to verify that such a random variable Y has a density that is, up to location and scale change, of the form (6.1).

There is yet another route that will lead to densities of the form (6.2), the original Azzalini form of the skew-normal. Loperfido (2002) (see also Roberts, 1966) observes that if $(W_1, W_2)^T$ has a bivariate normal density with standard normal marginals and correlation ρ then the random variable

$$Z = \max(W_1, W_2), \tag{6.5}$$

has a skew-normal density of the form (6.2) with $\alpha_1 = \sqrt{(1-\rho)/(1+\rho)}$.

It is not possible to identify the two parameter skew-normal density (6.1) with maxima of dependent normal random variables. Roberts (1966) does show however that if one of the W_i's in (6.5) is allowed to have a non-zero mean or equivalently if we consider

$$\tilde{Z} = \max(W_1, W_2 + \delta), \tag{6.6}$$

for some $\delta \in \mathbb{R}$, the random variable $\tilde{Z}$ will have a density which is a $\frac{1}{2} : \frac{1}{2}$ mixture of two two-parameter skew-normal densities. More generally, Cain (1994) considered the distribution of the minimum of two bivariate normal random variables with mean vector $\boldsymbol{\mu} = (\mu_1, \mu_2)^T$ and covariance matrix

$$\begin{pmatrix} \sigma_1^2 & \rho\sigma_1\sigma_2 \\ \rho\sigma_1\sigma_2 & \sigma_2^2 \end{pmatrix}. \tag{6.7}$$

The resulting distribution for the minimum was found to be a $\frac{1}{2} : \frac{1}{2}$ mixture of two skew-normal densities. When, in fact, the two variables have the same mean and standard deviation, the distribution of the minimum is skew-normal $SN(0, -\sqrt{(1-\rho)/(1+\rho)})$. Analogously, the distribution of the maximum of two identically distributed normal random variables is $SN(0, \sqrt{(1-\rho)/(1+\rho)})$.

6.3 Estimation for the Skew-Normal Distribution

Azzalini (1985) investigated the statistical aspects of the skew-normal distribution (6.2) with skewing parameter α. He considered a family of distributions generated by the linear transformation

$$Y = a_1 + a_2 Z, \quad a_2 > 0, \tag{6.8}$$

where Z is distributed as $SN(0, \alpha)$. He noticed that the information matrix corresponding to this model became singular as $|\alpha| \to 0$. He suggested that the following "centered" reparameterization be used:

$$Y = \theta_1 + \theta_2(Z - \mathrm{E}(Z))/\sqrt{\mathrm{Var}(Z)}. \tag{6.9}$$

In this case as α tends to zero, the information matrix tends to a diagonal matrix with diagonal elements $(\sigma^2, \sigma^2/2, 6)$. When $\mathrm{E}(Z)$ and $\sqrt{\mathrm{Var}(Z)}$ are both unknown, the sample mean and standard deviation are used to center and scale the data.

Liseo (1990) in his study of the skew-normal distribution noted the "strange" behavior of this class of densities from the point of view of likelihood-based inference for the model in (6.8). Pewsey (2000) formally addressed various issues related to inference for the Azzalini skew-normal model and he explained why the direct parameterization in (6.8) should not be used. He based his argument on redundancy of the parameterization as identified by Catchpole and Morgan (1997). For example, the mean of the random variable in (6.8) is a function of all three parameters whereas the mean for the random variable in (6.9) is a function of one parameter; see Pewsey (2000, p. 860). According to Theorem 2 in Catchpole and Morgan's paper, the likelihood based on (6.8) must contain a flat ridge, and hence no unique solution to the likelihood equations exists. These problems were observed by Arnold, Beaver, Groeneveld, and Meeker (1993) and Arnold and Beaver (2000a) in numerical computations for bivariate skew-normal models similar to (6.8) with a location vector $\boldsymbol{\mu}$ and a scale matrix Σ, and the three additional parameters $\boldsymbol{\alpha} = (\alpha_0, \alpha_1, \alpha_2)^T$.

Using Azzalini (1985)'s form (6.9) of the centered parameterization denoted by $SNC(\theta_1, \theta_2, \gamma_1)$, where γ_1 is the third central moment, the moment estimators are simply

$$\tilde{\theta}_1 = \bar{Y},\ \tilde{\theta}_2 = S,\ \tilde{\gamma}_1 = M_3/S^3, \tag{6.10}$$

where $\bar{Y}$ is the sample mean of a sample $Y_1, \ldots, Y_n$, S is the sample standard deviation, and M_3 is the sample third central moment. Maximum likelihood estimates for the centered model are well behaved except in situations where n is small and $|\alpha|$ is large since the maximum of the likelihood may occur for a boundary value of γ_1, indicating that a folded normal ($|\alpha| \to \infty$) is the most likely model that generated the data.

6.4 Other Univariate Skewed Distributions

Non-normal variants of the modeling scenarios discussed in Section 6.2 can be readily described. In the hidden truncation set-up we could begin with two independent random variables U, V with corresponding density functions $\psi_1(u)$ and $\psi_2(v)$ and distribution functions $\Psi_1(u)$ and $\Psi_2(v)$, not necessarily normal. If we focus on the conditional density of U given that $\alpha_0 + \alpha_1 U > V$, we find that it is of the form

$$f(u; \alpha_0, \alpha_1) = \psi_1(u)\Psi_2(\alpha_0 + \alpha_1 u)/\mathrm{P}(\alpha_0 + \alpha_1 U > V). \tag{6.11}$$

The model (6.1) is clearly recognizable as a special case here. If $\alpha_0 = 0$ and ψ_1 and ψ_2 are symmetric densities then (6.11) simplifies to yield

$$f(u; \alpha_1) = 2\psi_1(u)\Psi_2(\alpha_1 u). \tag{6.12}$$

Densities of the form (6.12) have been discussed by Azzalini (1985). They are a natural generalization of the original Azzalini skew-normal density. When $\alpha_0 \neq 0$, the denominator in (6.11) is typically difficult to evaluate. The exceptions are cases in which ψ_1 and ψ_2 are both normal or are both of the Cauchy form, see Arnold and Beaver (2000b) for discussion of the Cauchy case.

A non-normal version of the additive component construction is as follows. We begin with two independent random variables Y_0, Y_1 with corresponding densities ϕ_0 and ϕ_1 and distributions Φ_0 and Φ_1. For a fixed real number c, we define $Y_0(c)$ to be the random variable Y_0 truncated below at c. Then we define our skewed random variable by

$$Z = \nu_0 + \nu_1 Y_0(c) + \nu_2 Y_1. \tag{6.13}$$

The density function of Z is given by

$$f_Z(z; c, \nu_0, \nu_1, \nu_2) = \int_c^\infty \frac{1}{\nu_2}\phi_1\left(\frac{z - \nu_0 - \nu_1 y_0}{\nu_2}\right)\phi_0(y_0)dy_0/\bar{\Phi}_0(c), \tag{6.14}$$

where $\bar{\Phi}_0(c) = 1 - \Phi_0(c)$. The integration in (6.14) can be implemented in some special cases to yield an analytic expression for this density. Arnold and Beaver (2000b, 2000c) discussed the Cauchy, Laplace, and logistic cases.

6.5 Multivariate Skewed Distributions

We turn now to consider multivariate variants of the skewed univariate distributions introduced in Sections 6.2 and 6.4. The first multivariate skewed normal distribution discussed in the literature was an extension of the Azzalini model (6.2) which could be interpreted as hidden truncation with a zero threshold, see Azzalini and Dalla Valle (1996). A slight variant on this construction, introduced in Arnold and Beaver (2000a), involves a general threshold instead of 0. It is this model that we will now consider. We begin with $p+1$ iid standard normal random variables $W_1, \ldots, W_p, U$. Consider the conditional distribution of $\mathbf{w} = (W_1, \ldots, W_p)^T$ given that $\alpha_0 + \boldsymbol{\alpha}_1^T \mathbf{w} > U$, for some fixed $\alpha_0 \in \mathbb{R}$ and $\boldsymbol{\alpha}_1 \in \mathbb{R}^p$. The resulting density takes the form:

$$f(\mathbf{w}; \alpha_0, \boldsymbol{\alpha}_1) = \left(\prod_{i=1}^{p} \phi(w_i) \right) \Phi(\alpha_0 + \boldsymbol{\alpha}_1^T \mathbf{w}) / \Phi \left(\frac{\alpha_0}{\sqrt{1 + \boldsymbol{\alpha}_1^T \boldsymbol{\alpha}_1}} \right). \quad (6.15)$$

Azzalini and Dalla Valle (1996) considered the case in which $\alpha_0 = 0$. We then introduce location and scale parameters to arrive at the full p-dimensional skew-normal model, i.e., we consider random vectors of the form

$$\mathbf{x} = \boldsymbol{\mu} + \Sigma^{1/2} \mathbf{w}, \quad (6.16)$$

where $\boldsymbol{\mu} \in \mathbb{R}^p, \Sigma^{1/2}$ is positive definite, and $\mathbf{w}$ has density (6.15). It is at first glance remarkable that if $\mathbf{x}$ has a distribution of the form (6.16) then all of its marginal densities and all of its conditional densities are of the same form. This becomes more plausible if we consider the following alternative route to the model (6.16) which clearly involves hidden truncation. Consider $(Y_0, Y_1, \ldots, Y_p)^T$ with a $(p+1)$-dimensional classical multivariate normal distribution. The joint distribution of $(Y_1, \ldots, Y_p)^T$ given $Y_0 > c$ is of the form (6.16).

An additive component construction of such skewed distributions was suggested by Arnold and Beaver (2003). Consider $X_0, X_1, \ldots, X_p$ iid standard normal random variables. For $c \in \mathbb{R}$, define $X_0(c)$ to be X_0 truncated below at c. Now define $\mathbf{y}$ by

$$Y_i = \delta_i X_0(c) + \sqrt{1 - \delta_i^2} X_i, \ |\delta_i| < 1, \ \ i = 1, \ldots, p. \quad (6.17)$$

By considering the joint moment generating function (mgf) of $\mathbf{y}$, we may verify that this route also leads eventually to the model (6.16).

Jones (2002) described yet another way to arrive at the model (6.16).

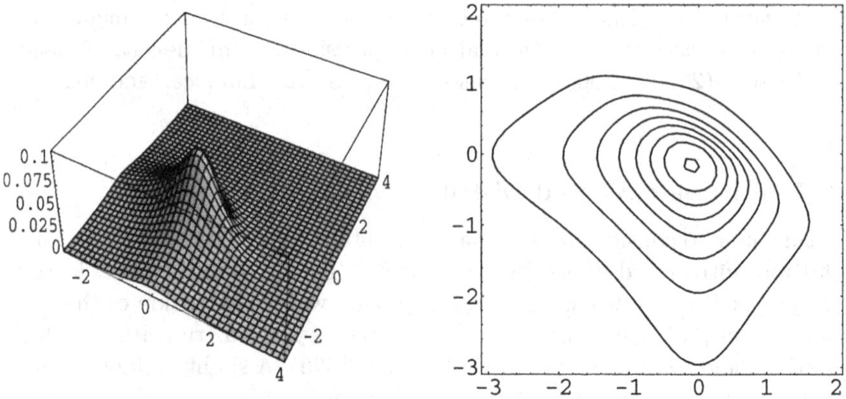

Figure 6.1 *Skewed bivariate Cauchy distribution based upon hidden truncation with* $\alpha_0 = 1$, $\boldsymbol{\alpha}_1 = (-2, -3)^T$*: 3-dimensional plot (left) and contour plot (right).*

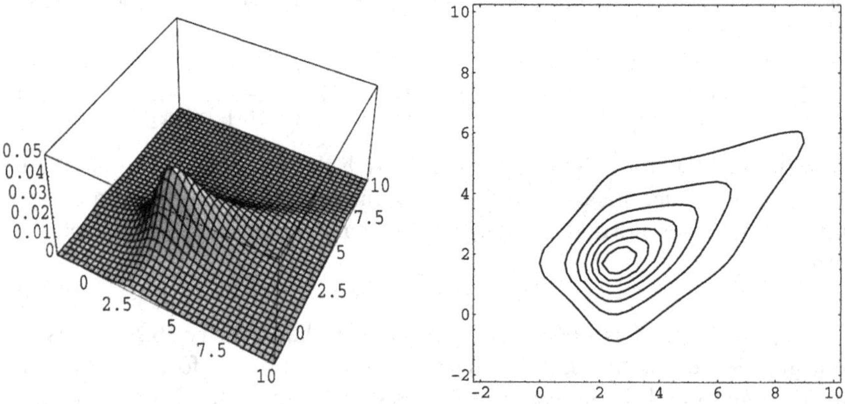

Figure 6.2 *Skewed bivariate Cauchy distribution based upon additive components with* $(c, \delta_1, \delta_2) = (4, 1/2, 1/3)$*: 3-dimensional plot (left) and contour plot (right).*

Using his approach we begin with a classical p-variate normal distribution for $(W_1, \ldots, W_p)^T$. We then replace the marginal distribution of W_1 by a skewed normal density, i.e., replace W_1 by $\tilde{W}_1 = W_1 + kZ_0(c)$, where Z_0 is a standard normal random variable independent of W_1. Then we define the conditional distribution of $(\tilde{W}_2, \ldots, \tilde{W}_p)^T$ given $\tilde{W}_1$ to be the same as the original conditional distribution of $(W_2, \ldots, W_p)^T$ given W_1. Once more, the model (6.16) appears in the end.

The hidden truncation paradigm, the additive component paradigm, and the Jones paradigm do lead to different distributions if we begin with non-normal components. Representative 3-dimensional and contour plots of two different forms of skewed bivariate Cauchy distributions are shown in Figures 6.1 and 6.2.

In the present chapter, our major concern is with skewed distributions with elliptical components rather than multivariate normal components. In this manner we will be dealing with a flexible family of models that we hope will be reasonably tractable and which will include skew-normal models as special cases. To this end, we need to introduce quite general multivariate skewed distributions involving dependent components, since elliptical distributions only have independent marginals in special normal cases. We will continue to use the hidden truncation and additive component constructions but they will be slightly generalized. We will also consider the Jones (2002) construction in the full generality introduced by Jones.

6.6 General Multivariate Skewed Distributions

At this point the general nature of the constructions to be used is well understood so that only a brief description of the general models is required.

6.6.1 Hidden Truncation Paradigm

Let $\mathbf{w} = (W_1, \ldots, W_p)^T$ have a joint density $\psi(\mathbf{w})$ and assume that W_0 is independent of $\mathbf{w}$ and has distribution function $\Psi_0(w_0)$. We then consider the conditional distribution of $\mathbf{w}$ given $A = \{\alpha_0 + \boldsymbol{\alpha}_1^T\mathbf{w} > W_0\}$. The resulting joint density is given by

$$f_{\mathbf{w}|A}(\mathbf{w}) = \psi(\mathbf{w})\Psi_0(\alpha_0 + \boldsymbol{\alpha}_1^T\mathbf{w})/\mathrm{P}(A). \tag{6.18}$$

The skew-normal model (6.15) is evidently a special case of (6.18) in which $\psi(\mathbf{w})$ is assumed to have independent standard normal marginals.

6.6.2 Hidden Truncation, More General

Let $(W_0, W_1, \ldots, W_p) = (W_0, \mathbf{w}^T)$ and denote the conditional distribution function of W_0 given $\mathbf{w} = \mathbf{w}^*$ by $\Psi_{\mathbf{w}^*}(w_0)$ and the marginal density of $\mathbf{w}$ by $\psi(\mathbf{w})$. Now once more, consider the conditional distribution of $\mathbf{w}$ given $A = \{\alpha_0 + \boldsymbol{\alpha}_1^T\mathbf{w} > W_0\}$. The resulting joint density is of the form

$$f_{\mathbf{w}|A}(\mathbf{w}) = \psi(\mathbf{w})\Psi_{\mathbf{w}}(\alpha_0 + \boldsymbol{\alpha}_1^T\mathbf{w})/\mathrm{P}(A). \tag{6.19}$$

It may be noted that (6.19) is a valid density for any indexed family of distributions $\{\Psi_{\mathbf{w}} : \mathbf{w} \in \mathbb{R}^p\}$. However, the apparent gain in generality is illusory, since they can always be interpreted as conditional distributions of some random variable W_0 given $\mathbf{w} = \mathbf{w}^*$.

6.6.3 Additive Component Paradigm

Let $\mathbf{w} = (W_1, \ldots, W_p)^T$ have a joint density $\psi(\mathbf{w})$ and assume that W_0 is independent of $\mathbf{w}$. Define $W_0(c)$ to be W_0 truncated below at c. Now define $\mathbf{y}$ by

$$Y_i = \alpha_i W_0(c) + \beta_i W_i, \;\; \alpha_i, \beta_i \in \mathbb{R}, \; i = 1, \ldots, p. \tag{6.20}$$

The model (6.17) is evidently a special case of (6.20) in which the W_i's are assumed to all be independent standard normal variables.

6.6.4 Additive Component, More General

Let $(W_0, W_1, \ldots, W_p)^T$ be a $(p+1)$ dimensional random vector, no longer assuming W_0 is independent of $(W_1, \ldots, W_p)$. Again define $W_0(c)$ to be W_0 truncated below at c and define Y_i as in (6.20).

6.6.5 The Jones Construction

As proposed in Jones (2002), we begin with $(W_1, \ldots, W_p)$ having joint density $\phi(\mathbf{w})$. Denote the marginal density of W_1 by $\phi_1(w_1)$ and the conditional density of $W_2, \ldots, W_p$ given $W_1 = w_1$ by $\psi_{w_1}(w_2, \ldots, w_p)$. Now replace $\psi_1(w_1)$ by a skewed version of itself, say $\tilde{\psi}_1(w_1)$, perhaps based on hidden truncation or on an additive component construction. Our skewed p-dimensional density is then given by

$$\tilde{\psi}(\mathbf{w}) = \tilde{\psi}_1(w_1)\psi_{w_1}(w_2, \ldots, w_p). \tag{6.21}$$

We thus keep the original conditional structure of $W_2, \ldots, W_p$ given W_1 and merely replace the marginal density of W_1 by a skewed version. Note that in fact we could more generally, and as Jones suggests, partition $\mathbf{w}$ into two subvectors $\mathbf{w}_1$ and $\mathbf{w}_2$ and replace the density of $\mathbf{w}_1$ by a skewed version while retaining the original conditional distribution of $\mathbf{w}_2$ given $\mathbf{w}_1 = \mathbf{w}_1^*$.

In all of these constructions we finish up by introducing location and scale parameters using (6.16), i.e., we define

$$\mathbf{x} = \boldsymbol{\mu} + \Sigma^{1/2}\mathbf{w}.$$

Of course if $\psi(\mathbf{w})$ in the scenarios described in Subsections 6.6.1, 6.6.3, and 6.6.5 has independent standard normal components, we will be led to the usual multivariate skew-normal distribution defined in (6.15)–(6.16). Otherwise, all of these constructions will lead to different models. Since independent standard normal components of $\mathbf{w}$ lead to a spherically symmetric joint density, it is quite natural to consider the use of general elliptically contoured densities in their place.

6.7 Skew-Elliptical Distributions

Azzalini and Capitanio (1999) suggested skew-elliptical densities as alternatives to their skew-normal distribution. Subsequently, Branco and Dey (2001) provided a detailed discussion of such densities. The Branco-Dey development may be identified with our general truncation formulation in Subsection 6.6.2, although they do not describe it in those terms. The description of the skew-elliptical distribution is conveniently simplified by considering skew-spherical distributions and finishing up by applying a transformation of the form (6.16) to take us to the corresponding skew-elliptical model. A caveat is in order, if the reader is familiar with the Branco-Dey development. Our discussion will, to a great extent, be equivalent to theirs with $\boldsymbol{\mu} = \mathbf{0}$ and $\Sigma = I_p$. However their "truncation" parameter $\boldsymbol{\alpha}$ is closely related to but not identical to our α_1. In addition, the α_0 appearing in our discussion does not appear in theirs; equivalently, we can say that it is set equal to zero.

Recall that a random vector $\mathbf{w}$ of dimension p is spherically symmetric if it admits the representation

$$\mathbf{w} = Z\mathbf{r}, \tag{6.22}$$

where the random variable Z and the random vector $\mathbf{r} = (R_1, \ldots, R_p)^T$ are independent. The random variable Z is positive with probability one and has density function $f_Z(z)$. The random vector $\mathbf{r}$ is assumed to be uniformly distributed over the unit p-sphere. The joint density of $\mathbf{w}$ will then be of the form

$$f_{\mathbf{w}}(\mathbf{w}) = g^{(p)}(\mathbf{w}^T\mathbf{w}), \tag{6.23}$$

where $g^{(p)}$ satisfies

$$\int_0^\infty u^{\frac{p}{2}-1} g^{(p)}(u)du = \Gamma\left(\frac{p}{2}\right)/(2\pi^{p/2}). \tag{6.24}$$

Of course $g^{(p)}$ is determined by the density of Z. In fact,

$$f_Z(z) = \frac{2\pi^{p/2}}{\Gamma(p/2)} z^{p-1} g^{(p)}(z^2). \tag{6.25}$$

It is evident from the representation (6.22) that spherically symmetric random vectors have spherically symmetric marginal densities. Moreover, inspection of (6.23) leads to the conclusion that they will have spherically symmetric conditional densities also.

Using the general hidden truncation paradigm described in Subsection 6.6.1, we begin with a $(p+1)$-dimensional spherically symmetric random vector $(W_0, W_1, \ldots, W_p) = (W_0, \mathbf{w}^T)$. The joint density of $(W_0, \mathbf{w}^T)^T$ is given by

$$f_{W_0,\mathbf{w}}(w_0, \mathbf{w}) = g^{(p+1)}(w_0^2 + \mathbf{w}^T\mathbf{w}), \tag{6.26}$$

where $g^{(p+1)}$ is related to f_Z as in (6.25). The marginal densities of W_0 and

$\mathbf{w}$ are then given by

$$f_{W_0}(w_0) = g^{(1)}(w_0^2), \tag{6.27}$$

and

$$f_{\mathbf{w}}(\mathbf{w}) = g^{(p)}(\mathbf{w}^T\mathbf{w}). \tag{6.28}$$

The conditional density of W_0 given $\mathbf{w} = \mathbf{w}^*$ may be denoted by

$$f_{\mathbf{w}^*}(w_0) = \frac{g^{(p+1)}(w_0^2 + \mathbf{w}^{*T}\mathbf{w}^*)}{g^{(p)}(\mathbf{w}^{*T}\mathbf{w}^*)}. \tag{6.29}$$

The symmetry of this conditional density is evident in (6.29). We will denote the conditional distribution function of W_0 given $\mathbf{w} = \mathbf{w}^*$ by $F_{\mathbf{w}^*}(w_0)$.

Now for $\alpha_0 \in \mathbb{R}$ and $\boldsymbol{\alpha}_1 \in \mathbb{R}^p$, consider the conditional distribution of $\mathbf{w}$ given the event $A = \{\alpha_0 + \boldsymbol{\alpha}_1^T\mathbf{w} > W_0\}$. By conditioning on $\mathbf{w} = \mathbf{w}^*$ we can conclude from (6.19) that

$$f_{\mathbf{w}|A}(\mathbf{w}) = g^{(p)}(\mathbf{w}^T\mathbf{w})F_{\mathbf{w}}(\alpha_0 + \boldsymbol{\alpha}_1^T\mathbf{w})/\mathrm{P}(A), \tag{6.30}$$

where the density corresponding to $F_{\mathbf{w}}$ is given in (6.29). Applying transformations of the form (6.12) to (6.30) we arrive at a general family of skew-elliptical densities, which includes the Azzalini and Branco-Dey models as special cases corresponding to the choice $\alpha_0 = 0$. The denominator $\mathrm{P}(A)$ in (6.30) can be troublesome to evaluate, though for many applications precise knowledge of this awkward normalizing constant is not crucial. A variety of estimation strategies can be used to sidestep this problem.

Of course, Branco and Dey (2001) did not have to deal with this normalization problem. This is because when $\alpha_0 = 0$, $\mathrm{P}(A) = 1/2$. This is analogous to the simplification encountered in going from (6.1) to (6.2), in the univariate case, also when $\alpha_0 = 0$. The reason is the same: symmetry. In our present case the event A can be written when $\alpha_0 = 0$, as $\boldsymbol{\alpha}_1^T\mathbf{w} - W_0 > 0$. But since $(W_0, \mathbf{w}^T)$ is spherically symmetric the random variable $\boldsymbol{\alpha}_1^T\mathbf{w} - W_0$ is symmetric, for any $\boldsymbol{\alpha}_1$, and so $\mathrm{P}(A) = 1/2$, trivially.

Some representative examples of densities of the form (6.30), with $\alpha_0 = 0$, are described in Branco and Dey (2001). As a simple example, if we take

$$f_Z(z) = \frac{1}{B(p/2, 1/2)} z^{p/2-1}(1+z)^{-(p+1)/2}, \tag{6.31}$$

in the representation in (6.22) with $B(p/2, 1/2) = \Gamma(p/2)\Gamma(1/2)/\Gamma((p+1)/2)$, our final skewed version of the spherically symmetric Cauchy distribution is given by

$$f_{\mathbf{w}}(\mathbf{w}) = c(p)(1+\mathbf{w}^T\mathbf{w})^{-(p+1)/2}\Psi(\alpha_0+\boldsymbol{\alpha}_1^T\mathbf{w})/\Psi\left(\frac{\alpha_0}{\sqrt{1+\boldsymbol{\alpha}_1^T\boldsymbol{\alpha}_1}}\right), \tag{6.32}$$

where $c(p) = \Gamma((p+1)/2)/\pi^{(p+1)/2}$ and $\Psi(u) = \frac{1}{2} + \frac{1}{\pi}\operatorname{Arctan}(u)$.

Of course we can arrive at simpler skew-elliptical densities by using the basic hidden truncation paradigm in Subsection 6.6.1. For it, we begin

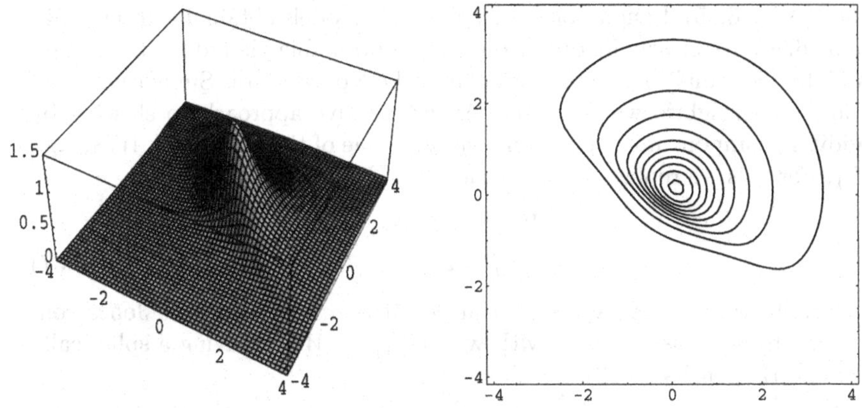

Figure 6.3 *Skewed bivariate Cauchy distribution with* $\alpha_0 = 1$, $\boldsymbol{\alpha}_1 = (2,3)^T$*: 3-dimensional plot (left) and contour plot (right).*

with $\mathbf{w}$ spherically symmetric of dimension p with density (6.23) and W_0 independent of $\mathbf{w}$ with a quite arbitrary distribution function $\Psi_0(w_0)$. From (6.18) we will then have a skewed version of the form

$$f_{\mathbf{w}|A}(\mathbf{w}) = g^{(p)}(\mathbf{w}^T\mathbf{w})\Psi_0(\alpha_0 + \boldsymbol{\alpha}_1^T\mathbf{w})/\mathrm{P}(A). \tag{6.33}$$

As an example of these types of skewed densities, suppose that (W_1, W_2) has a bivariate spherically symmetric Cauchy density and that Ψ_0 corresponds to an independent Cauchy variable. Plots for such a skewed density with $\alpha_0 = 1$ and $\boldsymbol{\alpha}_1 = (2,3)^T$ are shown in Figure 6.3. Note that in Arnold and Beaver (2002, p. 21) it is incorrectly stated that (6.30) with $\alpha_0 = 0$ is a special case of (6.33).

Now consider the additive component approach to skewing spherically symmetric densities. The general additive component approach, from Subsection 6.6.4, suggests consideration of $(W_0, W_1, \ldots, W_p) = (W_0, \mathbf{w}^T)$ having a $(p+1)$-dimensional spherically symmetric density, i.e., as in (6.26)

$$f_{W_0,\mathbf{w}}(w_0, \mathbf{w}) = g^{(p+1)}(w_0^2 + \mathbf{w}^T\mathbf{w}). \tag{6.34}$$

Now we define $W_0(c)$ to be W_0 truncated below at c and define a new random vector $\mathbf{y}$ by

$$Y_i = \alpha_i W_0(c) + \beta_i W_i,\ \alpha_i, \beta_i \in \mathbb{R},\ i = 1, \ldots, p, \tag{6.35}$$

as in (6.20). The paradigm described in Subsection 6.6.3 is of the same form except that we assume the joint density of $(W_0, \mathbf{w}^T)$ is given by

$$f_{W_0,\mathbf{w}}(w_0, \mathbf{w}) = g^{(p)}(\mathbf{w}^T\mathbf{w})\psi_0(w_0), \tag{6.36}$$

where ψ_0 is an arbitrary density function. Except in the normal case, i.e., $g^{(p)}(u) = (2\pi)^{-p/2}e^{-u/2}$, it will not be easy to determine the distribution of the Y_i's in (6.35). In non-normal settings, the models obtainable from (6.35) and (6.36) are clearly distinct from those obtainable via hidden truncation and further analytic investigation may be appropriate. Simulation is of course a straightforward matter. An alternative approach to skewing by additive components involves only skewing one of the coordinate W_i's, e.g., W_1. For this we replace the definition (6.35) for $\mathbf{y}$ by

$$\begin{aligned} Y_1 &= \alpha_1 W_0(c) + \beta_1 W_1, \ \alpha_1, \beta_1 \in \mathbb{R}, \\ Y_j &= W_j, \ j = 2, \ldots, p. \end{aligned} \tag{6.37}$$

Finally we turn to skew-elliptical densities arising from the Jones construction. For this we begin with $\mathbf{w} = (W_1, \ldots, W_p)^T$ having a spherically symmetric density (6.23),

$$f_{\mathbf{w}}(\mathbf{w}) = g^{(p)}(\mathbf{w}^T \mathbf{w}). \tag{6.38}$$

Now partition $\mathbf{w}$ as $(\mathbf{w}_1^T, \mathbf{w}_2^T)^T$ where $\mathbf{w}_1$ is of dimension p_1 and $\mathbf{w}_2$ is of dimension $p - p_1$. The approach involves constructing a new joint density which retains the conditional density of $\mathbf{w}_2$ given $\mathbf{w}_1$ implied by (6.38), but replaces the marginal density of $\mathbf{w}_1$ by a skewed version. For example, we might replace $f_{\mathbf{w}_1}(\mathbf{w}_1)$ by a hidden truncation skewed version as given by (6.30) or (6.33), but now of dimension p_1 rather than p. This will lead to skewed densities of the form:

$$f_{\mathbf{w}}(\mathbf{w}) = g^{(p)}(\mathbf{w}^T \mathbf{w}) F_{\mathbf{w}_1}(\alpha_0 + \tilde{\boldsymbol{\alpha}}_1^T \mathbf{w}_1)/\mathrm{P}(A^*), \tag{6.39}$$

and

$$f_{\mathbf{w}}(\mathbf{w}) = g^{(p)}(\mathbf{w}^T \mathbf{w}) \Psi_0(\alpha_0 + \tilde{\boldsymbol{\alpha}}_1^T \mathbf{w}_1)/\mathrm{P}(A^*), \tag{6.40}$$

where $A^* = \alpha_0 + \tilde{\boldsymbol{\alpha}}_1^T \mathbf{w}_1 > W_0$ and $F_{\mathbf{w}_1}$ denotes the conditional distribution of W_0 given $\mathbf{w}_1 = \mathbf{w}_1^*$. Jones (2002) gives some examples of skewed densities constructed using his paradigm, although the skewing mechanism he applies to the marginal density of $\mathbf{w}_1$ is typically not of the hidden truncation type.

6.8 Discussion

Arnold and Beaver (2000a) have argued that hidden truncation is quite likely to be encountered in data analysis. Models that incorporate hidden truncation will then be desirable additions to the modeler's tool-kit. The final hidden truncation models involve a plethora of parameters $\boldsymbol{\mu}, \Sigma, \alpha_0$ and $\boldsymbol{\alpha}_1$, or $\tilde{\boldsymbol{\alpha}}_1$ in the case of (6.39) and (6.40), and this will predictably make estimation problematic in high dimensional cases. More work is needed on estimation for non-normal skew-elliptical models.

CHAPTER 7

From Symmetric to Asymmetric Distributions: A Unified Approach

Reinaldo B. Arellano-Valle and Guido E. del Pino

7.1 Introduction

Asymmetric (skewed) distributions are useful in probability modeling, statistical analysis, and robustness studies. This chapter presents a unified approach to generating them, starting from a given family of symmetric distributions, either univariate or multivariate. The emphasis here lies on general methods to achieve the following aims:

- Define a skewed distribution.
- Find its probability density function (pdf).
- Provide a stochastic representation if possible.
- Simulate random variables from the given distribution.

Location and scale families occupy a predominant place in statistical modeling and analysis, and this will also be the case here. Any such family may be generated by applying the following steps:

- Pick a cumulative distribution function (cdf) F and let X be a random variable with $X \sim F$.
- Let $Y = \theta_0 + \theta_1 X$, where $-\infty < \theta_0 < \infty$, $\theta_1 > 0$.
- Let F_{θ_0,θ_1} be the cdf of Y.
- Define the family $\mathcal{F} = \{F_{\theta_0,\theta_1} : -\infty < \theta_0 < \infty,\ \theta_1 > 0\}$.
- With this notation, the generating distribution F is identified with $F_{0,1}$.

When the distributions in $\mathcal{F}$ are symmetric, θ_0 may always be chosen as the center of symmetry, which is equivalent to F being symmetric about the origin. In order to simplify the writing, a symmetric distribution will be understood as symmetric about the origin, unless explicitly stated otherwise.

A key idea in analyzing departures from a given location and scale family of symmetric distributions is to first obtain a one-parameter family containing $F_{0,1}$, where the parameter represents the degree of skewness. Each value of this skewness parameter defines a generator for a new location and scale family. One aim of this chapter is to present alternative ways to create these parametric families for the skewed generator.

Before proceeding to the main part, a few words about other approaches are in order. The idea of having a family of distributions that is both analytically tractable and able to accommodate practical values of skewness and kurtosis goes back to Karl Pearson. The famous family of Pearson curves contains the normal distributions, which are characterized as the only ones within the family with zero skewness and kurtosis. Pearson curves are not very popular today because they are not easily handled and the original estimation method based on matching the first four empirical moments to their theoretical counterparts leads generally to inefficient estimators, as pointed out by Sir Ronald Fisher.

A traditional approach to dealing with skewed responses in regression problems is to work with a transformed response variable, but linearity in the transformed scale is contradictory with linearity in the original one. The class of generalized linear models (see McCullagh and Nelder, 1989) does not have this drawback, since transformations are applied to the parameters instead of the responses. For these models the distribution of the response variable must belong to a linear exponential family.

Normal distributions are symmetric and belong both to a location and scale family and to an exponential family. From an inferential viewpoint it would be attractive to look for skewed distributions sharing this nice property. Unfortunately, the only such distributions are those corresponding to $c \log Y$, where Y follows a gamma distribution with fixed shape parameter and $c \neq 0$; see, e.g., Lehmann and Casella (1998, p. 32). Moreover, this second class does not contain any symmetric distributions.

To illustrate a simple way to build a parametric family of skewed distributions, consider the following model: the mixture of a symmetric distribution F_0 and an asymmetric distribution F_1, i.e., $F_\alpha = (1-\alpha)F_0 + \alpha F_1$, $0 \leq \alpha \leq 1$. This parametric family may only reflect skewness "in the direction of F_1." To get skewness in the opposite direction the parametric family may be enlarged by defining $X \sim F_\alpha \iff -X \sim F_{-\alpha}$. The extended family $\mathcal{F} = \{F_\alpha : -1 \leq \alpha \leq 1\}$ satisfies the reflection property: $X \sim F$, $F \in \mathcal{F}$, implies that the distribution of $-X$ also belongs to $\mathcal{F}$. A weakness of this parametric model is the arbitrariness in the choice of F_1.

An outline of this chapter follows. The first two sections deal with univariate distributions, while the remaining ones deal with various multivariate cases. Section 7.2 shows how a number of known classes of skewed distributions may be recovered and extended through the probability modeling of the absolute value and the sign of any given random variable. The results obtained depend on the type of conditioning employed. Conditioning on absolute values is seen to lead naturally to skewing functions, while conditioning on signs accommodates the skew-power distribution of Fernández, Osiewalsky, and Steel (1995), its extension by Fernández and Steel (1998), and the epsilon-skew-normal distribution of Mudholkar and Hutson (2000). Section 7.3 studies the important case where the skewing function is determined by conditioning on a latent random variable.

In Section 7.4 it is argued that the concept of skewness is not a well defined one, but rather depends on the desired concept of symmetry. Although the mathematical idea of invariance under group of transformations is used, this chapter only uses three such groups, which are discussed in this section. The advantage of this viewpoint is that many scattered facts may be seen as special cases of a general property. The results of Section 7.5 are applied in the next section to review the $\mathcal{C}$-class studied in Arellano-Valle, del Pino, and San Martín (2002) from a group theoretic viewpoint. The basic fact is that this class coincides with the symmetric distributions relative to the group of coordinate reflections. In that same section the results of Section 7.3 are applied to construct the pdf of a skewed distribution. Section 7.6 reviews a stochastic representation of skewed distributions, in the context of the class of sign invariant distributions, which is then applied to compute moments and to prove a property of the square of a skewed random variable. If X has a symmetric pdf and Y has a skewed pdf obtained from that of X, then Y^2 and X^2 have the same distribution. Section 7.7 provides a streamlined alternative derivation for the pdf of the multivariate skew-normal distribution, while the last section presents a canonical form for skew-elliptical distributions.

7.2 Signs, Absolute Values, and Skewed Distributions on the Real Line

The key for understanding skewed distributions on the real line lies in the simple fact that any real number x may be decomposed into its absolute value $t = |x|$ and its sign $w = \mathrm{sgn}(x)$, that is,

$$w = \begin{cases} 1, & \text{for } x > 0, \\ -1, & \text{for } x < 0. \end{cases}$$

We will use the symbol $\perp\!\!\!\perp$ to denote independence (or conditional independence when accompanied by the symbol $|$). In view of the one-to-one correspondence between x and the pair $(t, w) = (|x|, \mathrm{sgn}(x))$, any distribution of X transfers uniquely into the joint distribution of $(T, W) = (|X|, \mathrm{sgn}(X))$ and vice versa. To simplify the discussion we will assume throughout that X has an absolutely continuous distribution with pdf f. This implies that T is also absolutely continuous, while W is, of course, discrete. This mixed type of random variables suggests that their joint distribution be studied in terms of conditional distributions. There are then two options:

(a) Conditioning on the sign of a random variable

One must specify the conditional densities of $T|W = 1$ and $T|W = -1$, together with the marginal distribution of W, which is completely determined by $\gamma = \mathrm{P}(W = 1)$. For a symmetric distribution, the two conditional densities are identical (that is, $T \perp\!\!\!\perp W$) and $\mathrm{P}(W = 1) = 1/2$.

Denote by F_w the distribution of $T|W = w$, and let T_w be any random variable satisfying $T_w \sim F_w$, where $w = 1, -1$. Since their joint distribution is irrelevant, one may assume $T_1 \perp\!\!\!\perp T_{-1}$, which is useful for simulation. The distributions F_1 and F_{-1} of T_1 and T_{-1}, respectively, may be indexed by separate parameters, but they will typically have some dependence upon $\gamma = \mathrm{P}(W = 1)$.

This framework accommodates the skew-power distribution of Fernández, Osiewalsky, and Steel (1995) and its extension by Fernández and Steel (1998). It also includes the epsilon-skew-normal distribution of Mudholkar and Hutson (2000), and the general family of asymmetric pdf's introduced by Arellano-Valle, Gómez, and Quintana (2004a) as a special case. It may be verified that in all these papers, F_1 and F_{-1} are chosen to be elements of the same scale family, where the ratio of the corresponding scale parameters is chosen as $\frac{\gamma}{1-\gamma}$.

(b) Conditioning on the absolute value of a random variable

Since the sign W is a binary random variable, the conditional distributions are determined by

$$\gamma(t) = \mathrm{P}(W = 1|T = t) = \mathrm{P}(X > 0||X| = t).$$

Since these probabilities may be specified in an arbitrary fashion, the function $\gamma(t)$ needs only satisfy $0 \leq \gamma(t) \leq 1$ for all $t > 0$. For a symmetric random variable X, all the conditional distributions of $W||X| = t$ are uniform, that is, $\gamma(t) = 1/2$, for all $t > 0$.

The function γ is closely connected to the skewing function discussed in Azzalini and Capitanio (1999), Genton and Loperfido (2002), and Wang, Boyer, and Genton (2004a). In fact, for any given distribution F_T, there is a unique symmetric distribution F satisfying $|S| \stackrel{d}{=} T = |X|$ and $S \sim F$. In the absolutely continuous case the corresponding densities f_T and f satisfy $f_T(t) = 2f_S(t), t > 0$. Then

$$\pi(x) = \gamma(x),\ x \geq 0 \text{ and } \pi(x) = 1 - \gamma(-x),\ x < 0.$$

Since the functions γ and π are in one-to-one correspondence with each other, they provide equivalent representations of the conditional distributions of the sign of X given its absolute value. The arbitrariness of $\gamma(t)$ is equivalent to the result of Wang *et al.* (2004a) in terms of the skewing function. The representation $X = WT$ allows a direct proof of other results in that paper. Indeed, $|X| = |W|T = T$, which implies $|X| \stackrel{d}{=} |S|$ and so $h(X) \stackrel{d}{=} h(S)$ for any even function h. In particular, $X^2 \stackrel{d}{=} S^2$; see also Wang *et al.* (2004b). Equation (2) in Wang *et al.* (2004a) is obtained by writing $f_X = g$, $f_S = f$ and $f_T = \overline{f}$. Then $\overline{f}(t) = 2f(t)$, $t \geq 0$, and

$$g(x) = \overline{f}(x)\pi(x) = 2f(x)\pi(x), \tag{7.1}$$

which agrees with Equation (2) in Wang *et al.* (2004a).

A one-parameter family of skewed distributions may be defined by f and a one-parameter family of skewing functions. If one wants the family of skewed distributions to be closed under reflections about the origin, a natural choice of the parameter is $\alpha \in \mathbb{R}$, with $\alpha = 0$ corresponding to the symmetric density f and imposing the condition

$$-X \sim P_{-\alpha} \Longleftrightarrow X \sim P_{\alpha}.$$

Since $-X = (-W)T$, this translates into

$$\pi_{-\alpha}(x) = 1 - \pi_{\alpha}(x), \ \alpha < 0. \tag{7.2}$$

Thus one needs only to specify the function π_α for $\alpha > 0$. An example of (7.2) is obtained by picking the cdf G of any *symmetric* random variable and taking $\pi_\alpha(x) = G(\alpha x)$. For $\alpha > 0$, the right-hand side may be interpreted as a a scale family of distributions generated by G, with scale parameter α^{-1}. For the skew-normal distribution $G = F$, but this equality need not hold in general; see Genton and Loperfido (2002), Ma and Genton (2004), and Wang *et al.* (2004a).

7.3 Latent Variables, Selection Models, and Skewed Distributions on the Real Line

The following proposition is a powerful tool to compute the pdf of skewed distributions and it generalizes Theorem 5.1 in Arellano-Valle *et al.* (2002). A useful viewpoint is to interpret it as a way of generating skewing functions. The proposition is stated for an arbitrary dimension, but it will be used here for dimension 1. The proof follows easily by applying the definition of conditional distributions.

Proposition 7.3.1 Let $\mathbf{x}$ be a random vector with pdf f. Then, for any random vector $\mathbf{r}$ and a set A, the conditional pdf of $\mathbf{x}$ given the event $\mathbf{r} \in A$ is

$$g(\mathbf{x}|A) = f(\mathbf{x})\frac{\mathrm{P}(\mathbf{r} \in A|\mathbf{x})}{\mathrm{P}(\mathbf{r} \in A)}. \tag{7.3}$$

The underlying probability model implicit in Proposition 7.3.1 may be interpreted as a selection model, where $\mathbf{x}$ can only be observed provided the event $\mathbf{r} \in A$ occurs. If f is a symmetric pdf, then (7.3) corresponds to the skewing function

$$\pi(\mathbf{x}) = \frac{\mathrm{P}(\mathbf{r} \in A|\mathbf{x})}{2\mathrm{P}(\mathbf{r} \in A)}.$$

For $\mathrm{P}(\mathbf{r} \in A) = 1/2$ this simplifies to

$$\pi(\mathbf{x}) = \mathrm{P}(\mathbf{r} \in A|\mathbf{x}).$$

When $A \subseteq \mathbb{R}$, the order structure becomes relevant and a natural choice is $A = (c, \infty)$. Intuitively, the conditional density $g(x|R > c)$ is positively skewed when R and X are positively associated. For $c = 0$ and R symmetric, $\pi(x) = \mathrm{P}(R > 0|X = x)$.

A simple model that assures a positive association between R and X is $R = h(X, Z; \alpha)$, where the function h is strictly increasing in its first argument and Z is independent of X. This independence assumption can be weakened as discussed in Section 7.8. The skew-normal distribution is obtained by choosing $h(x, z; \alpha) = \alpha x - z$, which implies $\pi(x) = F_Z(\alpha x)$. A similar representation holds for its extensions; see, e.g., Arellano-Valle *et al.* (2002), Wang *et al.* (2004a), and Arellano-Valle and Genton (2003).

7.4 Symmetry, Invariance, and Skewness

In general, skewness is essentially defined as lack of symmetry. Although the concept of symmetry in the real line is clear, this is not so in higher dimensions and one has to be explicit about what is meant by being symmetric. A mathematical concept associated with symmetry is that of invariance under a group of transformations. Below is a glossary, followed by three particular examples that are the only ones relevant for this chapter.

7.4.1 Glossary and Basic Facts

The application of invariance to statistics is covered in several books, e.g., Lehmann and Casella (1998), and Eaton (1983). Invariance involves a group of transformations $\mathcal{G}$ acting on the sample space $\mathcal{X}$. A glossary and brief account of the main facts are reviewed below:

- A random vector has a symmetric distribution relative to $\mathcal{G}$ if $G(\mathbf{x}) \stackrel{d}{=} \mathbf{x}$ for all $G \in \mathcal{G}$.
- A function h defined on $\mathcal{X}$ is symmetric relative to $\mathcal{G}$ if $h(G(\mathbf{x})) = h(\mathbf{x})$ for any $\mathbf{x} \in \mathcal{X}$ and $G \in \mathcal{G}$. If the random vector $\mathbf{x}$ has a pdf f, its distribution is symmetric if, and only if, f is symmetric.
- $\mathbf{t}(\mathbf{x})$ is maximal invariant if any symmetric function h has the form $h(\mathbf{x}) = v(\mathbf{t}(\mathbf{x}))$ for some function v.
- Given $\mathbf{x}_0$, the set $\{G(\mathbf{x}_0) : G \in \mathcal{G}\}$ is called the orbit that contains $\mathbf{x}_0$. This orbit is determined by the value $\mathbf{t}(\mathbf{x}_0)$ of the maximal invariant.

Another equivalent condition for a distribution to be symmetric is that the conditional distributions of $\mathbf{x}$ given $\mathbf{t} = \mathbf{t}_0$ are uniform over the orbit $\{\mathbf{x} : \mathbf{t}(\mathbf{x}) = \mathbf{t}_0\}$.

7.4.2 Three Groups of Transformations

The following three groups appear in applications to skewed distributions:

($\mathcal{G}_{ps}$) *Polar symmetry:* It consists of the identity $G(\mathbf{x}) = \mathbf{x}$ and the reflection about the origin, i.e., $G(\mathbf{x}) = -\mathbf{x}$. The orbits are $\{\mathbf{x}, -\mathbf{x}\}$ and may be identified with the maximal invariant. In the one-dimensional case the latter may be chosen as $t = |x|$ and the orbit may be written as $\{-t, t\}$. On the other hand, the elements of $\mathcal{G}_{ps}$ correspond to multiplication of the vector $\mathbf{x}$ by 1 or by -1. Therefore $\mathcal{G}_{ps}$ is in one-to-one correspondence with the multiplicative group $\mathcal{W} = \{1, -1\}$. The symmetric distributions associated with this group are the *polar symmetric* distributions, whose class is denoted here by $\mathcal{PS}$.

($\mathcal{G}_{si}$) *Coordinate reflections*: The transformation G acts on the entries of each vector by leaving them unchanged or reversing their signs. If $\mathbf{x} = (x_1, \ldots, x_p)^T$ the maximal invariant is $\mathbf{t} = |\mathbf{x}| = (|x_1|, \ldots, |x_p|)^T$. If $\mathrm{sgn}(\mathbf{x}) = \mathbf{w} = (w_1, \ldots, w_p)^T$, where w_i is the sign of x_i, then $\mathbf{x}$ is the pointwise multiplication of $\mathbf{t}$ and $\mathbf{w}$. Thus $\mathcal{G}_{si}$ is in one-to-one correspondence with the set $\mathcal{W}^p = \{1, -1\}^p$, which is a group under pointwise multiplications. The symmetry concept associated with $\mathcal{G}_{si}$ leads to the $\mathcal{C}$-class of distributions studied in Arellano-Valle *et al.* (2002). Since the orbit has 2^p points, a natural extension of the skewing function is the probability function associated with the conditional distribution of the 2^p points satisfying $|\mathbf{x}| = \mathbf{t}$. The symmetric distributions associated with this group are the *sign invariant* distributions, whose class is denoted here by $\mathcal{SI}$.

($\mathcal{G}_{ss}$) *Spherical symmetry:* $G(\mathbf{x}) = P\mathbf{x}$, where P is an orthogonal matrix. This means essentially that there is invariance under rotations. A maximal invariant is the distance $\|\mathbf{x}\|$ of $\mathbf{x}$ to the origin, or any one-to-one function of it, like $\|\mathbf{x}\|^2$. The orbits are spheres and any point in the orbit is uniquely determined by the rotation required to move a reference point into it. Another characterization is given by the unit vector $\frac{\mathbf{x}}{\|\mathbf{x}\|}$ in the unit sphere. The symmetric distributions associated with this group are the *spherical* distributions, whose class is denoted by $\mathcal{SS}$.

In dimension 1, the groups $\mathcal{G}_{ps}, \mathcal{G}_{si}$, and $\mathcal{G}_{ss}$ are identical, and therefore $\mathcal{SS} = \mathcal{SI} = \mathcal{PS}$. In higher dimensions ($p > 1$), $\mathcal{G}_{ps} \subset \mathcal{G}_{si} \subset \mathcal{G}_{ss}$, implying that the symmetry condition is strongest for $\mathcal{G}_{ss}$ and weakest for $\mathcal{G}_{ps}$. In other words,

$$\mathcal{SS} \subset \mathcal{SI} \subset \mathcal{PS}, \text{ for } p > 1.$$

7.4.3 Conditional Representations

In these three groups, a one-to-one correspondence may be established between $\mathbf{x}$ and $(\mathbf{t}, \mathbf{w})$, where the maximal invariant $\mathbf{t}$ determines the orbit and $\mathbf{w} \in \mathcal{W}$ determines a unique point within this orbit. Furthermore, $\mathbf{w}$ is in one-to-one correspondence with a transformation G in the corresponding group. By employing the standard marginal/conditional decomposition, one gets two alternative descriptions:

(a) The marginal distribution of $\mathbf{w}$ and the conditional densities of $\mathbf{t}$ given $\mathbf{w}$, for $\mathbf{w} \in \mathcal{W}$. Symmetric distributions are characterized by $\mathbf{w} \perp\!\!\!\perp \mathbf{t}$ and a uniform distribution of $\mathbf{w}$.

(b) The marginal distribution of $\mathbf{t}$ together with either (i) the conditional probability distributions of $\mathbf{x}$ on the orbits (which are uniform for symmetric distributions) or (ii) the conditional probability distributions of $\mathbf{w}$ given $\mathbf{t}$ (which are uniform and independent of $\mathbf{t}$ for symmetric distributions). Although (i) and (ii) are equivalent, they lead to different classes of skewed distributions:

   (i) Choose an arbitrary distribution on each orbit. For the polar group $\mathcal{G}_{ps}$, each orbit consists of two points. Letting $\pi(\mathbf{x}_0) = \mathrm{P}(\mathbf{x} = \mathbf{x}_0 | \mathbf{t} = \mathbf{t}(\mathbf{x}_0))$ it follows that $\mathrm{P}(\mathbf{x} = -\mathbf{x}_0 | \mathbf{t} = \mathbf{t}(\mathbf{x}_0)) = 1 - \pi(\mathbf{x}_0)$. Then $\pi(\mathbf{x})$ can be recognized as the skewing function introduced in Genton and Loperfido (2002). The symmetry concept associated with $\mathcal{G}_{si}$ leads to the $\mathcal{C}$-class of distributions studied in Arellano-Valle *et al.* (2002) that are denoted here by $\mathcal{SI}$ and will be reviewed in Section 7.5. Since the orbit has now 2^p points, a natural extension of the skewing function would be the probability function of the conditional distribution of the 2^p points satisfying $|\mathbf{x}| = \mathbf{t}$.

   (ii) Choose a single non-uniform distribution on $\mathcal{G}$ (or on $\mathcal{W}^p$). This means that the maximal invariant and the random transformations are independent. For the spherical distributions, this conditional representation recovers the skewed distributions of Fernández *et al.* (1995).

7.4.4 Density or Probability Functions for the Maximal Invariant

The relationships between a symmetric pdf and the pdf of the absolute values, or the pdf of the distance to the origin, can be viewed as special cases of the following problem: *Express the pdf of the maximal invariant in terms of the symmetric pdf.*

When the orbits are finite sets of m elements, the pdf $\bar{f}$ of $\mathbf{t}(\mathbf{x})$ is given by

$$\bar{f}(\mathbf{t}_0(\mathbf{x})) = \sum\nolimits_{\mathbf{t}(\mathbf{x})=\mathbf{t}_0} f(\mathbf{x}),$$

and, therefore, the conditional distribution of $\mathbf{x}$ given an orbit is

$$\mathrm{P}(\mathbf{x} = \mathbf{x}_0 | \mathbf{t}(\mathbf{x}) = \mathbf{t}_0) = \frac{f(\mathbf{x}_0)}{\sum\nolimits_{\mathbf{t}(\mathbf{x})=\mathbf{t}_0} f(\mathbf{x})}.$$

For a symmetric pdf f these expressions reduce to $\bar{f}(\mathbf{t}(\mathbf{x})) = mf(\mathbf{x})$ and $\mathrm{P}(\mathbf{x} = \mathbf{x}_0 | \mathbf{t}(\mathbf{x}) = \mathbf{t}_0) = 1/m$, respectively.

To tackle a more general case, consider a given distribution $\bar{F}$ of the maximal invariant $\mathbf{t}(\mathbf{x})$. Then there is a unique symmetric distribution F_0 satisfying

$$\mathbf{s} \sim F_0 \text{ implies } \mathbf{t}(\mathbf{s}) \stackrel{d}{=} \mathbf{t}(\mathbf{x}).$$

If $\mathbf{s}$ has a pdf f_0, so will $\mathbf{t}$ (with respect to a suitable dominating measure) and

$$\bar{f}(\mathbf{t}(\mathbf{s})) = c(\mathbf{t}) f_0(\mathbf{s}).$$

The densities f_0 and $\bar{f}$ thus differ only by a multiplicative factor $c(\mathbf{t})$, which somehow represents the size of the orbit. When the orbits have m elements, as in $\mathcal{G}_{ps}$ and $\mathcal{G}_{si}$, then $c(\mathbf{t}) = m$.

For a finite group $\mathcal{G}$, the uniform probability distribution just assigns equal probabilities to its elements. For a compact group $\mathcal{G}$ there exists a normalized invariant measure, which is unique. This measure defines the uniform distribution on this group. The following proposition shows this can be applied to get symmetric distributions relative to $\mathcal{G}$.

Proposition 7.4.1 Let $\mathbf{x}$ be an arbitrary random vector with values in $\mathcal{X}$ and let $\mathcal{G}$ be a compact group of transformations on $\mathcal{X}$. If G is uniformly distributed on $\mathcal{G}$ and $G \perp\!\!\!\perp \mathbf{x}$, then $\mathbf{y} = G(\mathbf{x})$ has a symmetric distribution relative to $\mathcal{G}$.

When the orbits are not discrete, some mathematical subtleties arise, but the ideas remain similar. Let us discuss the case of spherical distributions, where the factor m is replaced by an analog of the Jacobian. Let $r = \|\mathbf{x}\|$, $t = r^2$, and let $A_p(r)$ be the $(p-1)$-dimensional area of the sphere of radius $r \in \mathbb{R}_+$. Then f and $\bar{f}$ are connected through $\bar{f}(t(\mathbf{x})) = c(t) f(\mathbf{x})$, with

$$c(t) = \frac{\pi^{\frac{p}{2}}}{\Gamma(\frac{p}{2})} t^{\frac{p}{2}-1} = \left[\frac{2\pi^{\frac{p}{2}}}{r\Gamma(\frac{p}{2})}\right] r^{p-1} = \frac{A_p(r)}{2r}.$$

Furthermore, the uniform distribution on the unit sphere is directly linked to the uniform distribution on $\mathcal{G}_{ss}$ and the conditional distribution of $\mathbf{x}$ given the orbit $\|\mathbf{x}\|^2 = t$ is determined by that of $\frac{\mathbf{x}}{\|\mathbf{x}\|}$, which is also a distribution on the unit sphere, which may depend upon t. For an elegant presentation of invariance computations the reader is referred to Eaton (1983).

7.5 The $\mathcal{SI}$ Class of Sign Invariant Distributions

7.5.1 Examples and Main Results

The class of symmetric distributions over $\mathbb{R}$, associated with the group $\mathcal{G}_{si}$ of coordinate reflections, will be denoted here by $\mathcal{SI}^p$ (from sign invariant), where p may be omitted if it is clear from the context. It turns out that $\mathcal{SI}$ is identical to the $\mathcal{C}$-class studied in Arellano-Valle *et al.* (2002) and so an application of the general results of Section 7.4 allows most of the nice properties of this family to be easily obtained. For simplicity in writing we abuse notation and write $\mathbf{x} \sim \mathcal{SI}^p$ if $\mathbf{x} \sim F$, with $F \in \mathcal{SI}^p$. The uniform distribution on $\mathcal{W}^p = \{-1, 1\}^p$ is denoted by $\mathcal{U}_p$.

Some situations where $\mathbf{x} \sim \mathcal{SI}^p$ holds are:

- The components $X_1, \ldots, X_p$ of $\mathbf{x}$ are independent and symmetric random variables.
- $\mathbf{x}_0 \in \mathcal{SI}^q, \mathbf{x}_1 \in \mathcal{SI}^{p-q}$ and $\mathbf{x}_0 \perp\!\!\!\perp \mathbf{x}_1$.
- $\mathbf{x}$ follows a spherical distribution centered at $\mathbf{0}$, with a density generator h.
- $\mathbf{x}$ follows an elliptically contoured distribution $EC_p(\boldsymbol{\mu}, \Sigma, h)$ with $\boldsymbol{\mu} = \mathbf{0}$ and Σ diagonal, where $\mathbf{x} \sim EC_p(\boldsymbol{\mu}, \Sigma, h)$ means that $\mathbf{y} = \Sigma^{-1/2}(\mathbf{x} - \boldsymbol{\mu})$ follows a spherical distribution centered at $\mathbf{0}$ and with density generator h.

The function h determines the pdf of $R = \|\mathbf{y}\|$. Linear transformations of $\mathbf{x}$, in particular the marginals, are also elliptical. The same holds for conditional distributions, but the new function h will generally depend on the fixed values of the conditioned variables.

Following the notation of Section 7.4, any $\mathbf{x} \in \mathbb{R}^p$ may be written as the pointwise multiplication of $\mathbf{t} = |\mathbf{x}| = (|x_1|, \ldots, |x_p|)^T$ and $\mathbf{w} = \text{sgn}(\mathbf{x}) = (\text{sgn}(x_1), \ldots, \text{sgn}(x_p))^T \in \mathcal{W}^p$. The vector $\mathbf{t}$ is the maximal invariant under $\mathcal{SI}^p$. Since the sets $\mathcal{G}_{si}$ and $\mathcal{W}^p$ are in one-to-one correspondence through $G_w(\mathbf{x}) = (w_1 x_1, \ldots, w_p x_p)$ and they are finite, uniformity on $\mathcal{SI}^p$ is equivalent to uniformity on $\mathcal{W}^p$. In fact, these two sets are isomorphic groups under the operation of pointwise multiplication for $\mathcal{W}^p$. This means that $G_{w_1}(G_{w_2}(\mathbf{x})) = G_{w_3}(\mathbf{x})$, where $\mathbf{w}_3$ is the pointwise multiplication of $\mathbf{w}_1$ and $\mathbf{w}_2$.

Proposition 7.5.1 The following conditions are equivalent for $\mathbf{x} \in \mathcal{SI}^p$:

(a) $|\mathbf{x}| \perp\!\!\!\perp \text{sgn}(\mathbf{x})$ and $\text{sgn}(\mathbf{x}) \sim \mathcal{U}_p$. The second condition is equivalent to $\text{sgn}(X_i) \overset{iid}{\sim} \mathcal{U}_1$, $i = 1, \ldots, p$.

(b) There exists a regular family of probability distributions of $(\mathbf{x} \mid |\mathbf{x}| = \mathbf{t}_0)$ consisting of discrete uniform distributions on $O_{\mathbf{t}_0} = \{\mathbf{x} : |\mathbf{x}| = \mathbf{t}_0\}$.

(c) There exists a positive random vector $\mathbf{t}$ and $\mathbf{w} \sim \mathcal{U}_p$, with $\mathbf{w} \perp\!\!\!\perp \mathbf{t}$, such that

$$(X_i, i = 1, \ldots, p) \overset{d}{=} (W_i T_i, i = 1, \ldots, p). \tag{7.4}$$

(d) $\mathbf{x} = G(\mathbf{t})$, where G is a random transformation on $\mathcal{G}_{si}$.

Applying Proposition 7.4.1 to the group $\mathcal{W}^p$ one gets:

Proposition 7.5.2 Let $\mathbf{s} \sim \mathcal{U}_p$ and let $\mathbf{z}$ be an arbitrary random vector with values on $\mathcal{W}^p$, with $\mathbf{z} \perp\!\!\!\perp \mathbf{s}$. Then $(Z_1 S_1, \ldots, Z_p S_p)^T \sim \mathcal{U}_p$.

The following proposition shows that the class $\mathcal{SI}$ is closed under the following: marginalization, conditioning, and disjoint linear combinations.

Proposition 7.5.3 For any $\mathbf{x} \in \mathcal{SI}^p$ and $I \subset \{1, \ldots, p\}$, with q elements, denote by $\mathbf{x}_I$ the array $(X_i : i \in I)$. Then:

(i) The marginal distribution of $\mathbf{x}_I$ belongs to $\mathcal{SI}^q$.

(ii) The conditional distribution of $(\mathbf{x}_I \mid \mathbf{x}_J)$ belongs to $\mathcal{SI}^q$ for all $J \subset I^c$.

(iii) Let $(I_i, i = 1, \ldots, r)$ be a partition of $\{1, \ldots, p\}$. Then

$$\mathbf{y} = \left(\sum\nolimits_{j \in I_i} a_{ij} X_j, i = 1, \ldots, r\right)^T \sim \mathcal{SI}^r. \tag{7.5}$$

Since each orbit $O_{\mathbf{t}_0}$ now has 2^p points, a natural extension of the skewing function would be the probability function induced by the conditional distribution of the 2^p points $\mathbf{x}$ that satisfy $|\mathbf{x}| = \mathbf{t}_0$.

7.5.2 Application to the Density Formula for a Skewed Distribution

Applying Proposition 7.3.1 with $\mathbf{r} \in A$ replaced by $\mathbf{x}_0 > \mathbf{0}$, where $\mathbf{x}_0 \in \mathcal{SI}^p$ one gets:

Proposition 7.5.4 Assume $\mathbf{x}_0 \in \mathcal{SI}^p$. Let f be the pdf of a random vector $\mathbf{z}$ in $\mathbb{R}^p$ and let g be the conditional density of $\mathbf{z}$ given $\mathbf{x}_0 > \mathbf{0}$. Then

$$g(\mathbf{z}) \;=\; 2^p\, f(\mathbf{z})\, \mathrm{P}(\mathbf{x}_0 > \mathbf{0} \mid \mathbf{z}). \tag{7.6}$$

Next, we consider a dimension reduction. When $\mathbf{x}_0 \perp\!\!\!\perp \mathbf{z}$ given $\mathbf{u}(\mathbf{z})$ holds, the conditional distribution of $\mathbf{x}_0$ given $\mathbf{z}$ depends only on $\mathbf{u}(\mathbf{z})$. A maximal reduction corresponds to a scalar $u(\mathbf{z})$, and, in particular, a linear combination $\boldsymbol{\alpha}^T \mathbf{z}$. An example is provided by the location and scale family:

Proposition 7.5.5 Let X_0 be a real symmetric random variable and assume that the conditional distributions of X_0 given $\mathbf{z}$ belong to a common location and scale family, with standardized cdf H, and with location and scale parameters $\theta_0(\mathbf{z})$ and $\theta_1(\mathbf{z})$ respectively. Then the skewing function is

$$\mathrm{P}(X_0 > 0 \mid \mathbf{z}) = H\left(\frac{\theta_0(\mathbf{z})}{\theta_1(\mathbf{z})}\right). \tag{7.7}$$

Thus $X_0 \perp\!\!\!\perp \mathbf{z}$ given $u(\mathbf{z})$ holds with

$$u(\mathbf{z}) = \frac{\theta_0(\mathbf{z})}{\theta_1(\mathbf{z})}. \tag{7.8}$$

Multiplication of X_0 by any constant factor leaves (7.7) invariant, and so there is no loss of generality in assuming that the latent random variable X_0 has a unit variance.

Proof. $\mathrm{P}(X_0 > 0|\mathbf{z}) = \mathrm{P}(\frac{X_0-\theta_0(\mathbf{z})}{\theta_1(\mathbf{z})} > \frac{0-\theta_0(\mathbf{z})}{\theta_1(\mathbf{z})}) = H(\frac{\theta_0(\mathbf{z})}{\theta_1(\mathbf{z})})$ by the symmetry of X_0. Finally, multiplication of X_0 by c generates new parameters $c\theta_0(\mathbf{z})$ and $c\theta_1(\mathbf{z})$, whose ratio equals $u(\mathbf{z})$. □

For the scalar case, let $\mathbf{u}(\mathbf{z}) = u(\mathbf{z}) \in \mathbb{R}$. In order that $\pi(\mathbf{z}) = G(u(\mathbf{z}))$ be a skewing function, the condition $\pi(u(-\mathbf{z})) = 1-\pi(u(\mathbf{z}))$ must be satisfied. If G is the cdf of a continuous symmetric distribution, this condition is equivalent to

$$u(\mathbf{z}) \text{ is an antisymmetric function.} \tag{7.9}$$

The general skew-normal distribution defined in Arellano-Valle, Gómez, and Quintana (2004b) falls into this framework, with

$$u(z) = \frac{\lambda_1 z}{\sqrt{1+\lambda_2 z^2}}.$$

For the multivariate skew-normal distribution treated by Azzalini and his collaborators, $u(\mathbf{z}) = \boldsymbol{\alpha}^T\mathbf{z}$ for some vector $\boldsymbol{\alpha}$. The antisymmetry assumption (7.9) is also used in Azzalini and Capitanio (1999), Section 7, Lemma 1, to introduce an extension to elliptical distributions; see also Genton and Loperfido (2002), Azzalini and Capitanio (2003), and Wang *et al.* (2004a).

7.6 A Stochastic Representation Associated with the $\mathcal{SI}$ Class

In the previous sections, the conditional distribution of one random variable X_1, given that another random variable X_0 was positive, was shown to be a way of generating skewed distributions. This idea is here extended to two random vectors $\mathbf{x}_1$ and $\mathbf{x}_0$, where $\mathbf{x}_0 > \mathbf{0}$ is understood to be coordinatewise. The main result is a stochastic representation of the conditional distribution under

$$\mathbf{x}_1 \perp\!\!\!\perp \mathrm{sgn}(\mathbf{x}_0) \mid |\mathbf{x}_0|, \tag{7.10}$$

a condition that is satisfied under

$$(\mathbf{x}_0^T, \mathbf{x}_1^T)^T \in \mathcal{SI}^{q+r}. \tag{7.11}$$

Proposition 7.6.1 Under (7.10), the following equivalent properties hold:

$$\left(\begin{array}{c}\mathbf{x}_0\\ \mathbf{x}_1\end{array}\right)\Bigg|\, \mathbf{x}_0 > \mathbf{0} \;\stackrel{d}{=}\; \left(\begin{array}{c}|\mathbf{x}_0|\\ \mathbf{x}_1\end{array}\right). \tag{7.12}$$

$$h(\mathbf{x}_0, \mathbf{x}_1) \mid \mathbf{x}_0 > \mathbf{0} \;\stackrel{d}{=}\; h(|\mathbf{x}_0|, \mathbf{x}_1), \text{ for any function } h. \tag{7.13}$$

7.6.1 *Moments of a Multivariate Skewed Distribution*

Let

$$\mathbf{x} = \begin{pmatrix} \mathbf{x}_0 \\ \mathbf{x}_1 \end{pmatrix}, \ \mathbf{v} = \begin{pmatrix} |\mathbf{x}_0| \\ \mathbf{x}_1 \end{pmatrix}, \ \mathbf{t} = \begin{pmatrix} |\mathbf{x}_0| \\ |\mathbf{x}| \end{pmatrix}, \ \mathbf{w} = \begin{pmatrix} \text{sgn}(\mathbf{x}_0) \\ \text{sgn}(\mathbf{x}) \end{pmatrix}, \quad (7.14)$$

where $\mathbf{x}_0$ and $\mathbf{x}_1$ are random vectors of lengths q and r, respectively, and $\mathbf{x} \in \mathcal{SI}^{q+r}$. Let $\tilde{\mathbf{z}} = A|\mathbf{x}_0| + B\mathbf{x}_1$. The moments of $\tilde{\mathbf{z}}$ can be written in terms of linear combinations of the moments of $\mathbf{v}$ in (7.14). Letting

$$c(k_{q+1}, \ldots, k_{q+r}) = \mathrm{E}\left(\prod_{i=q+1}^{q+r} W_i^{k_i}\right),$$

the independence of the signs implies that $c(k_{q+1}, \ldots, k_{q+r}) = 1$ if $k_{q+1}, \ldots, k_{q+r}$ are all even, and it is zero otherwise. On the other hand, the independence of signs and absolute values yields

$$\mathrm{E}\left(\prod_{i=1}^{q+r} V_i^{k_i}\right) = \mathrm{E}\left(\prod_{i=1}^{q+r} T_i^{k_i}\right) c(k_{q+1}, \ldots, k_{q+r}), \quad (7.15)$$

which reduces the problem to the computation of the moments of $\mathbf{t}$, making the distinction between the individual dimensions q and r irrelevant. In what follows we denote $q + r$ by p.

Since $\mathbf{x} \in \mathcal{SI}^p$, the exchangeability of $X_1, \ldots, X_p$ is equivalent to that of $T_1, \ldots, T_p$, a condition that generates some savings in computations. Denoting by $(l_i,\ i = 1, \ldots, s)$ the ordered array of the positive components of $(k_i,\ i = 1, \ldots, p)$, we have

$$\mathrm{E}\left(\prod_{i=1}^{p} T_i^{k_i}\right) = \mathrm{E}\left(\prod_{i=1}^{s} T_i^{l_i}\right).$$

When $\mathbf{x}$ has a spherical distribution, it admits the well-known stochastic representation $\mathbf{x} \overset{d}{=} R\mathbf{u}$, where $R \perp\!\!\!\perp \mathbf{u}$, $R > 0$, and $\mathbf{u}$ is uniformly distributed on the unit sphere in $\mathbb{R}^p$. It follows that $|\mathbf{x}| = R|\mathbf{u}|$, $R \perp\!\!\!\perp |\mathbf{u}|$ and

$$\mathrm{E}\left(\prod_{i=1}^{s} T_i^{l_i}\right) = \mathrm{E}\left(R^l\right) \mathrm{E}\left(\prod_{i=1}^{s} |U_i|^{l_i}\right), \quad (7.16)$$

where $l = \sum_{i=1}^{s} l_i$. Letting Q and $\mathbf{u}$ be independent, with $Q^2 \sim \chi_k^2$ and $\mathbf{u} \sim S_p(1)$, it is well-known that $\mathbf{n} = Q\mathbf{u} \sim N_p(\mathbf{0}, I_p)$ and from this

$$\mathrm{E}\left(\prod_{i=1}^{s} T_i^{l_i}\right) = \frac{\mathrm{E}(R^l)}{\mathrm{E}(Q^l)} \prod_{i=1}^{s} \mathrm{E}\left((N_i^2)^{\frac{l_i}{2}}\right).$$

Finally, $(Q^2/2) \sim Gamma(p/2, 1)$ and $(N_i^2/2) \sim Gamma(1/2, 1)$ imply that

$$\mathrm{E}\left(\prod_{i=1}^{s} T_i^{l_i}\right) = \mathrm{E}(R^l)\frac{\Gamma(\frac{k}{2})}{\Gamma(\frac{k+l}{2})}\pi^{\frac{s}{2}}\prod_{i=1}^{s}\Gamma\left(\frac{l_i+1}{2}\right). \tag{7.17}$$

For recent results on the moments of a multivariate skew-normal distribution, see Genton, He, and Liu (2001).

7.6.2 Distribution of the Square of a Skewed Random Variable

The following proposition states that the distribution of the square of a skewed random variable is identical to that of the square of the original symmetric random variable, in the context of the $\mathcal{SI}$ class.

Proposition 7.6.2 Let $\mathbf{x} \in \mathcal{SI}^p$ and consider the linear combination $Y = \sum_{i=1}^{p} a_i X_i$. Let $\tilde{Y}$ be obtained by replacing a single component, say X_r, by its absolute value $|X_r|$. Then

$$\tilde{Y}^2 \stackrel{d}{=} Y^2. \tag{7.18}$$

Proof: By symmetry one may assume $r = 1$. Then $(a_1X_1, \sum_{i=2}^{p} a_iX_i)^T \in \mathcal{SI}^2$. By a simple redefinition of the terms involved, one needs only to give a proof for $p = 2$ and $a_1 = a_2 = 1$. In other words, one must prove that under $(X_1, X_2)^T \in \mathcal{SI}^2$

$$(|X_1| + X_2)^2 \stackrel{d}{=} (X_1 + X_2)^2.$$

Expanding these two expressions, conditioning on $|X_1| = t_1$ and $|X_2| = t_2$, and using the independence of signs and absolute values, one is led to the simple comparison

$$t_1^2 + t_2^2 + 2t_1t_2W_2 \text{ vs. } t_1^2 + t_2^2 + 2t_1t_2W_1W_2.$$

The desired result follows immediately from $W_1W_2 \stackrel{d}{=} W_1 \sim \mathcal{U}_1$. □

When more than one component is replaced by its absolute value in the linear combination, the proposition no longer holds. A simple counterexample is obtained by examining the particular case $\mathbf{x} \sim \mathcal{U}_3$.

7.7 Application to the Multivariate Skew-Normal Distribution

This section applies some of the general results in the previous sections to take a fresh look at the distribution of the skewed multivariate normal distribution. The original definition of Azzalini and Dalla Valle (1996) is a natural application of the stochastic representation (7.13), which is the distribution obtained replacing X_0 by $|X_0|$ in

$$\mathbf{z} = h(X_0, \mathbf{x}_1) = \boldsymbol{\delta} X_0 + \Delta \mathbf{x}_1, \tag{7.19}$$

where

$$|\delta_j| < 1, i = 1, \dots, p, \text{ and } \Delta = \text{diag}\{(1-\delta_1)^{\frac{1}{2}}, \dots, (1-\delta_p)^{\frac{1}{2}}\}, \quad (7.20)$$

$$X_0 \sim N(0,1), \ \mathbf{x}_1 \sim N_p(\mathbf{0}, \Psi), \ \mathbf{x}_1 \perp\!\!\!\perp X_0. \quad (7.21)$$

The notation $D = \text{diag}\{d_1, \dots, d_p\}$ means that D is a $p \times p$ diagonal matrix with diagonal elements $d_1, \dots, d_p$. Applying (7.13) in one dimension, all univariate marginals are seen to be skew-normal. Quoting Azzalini and Dalla Valle (1996), the computations to get the pdf

$$g(\mathbf{z}) = 2\phi_p(\mathbf{z}; \mathbf{0}, \Omega)\ \Phi(\boldsymbol{\alpha}^T\mathbf{z}), \ \mathbf{z} \in \mathbb{R}^p, \quad (7.22)$$

are "trivial but lengthy," the second part being clear by looking at the appendix in that paper.

First of all, $\Phi(\boldsymbol{\alpha}^T\mathbf{z})$ is a valid skewing function and can be identified with $\text{P}(X_0 > 0|\mathbf{z})$. Since the marginal of $\mathbf{z}$ is already specified, the probabilistic definition is completed by specifying the conditional distributions of X_0 given $\mathbf{z}$ (or the joint distribution of X_0 and $\mathbf{z}$).

From Propositions 7.5.4 and 7.5.5, (7.22) can be identified with (7.6), under the condition (7.8). Since the conditional distributions of $\mathbf{x}_0$ given $\mathbf{z}$ are normal, with a mean linear in $\mathbf{z}$ and with constant variance, this implies that $\mathbf{x}_0$ and $\mathbf{z}$ are jointly normal.

Now write $(\mathbf{x}_0^T, \mathbf{z}^T)^T \sim N_{q+p}(\boldsymbol{\mu}, \Sigma)$, and partition Σ as

$$\Sigma_{00} = \text{Cov}\,(\mathbf{x}_0, \mathbf{x}_0), \ \Sigma_{10} = \text{Cov}\,(\mathbf{z}, \mathbf{x}_0), \text{ and } \Sigma_{11} = \text{Cov}\,(\mathbf{z}, \mathbf{z}).$$

The form of (7.22) indicates that $\boldsymbol{\mu} = \mathbf{0}$ and $\Sigma_{11} = \Omega$. In fact, $\mathbf{x}_0$ is a random variable, and so Σ_{00} is a scalar, and may be set to 1 without any loss of generality. Specifying Σ_{10} fully determines the joint distribution. In particular, $\mathbf{x}_0|\mathbf{z} \sim N(\mathbf{a}^T\mathbf{z}, \omega)$, with $\mathbf{a}^T\mathbf{z} = \Sigma_{10}^T\Sigma_{11}^{-1}\mathbf{z}$, $\omega = \Sigma_{00} - \Sigma_{10}^T\Sigma_{11}^{-1}\Sigma_{10}$. Applying this general expression with $\Sigma_{10} = \boldsymbol{\delta}$, one gets $\mathbf{a}^T\mathbf{z} = \boldsymbol{\delta}^T\Omega^{-1}\mathbf{z}$ and $\omega = \text{Var}(X_0) - \boldsymbol{\delta}^T\Omega^{-1}\boldsymbol{\delta} = 1 - \boldsymbol{\delta}^T\Omega^{-1}\boldsymbol{\delta}$. From (7.7) it follows that $\boldsymbol{\alpha} = \omega^{-1}\mathbf{a}$ and finally

$$\boldsymbol{\alpha} = \frac{1}{(1 - \boldsymbol{\delta}^T\Omega^{-1}\boldsymbol{\delta})^{\frac{1}{2}}}\,\Omega^{-1}\boldsymbol{\delta},$$

in agreement with (5) in Azzalini and Capitanio (1999).

The basic results in Azzalini and Dalla Valle (1996) may be easily obtained by applying standard results for the Bayesian linear regression model

$$\mathbf{y} = A\boldsymbol{\beta} + \boldsymbol{\epsilon}, \ \boldsymbol{\epsilon} \sim N_n(\mathbf{0}, V), \text{ with prior } \boldsymbol{\beta} \sim N_q(\mathbf{0}, B),$$

where the posterior for $\boldsymbol{\beta}$ is well-known to be

$$\boldsymbol{\beta}|\mathbf{y} \sim N_q\left((B^{-1} + A^TV^{-1}A)^{-1}A^TV^{-1}\mathbf{z},\ (B^{-1} + A^TV^{-1}A)^{-1}\right).$$

The vector $\boldsymbol{\delta}$ appearing in (2.4) of Azzalini and Dalla Valle (1996), i.e., our (7.19), coincides with Σ_{01}. Applying the general results with $n = p$,

$\mathbf{y} = \mathbf{z}$, $A = \boldsymbol{\delta}$, $\boldsymbol{\epsilon} = \mathbf{y}$, $V = \Delta\Psi\Delta$, $B = 1$, and writing $\boldsymbol{\lambda} = \Delta^{-1}\boldsymbol{\delta}$, one gets $A^T V^{-1} A = \boldsymbol{\lambda}^T \Psi^{-1} \boldsymbol{\lambda}$ and $A^T V^{-1}\mathbf{z} = \boldsymbol{\delta}^T \Delta^{-1}\Psi$, showing that

$$\boldsymbol{\alpha} = \frac{1}{(1 + \boldsymbol{\lambda}^T \Psi^{-1}\boldsymbol{\lambda})^{\frac{1}{2}}} \Delta^{-1}\Psi^{-1}\boldsymbol{\lambda},$$

in agreement with (2.4) in Azzalini and Dalla Valle (1996). When the design matrix A is diagonal, the components of $\mathbf{z}$ are independent and follow skew-normal distributions. This agrees with the model discussed in Sahu *et al.* (2003).

7.8 A Canonical Form for Skew-Elliptical Distributions

The replacement of the multivariate normal by an elliptical distribution in a number of statistical models has become quite popular. In the context of skewed distributions, this has been hinted at by Azzalini and Capitanio (1999, Section 7), investigated by Branco and Dey (2001), and there is further unpublished work by Arellano-Valle and Genton (2003) along these lines. The discussion presented here tackles the problem starting from a general formulation and showing how this may be reduced to a canonical form, which simplifies the computations and provides some geometric intuition.

As an initial point assume that the representation

$$\mathbf{x} = (\mathbf{x}_0^T, \mathbf{x}_1^T)^T \in \mathcal{SI}^{q+p}, \ \ \mathbf{z} \ = \ A\mathbf{x}_0 + B\mathbf{x}_1 \tag{7.23}$$

holds; where A and B are arbitrary matrices with dimensions $p \times q$ and $p \times p$, respectively, B is non-singular and

$$(\mathbf{x}_0^T, \mathbf{x}_1^T)^T \sim EC_{q+p}(\mathbf{0}, \Sigma, h).$$

This last condition, together with the first condition in (7.23), forces Σ to be diagonal. The arbitrariness of A and B allows the diagonal elements of Σ to be absorbed into these two matrices and, consequently, Σ may be taken as the identity matrix I_{q+p}. Thus $\mathbf{x}$ may be assumed to follow a spherical distribution.

From Proposition 7.5.4, the density of $(\mathbf{z}|\mathbf{x}_0 > \mathbf{0})$ is

$$g(\mathbf{z}) \ = \ 2^q \, f(\mathbf{z}) \, \mathrm{P}\left(\mathbf{x}_0 > \mathbf{0} \mid \mathbf{z}\right),$$

and so the problem reduces to finding the conditional density of $\mathbf{x}_0$ given $\mathbf{z}$. It is clear that this density does not determine A, B, and Σ uniquely. To simplify the computations and to avoid a lack of identifiability, it is useful to find a canonical form.

Letting $\mathbf{s} = B^{-1}\mathbf{z}$ and $C = B^{-1}A$, $\mathrm{P}(\mathbf{x}_0 > \mathbf{0}|\mathbf{z}) = \mathrm{P}(\mathbf{x}_0 > \mathbf{0}|\mathbf{s})$, where s satisfies

$$\mathbf{s} = C\mathbf{x}_0 + \mathbf{x}_1, \ \ \mathbf{x} = (\mathbf{x}_0^T, \mathbf{x}_1^T)^T \text{ spherically distributed.} \tag{7.24}$$

But this may be viewed as a Bayesian regression problem, where the prior coincides with the distribution of $\mathbf{x}_0$. The posterior distribution is also spherical, with location and dispersion respectively

$$(I + C^T C)^{-1} C^T \mathbf{s} \text{ and } (I + C^T C)^{-1}, \tag{7.25}$$

while its (conditional) generator is

$$h^{(q)}_{v(s)}(u) = \frac{h^{(p+q)}(u + v(\mathbf{s}))}{h^{(p)}(v(\mathbf{s}))}, \quad \text{with} \quad v(\mathbf{s}) = \mathbf{s}^T (I + C^T C)^{-1} \mathbf{s}, \tag{7.26}$$

where $h^{(p+q)}$ is the generator of the distribution of $\mathbf{x}$ and $h^{(p)}$ is the (marginal) generator corresponding to the distribution of $\mathbf{x}_1$.

Let r be the rank of C and consider the singular value decomposition $C = UDV^T$, where U, D, and V have orders $p \times r$, $r \times r$, and $q \times r$, respectively; U and V have orthogonal columns of length 1; and $D = \text{diag}\{d_1, \ldots, d_r\}$ is a positive diagonal matrix. Substituting this into (7.24), one gets $\mathbf{s} = UDV^T \mathbf{x}_0 + \mathbf{x}_1$. Let now N be any orthogonal matrix such that $\mathbf{y} = N\mathbf{x}$ may be partitioned as $(\mathbf{y}_0^T, \mathbf{y}_1^T)^T$, with $\mathbf{y}_0 = V^T \mathbf{x}_0$. Then $\mathbf{y} \stackrel{d}{=} \mathbf{x}$ and, in particular, $\mathbf{y}_0 \stackrel{d}{=} \mathbf{x}_0$. With this normalization, (7.24) further simplifies to

$$\mathbf{s} = UD\mathbf{y}_0 + \mathbf{y}_1.$$

From (7.25), the distribution of $(\mathbf{y}_0|\mathbf{s})$ is spherical, with location $\boldsymbol{\mu} = (I + D^2)^{-1} D\mathbf{t}$ and dispersion matrix $\Sigma = (I + D^2)^{-1}$, where $\mathbf{t} = U^T \mathbf{s}$. The corresponding entries are $\mu_i = \frac{d_i t_i}{1+d_i^2}$ and $\sigma_i^2 = \frac{1}{(1+d_i^2)}$. Then

$$\frac{\mu_i}{\sigma_i} = \frac{d_i t_i}{(1 + d_i^2)^{\frac{1}{2}}}.$$

Since the spherical posterior distribution belongs to $\mathcal{SI}^q$,

$$\mathrm{P}(Y_i > 0,\ i = q - r + 1, \ldots, q \mid Y_1 > 0, \ldots, Y_r > 0, \mathbf{t}) = 2^{-(q-r)}.$$

Hence, the original problem reduces to computing the probability

$$\begin{aligned} \mathrm{P}(Y_i > 0,\ i = 1, \ldots, r \mid \mathbf{t}) &= \mathrm{P}\left(\frac{Y_i - \mu_i}{\sigma_i} > \frac{-d_i t_i}{(1 + d_i^2)^{\frac{1}{2}}},\ i = 1, \ldots, r \mid \mathbf{t} \right) \\ &= \mathrm{P}\left(\frac{Y_i - \mu_i}{\sigma_i} \leq \frac{d_i t_i}{(1 + d_i^2)^{\frac{1}{2}}},\ i = 1, \ldots, r \mid \mathbf{t} \right), \end{aligned}$$

by symmetry. The extreme case $r = 1$ corresponds to Azzalini and Capitanio's (1999) model, while $r = q = p$ corresponds to the model discussed in Sahu *et al.* (2003). In the normal case, the components Y_i are independent, which greatly simplifies the computations.

Acknowledgments

This work has been partially financed by Fondecyt Project 1030801 and 1040865.

CHAPTER 8

Skewed Link Models for Categorical Response Data

Ming-Hui Chen

8.1 Introduction

This chapter discusses the general univariate and multivariate skewed link models for categorical response data. Both independent and correlated categorical (binary or ordinal) responses will be considered. The multivariate skewed link models are particularly useful in modeling correlated categorical response data, since they allow each binary or ordinal component to have different skewness and heaviness of tails of links, and at the same time they capture the correlation among binary and/or ordinal responses.

The analysis of binary and ordinal response data arises in a variety of disciplines such as biometrics, econometrics, and social sciences. Amemiya (1981) provided a comprehensive review on the development of qualitative response models. His review mainly focuses on the application in the field of econometrics. He summarized the properties of the link functions for univariate dichotomous models and discussed the linear relationship between the regression coefficients under different links. However, the link functions considered in his review include only the probit, logit, and linear probability link. Albert and Chib (1993) illustrated the Bayesian approach by extending the probit link to the t-link, and also incorporated these links into hierarchical models. By introducing a skewed distribution for the underlying latent variable, Chen, Dey, and Shao (1999) proposed a class of asymmetric link models for binary response data. In this chapter, we discuss and address several important issues in modeling categorical response data, such as the importance of links in fitting categorical response data, the general models with flexible parametric and nonparametric links for independent or correlated binary, ordinal, or mixed binary and ordinal, and discrete choice response data, the conditions for the properiety of the posteriors with improper uniform priors for the regression coefficients, and the general Bayesian criteria for model assessment and model diagnostics for categorical response data.

8.2 Preliminaries

Generalized linear models (see, for example, McCullagh and Nelder, 1989) allow for the treatment of regression problems in which the distribution of the response variable can be chosen as a one parameter exponential family. One of the commonly used models is the logistic regression model for modeling independent binary response data. The logit link in binomial regression gives a simple interpretation of the regression parameters, because in this case, we obtain a linear model for the natural parameter of the underlying exponential family. For example, the logit link allows for a simple representation of the odds ratio, which helps in the interpretation of the results. An excellent reference on the ordinal regression model is McCullagh (1980). Commonly used links such as the logit and probit links for binary or ordinal response data models, however, do not always provide the best fit available for a given data set. In this case, the link could be misspecified, which can yield substantial bias in the mean response estimates; see Czado and Santner (1992). In particular, when the probability of a given binary response approaches to zero at a different rate than it approaches to one, the symmetric link, such as the logit or probit, is inappropriate. The most intuitive approach to guard against such a misspecification is to embed the symmetric links, such as the logit or probit, into a wide parametric class of links. Many such parametric classes for binary response data have been proposed in the literature. Aranda-Ordaz (1981), Guerrero and Johnson (1982), Morgan (1983), and Whittmore (1983) proposed one-parameter families, while Prentice (1976), Pregibon (1980), Stukel (1988), and Czado (1992, 1994a) considered two-parameter families. Other related work includes Czado (1994b), and Basu and Mukhopadhyay (2000). Chen *et al.* (1999) introduced a new class of skewed link models for binary response data by using a latent variable approach of Albert and Chib (1993), where the underlying latent variable has a mixed effects model structure.

There is an extensive literature on methods for modeling correlated categorical (binary or ordinal) data using symmetric links. Prentice (1988) provided a comprehensive review of various modeling strategies using generalized linear regression analysis of correlated binary data with covariates associated with each binary response. Chib and Greenberg (1998) used the multivariate probit (MVP) model for correlated binary data, while Chen and Dey (1998) considered general scale mixture of multivariate normal (SMMVN) link functions for longitudinal binary responses. Moreover, Chen and Dey (2000a, 2000b) considered generalized linear models with SMMVN link functions to model correlated binary response data and ordinal response data, and Chen (1998) used the SMMVN links for modeling correlated mixed binary and ordinal response data, and the properiety of the resulting posterior distribution was studied in Chen and Shao (1999a, 1999b). In addition, Chen and Dey (2003) used multivariate logistic regression models to incorporate correlation among binary response data and investigated the role of correlation in Bayesian model comparisons in detail.

Model assessment is a crucial part of statistical analysis. Due to recent computational advances, sophisticated techniques for Bayesian model assessment are becoming increasingly popular. We have seen a surge in the statistical literature on Bayesian methods for model assessment, including George and McCulloch (1993), Madigan and Raftery (1994), Bernardo and Smith (1994), Kass and Raftery (1995), Raftery, Madigan, and Volinsky (1995), George, McCulloch, and Tsay (1996), Raftery, Madigan, and Hoeting (1997), and Clyde (1999). The scope of Bayesian model assessment is quite broad, and can be investigated via model diagnostics, goodness of fit measures, or posterior model probabilities (or Bayes factors). A comprehensive account of model diagnostics and related methods for model assessment is given in Geisser (1993) and many references therein.

Many of the proposed Bayesian methods for model comparison usually rely on posterior model probabilities or Bayes factors, and it is well-known that to use these methods, proper prior distributions are needed except on common parameters. It is usually a major task to specify prior distributions for all models under consideration, especially if the model space is large. This issue has been discussed in detail by several authors including Ibrahim and Laud (1994), Laud and Ibrahim (1995), and Chen, Ibrahim and Yiannoutsos (1999). In addition, it is well-known that Bayes factors and posterior model probabilities are generally sensitive to the choices of prior parameters, and thus one cannot simply select vague proper priors to get around the elicitation issue. Thus, computing these quantities can become a monumental chore if informative prior distributions are difficult to specify. Alternatively, criterion-based methods can be attractive in the sense that they do not require proper prior distributions in general, and thus have an advantage over posterior model probabilities in this sense. Several recent papers advocating the use of Bayesian criteria for model assessment include Ibrahim and Laud (1994), Laud and Ibrahim (1995), Gelman, Meng, and Stern (1996), Gelfand and Ghosh (1998), and Ibrahim, Chen, and Sinha (2001a). However, criterion-based methods for categorical data pose a challenge due to the discreteness of the data structure, and the properties attributable to discrete data models. In this chapter, we will consider several general Bayesian criteria for model assessment for categorical data, such as the weighted L measure, the Conditional Predictive Ordinate (CPO), and the Deviance Information Criterion (DIC) for categorical response data.

8.3 Importance of Links in Fitting Categorical Response Data

To fit categorical response data, it is important to investigate (i) whether there is a relationship between the regression coefficients under different links, and (ii) whether the choice of links is important in predicting categorical responses. We will discuss each of these two issues in detail.

8.3.1 Relationship between Regression Coefficients under Different Links

One common thought is that there is a linear relationship between the regression coefficients under two different links. If such a relationship exists, it would be convenient to use the estimates of the regression coefficients under one link to get the quick estimates of the regression coefficients under another link.

To examine this issue, we consider the bottle return data given in Neter, Kutner, Nachtsheim, and Wasserman (1996, p. 618), which was collected from a study on the effect of the size of the deposit level on the likelihood that a returnable one-liter soft-drink bottle will be returned. The number of bottles returned (Y_i) out of 500 sold (n_i) at each deposit level (x_i, in cents), $i = 1, \dots, 6$, was recorded.

To fit this data set, we consider a binomial regression model with the size of the deposit (x_i) as a covariate so that the probability for observing Y_i with deposit x_i is given by

$$f(y_i|\mathbf{x}_i, \boldsymbol{\beta}) = \binom{n_i}{y_i}[F(\mathbf{x}_i^T\boldsymbol{\beta})]^{y_i}[1 - F(\mathbf{x}_i^T\boldsymbol{\beta})]^{n_i-y_i},$$

where F is a cumulative distribution function (cdf), F^{-1} is called a link function, $\mathbf{x}_i = (1, x_i)^T$, and $\boldsymbol{\beta} = (\beta_1, \beta_2)^T$. We consider the probit, logit, and complementary log-log (C log-log) links, which are obtained by taking F to be the standard normal, logistic, and extreme value distribution functions, respectively. The maximum likelihood estimates $\hat{\boldsymbol{\beta}}$ of $\boldsymbol{\beta}$ are $(-1.2501, 0.0819)^T$, $(-2.0766, 0.1359)^T$, and $(-1.9195, 0.0935)^T$, and the resulting log likelihoods are -1530.5, -1531.4, and -1529.4 under the probit, logit, and C log-log links, respectively. Based on the log likelihood criterion, which is essentially equivalent to the AIC criterion (Akaike, 1973), the C log-log link fits the data slightly better than the other two links. Now, the question is whether those estimates of the regression coefficients are related or not.

To formally establish the relationship between regression coefficients, we consider Taylor expansion and quantile matching. Suppose we consider two links, F_1^{-1} and F_2^{-1}, corresponding to two respective cdf's F_1 and F_2. Let $\xi_j \sim F_j$. Also, let $\boldsymbol{\beta}_{F_j}$ denote the vector of regression coefficients under the link F_j for $j = 1, 2$. The Taylor expansion can be derived in the following way. Suppose $F_1(\mathbf{x}_i^T\boldsymbol{\beta}_{F_1}) = F_2(\mathbf{x}_i^T\boldsymbol{\beta}_{F_2})$. Then, we have

$\mathbf{x}_i^T\boldsymbol{\beta}_{F_1} = F_1^{-1}(F_2(\mathbf{x}_i^T\boldsymbol{\beta}_{F_2}))$. The first-order approximation gives the following relationship between the two sets of regression coefficients under F_1^{-1} and F_2^{-1} links:

$$\mathbf{x}_i^T\boldsymbol{\beta}_{F_1} \simeq F_1^{-1}(F_2(0)) + \frac{dF_1^{-1}(F_2(\xi))}{d\xi}\Big|_{\xi=0}\mathbf{x}_i^T\boldsymbol{\beta}_{F_2}.$$

When F_2^{-1} is logit and F_1^{-1} is probit, we obtain $\mathbf{x}_i^T\boldsymbol{\beta}_{F_1} \simeq 0.627\mathbf{x}_i^T\boldsymbol{\beta}_{F_2}$, while F_2^{-1} is C log-log and F_1^{-1} is logit, we have

$$\mathbf{x}_i^T\boldsymbol{\beta}_{F_1} \simeq 0.541 + 1.582\mathbf{x}_i^T\boldsymbol{\beta}_{F_2}.$$

The linear relationship between $\boldsymbol{\beta}_{F_1}$ and $\boldsymbol{\beta}_{F_2}$ can also be established by quantile matching. Let $\xi_j^{(q)} = F_j^{-1}(q)$ for $j = 1, 2$ and assume $\xi_1^{(q)} \simeq \lambda_1 + \lambda_2\xi_2^{(q)}$ for all $0 < q < 1$. Then, it can be shown that

$$\mathbf{x}_i^T\boldsymbol{\beta}_{F_1} \simeq \lambda_1 + \lambda_2\mathbf{x}_i^T\boldsymbol{\beta}_{F_2}.$$

By choosing q evenly spaced from 0.01 to 0.99, and fitting the linear regression model $\xi_1^{(q)} \simeq \lambda_1 + \lambda_2\xi_2^{(q)}$, the least squares method gives that $(\hat{\lambda}_1, \hat{\lambda}_2) = (0, 0.568)$ when F_2^{-1} is logit and F_1^{-1} is probit and $(\hat{\lambda}_1, \hat{\lambda}_2) = (0.760, 1.382)$ when F_2^{-1} is C log-log and F_1^{-1} is logit.

Now, for the bottle return data, under the logit and probit links, we obtain the ratio of the two estimated slopes (i.e., $\hat{\beta}_2$'s) is 0.0819/0.1359=0.603, which is between 0.568 and 0.627 based on quantile matching and Taylor expansion. For the estimated intercept, the linear approximations give $0.568\times(-2.0766) = -1.180$ and $0.627\times(-2.0766) = -1.302$ based on quantile matching and Taylor expansion, which are very close to -1.2501, the MLE of β_1, under the probit link. Similarly, under the C log-log and logit links, $0.1359/0.0935 = 1.453$, which is between 1.382 and 1.582 based on quantile matching and Taylor expansion, while for the estimated intercept, quantile matching gives the approximate intercept $0.760+1.382\times(-1.9195) = -1.893$ and Taylor expansion gives $0.541 + 1.582 \times (-1.9195) = -2.496$, which are close to -2.0766 under the logit link. These illustrations indicate that the relationships between the regression coefficients based on the MLE's are mostly between quantile matching and Taylor expansion. Thus, when the models under different links fit the data fairly well, the linear relationship between regression coefficients is maintained.

8.3.2 Prediction under Different Links

As discussed above, there exists, at least approximately, a relationship between the regression coefficients under two different links. Can this type of relationship warrant the goodness-of-fit of the binary or ordinal regression model under different links?

To address this issue, we first conduct a simulation study, in which the response is binary. Our data generation proceeds as follows. Step 1: we specify $\boldsymbol{\beta}$, a vector of the regression coefficients. Step 2: we select a probability p_i. Step 3: we compute $\mathbf{x}_i = (1, x_i)^T$ from a given link, say, F^{-1}, so that $\mathbf{x}_i^T\boldsymbol{\beta} = F^{-1}(p_i)$. Step 4: we generate $Y_i \sim b(n_i, p_i)$, which is the binomial distribution with p_i as the probability of "success" and n_i is a prespecified sample size for the repeated Bernoulli trials. We consider the following two simulations. Simulation I: the data are generated from the probit model with p_i's that are evenly spaced from 0.01 to 0.99; and Simulation II: the data are generated from the C log-log link model with p_i's that are evenly spaced from 0.01 to 0.99. For each simulation, we generate 500 data sets and for each dataset, we generate the binomial response variable Y_i, $i = 1, \ldots, 60$, where $Y_i \sim b(50, p_i)$. The true values of the regression coefficients are $\beta_1 = 0$ (the intercept) and $\beta_2 = 2$ (the slope) for both simulations. For each simulated data set, we fit the probit, logit, and C log-log links. Then, we compute the p-values for the Pearson's test. The definition of Pearson's test statistic can be found from, for example, Agresti (1990). Thus, the detail is omitted here for brevity. Figure 8.1 displays the boxplots of 500 p-values under each fitted link for both simulations. The results shown in Figure 8.1 are very interesting and informative. First, when the true link is the probit, which corresponds to Simulation I, there is almost no difference between the logit and probit links. However, the Pearson's test gives very small p-values for the asymmetric C log-log link model, which

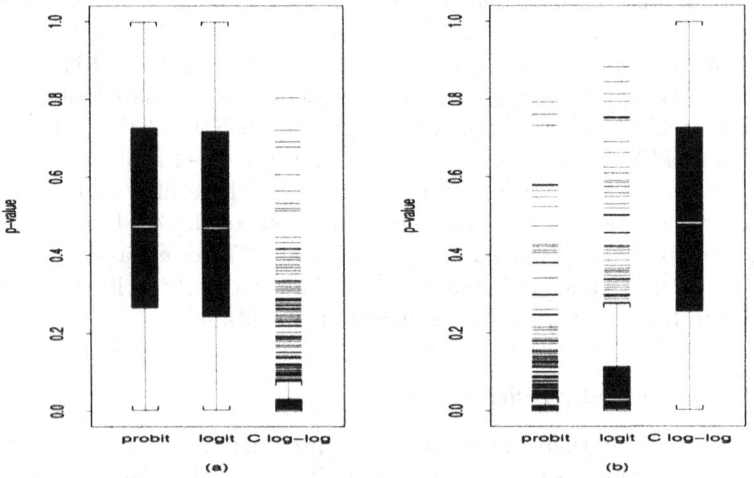

Figure 8.1 *Boxplot of p-values of the Pearson's test (plots (a) and (b) correspond to Simulations I and II).*

indicates that the asymmetric link is inadequate in fitting the data generated from the light tailed symmetric link. On the contrary, when the true link is asymmetric, both symmetric links (probit and logit) yield very small p-values, see Plot (b). But, the logit link fits the data slightly better than the probit. This simple simulation study empirically demonstrates that a wrong choice of link can lead to a poor fit.

Second, we consider the case where the response is ordinal. McCullagh and Nelder (1989) provided an excellent illustration to show the difference between skewed and symmetric links. They considered an ordinal (0-3) response along with the logit and complementary log-log links. Let $p_j = \mathrm{P}(Y = j|x, \beta)$ for $j = 0, 1, 2, 3$, where x is univariate. From Figures 5.1a and 5.1b in McCullagh and Nelder (1989, p. 152), it can be clearly seen that (i) when x is small, p_3 under the logit link is much larger than the one under the C log-log link when $\beta > 0$; and (ii) when x is large, the C log-log link yields a much larger p_3 than the logit link. Thus, a wrong choice of the links can lead to an underfit or overfit of the probability that Y falls in a particular category.

From these two simple illustrations, we can conclude that the choice of the links is important since a misspecified link can lead to a poor prediction.

8.4 General Skewed Link Models for Categorical Response Data

In this section, we discuss the general skewed link models for independent binary and/or ordinal response data, correlated binary or ordinal response data, and discrete choice data.

8.4.1 Independent Binary and/or Ordinal Regression Models

Suppose we observe a binary $(0-1)$ or ordinal $(0-(J-1))$ response from the i-th subject and let $\mathbf{x}_i = (x_{i1}, \ldots, x_{ip})^T$ be the corresponding $p \times 1$ vector of covariates for $i = 1, \ldots, n$. Note that x_{i1} may be 1, which corresponds to an intercept and also note that when $J = 2$, Y_i represents a binary response. Denote $\mathbf{y} = (Y_1, \ldots, Y_n)^T$ and assume that $Y_1, \ldots, Y_n$ are independent. Let $D = (n, \mathbf{y}, X)$ denote the observed data, where X is the $n \times p$ matrix with i-th row $\mathbf{x}_i^T$. Also let $\boldsymbol{\beta} = (\beta_1, \ldots, \beta_p)^T$ be a $p \times 1$ vector of regression coefficients. Then, we assume that Y_i takes values of 0, 1, ..., $J-1$ with probability

$$p_{ij} = F_{\boldsymbol{\delta}}(\gamma_{j+1} - \mathbf{x}_i^T\boldsymbol{\beta}) - F_{\boldsymbol{\delta}}(\gamma_j - \mathbf{x}_i^T\boldsymbol{\beta}), \tag{8.1}$$

for $j = 0, 1, \ldots, J-1$, where $F_{\boldsymbol{\delta}}$ is a cdf, which may depend on additional parameters $\boldsymbol{\delta}$, $F_{\boldsymbol{\delta}}^{-1}$ is called a link function, and $-\infty = \gamma_0 < \gamma_1 \leq \ldots \leq \gamma_{J-1} < \gamma_J = \infty$ are cutpoints dividing the real line into J intervals. Here, we assume either $\gamma_1 = 0$ or $\gamma_1 = F_{\boldsymbol{\delta}}^{-1}(0.5)$ to ensure identifiability; see Nandram and Chen (1996) or Chen and Dey (2000a) for a detailed explanation. We consider $\gamma_1 = F_{\delta}^{-1}(0.5)$ so that it always corresponds to the middle point (median) of the distribution function no matter whether F_{δ} is symmetric or asymmetric. Using (8.1), we have

$$f(y_i|\mathbf{x}_i, \boldsymbol{\beta}, \boldsymbol{\gamma}, \boldsymbol{\delta}) = F_{\boldsymbol{\delta}}(\gamma_{y_i+1} - \mathbf{x}_i^T\boldsymbol{\beta}) - F_{\boldsymbol{\delta}}(\gamma_{y_i} - \mathbf{x}_i^T\boldsymbol{\beta}), \quad (8.2)$$

where $\boldsymbol{\gamma} = (\gamma_2, \dots, \gamma_{J-1})^T$ denotes the vector of $J-2$ unknown cutpoints.

In (8.1) or (8.2), three types of links can be considered: (i) fixed link; (ii) parametric link; and (iii) nonparametric link. For (i), we simply take $F_{\boldsymbol{\delta}} = F$, where F is a known cdf. There are several commonly used fixed links, such as the probit ($F_{\boldsymbol{\delta}}(u) = \Phi(u)$, which is the standard normal $N(0,1)$ cdf), logit ($F_{\boldsymbol{\delta}}(u) = \frac{\exp(u)}{1+\exp(u)}$), C log-log link ($F_{\boldsymbol{\delta}}(u) = 1 - \exp(-\exp(u))$), and t_ν-link with a known ν ($F_{\boldsymbol{\delta}}(u) = \int_{-\infty}^{u} \frac{\Gamma(\frac{1}{2}(\nu+1))}{\sqrt{\pi\nu}\Gamma(\frac{1}{2}\nu)} \left[1 + w^2/\nu\right]^{-(\nu+1)/2} dw$). For (ii), we assume F takes the form:

$$F_{\boldsymbol{\delta}}(\xi) = \int F(\xi - \delta z) dG(z), \quad (8.3)$$

where F is the cdf of a symmetric distribution and G is the cdf of a skewed distribution. In (8.3), $-\infty < \delta < \infty$ is a skewness parameter. The model defined by (8.3) has several attractive properties. First, when either $\delta = 0$ or G is a degenerate distribution, the model in (8.3) reduces to a standard symmetric link model, since F corresponds to a symmetric distribution. Second, the skewness of $F_{\boldsymbol{\delta}}$ can be characterized by δ and G. Third, the heavy or light tailed link can be obtained by letting F be the cdf of a scale mixture of normal distributions with an unknown mixing distribution. With an unknown scale mixing distribution, the model in (8.3) is very flexible since the skewness and heaviness of the tails of the link can be governed by δ and an unknown scale mixing distribution. This is very important, since we do not know what kind of link is needed before data analysis.

Regarding the choice of G, it may be sufficient for most applications to take G to be the cdf of the half standard normal distribution with density function $g(\xi) = \frac{2}{\sqrt{2\pi}} e^{-\frac{\xi^2}{2}}$ ($\xi > 0$), or the cdf of an exponential distribution with mean of 1.

Note that the models given by (8.1) and (8.3) can be represented via latent variables. Specifically, we define

$$Y_i = j, \text{ if } \gamma_j \leq \xi_i < \gamma_{j+1}, \quad (8.4)$$

for $j = 0, 1, \dots, J-1$ and

$$\xi_i = \mathbf{x}_i^T\boldsymbol{\beta} + \delta z_i + \epsilon_i, \quad (8.5)$$

where $\epsilon_i \sim F$ and $z_i \sim G$. In the latent variable version of the general skewed link models, it is clear that $\gamma_0, \gamma_1, \dots, \gamma_{J-1}$ are cutpoints that divide the real line into J intervals. When $\mathbf{x}_i^T\boldsymbol{\beta}$ includes an intercept, it is more desirable to modify (8.5) as follows:

$$\xi_i = \mathbf{x}_i^T \boldsymbol{\beta} + \delta(z_i - \mu_G) + \epsilon_i, \tag{8.6}$$

where $\mu_G = \int z dG(z)$. From (8.6), it is easy to see that $\mathrm{E}(\xi_i) = \mathbf{x}_i^T \boldsymbol{\beta}$, and hence $\mathbf{x}_i^T \boldsymbol{\beta}$ controls the location of the latent variable ξ_i while δ purely serves as a skewness parameter.

We note that (8.4) and (8.6) are also scale invariant. To see this, we divide both sides of (8.4) and (8.6) by a scalar σ and write $\gamma_j^* = \gamma_j/\sigma$, $\xi_i^* = \xi/\sigma$, $\boldsymbol{\beta}^* = \boldsymbol{\beta}/\sigma$, δ^*/σ, and $\epsilon_i^* = \epsilon_i/\sigma$. Then, equivalent latent variable models can be obtained via the following two equations:

$$Y_i = j, \text{ if } \gamma_j^* \leq \xi_i^* < \gamma_{j+1}^*, \tag{8.7}$$

for $j = 0, 1, \ldots, J-1$ and

$$\xi_i^* = \mathbf{x}_i^T \boldsymbol{\beta}^* + \delta^*(z_i - \mu_G) + \epsilon_i^*, \tag{8.8}$$

where $\epsilon_i^* \sim F_\sigma$, $F_\sigma(u) = F(\sigma u)$, and $z_i \sim G$.

Now, suppose we consider a t_ν distribution with an unknown ν for ϵ_i. From (8.7) and (8.8), ν may primarily control the tail behavior of the link, but at the same time, it is also partially confounded with the scale of the latent variable ξ_i. That is, ν plays a dual role in the latent variable model. Thus, it is difficult to interpret the meaning of ν, and it may also cause slow convergence and poor mixing of the Gibbs sampling algorithm, which is used to sample from the posterior distribution in Bayesian inference, due to the weak identifiability of ν. To overcome such a difficulty in fitting a skewed t-link model, Bhaumik, Chen, and Dey (2004) proposed a novel *skewed generalized t-link model.* The probability density function (pdf) of the generalized t-distribution with degrees of freedom (ν_1, ν_2) is of the form:

$$f_{\nu_1,\nu_2}(u) = \frac{\Gamma\left(\frac{\nu_1+1}{2}\right)}{(\pi\nu_2)^{1/2}\Gamma\left(\frac{\nu_1}{2}\right)} \left(1 + \frac{u^2}{\nu_2}\right)^{-\frac{\nu_1+1}{2}}. \tag{8.9}$$

To fix the scaling problem in the latent variable model, we simply take $\nu_1 = \nu$ and $\nu_2 = 1$, and let $F(u) = \int_{-\infty}^{u} f_{\nu,1}(w)dw$. The skewed generalized t-link model has several advantages. First, the unknown parameter ν purely controls the heaviness of the tails of the link. Second, δ in (8.6) governs the skewness of the tails of the link. Third, both ν and δ are identifiable. Fourth, it facilitates a Gibbs sampling scheme that leads to accelerated convergence in carrying out Bayesian inference.

Finally, we briefly discuss another class of links for $F_{\boldsymbol{\delta}}$. Instead of assuming any parametric form of $F_{\boldsymbol{\delta}}$ in (8.1), the most general link can be obtained by taking a nonparametric specification of $F_{\boldsymbol{\delta}}$ through the Dirichlet process (DP) introduced by Ferguson (1973). The notable references in estimating nonparametric link functions include Newton, Czado, and Chappell (1996), Erkanli, Stangl, and Müller (1993), Basu and Mukhopadhyay

(2000), and Mukhopadhyay and Gelfand (1997). There are different approaches for considering DP mixture for categorical response models. One approach is to directly take $F_{\boldsymbol{\delta}} \sim DP(\nu F_0)$ in (8.1), where F_0 is a proper base probability distribution and ν is a precision parameter. It follows from Lo (1984) that any distribution of $\mathbb{R}^1$ can be approximated by a DP mixture of location and scale parameters of a normal distribution. Thus, a natural choice of F_0 is a normal distribution. Another way to specify $F_{\boldsymbol{\delta}}$ in (8.1) is to model $F_{\boldsymbol{\delta}}$ using a finite dimensional mixing distribution. One form of such a distribution function is a stick-breaking distribution of Ishwaran and James (2001), which can be expressed as $F_{\boldsymbol{\delta}}(\cdot) = \sum_{i=1}^{d} V_i 1_{Z_i}(\cdot)$, where d is fixed, Z_i are independent and identically distributed (iid) from F_0, and $\mathbf{v} = (V_1, \ldots, V_d)^T$ are independent of the Z_i's and distributed as a generalized Dirichlet distribution $\mathcal{GD}(\boldsymbol{\alpha}, \boldsymbol{\lambda})$ with $\boldsymbol{\alpha} = (\alpha_1, \ldots, \alpha_{d-1})^T$ and $\boldsymbol{\lambda} = (\lambda_1, \ldots, \lambda_{d-1})^T$. The definition of $\mathcal{GD}(\boldsymbol{\alpha}, \boldsymbol{\lambda})$ can be found in Ishwaran and James (2001) and thus the detail is omitted for brevity.

8.4.2 Correlated Binary and/or Ordinal Regression Models

When two or more binary or ordinal responses are taken from the same individual or subject at one time or over time, the responses are correlated within each subject and independent between subjects. There are three challenges in fitting these types of data: (i) modeling the skewness and heaviness of the tails of links; (ii) building the correlations among correlated categorical responses; and (iii) computations. One of the main difficulties for modeling multivariate categorical response data is that the skewness and heaviness of the tails of the link for one categorical response may be different than those for the others. To analyze correlated or longitudinal binary response data, Chib and Greenberg (1998) used the multivariate probit (MVP) model and Dey and Chen (2000) used the MVP and multivariate t-link models along with models proposed by Prentice (1988). For correlated ordinal response data, Chen and Dey (2000a) proposed the generalized linear models with scale mixture of multivariate normal (SM-MVN) link functions. However, the literature on the multivariate skewed link models for the correlated categorical responses is still quite sparse. In this subsection, we extend the model in (8.1) to the multivariate categorical responses.

To introduce our general multivariate link models, we need the following notation. Suppose we observe an ordinal (or binary) $(0 - (J_k - 1))$ response Y_{ik} on the i-th observations and the k-th variable and let $\mathbf{x}_{ik} = (x_{ik1}, \ldots, x_{ikp_k})^T$ be the corresponding $p_k \times 1$ vector of covariates for $i = 1, \ldots, n$ and $k = 1, \ldots, K$. Denote $\mathbf{y}_i = (Y_{i1}, \ldots, Y_{iK})^T$ and assume that $Y_{i1}, \ldots, Y_{iK}$ are dependent whereas $\mathbf{y}_1, \ldots, \mathbf{y}_n$ are independent. Let $D = (n, \mathbf{y}, X)$ denote the data, where $\mathbf{y} = (\mathbf{y}_1^T, \ldots, \mathbf{y}_n^T)^T$, $X = (X_1, \ldots, X_n)$, and $X_i = \text{diag}(\mathbf{x}_{i1}^T, \ldots, \mathbf{x}_{iK}^T)$ is a $K \times (p_1 + \ldots + p_K)$

matrix. Also let $\boldsymbol{\beta}_k = (\boldsymbol{\beta}_{k1}, \ldots, \boldsymbol{\beta}_{kp_k})^T$ be a $p_k \times 1$ vector of regression coefficients and $\boldsymbol{\beta} = (\boldsymbol{\beta}_1^T, \ldots, \boldsymbol{\beta}_K^T)^T$.

Let $A_{ik} = [\gamma_{kj} - \mathbf{x}_{ik}^T\boldsymbol{\beta}_k, \gamma_{k,j+1} - \mathbf{x}_{ik}^T\boldsymbol{\beta}_k)$ if $Y_{ik} = j$ for $j = 0, 1, \ldots, J_k - 1$ and $k = 1, \ldots, K$ and $\tilde{\mathbf{y}}_i = (Y_{i1}, \ldots, Y_{iK})^T$. Then, our general model for the multivariate categorical responses is given by

$$f(\mathbf{y}_i|X_i, \boldsymbol{\beta}, \boldsymbol{\gamma}, \boldsymbol{\delta}) = \mathrm{P}(\tilde{\mathbf{y}}_i = \mathbf{y}_i|X_i, \boldsymbol{\beta}, \boldsymbol{\gamma}, \boldsymbol{\delta}) = \int_{A_{i1} \times \cdots \times A_{iK}} dF_{\boldsymbol{\delta}}(\xi_1, \ldots, \xi_K), \tag{8.10}$$

where $F_{\boldsymbol{\delta}}$ is a K-dimensional cdf which may depend on the parameters $\boldsymbol{\delta}$, $-\infty = \gamma_{k0} < \gamma_{k1} = 0 \leq \gamma_{k2} \leq \cdots \leq \gamma_{k,J_k-1} < \gamma_{kJ_k} = \infty$ are the cutpoints, and $\boldsymbol{\gamma} = (\gamma_{12}, \ldots, \gamma_{1,J_1-1}, \ldots, \gamma_{K2}, \ldots, \gamma_{K,J_K-1})^T$ denotes the vector of unknown cutpoints.

Similar to the univariate case, (i) fixed, (ii) parametric, and (iii) nonparametric forms for $F_{\boldsymbol{\delta}}$ can be considered. Since for (i) and (iii), we can easily develop the multivariate version of the fixed and nonparametric links for a univariate binary or ordinal response, our focus here will be mainly on the development of an extension of (8.3) to the multivariate case. To this end, we consider

$$F_{\boldsymbol{\delta}}(\boldsymbol{\xi}) = \int F(\boldsymbol{\xi} - \boldsymbol{\delta}\circ(\mathbf{z} - \boldsymbol{\mu}_G)|R)dG(\mathbf{z}), \tag{8.11}$$

where $\boldsymbol{\delta} = (\delta_1, \ldots, \delta_K)^T$, $\mathbf{z} = (z_1, \ldots, z_K)^T$, $\boldsymbol{\mu}_G = (\mu_{G1}, \ldots, \mu_{GK})^T = \int \mathbf{z}dG(\mathbf{z})$, $\boldsymbol{\delta}\circ(\mathbf{z} - \boldsymbol{\mu}_G) = (\delta_1(z_1 - \mu_{G1}), \ldots, \delta_K(z_K - \mu_{GK}))^T$, and $\boldsymbol{\xi} = (\xi_1, \ldots, \xi_K)^T$. In (8.11), we take $F(\cdot|R)$ to be a K-dimensional SMMVN such that $F(\cdot|R)$ is the cdf of

$$N_K\Big(\mathbf{0}, \mathrm{diag}((s_1(\lambda_1))^{1/2}, \ldots, (s_K(\lambda_K))^{1/2}) \cdot R \cdot \mathrm{diag}((s_1(\lambda_1))^{1/2}, \ldots, (s_K(\lambda_K))^{1/2})\Big), \tag{8.12}$$

where R is a $K \times K$ correlation matrix, the λ_k are iid $\pi_k(\lambda_k)$, $s_k(\lambda_k)$ is a positive function of one-dimensional positive-valued scale mixing variable λ_k, and $\pi_k(\lambda_k)$ is a mixing distribution which is either discrete or continuous. In addition, we take $G(\mathbf{z}) = \prod_{k=1}^K G_k(z_i)$, where G_k is the cdf of a skewed distribution. We call the above proposed multivariate link as the skewed SMMVN link, which is abbreviated by SSMMVN. The SSMMVN link has several desirable features. First, it allows each binary or ordinal component to have a different skewness parameter δ_k and skewness distribution G_k. Second, each binary or ordinal component can have a different heaviness of the tails governed by $s_k(\lambda_k)$ and π_k. Third, the correlation between binary or ordinal components can be captured by the correlation matrix R. Thus, the SSMMVN link provides a great flexibility in controlling the skewness, heaviness of tails, and correlation for correlated binary and ordinal responses. In addition, after introducing latent variables $\boldsymbol{\xi}$ and $\mathbf{z}$, Gibbs sampling can be used to facilitate Bayesian computation or Monte Carlo EM for the SSMMVN models without any numerical evaluation of the multi-dimensional integrals in (8.10) and (8.11). Note that the multivariate extension of the skewed generalized t-link model can be obtained simply by taking

$$s_k(\lambda_k) = \lambda_k$$

and

$$\pi_k(\lambda_k) = \frac{1}{\Gamma(\nu_k/2)} \left(\frac{1}{2}\right)^{1/2} \lambda_k^{\frac{\nu_k}{2}-1} \exp(-\lambda_k/2).$$

Hence, the SSMMVN link models include the *multivariate skewed generalized t-link model* as a special case. Similar to (8.4) and (8.5) for the univariate skewed link model, the latent variable version of the SSMMVN link model given by (8.10) and (8.11) is given as follows:

$$Y_{ik} = j, \text{ if } \gamma_{kj} \leq \xi_{ik} < \gamma_{k,j+1}, \tag{8.13}$$

for $j = 0, 1, \ldots, J_k - 1$ and

$$\xi_{ik} = \mathbf{x}_{ik}^T \boldsymbol{\beta}_k + \delta_k(Z_{ik} - \mu_{G_k}) + \epsilon_{ik}, \tag{8.14}$$

where $Z_{ik} \sim G_k$ for $k = 1, \ldots, K$ and $\boldsymbol{\epsilon}_i = (\epsilon_{i1}, \ldots, \epsilon_{iK})^T \sim F(\cdot|R)$.

Finally, we mention that alternatively to (8.11), we can also directly model $F_{\boldsymbol{\delta}}(\boldsymbol{\xi})$ by using a multivariate skew-elliptical distribution given in Branco and Dey (2001), Sahu, Dey, and Branco (2003), and Ferreira and Steel (2003).

8.4.3 Discrete Choice Models

Discrete choice data often arise in various fields, especially in economics, transportation, marketing, health sciences, political science, psychology, and biology. See Chen (2001) for a comparative review. To model such data, we first introduce some notations. Suppose we observe a multinomial choice (1 through J) response Y_i on the i-th subject, and let $\mathbf{x}_{ij} = (x_{ij1}, \ldots, x_{ijp})^T$ be the corresponding p-dimensional vector of covariates for $j = 1, \ldots, J$ and $i = 1, \ldots, n$.

Let $\boldsymbol{\xi} = (\xi_1, \ldots, \xi_{J-1})^T$. Define $A_{ij} = \{\boldsymbol{\xi} : \ \xi_{ij} + (\mathbf{x}_{ij} - \mathbf{x}_{iJ})^T\boldsymbol{\beta} \geq \xi_{ij^*} + (\mathbf{x}_{ij^*} - \mathbf{x}_{iJ})^T\boldsymbol{\beta} \text{ for } j^* \neq j, \ \xi_{ij} > 0\}$ if $Y_i = j$ for $j < J$ and $\{\boldsymbol{\xi} : \ \xi_{ij} \leq 0, \ j = 1, \ldots, J-1\}$ if $Y_i = J$. Then, a general model for the discrete choice data is given by

$$f(y_i|\boldsymbol{\beta}, \boldsymbol{\delta}) = \mathrm{P}(Y_i = y_i|\boldsymbol{\beta}, \boldsymbol{\delta}) = \int_{A_{iy_i}} dF_{\boldsymbol{\delta}}(\boldsymbol{\xi}), \tag{8.15}$$

where $F_{\boldsymbol{\delta}}(\boldsymbol{\xi})$ is a $(J-1)$-dimensional cdf which may depend on the parameters $\boldsymbol{\delta}$. It can be easily observed that the general model in (8.15) has a great similarity to the model in (8.10) for correlated binary responses. Thus, we can directly use the SSMMVN link to model $F_{\boldsymbol{\delta}}(\boldsymbol{\xi})$ in (8.15). We note that for the model in (8.15), we can allow R in (8.11) to be a variance-covariance matrix with a fixed first diagonal element to ensure identifiability.

8.5 Bayesian Inference

In general, likelihood-based inference relies on asymptotics. In contrast, using a Bayesian approach, we can make exact inference without resorting to asymptotic calculations. Bayesian inference is also advantageous in the availability and flexibility of model building and data analysis tools. For example, in the Bayesian paradigm, model comparisons of nested or non-nested models are easily entertained via model selection criteria such as those which will be discussed in the next section. Exact computations of model selection criteria can be obtained via Markov chain Monte Carlo (MCMC) sampling. In the frequentist paradigm, there is no unified methodology for comparing non-nested models and comparisons of nested models usually require asymptotic arguments. Other data analysis tools such as predictive distributions and residuals can be more easily calculated for the models for categorical response data under the Bayesian paradigm. In addition, the Bayesian paradigm also enables us to incorporate prior information in a natural way, whereas the likelihood-based approach does not. Also, for many models, likelihood-based inference can be obtained as a special case of Bayesian inference with different types of noninformative priors such as an improper uniform prior. For example, with a uniform prior, the posterior mode corresponds to the maximum likelihood estimate.

To make Bayesian inference, we need to construct prior distributions for model parameters. In this regard, we consider both noninformative and informative priors. For noninformative priors, we consider improper uniform priors for regression coefficients $\boldsymbol{\beta}$ and cutpoints γ, proper but vague priors for link indexing parameters $\boldsymbol{\delta}$, and Jeffreys' prior for correlation matrix R. In many cancer and AIDS clinical trials, current studies often use treatments that are very similar or slight modifications of treatments used in previous studies. We refer to data arising from previous similar studies as historical data. To incorporate historical data into the prior distribution, we use the power priors proposed in Ibrahim and Chen (2000) and Chen and Shao (1999a).

The techniques developed in Chen and Shao (2001, 1999a, 1999b) and Chen *et al.* (1999) are ready to be used for characterizing the conditions for the properiety of the posterior distributions when we take noninformative priors. As an illustration, we characterize the conditions for the properiety of the posterior for the general model in (8.1) with a parametric link given by (8.3). Let $\tilde{\mathbf{x}}_i = -\mathbf{x}_i 1_{\{1 \leq Y_i \leq J-1\}}$, where the indicator function $1_{\{1 \leq Y_i \leq J-1\}} = 1$ if $1 \leq Y_i \leq J-1$ and 0 otherwise, and the vectors $\mathbf{x}_i^* = \mathbf{x}_i 1_{\{0 \leq Y_i \leq J-2\}}$, $\tilde{\mathbf{c}}_i = (1_{\{2 \leq Y_i\}}, \ldots, 1_{\{Y_i = J-1\}})^T$, $\mathbf{c}_i^* =$

$-(1_{\{1\leq Y_i\}},\ldots,1_{\{J-2\leq Y_i\}})^T 1_{\{0\leq Y_i\leq J-2\}}$, $\mathbf{g}_i = (\tilde{\mathbf{x}}_i, \mathbf{x}_i^*)^T$, $\mathbf{h}_i = (\tilde{\mathbf{c}}_i, \mathbf{c}_i^*)^T$, and the matrices $G = (\mathbf{g}_1^T,\ldots,\mathbf{g}_n^T)^T$, and $H = (\mathbf{h}_1^T,\ldots,\mathbf{h}_n^T)^T$. Then, we are led to the following proposition.

Proposition 8.5.1 Assume that the following conditions are satisfied: (i) (G, H) are of full rank; (ii) there exists a positive vector $\mathbf{b}$ such that $\mathbf{b}^T G = \mathbf{0}$ and $\mathbf{b}^T H \geq \mathbf{0}$; and (iii) the prior $\pi(\boldsymbol{\delta})$ is proper. Then, the posterior for the model in (8.1) with a parametric link given by (8.3) is proper.

The proof of Proposition 8.5.1 directly follows from Chen and Shao (2001, 1999a, 1999b). Thus, the detail is omitted here. The sufficient conditions for the other models discussed in the previous section can also be established.

8.6 Bayesian Model Assessment

In the context of model comparison, we are interested in three scenarios: (i) variable subset selection under a given link; (ii) link selection given a set of covariates; and (iii) simultaneous selection of variable subsets and links. As discussed earlier, when noninformative priors are used, criterion-based methods are more attractive. In this regard, the predictive-based weighted L measure (Chen, Dey, and Ibrahim, 2004), Conditional Predictive Ordinate (CPO) (Geisser and Eddy, 1979; Gelfand, Dey, and Chang, 1992), and the Deviance Information Criterion (DIC) (Spiegelhalter, Best, Carlin, and van der Linde, 2002) may be suitable for the categorical response data.

8.6.1 Weighted L Measure

The L measure, the quadratic loss version in particular, which was originally proposed by Ibrahim and Laud (1994) and Laud and Ibrahim (1995), and later extended by Gelfand and Ghosh (1998) and Ibrahim *et al.* (2001a), is a useful model assessment tool for continuous univariate data, multivariate data, as well as censored survival data. Consider the independent binary regression model, which is a special case of (8.2) with $J = 2$. Let $\mathbf{z} = (Z_1,\ldots,Z_n)^T$ denote a vector of future values of a replicate experiment. That is, $\mathbf{z}$ is a future response vector with the same sampling density as $[\mathbf{y}|\boldsymbol{\beta}]$. The L measure proposed by Gelfand and Ghosh (1998) and Ibrahim *et al.* (2001a) is defined as

$$L_\nu(\mathbf{y}) = \sum_{i=1}^{n} \mathrm{Var}(Z_i|\mathbf{x}_i, D) + \nu \sum_{i=1}^{n} (\mu_i - Y_i)^2, \tag{8.16}$$

where $\mu_i = \mathrm{E}(Z_i|\mathbf{x}_i, D)$, $\mathrm{Var}(Z_i|\mathbf{x}_i, D) = \mu_i(1-\mu_i)$, $0 < \nu < 1$, and the expectation is taken with respect to the predictive posterior distribution. The quantity ν plays an important role in (8.16). It can be interpreted as a weight term in the squared bias component of (8.16), and appears to have potential impact on the ordering of the models, as well as characterizing the properties of the L measure. It can be shown that the choice $\nu = 1/2$ yields attractive theoretical properties of the L measure (see Ibrahim *et al.*, 2001a).

Chen *et al.* (2004) proposed the weighted L measure, which extends the L measure in (8.16), defined as follows:

$$L_{\nu,w}(\mathbf{y}) = \sum_{i=1}^{n} w_i \text{Var}(Z_i|\mathbf{x}_i, D) + \nu \sum_{i=1}^{n} w_i(\mu_i - Y_i)^2, \tag{8.17}$$

where the weight w_i is taken to be

$$w_i = \left\{ F(w\mathbf{x}_i^T\hat{\boldsymbol{\beta}})[1 - F(w\mathbf{x}_i^T\hat{\boldsymbol{\beta}})] \right\}^{-1}, \tag{8.18}$$

for $i = 1, \ldots, n$, $0 \leq w \leq 1$, $\hat{\boldsymbol{\beta}}$ is an estimate of $\boldsymbol{\beta}$, such as the posterior mean of $\boldsymbol{\beta}$, and F is a known cdf. We see that (8.18) weighs (scales) each observation by the inverse of its sampling variance. This choice of the weight is attractive because when $w = 0$, $L_{\nu,w=0}(\mathbf{y}) = \{F(0)[1 - F(0)]\}^{-1} L_\nu(\mathbf{y})$, and hence the weighted L measure includes the unweighted L measure as a special case. In addition, w_i amounts to weighting the cases according to their covariate values. The simulation study conducted by Chen *et al.* (2004) shows that the weighted L measure outperforms the unweighted L measure in the context of variable subset selection for independent binary regression models. Based on our empirical results, a w such that $0.3 \leq w \leq 0.7$ is a desirable choice for model selection.

To develop the weighted L measure for our general multivariate skewed link models defined by (8.10) and (8.11), we first dichotomize the "ordinal" responses by defining

$$Y_{ik,j+1} = \begin{cases} 1, & \text{if } j = Y_{ik} \\ 0, & \text{if } j \neq Y_{ik}, \end{cases} \tag{8.19}$$

for $i = 1, \ldots, n$ and $k = 1, \ldots, K$. Let $\mathbf{y}_i^* = (Y_{i11}, \ldots, Y_{i1J_1}, \ldots, Y_{iK1}, \ldots, Y_{iKJ_K})^T$. Then, for each binary vector $\mathbf{y}_{ik} = (Y_{ik1}, \ldots, Y_{ikJ_k})^T$, there is only one component with a value of 1, and for the entire $\left(\sum_{k=1}^K J_k\right)$-dimensional vector, there are exactly K components with values of 1. Let $\boldsymbol{\theta}$ denote the vector of all model parameters, which may include $\boldsymbol{\beta}$, $\boldsymbol{\gamma}$, $\boldsymbol{\delta}$, $\boldsymbol{\nu} = (\nu_1, \ldots, \nu_K)^T$, and R. Denote $\Sigma(W_iX_i, \boldsymbol{\theta}^*) = \text{Var}(\mathbf{y}_i^*|W_iX_i, \boldsymbol{\theta}^*)$ to be the sampling variance-covariance matrix of $\mathbf{y}_i^*$ given W_iX_i and $\boldsymbol{\theta} = \boldsymbol{\theta}^*$, where $W_i = \text{diag}(w_{i1}, \ldots, w_{iK})$. This construction of the weight matrix W_i allows each vector of covariates $\mathbf{x}_{ik}$ to be weighted differently for each component of binary or ordinal responses and for each observation. Of course, the simplest version of the weight matrix is $W_i = wI_K$, which may be sufficient for most applications. It is easy to show that the mean of Y_{ikj}

given $\mathbf{x}_{ik}$ and $\boldsymbol{\theta}$ is

$$\mu_{ikj}(\mathbf{x}_{ik}|\boldsymbol{\theta}) = \mathrm{E}(Y_{ikj}|\mathbf{x}_{ik}, \boldsymbol{\theta}) = \int_{A_{ikj}} dF_{\boldsymbol{\delta}}(\xi_k), \tag{8.20}$$

where $A_{ikj} = [\gamma_{kj} - \mathbf{x}_{ik}^T\boldsymbol{\beta}_k, \gamma_{k,j+1} - \mathbf{x}_{ik}^T\boldsymbol{\beta}_k)$, the variance of Y_{ikj} given $\mathbf{x}_{ik}$ and $\boldsymbol{\theta}$ is

$$\mathrm{Var}(Y_{ikj}|\mathbf{x}_{ik}, \boldsymbol{\theta}) = \mu_{ikj}(\mathbf{x}_{ik}|\boldsymbol{\theta})[1 - \mu_{ikj}(\mathbf{x}_{ik}|\boldsymbol{\theta})], \tag{8.21}$$

the covariance of Y_{ikj} and $Y_{ikj'}$ for $j \neq j'$ given $\mathbf{x}_{ik}$ and $\boldsymbol{\theta}$ is

$$\mathrm{Cov}(Y_{ikj}, Y_{ikj'}|\mathbf{x}_{ik}, \boldsymbol{\theta}) = -\mu_{ikj}(\mathbf{x}_{ik}|\boldsymbol{\theta})\mu_{ikj'}(\mathbf{x}_{ik}|\boldsymbol{\theta}), \tag{8.22}$$

and the covariance of Y_{ikj} and $Y_{ik'j'}$ for $k \neq k'$ given $\mathbf{x}_{ik}$, $\mathbf{x}_{ik'}$, and $\boldsymbol{\theta}$ is

$$\begin{aligned} &\mathrm{Cov}(Y_{ikj}, Y_{ik'j'}|\mathbf{x}_{ik}, \boldsymbol{\theta}) \\ &= \int_{A_{ikj}\times A_{ik'j'}} dF_{\boldsymbol{\delta}}(\xi_k, \xi_{k'}) - \mu_{ikj}(\mathbf{x}_{ik}|\boldsymbol{\theta})\mu_{ik'j'}(\mathbf{x}_{ik'}|\boldsymbol{\theta}). \end{aligned} \tag{8.23}$$

The variance-covariance matrix $\Sigma(W_iX_i, \boldsymbol{\theta}^*)$ can now be computed via (8.20), (8.21), (8.22), and (8.23). Since $\mathbf{y}_i^*$ is the dichotomized vector of ordinal responses, it can be shown that the variance-covariance matrix $\Sigma(W_iX_i, \boldsymbol{\theta}^*)$ is of rank $\sum_{k=1}^K (J_k - 1)$. To overcome this difficulty, we need a dimension reduction technique. To this end, let $B = \mathrm{diag}(B_1, \ldots, B_K)$, where B_k is any $(J_k - 1) \times J_k$ constant matrix of rank $J_k - 1$. Then,

$$\mathrm{Var}(B\mathbf{y}_i^*|W_iX_i, \boldsymbol{\theta}^*) = B\Sigma(W_iX_i, \boldsymbol{\theta}^*)B^T,$$

is of full rank. Following Chen *et al.* (2004), the weighted L measure for the SSMMVN link models under the weighted quadratic loss is given by

$$\begin{aligned} L_{\nu,W}^{SSMMVN}(\mathbf{y}, B) = \sum_{i=1}^n \Big\{ &\mathrm{E}\left[\mathrm{tr}\left(\left(B\Sigma(W_iX_i, \boldsymbol{\theta}^*)B^T\right)^{-1} B\Sigma(X_i, \boldsymbol{\theta})B^T\right)|D\right] \\ &+ \mathrm{E}\left[(B\boldsymbol{\mu}_i(\boldsymbol{\theta}))^T(B\Sigma(W_iX_i, \boldsymbol{\theta}^*)B^T)^{-1}B\boldsymbol{\mu}_i(\boldsymbol{\theta})|D\right] \\ &- \left[(\mathrm{E}\left[B\boldsymbol{\mu}_i(\boldsymbol{\theta})|D\right])^T (B\Sigma(W_iX_i, \boldsymbol{\theta}^*)B^T)^{-1}\mathrm{E}\left[B\boldsymbol{\mu}_i(\boldsymbol{\theta})|D\right]\right] \\ &+ \nu\Big[(\mathrm{E}\left[B\boldsymbol{\mu}_i(\boldsymbol{\theta})|D\right] - B\mathbf{y}_i^*)^T (B\Sigma(W_iX_i, \boldsymbol{\theta}^*)B^T)^{-1} \\ &\quad \times (\mathrm{E}\left[B\boldsymbol{\mu}_i(\boldsymbol{\theta})|D\right] - B\mathbf{y}_i^*)\Big]\Big\}, \end{aligned} \tag{8.24}$$

where the expectation is taken with respect to the posterior distribution of $\boldsymbol{\theta}$, $\boldsymbol{\mu}_i(\boldsymbol{\theta}) = (\mu_{i11}(\mathbf{x}_{i1}|\boldsymbol{\theta}), \ldots, \mu_{i1J_1}(\mathbf{x}_{i1}|\boldsymbol{\theta}), \ldots, \mu_{iK1}(\mathbf{x}_{iK}|\boldsymbol{\theta}), \ldots, \mu_{iKJ_K}(\mathbf{x}_{iK}|\boldsymbol{\theta}))^T$, $W = (W_1, \ldots, W_n)^T$, $\Sigma(X_i, \boldsymbol{\theta})$ is similar to $\Sigma(W_iX_i, \boldsymbol{\theta}^*)$ by replacing W_iX_i and $\boldsymbol{\theta}^*$ with X_i and $\boldsymbol{\theta}$, $\mathbf{y}_i^*$ is the observed value of the dichotomized binary vector, and $\boldsymbol{\theta}^*$ can be the posterior mean of $\boldsymbol{\theta}$. From (8.24), it appears that $L_{\nu,W}^{SSMMVN}(\mathbf{y}, B)$ depends on the choice of B. But, $L_{\nu,W}^{SSMMVN}(\mathbf{y}, B)$ is indeed invariant in B. We formally state this result in the next proposition.

Proposition 8.6.1 Let B_k^* be any $(J_k - 1) \times J_k$ constant matrix of rank $J_k - 1$ and write $B^* = \text{diag}(B_1^*, \ldots, B_K^*)$. Then,

$$\text{Var}(B^* \mathbf{y}_i^* | W_i X_i, \boldsymbol{\theta}^*) = B^* \Sigma(W_i X_i, \boldsymbol{\theta}^*) B^{*T}$$

is of full rank and

$$L_{\nu,W}^{SSMMVN}(\mathbf{y}, B) = L_{\nu,W}^{SSMMVN}(\mathbf{y}, B^*).$$

Similar to Chen *et al.* (2004), the proof of Proposition 8.6.1 directly follows from the matrix algebra and the detail is thus omitted. It can be seen from (8.20), (8.21), (8.22), and (8.23) that the computation of $L_{\nu,W}^{SSMMVN}(\mathbf{y}, B)$ involves at most two-dimensional integrals and in the class of SSMMVN links, those two-dimensional integrals can be written as the functions of bivariate normal cdf's. Thus, $L_{\nu,W}^{SSMMVN}(\mathbf{y}, B)$ is an attractive model assessment tool for multivariate categorical response data.

8.6.2 *Conditional Predictive Ordinate*

The Conditional Predictive Ordinate (CPO) statistic is a very useful model assessment tool that has been widely used in the statistical literature under various contexts. For a detailed discussion of the CPO statistic and its applications to model assessment, see Geisser (1993), Gelfand *et al.* (1992), and Dey, Chen, and Chang (1997). Here, we mainly focus on the development of CPO for the general model given by (8.10). Analogous to leave-one-out cross-validation methods, for the i-th observation, the CPO statistic is defined as

$$\text{CPO}_i = f(\mathbf{y}_i | X_i, D^{(-i)}) = \int f(\mathbf{y}_i | X_i, \boldsymbol{\beta}, \boldsymbol{\gamma}, \boldsymbol{\delta}) \pi(\boldsymbol{\theta} | D^{(-i)}) \, d\boldsymbol{\theta}, \qquad (8.25)$$

where $f(\mathbf{y}_i | X_i, \boldsymbol{\beta}, \boldsymbol{\gamma}, \boldsymbol{\delta})$ is defined in (8.10), $\boldsymbol{\theta}$ is the vector of all unknown model parameters, $D^{(-i)}$ denotes the data with the i-th case deleted, and $\pi(\boldsymbol{\theta} | D^{(-i)})$ is the posterior density of $\boldsymbol{\theta}$ based on the data $D^{(-i)}$. From (8.25), we see that CPO_i is the marginal posterior predictive density of $\mathbf{y}_i$ given X_i and $D^{(-i)}$, and can be interpreted as the height of this marginal density at $\mathbf{y}_i$. Thus, large values of CPO_i imply a better fit of the model.

For comparing two competing models, we examine the CPO_i's under both models. The observation with a larger CPO value under one model will support that model over the other. Therefore, a plot of CPO_i's under both models against observation number should reveal that the better model has the majority of its CPO_i's above those of the poorer fitting model. In comparing several competing models, the CPO_i values under all models can be plotted against the observation number in a single graph.

An alternative to CPO plots is the summary statistic called the logarithm of the pseudo-marginal likelihood (LPML), see Geisser and Eddy (1979), defined as

$$\text{LPML} = \sum_{i=1}^{n} \log(\text{CPO}_i). \tag{8.26}$$

We select the model which yields the largest LPML.

As discussed in Chen, Shao, and Ibrahim (2000), CPO_i can be computed via the following identity:

$$\begin{aligned}\text{CPO}_i =& f(\mathbf{y}_i|X_i, D^{(-i)}) \\ =& \left\{ \int \frac{1}{f(\mathbf{y}_i|X_i, \boldsymbol{\beta}, \boldsymbol{\gamma}, \boldsymbol{\delta})} \pi(\boldsymbol{\theta}|D) d\boldsymbol{\theta} \right\}^{-1},\end{aligned} \tag{8.27}$$

where $\pi(\boldsymbol{\theta}|D)$ is the posterior density of $\boldsymbol{\theta}$ based on the whole data D. Therefore, CPO_i can be computed using an MCMC sample of $\boldsymbol{\theta}$ from the joint posterior $\pi(\boldsymbol{\theta}|D)$. Compared to the weighted L measure, CPO_i is computationally more expensive, since for each MCMC sample value of $(\boldsymbol{\beta}, \boldsymbol{\gamma}, \boldsymbol{\delta})$, it is required to evaluate $f(\mathbf{y}_i|X_i, \boldsymbol{\beta}, \boldsymbol{\gamma}, \boldsymbol{\delta})$. Due to the complexity of the general model (8.10), an analytical evaluation of $f(\mathbf{y}_i|X_i, \boldsymbol{\beta}, \boldsymbol{\gamma}, \boldsymbol{\delta})$ does not appear possible. However, the efficient Monte Carlo methods developed by Chen and Dey (2000a) and Chen and Shao (1998) may be used for computing $f(\mathbf{y}_i|X_i, \boldsymbol{\beta}, \boldsymbol{\gamma}, \boldsymbol{\delta})$.

8.6.3 Deviance Information Criterion

Again let $\boldsymbol{\theta}$ denote the vector of all unknown model parameters. The criterion DIC, proposed by Spiegelhalter *et al.* (2002), is given by

$$\text{DIC} = D(\bar{\boldsymbol{\theta}}) + 2p_D, \tag{8.28}$$

where $p_D = \overline{D(\boldsymbol{\theta})} - D(\bar{\boldsymbol{\theta}})$, $\bar{\boldsymbol{\theta}} = \text{E}(\boldsymbol{\theta}|D)$, and $\overline{D(\boldsymbol{\theta})} = \text{E}(D(\boldsymbol{\theta})|D)$.

For the general model given by (8.10), where $f(\mathbf{y}_i|X_i, \boldsymbol{\beta}, \boldsymbol{\gamma}, \boldsymbol{\delta})$ is defined in (8.10), we may take $D(\boldsymbol{\theta})$ to be the unstandardized deviance, defined as

$$D(\boldsymbol{\theta}) = -2 \sum_{i=1}^{n} \log f(\mathbf{y}_i|X_i, \boldsymbol{\beta}, \boldsymbol{\gamma}, \boldsymbol{\delta}). \tag{8.29}$$

Similar to CPO_i and LPML, to compute DIC, we need to evaluate $f(\mathbf{y}_i|X_i, \boldsymbol{\beta}, \boldsymbol{\gamma}, \boldsymbol{\delta})$, which is computationally expensive.

8.7 Bayesian Model Diagnostics and Outlier Detection

Model diagnostics and outlier detection are another aspect of model assessment. In this regard, we discuss three Bayesian diagnostic methods.

8.7.1 Bayesian Latent Residuals

Albert and Chib (1995) proposed Bayesian latent residuals for binary response regression models. Chen and Dey (2000a) generalized the univariate Bayesian residuals of Albert and Chib (1995) for correlated ordinal data. Following Chen and Dey (2000a), and using (8.13) and (8.14) for the SSMMVN link models, we define the Bayesian latent residuals as

$$\epsilon_{ik}^* = \frac{\xi_{ik} - \mu_{ik}}{\sigma_{ik}}, \tag{8.30}$$

where $\mu_{ik} = \mathrm{E}(\xi_{ik}|D)$ and $\sigma_{ik}^2 = \mathrm{Var}(\xi_{ik}|D)$, that is, μ_{ik} and σ_{ik}^2 are the posterior mean and variance of ξ_{ik}, for $k = 1, \ldots, K$, $i = 1, \ldots, n$. Note that μ_{ik} and σ_{ik} can simply be calculated by using the readily available MCMC samples of ξ_{ik}'s generated by the MCMC sampling algorithm. Therefore, no additional MCMC samples are needed in order to obtain the latent residuals ϵ_{ik}^*'s.

Based on the Bayesian latent residuals ϵ_{ik}'s, we can use boxplots of the posterior distributions of the ϵ_{ik}^*'s to detect outlying observations. Alternatively, we can calculate $\mathrm{P}(|\epsilon_{ik}^*| \geq K^*|D)$ and plot $\mathrm{P}(|\epsilon_{ik}^*| \geq K^*|D)$ versus $\mathrm{E}(\mathbf{x}_{ik}^T\boldsymbol{\beta}_k|D)$, where the expectation is taken with respect to the posterior distribution $\pi(\boldsymbol{\theta}|D)$. Following Albert and Chib (1995), $K^* = 2$ may be sufficient for most applications. Bayesian latent residuals are advantageous due to their computational simplicity.

8.7.2 Bayesian CPO-Based Residuals

For the SSMMVN link models given by (8.10) and (8.11), following the notations used in Section 8.6.1, we develop the Bayesian residual via the dichotomized vector $\mathbf{y}_i^*$ of $\mathbf{y}_i$. Let $\mathbf{z}_i^*$ denote a vector of the future value of a replicate experiment. That is, $\mathbf{z}_i^*$ has the same sampling distribution as $\mathbf{y}_i^*$. Then, the Bayesian residual is defined as

$$\mathbf{d}_i = \left[\mathrm{Var}(B\mathbf{z}_i^*|X_i, D^{(-i)})\right]^{-1/2} [B\mathbf{y}_i^* - \mathrm{E}(B\mathbf{z}_i^*|X_i, D^{(-i)})], \tag{8.31}$$

where the variance and expectation are taken with respect to the posterior predictive distribution given the data $D^{(-i)}$, and $\mathbf{y}_i^*$ is the observed value of the dichotomized vector. Again, let $\boldsymbol{\theta}$ denote the vector of all unknown model parameters. Then, we have

$$\mathrm{E}(B\mathbf{z}_i^*|X_i, D^{(-i)}) = \mathrm{CPO}_i B \int \frac{\boldsymbol{\mu}_i(\boldsymbol{\theta})}{f(\mathbf{y}_i|X_i, \boldsymbol{\theta})} \pi(\boldsymbol{\theta}|D) d\boldsymbol{\theta}, \tag{8.32}$$

where $\boldsymbol{\mu}_i(\boldsymbol{\theta})$ and $f(\mathbf{y}_i|X_i,\boldsymbol{\theta}) = f(\mathbf{y}_i|X_i,\boldsymbol{\beta},\boldsymbol{\gamma},\boldsymbol{\delta})$ are defined in (8.24) and (8.10), respectively, and CPO_i is given by (8.25). Similarly, we obtain

$$\begin{aligned}&\text{Var}(B\mathbf{z}_i^*|X_i,D^{(-i)})\\ =&\text{CPO}_i B\Big\{\text{E}\Big[\frac{\Sigma(X_i,\boldsymbol{\theta})}{f(\mathbf{y}_i|X_i,\boldsymbol{\theta})}\Big|D\Big]+\text{E}\Big[\frac{\boldsymbol{\mu}_i(\boldsymbol{\theta})\boldsymbol{\mu}_i^T(\boldsymbol{\theta})}{f(\mathbf{y}_i|X_i,\boldsymbol{\theta})}\Big|D\Big]\\ &-\text{CPO}_i\text{E}\Big(\frac{\boldsymbol{\mu}_i(\boldsymbol{\theta})}{f(\mathbf{y}_i|X_i,\boldsymbol{\theta})}\Big|D\Big)\text{E}\Big(\frac{\boldsymbol{\mu}_i(\boldsymbol{\theta})}{f(\mathbf{y}_i|X_i,\boldsymbol{\theta})}\Big|D\Big)^T\Big\}B^T,\end{aligned} \tag{8.33}$$

where $\Sigma(X_i,\boldsymbol{\theta})$ is defined in (8.24). As discussed in Chen and Deely (1996), the Bayesian residual defined by (8.31) is analogous to the Studentized residual. However, $\mathbf{d}_i$ is no longer invariant in B. With the sacrifice of the sign of the residual, we can modify (8.31) as follows:

$$\begin{aligned}d_i^* =&\Big\{[B\mathbf{y}_i^*-\text{E}(B\mathbf{z}_i^*|X_i,D^{(-i)})]^T\left[\text{Var}(B\mathbf{z}_i^*|X_i,D^{(-i)})\right]^{-1}\\ &\times[B\mathbf{y}_i^*-\text{E}(B\mathbf{z}_i^*|X_i,D^{(-i)})]\Big\}^{1/2}.\end{aligned} \tag{8.34}$$

Then, it can be shown that d_i^* is invariant in B. The Bayesian residual $\mathbf{d}_i$, or d_i^*, is attractive for assessing the fit of the SSMMVN link model to the correlated categorical data. A large value of d_i^* indicates a poor fit. However, both $\mathbf{d}_i$ and d_i^* are expensive to compute, which may be a drawback of this method.

8.7.3 Observationwise Weighted L Measure

For the independent binary response model, we let

$$L_{\nu,w,i}(\mathbf{y}) = w_i\text{Var}(Z_i|\mathbf{x}_i,D)+\nu w_i(\mu_i-Y_i)^2$$

which is the i-th term inside of the summation in (8.17). For the SSMMVN link models, we write

$$\begin{aligned}L_{\nu,W,i}^{SSMMVN}(\mathbf{y},B) =& \text{E}\left[\text{tr}\left(\left(B\Sigma(W_iX_i,\boldsymbol{\theta}^*)B^T\right)^{-1}B\Sigma(X_i,\boldsymbol{\theta})B^T\right)|D\right]\\ &+\text{E}\left[(B\boldsymbol{\mu}_i(\boldsymbol{\theta}))^T(B\Sigma(W_iX_i,\boldsymbol{\theta}^*)B^T)^{-1}B\boldsymbol{\mu}_i(\boldsymbol{\theta})|D\right]\\ &-\left[(\text{E}\left[B\boldsymbol{\mu}_i(\boldsymbol{\theta})|D\right])^T(B\Sigma(W_iX_i,\boldsymbol{\theta}^*)B^T)^{-1}\text{E}\left[B\boldsymbol{\mu}_i(\boldsymbol{\theta})|D\right]\right]\\ &+\nu\Big[(\text{E}\left[B\boldsymbol{\mu}_i(\boldsymbol{\theta})|D\right]-B\mathbf{y}_i^*)^T(B\Sigma(W_iX_i,\boldsymbol{\theta}^*)B^T)^{-1}\\ &\times(\text{E}\left[B\boldsymbol{\mu}_i(\boldsymbol{\theta})|D\right]-B\mathbf{y}_i^*)\Big],\end{aligned}$$

which is the i-th term inside of the summation in (8.24). Then, $L_{\nu,w,i}(\mathbf{y})$ or $L_{\nu,W,i}^{SSMMVN}(\mathbf{y},B)$ describes how much the i-th observation supports the model. A large value of $L_{\nu,w,i}(\mathbf{y})$ or $L^{SSMMVN}_{\nu,W,i}(\mathbf{y},B)$ is an indication of an abnormal or outlying observation. This approach is attractive, since it captures the observationwise predictive variation and biasedness. In addition, both $L_{\nu,w,i}(\mathbf{y})$ and $L^{SSMMVN}_{\nu,W,i}(\mathbf{y},B)$ are very easy to compute. Thus, they are potentially very useful in analyzing categorical response data.

8.8 Concluding Remarks

To analyze categorical response data, the choice of links is important. As discussed in Chen *et al.* (1999), the symmetric links may be inadequate to fit the data when the probability of a given binary response approaches to zero at a different rate than it approaches to one. Thus, the problems of interest are how (a) to explore the relationship between various covariates and the categorical outcome measure, (b) to identify the important covariates, (c) to estimate quantities of interest, e.g., multinomial probabilities or a function of multinomial probabilities, (d) to identify which model fits the data best, and (e) to assess the goodness of fit.

In this chapter, we have developed the general skewed link models for independent and correlated binary and/or ordinal response data, and discrete choice data. The SSMMVN link models are attractive and useful for multivariate categorical response data as these models provide a great flexibility in controlling the skewness, heaviness of tails, and correlation for correlated binary and ordinal responses. In addition, we have developed several general Bayesian criteria for model assessment and several methods for model diagnostics for multivariate categorical response data. The extension of the SSMMVN links models can also be obtained in order to fit dynamic correlated binary or multinomial data, e.g., Fahrmeir and Tutz (1997), Bhaumik *et al.* (2004).

CHAPTER 9

Skew-Elliptical Distributions in Bayesian Inference

Brunero Liseo

9.1 Introduction

In this chapter we review recent developments in inference procedures based on skew-elliptical (*SE*, henceforth) distributions and discuss their impact in Bayesian analysis.

Non-symmetric distributions can be exploited in Bayesian inference both for modeling prior beliefs and for theoretical representation of skewed observables. In the former case, these distributions may represent an honest mathematical formalization of genuine asymmetric information about a vector of parameters of interest; also, they arise in the context of hierarchical linear models when the elicitation process on the location parameters involves linear constraints (O'Hagan and Leonard, 1976; Liseo and Loperfido, 2003a). In the latter case, *SE* distributions play the same role as they do in classical inference. However, in a frequentist or likelihood setup, inferential procedures for *SE* classes of densities usually pose challenging problems; as a result, it is not surprising that the literature on skewed distributions has experienced an explosion of papers dealing with distributional properties of these families while devoting minor attention to inferential problems.

On the other hand, the use of Monte Carlo Markov Chains algorithms makes the Bayesian approach to *SE* distributions quite appealing, so that the use of this class of distributions has become popular in Bayesian linear regression, generalized linear models, time series, and survival analysis (Sahu, Dey, and Branco, 2003).

Finally, the use of *SE* classes of densities poses interesting questions in the objective Bayes methodology. For example, Liseo and Loperfido (2003b) have shown that, in the skew-normal case, Jeffreys' prior for the skewness parameter is proper: this represents the second example in the literature of a proper default prior for a parameter with unbounded support; see Berger, Oliveira, and Sansó (2001) for another example.

The chapter is organized as follows. In Section 9.2, we illustrate the use of skewed distributions to describe prior beliefs about location parameters in different contexts. We provide closed-form expressions for the posterior distributions of the parameters of interest and some alternative Monte Carlo Markov Chains strategies for inference. In Section 9.3, we assume that our data come from some skew-elliptical distributions and illustrate how to implement simulation-based Bayesian inference techniques for independent and identically distributed (iid) data and time series. Finally, Section 9.4 is devoted to the illustration of objective Bayesian methods with skew-normal distributions.

9.2 Skewed Prior Distributions for Location Parameters

This section illustrates the role of skewed distributions in modeling prior beliefs about the parameters in a statistical model. We confine ourselves to the analysis of location parameters; nothing in the mathematical nature of the skewed distributions prevents us from using them in more general scenarios. However, they play a specific role when one wants to stress the asymmetric nature of the information about some quantity that is "symmetric," at least in a topological sense. We discuss the use of skewed priors in two examples: hierarchical linear models with linear constraints (see Section 9.2.1) and robustness analysis of the Bayes risk of linear Bayes rules (see Section 9.2.2).

Up to now, mainly for computational reasons, much of the use of skewed priors has been restricted to the skew-normal distribution and applications of more general skew-elliptical priors have been quite rare: see Section 9.2.3 for more on this issue.

9.2.1 Hierarchical Models with Linear Constraints

Consider the following simple scenario (O'Hagan and Leonard, 1976). Suppose we have an iid sample from a Gaussian distribution $N(\theta, \psi^2)$ with ψ^2 known, and we need to elicit a prior on θ. We use a hierarchical conjugate prior, that is,

- $\theta \mid \theta_0 \sim N(\theta_0, \sigma^2)$,
- $\theta_0 \sim N(\mu, \omega^2)$.

Moreover, we assume that, at the highest level of the hierarchy, there is a linear constraint, e.g., $\theta_0 > \mu$. This context can arise quite commonly in practice: instead of cutting the parameter space, excluding *a priori* some values for θ, one postpones such a cut to the highest level of the hierarchical model, this way penalizing, to a lesser extent, small values of the location parameter.

Standard algebra shows that the marginal prior distribution of θ is then skew-normal; more precisely

$$\theta \sim SN\left(\mu, \sqrt{\sigma^2 + \omega^2}, \frac{\omega}{\sigma}\right).$$

The skewing factor depends on the ratio of the two scale parameters: the larger is ω^2, the variance of θ_0, compared to σ^2, the variance of θ, the larger is the probability that θ_0 (the mean of θ) assumes large values and the larger is the amount of skewness induced in the law of θ.

Many generalizations of this result are possible: one can consider a multivariate scenario, and a more general system of linear constraints can be assumed on the hyperparameter. Liseo and Loperfido (2003a) consider the following framework: suppose we observe data from a random vector $\mathbf{x} \sim N_p(\boldsymbol{\theta}, \Psi)$, where Ψ is assumed to be known; conditionally on a hyperparameter $\boldsymbol{\theta}_0$, the prior distribution on $\boldsymbol{\theta}$ is also multivariate normal with mean $\boldsymbol{\theta}_0$ and known covariance matrix Σ. At the third level of the hierarchy, $\boldsymbol{\theta}_0$ follows a $N_p(\boldsymbol{\mu}, \Omega)$ distribution, with the additional prior information on $\boldsymbol{\theta}_0$ which comes in the form of k $(1 \leq k \leq p)$ linear constraints. These can be formulated in a matrix form as

$$C\boldsymbol{\theta}_0 + \mathbf{d} \leq \mathbf{0},$$

where C is a full rank $k \times p$ matrix and $\mathbf{d} \in \mathbb{R}^k$. The following theorem provides the closed-form expression of the marginal prior distribution of $\boldsymbol{\theta}$:

Theorem 9.2.1 (Liseo and Loperfido, 2003a) The marginal density of $\boldsymbol{\theta}$ is given by

$$\frac{\Phi_k\left(\mathbf{0}; C\Delta\left(\Sigma^{-1}\boldsymbol{\theta} + \Omega^{-1}\boldsymbol{\mu}\right) + \mathbf{d}, C\Delta C^T\right)}{\Phi_k\left(\mathbf{0}; C\boldsymbol{\mu} + \mathbf{d}, C\Omega C^T\right)} \phi_p\left(\boldsymbol{\theta}, \boldsymbol{\mu}, \Sigma + \Omega\right), \tag{9.1}$$

where ϕ_k and Φ_k denote, as usual, the pdf and the cumulative distribution function (cdf) of a k-dimensional standard normal distribution and $\Delta^{-1} = \Sigma^{-1} + \Omega^{-1}$.

Proof. See the Appendix. □

For $k = 1$, the constraints turn into a simple linear one, say $\boldsymbol{c}^T\boldsymbol{\theta}_0 + d \leq 0$, and the probability density function (pdf) of $\boldsymbol{\theta}$ is a slight generalization of the Azzalini and Dalla Valle (1996) skew-normal distribution, as it has been proposed by Arnold and Beaver (2000a). When $k = p$, i.e., the number of constraints equals the dimension of the parameter, expression (9.1) reduces to another type of multivariate skew-normal distribution, namely, the one proposed by Domínguez-Molina, González-Farías, and Gupta (2003). This distribution can be considered as a generalization of the one proposed by Branco and Dey (2001). The generalization mainly consists of allowing the normalizing factor in the denominator to be different from 2^{-p} and in accommodating for a non-diagonal matrix D.

As suggested above, the skewed prior (9.1) can be easily used in conjunction with multivariate Gaussian data. Suppose we observe $\mathbf{x} \sim N_p(\boldsymbol{\theta}, \Psi)$, with known Ψ and let the prior on $\boldsymbol{\theta}$ be as in (9.1), then the following result holds.

Theorem 9.2.2 Let

$$\begin{aligned} \mathbf{x} \mid \boldsymbol{\theta}, \boldsymbol{\theta}_0 &\sim N_p(\boldsymbol{\theta}, \Psi), & \boldsymbol{\theta} \mid \boldsymbol{\theta}_0 \sim N_p(\boldsymbol{\theta}_0, \Sigma), \\ \boldsymbol{\theta}_0 &\sim N_p(\boldsymbol{\mu}, \Omega), & C\boldsymbol{\theta}_0 + \mathbf{d} \leq \mathbf{0}. \end{aligned}$$

Then the posterior distribution of $\boldsymbol{\theta}$, upon observing $\mathbf{x} = \mathbf{x}^*$ is given by

$$f(\boldsymbol{\theta}|\mathbf{x}^*) = \frac{\Phi_k\left(\mathbf{0}; C\Delta_{\mathbf{x}}\left(\Sigma_{\mathbf{x}}^{-1}\boldsymbol{\theta} + \Omega^{-1}\boldsymbol{\mu}_{\mathbf{x}}\right) + \mathbf{d}_{\mathbf{x}}, C\Delta_{\mathbf{x}}C^T\right)\phi_p\left(\boldsymbol{\theta}; \boldsymbol{\mu}_{\mathbf{x}}, \Sigma_{\mathbf{x}} + \Omega\right)}{\Phi_k\left(\mathbf{0}; C\boldsymbol{\mu}_{\mathbf{x}} + \mathbf{d}_{\mathbf{x}}, C\Omega C^T\right)}$$

where

$$\begin{aligned} \boldsymbol{\mu}_{\mathbf{x}} &= \boldsymbol{\mu} + \Sigma\left(\Sigma + \Psi\right)^{-1}\mathbf{x}^*, & \Sigma_{\mathbf{x}} = \Sigma - \Sigma\left(\Sigma + \Psi\right)^{-1}\Sigma, \\ \Delta_{\mathbf{x}} &= \left(\Sigma_{\mathbf{x}}^{-1} + \Omega^{-1}\right)^{-1}, & \mathbf{d}_{\mathbf{x}} = \mathbf{d} - C\Sigma\left(\Sigma + \Psi\right)^{-1}\mathbf{x}^*. \end{aligned}$$

Proof. See Liseo and Loperfido (2003a). □

Although these results hold for the case of known covariance matrices, they can be easily extended to the case of unknown covariance matrices, using Monte Carlo Markov Chains methods. The conditional expression of the prior on $\boldsymbol{\theta}$ makes this problem *one line code more difficult* than usual hierarchical normal linear models, and the implementation of these kinds of models with user-friendly software packages like BUGS is not difficult.

The above theorems show that skewed priors naturally arise in the elicitation of linear constraint when dealing with Gaussian linear models. For other elliptical distributions, analogous results are not available so far.

9.2.2 Efficiency of Linear Bayes Rules with Skewed Priors

A different use of skewed priors is described in Mukhopadhyay and Vidakovic (1995). The authors conduct a robust Bayesian analysis to exploit the efficiency of linear Bayes rules compared to exact Bayes rules in some simple settings, under squared error loss.

Consider the following situation: let X be a $N(\theta, 1)$ random variable, and suppose that the prior distribution for θ, $\pi(\theta)$, say, belongs to the following class Γ of standard skew-normal priors:

$$\Gamma = \{\pi_\alpha : \pi_\alpha(\theta) = 2\phi(\theta)\Phi(\alpha\theta), \alpha \in [0, \alpha_{\max}]\}.$$

For each fixed π_α, the exact Bayes rule for this problem is of course the posterior mean of θ, say $\hat{\theta}_B$, whose Bayes risk

$$r(\pi_\alpha) = \int_{\mathcal{X}}\int_{\theta}\left(\theta - \hat{\theta}_B\right)^2 \pi_\alpha(\theta) f(x \mid \theta)\, dx d\theta,$$

where f is the likelihood arising from an iid sample from $N(\theta, 1)$, can be calculated using a well-known result in Brown (1971), and equals

$$r(\pi_\alpha) = 1 - \mathcal{I}(m_\alpha(x)),$$

where m_α is the marginal density of the data under the prior π_α and the Fisher information

$$\mathcal{I}(f) = \int \frac{f'(x)^2}{f(x)} dx.$$

On the other hand, in our setting, it can be easily shown that the linear Bayes rule can be written as

$$\delta_L(x) = \frac{x\,V_\pi + \mu_\pi r(\pi, x)}{V_\pi + r(\pi, x)},$$

where μ_π and V_π denote the prior mean and variance, respectively, and $r(\pi, x)$ is the Bayes risk for the decision x, which in this case is equal to unity. Also, the Bayes risk of the above linear Bayes rule, denoted by $r_L(\pi_\alpha)$, equals $V_\pi^{-1} + r(\pi, x)^{-1}$. Mukhopadhyay and Vidakovic (1995) consider the relative efficiency of $\delta_L(x)$, by computing, numerically, the ratio $r_L(\pi_\alpha)/r(\pi_\alpha)$, as π_α varies over Γ. They conjecture, without a formal proof, that the above ratio is decreasing in α, but the efficiency is still as large as 0.974 for $\alpha = 1,000$.

Such results may be very important in convincing statisticians to use more complex priors, including some degree of skewness, and, maybe, heavier tails. In this respect, more general classes of priors need to be studied; some preliminary results are given in Loperfido (2003).

9.2.3 Heavy Tail Priors

In the previous subsection we have mainly considered skew-normal type priors. It is well-known (see, for example, Berger, Rios Insua, Ruggeri, 2000), that Gaussian-like tail priors for location (or regression) parameters may induce a lack of robustness. To avoid this, it would be helpful to adopt priors with heavier tails, e.g., skew-t distributions, either in the form proposed by Azzalini and Capitanio (2003) or in the form of Sahu *et al.* (2003).

For example, the Azzalini and Capitanio (1999) version of the multivariate skew-t distribution can be represented as a scale mixture of their own version of the multivariate skew-normal distribution. More precisely, let $\mathbf{z} \sim SN_p(\mathbf{0}, \Omega, \boldsymbol{\alpha})$ and $V \sim \chi^2_\nu/\nu$, independent of $\mathbf{z}$; then, for some vector $\boldsymbol{\xi}$, the random vector $\mathbf{y} = \boldsymbol{\xi} + V^{-1/2}\mathbf{z}$ has a skew-t distribution; this fact can be easily implemented in a Gibbs sampling algorithm. Similar results hold for the Sahu *et al.* (2003) version of the skew-t distribution, obtained as a scale mixture of their own definition of skew-normal distribution.

We conclude this section with an illustration of the use of a skew-t prior for a normal linear model.

Example 1. Consider a vector $\mathbf{x} \sim N_p(\boldsymbol{\theta}, I_p)$. The p-dimensional vector of parameters $\boldsymbol{\theta}$ is given a multivariate skew-t prior, in the sense of Sahu *et al.* (2003), in the following way: given a random variable W and a random vector $\mathbf{z}$, $\boldsymbol{\theta} \mid \mathbf{z} = \mathbf{z}^*, W = w$ has a p-dimensional normal distribution with mean $D\mathbf{z}^*$ and covariance matrix $(\tau^2/w)I_p$; here D is the diagonal matrix introducing skewness. For simplicity, we assume that D is known, but a $N_p(\mathbf{0}, \Gamma)$ distribution could be assumed for its non-zero elements with a slight increase in computation. The vector $\mathbf{z}$ is also p-dimensional multivariate standard normal restricted to the positive orthant. Finally W has a $Gamma(\nu/2, \nu/2)$ distribution, and ν is a hyperparameter controlling the tails of the prior: we assume here that ν is fixed, but generalizations are possible.

Using standard calculations for normal models, one can see that the full conditionals have the following forms:

- $\boldsymbol{\theta} \mid \mathbf{z}^*, w, D, \mathbf{x}^* \sim N_p\left(\frac{\tau^2\mathbf{x}^* + wD\mathbf{z}^*}{\tau^2 + w}, \frac{\tau^2}{\tau^2+w} I_p\right)$,
- $\mathbf{z} \mid \boldsymbol{\theta}, w, D, \mathbf{x}^* \sim N_p^+\left(\left(I_p + \frac{w}{\tau^2}D^2\right)^{-1} \frac{w}{\tau^2} D\boldsymbol{\theta}, \left(I_p + \frac{w}{\tau^2}D^2\right)^{-1}\right)$,
where '+' represents the positive orthant of $\mathbb{R}^p$,
- the density of $W \mid \mathbf{z}^*, \boldsymbol{\theta}, D, \mathbf{x}^*$ is proportional to
$w^{\frac{\nu+p}{2}-1} \exp\left\{-\frac{w}{2}\left[\nu + \frac{1}{\tau^2}(\boldsymbol{\theta} - D\mathbf{z}^*)^T(\boldsymbol{\theta} - D\mathbf{z}^*)\right]\right\}$,
which is log-concave for all $\nu > 2$.

The above model (and various generalizations of it) can be easily implemented in user-friendly softwares like BUGS. This example is a simple illustration of the dramatic impact of simulation-based methods in Bayesian inference: we have been able to use a rather sophisticated prior distribution for $\boldsymbol{\theta}$ in a rather elementary way.

9.3 Skew-Elliptical Likelihood

In this section, we review the Bayesian approach to modeling and inference when the data can be assumed to follow an SE distribution. Before going into details, a premise is necessary. Since the pioneering work of Azzalini (1985) it is well-known that inference procedures are problematic with the skew-normal distribution; see, for example, Liseo (1990), Azzalini and Capitanio (1999), and Pewsey (2000). Moreover, the difficulties increase with higher generality of SE distributions. This has motivated considerable effort in exploiting the probabilistic properties of the SE class and the inferential aspects have received, in some sense, less attention. In particular, likelihood-based inference techniques pose several problems and, with the exception of normal and t cases, little is known about how to make inference on SE models. Moreover, even in the above-mentioned special cases, computations are far from trivial and anomalies of the maximum likelihood estimator are quite common. In Section 9.3.1, we first briefly illustrate the basic inferential problems which arise with the skew-normal distributions and its generalizations. In Section 9.3.2 we describe how the above problems can be generally tackled within a Bayesian framework; in fact, Monte Carlo Markov Chains methodology provides a relatively simple numerical tool that allows us to bypass the analytical difficulties inherent in the likelihood approach. We conclude the section with a brief account of several Bayesian uses of the SE model. Further insights on these aspects can be found in the second part of the book.

9.3.1 Inferential Problems

Consider the simplest example of an SE model, namely, the one parameter standard univariate skew-normal distribution with density

$$f(x;\alpha) = 2\phi(x)\Phi(\alpha x), \qquad x \in \mathbb{R},\ \alpha \in \mathbb{R}.$$

A sample of n iid observations from the above density provides the likelihood function $L(\alpha) \propto \prod_{j=1}^{n} \Phi(\alpha x_j)$. When the x_j's have all the same sign, it is readily seen that $L(\alpha)$ is a monotonic function of α, and the maximum likelihood estimate is infinite!

More general forms of SE distributions do not avoid this basic problem, which can be detected only in relatively simple settings but not for the general SE class. Consider, for example, the multivariate skew-normal case when the location parameter $\boldsymbol{\xi}_0$ is known. The likelihood function associated to an n-dimensional sample of iid replications of an $SN_p(\boldsymbol{\xi}_0, \Omega, \boldsymbol{\alpha})$ random vector is given by

$$L\left(\boldsymbol{\alpha}, \Omega\right) \propto \prod_{i=1}^{n} \phi_p(\mathbf{x}_i|\Omega) \prod_{i=1}^{n} \Phi(\boldsymbol{\alpha}^T \mathbf{x}_i). \tag{9.2}$$

In this case it is possible to characterize the samples which give an infinite maximum likelihood estimate of $\boldsymbol{\alpha}$: it suffices to show that there exists a vector $\boldsymbol{\alpha}_0$ such that $\boldsymbol{\alpha}_0^T \mathbf{x}_i > 0$ for all $\mathbf{x}_i$'s. No other simple characterizations of infinite maximum likelihood estimates are available for more general SE distributions. As an example of a data set which provides infinite maximum likelihood estimates, consider the one discussed in Azzalini and Capitanio (1999). This consists of several physical characteristics for 202 Australian athletes in different disciplines. For simplicity – but the phenomenon is more general – we restrict our attention to the bivariate case and consider only the two variables *weight* and *height* observed on female "Row" athletes ($n = 22$). A graphical analysis of the data set, not reported here, shows a substantial degree of skewness and the use of the SN_2 model seems appropriate. However, when the location parameters are *estimated* with the known values of the Australian population (mean height = 161.4 cm, mean weight = 67.0 kg), the maximum likelihood estimates of the two-dimensional skewness parameter turns out to be $\hat{\boldsymbol{\alpha}} = (+\infty, +\infty)^T$. In this example it is easy to check that the problem arises from the fact that all the 22 athletes have heights and weights greater than the respective population averages. Then it should be more correct to estimate the population vector mean from the data; this way one could take into account the fact that athletes should be considered *outliers* within the entire population. However, this example should sound as a warning: similar remedies are not always easy to find and the estimation of $\boldsymbol{\alpha}$ may be potentially problematic in real data applications. We will return to this example below.

There are a few papers dealing with the inferential problems of the skew-normal distribution (and generalizations) from the classical perspective. We suggest to read Azzalini (1985), Pewsey (2000), and Sartori (2003). In Section 9.4 we will discuss in some detail these inferential problems from a default Bayesian point of view.

9.3.2 Regression Models with SE Errors

In this section we describe Bayesian inferential procedures for the general SE class of distributions using Monte Carlo Markov Chains (MCMC) tools. This section is mainly based on Sahu *et al.* (2003): consequently, methods are described having in mind their definition of SE densities. Adaptations to other variants, like the one defined in Azzalini and Capitanio (2003), are not difficult to implement.

Consider a sample $\mathbf{y}_1, \ldots, \mathbf{y}_n$ of random vectors with p-dimensional density

$$f(\mathbf{y}; \boldsymbol{\mu}, \Sigma, D, g_0^{(p)}) = 2^p f_{\mathbf{y}}\left(\mathbf{y}; \boldsymbol{\mu}, \Sigma + D^2, g_0^{(p)}\right) \mathrm{P}(\mathbf{v} > 0), \qquad (9.3)$$

where f is the pdf of some p-dimensional elliptically contoured distribution,

$$\mathbf{v} \sim EC_p\left(D(\Sigma + D^2)^{-1}(\mathbf{y} - \boldsymbol{\mu}), I_p - D(\Sigma + D^2)^{-1}D, g^{(p)}_{q(\mathbf{y}-\boldsymbol{\mu})}\right),$$

and $g_a^{(p)}$ represents the density generator of the elliptically contoured random vector, that is,

$$g_a^{(p)}(u) = \frac{\Gamma(p/2)}{\pi^{p/2}} \frac{g(a+u; 2p)}{\int_0^\infty r^{m/2-1}\, g(a+r; 2p)\, dr}, \qquad a > 0.$$

Finally, $g(v;p)$ is a positive non-increasing function of a positive argument v and such that the integral over $\mathbb{R}_+$, $\int v^{p/2-1} g(v;p) dv$, is finite. Also, $\boldsymbol{\mu}$ and Σ are the location and scale parameters, D is a diagonal matrix, say $D = \mathrm{diag}(\delta_1, \ldots, \delta_p)$, and $q(\mathbf{y} - \boldsymbol{\mu})$ is the quadratic form

$$q(\mathbf{y} - \boldsymbol{\mu}) = (\mathbf{y} - \boldsymbol{\mu})^T (\Sigma + D^2)^{-1} (\mathbf{y} - \boldsymbol{\mu}).$$

The location vector $\boldsymbol{\mu}$ may also contain a regression structure so that our observations are independent but not identically distributed with, for the i-th observation,

$$\boldsymbol{\mu}_i = X_i^T \boldsymbol{\beta},$$

where $\boldsymbol{\beta}$ is a k-dimensional vector of unknown coefficients and X_i is a $k \times p$

design matrix. We must recall that the density of the $\mathbf{y}_i$ can be interpreted as the density of $\boldsymbol{\epsilon} + D\mathbf{z}|\mathbf{z} > \mathbf{0}$, where

$$\begin{pmatrix} \boldsymbol{\epsilon} \\ \mathbf{z} \end{pmatrix} \sim EC_{2p}\left(\begin{pmatrix} \boldsymbol{\mu} \\ \mathbf{0} \end{pmatrix}, \begin{pmatrix} \Sigma & O \\ O & I_p \end{pmatrix}, g_0^{2p} \right),$$

that is the vector $(\boldsymbol{\epsilon}, \mathbf{z})^T$ has a $2p$-dimensional elliptically contoured distribution.

The direct analysis of the likelihood function of such models is of course prohibitive: the multivariate cdf which appears in the expression of the density (9.3) can be even harder to evaluate. However, within the MCMC framework, one can exploit the conditional nature of the SE class by noticing that

$$\boldsymbol{\epsilon} + D\mathbf{z} \mid \mathbf{z} = \mathbf{z}^* \sim EC_p\left(\boldsymbol{\mu} + D\mathbf{z}^*, \Sigma, g^{(p)}_{\tilde{q}(\mathbf{z}^*)}\right),$$
$$\mathbf{z} \sim EC_p^+\left(\mathbf{0}, I_p, g_0^{(p)}\right),$$

with $\tilde{q}(\mathbf{z}^*) = \mathbf{z}^{*T}\mathbf{z}^*$.

The ease of implementation now depends on the particular elliptically contoured distribution adopted. Here we describe the case of the multivariate skew-t distribution: this context includes, as a special case, the multivariate skew-normal model. Suppose henceforth that

$$g(u, 2p) = \left(1 + \frac{u}{\nu}\right)^{\frac{\nu+2p}{2}},$$

where ν is some integer representing the degrees of freedom. Because of well-known standard results about the representation of t distributions as a scale mixture of normal ones (see, for example, Dickey, 1968), this is equivalent to saying that each observation $\mathbf{y}_i$, $i = 1, \ldots, n$, can be represented in the following way:

$$\begin{aligned} \mathbf{y}_i \mid \mathbf{z}_i = \mathbf{z}_i^*, \boldsymbol{\beta}, X_i, \Sigma, D, W_i = w_i &\sim N_p\left(X_i^T\boldsymbol{\beta} + D\mathbf{z}_i^*, \frac{\Sigma}{w_i}\right), \\ \mathbf{z}_i &\sim N_p^+(\mathbf{0}, I_p), \\ W_i &\sim Gamma(\frac{\nu}{2}, \frac{\nu}{2}). \end{aligned}$$

Prior distributions are chosen in a conjugate way in order to simplify Monte Carlo Markov Chains techniques. Following Sahu *et al.* (2003), we assume that

$$\begin{aligned} \boldsymbol{\beta} &\sim N_k(\boldsymbol{\beta}_0, \Lambda), \\ \Sigma^{-1} &\sim W_p(2\Omega, 2r), \\ \boldsymbol{\delta} &\sim N_p(\mathbf{0}, \Gamma), \\ \nu &\sim Gamma(\xi, \psi)I_{(\nu>2)}(\nu), \end{aligned}$$

where $\boldsymbol{\beta}_0, \Lambda, r, \Omega, \Gamma, \xi, \psi$ are fixed hyperparameters, and W_p denotes the Wishart distributions for $p \times p$ positive semidefinite matrices. This setup can also be used for skew-normal distributions: it suffices to set the W_i's equal to 1 and to omit the last line in the two above schemes. Within this construction, it is easy to derive complete conditionals for all the parameters, see Sahu *et al.* (2003) for details.

9.3.3 Some Applications

As discussed in the previous sections, standard Bayesian regression models can be easily generalized to include skew-elliptical errors. As a consequence, skewed Bayesian regression models have been recently used in a variety of disciplines: the reader will find a detailed account of these applications in the second part of this book. Here we merely sketch some of them.

A very natural context for the use of skew-normal or t distributions arises in the so-called *stochastic frontier models* that are used to evaluate the efficiency of some economic agents, such as firms or countries. The basic idea is that the observed production of a single unit cannot exceed the (latent) potential production, which is named *frontier*; the difference between the frontier and the actual production can be used as a measure of inefficiency. Consider then the model

$$Y_i = h(\mathbf{x}_i, \boldsymbol{\beta}) + \epsilon_i - Z_i, \qquad i = 1, \ldots, n, \tag{9.4}$$

where Y_i is the logarithm of the output variable for the i-th unit, $\mathbf{x}_i$ is a vector of covariates, $\boldsymbol{\beta}$ represents the unknown coefficients, and h is a smooth function. Also, ϵ_i denotes the error term, which is usually given a symmetric, zero-mean distribution, while Z_i, representing the inefficiency, is a positive random variable. It can be easily demonstrated that, when $\epsilon_i \sim N(0, \sigma_\epsilon^2)$ and $Z_i \sim N(0, \sigma_z^2)$ is restricted to be positive, then the difference is

$$V_i = \epsilon_i - Z_i \sim SN\left(0, \sqrt{\sigma_z^2 + \sigma_\epsilon^2}, -\frac{\sigma_z/\sigma_\epsilon}{\sqrt{\sigma_z^2 + \sigma_\epsilon^2}}\right).$$

Then, a stochastic frontier model can be represented as a standard regression model with skew-normal errors. The extension to skew-t errors with ν degrees of freedom can be easily accommodated by assuming that the difference $\epsilon_i - Z_i$ in (9.4), is replaced by the quantity $(\epsilon_i - Z_i)/\sqrt{W}$, where $W \sim Gamma(\nu/2, \nu/2)$. For this model, Tancredi (2003) describes a full likelihood approach, although a Monte Carlo Markov Chains Bayesian analysis seems straightforward; see also Koop, Osiewalski, Steel, and van den Broeck (1994) for related work in this subject.

In an unpublished Ph.D. thesis at ISDS, Duke University, Lietchy (2003) uses a generalized version of multivariate skew-normal distributions to explore the influence of third moments (i.e., skewness) in the construction of optimal Markowitz's portfolios, which are usually based only on means and variances.

A further topic where the presence of skewness can be important is the analysis of returns in financial time series. Various empirical studies (Mills, 1995) have shown that returns distributions tend to be negatively skewed. Peiró (1999) discusses the problem in the context of time series analysis where the dependence among data makes the usual Mardia's test for asymmetry inapplicable. Liseo and Macaro (2003) consider the use of a $GARCH(1,1)$ model with innovation terms being skew-normal random variables. Since arbitrage constraints force the mean of the process to be zero, they use error terms with general $SN(\mu, \tau, \alpha)$ under the constraint that

$$\mu = \mu(\alpha, \tau) = -\sqrt{\frac{2}{\pi}} \frac{\alpha \tau}{\sqrt{1+\alpha^2}}. \tag{9.5}$$

In detail, the Bayesian model is the following. For $t = 1, \ldots, T$, let $Y_t = \sigma_t X_t$ where the conditional variances

$$\sigma_t^2 = \alpha_0 + \alpha_1 Y_{t-1}^2 + \beta_1 \sigma_{t-1}^2,$$

follow a $GARCH(1,1)$ structure, and the innovation terms X_t are iid SN, that is,

$$X_t \sim SN(\mu(\alpha, \tau), \tau(\alpha), \alpha),$$

where $\mu(\alpha, \tau)$ is given by (9.5). In order to maintain the actual meaning of variance for σ_t^2, the scale parameter of X_t is set to

$$\tau(\alpha) = \left(1 - \frac{2\alpha^2}{\pi(1+\alpha^2)}\right)^{-1}.$$

Note that, whereas the innovation terms X_t's follow an SN distribution, the random variables Y_t's have a distribution that is analytically intractable and can only be handled via a Monte Carlo Markov Chains approach. Liseo and Macaro (2003) compare the prediction performances of skewed $GARCH$ models with those of regular $GARCH$ and $EGARCH$ models.

Finally, Kim and Mallick (2004) use skew-normal random processes for Bayesian spatial prediction.

9.4 Objective Bayesian Analysis of the Skew-Normal Model

In Section 9.3.1 we have already mentioned the unusual behavior of the likelihood function when we try to make inference on the shape parameter of the skew-normal distribution. In this section we describe in some detail an objective Bayesian approach which represents a possible solution to these problems. The skew-normal model represents an interesting case where classical inference is problematic and objective Bayes methods (which usually give, at least for point estimation, similar results) provide strongly disagreeing solutions.

Although we conjecture that analogous conclusions hold for the general SE class, complete results are available only for the skew-normal case. In Section 9.4.1 we give a detailed account of the scalar case. In Section 9.4.2 we sketch the currently available extensions to the multivariate case. Most of the results in this section are based on Liseo and Loperfido (2003b).

9.4.1 The Scalar Case

Here we assume to observe an iid random sample $\mathbf{x} = (X_1, \ldots, X_n)^T$ from the scalar skew-normal density $SN(\xi, \tau, \alpha)$. As discussed in Section 9.3.1, in the standard case ($\xi = 0, \tau = 1$) this model can give infinite maximum likelihood estimates with positive sampling probability for all $\alpha \neq 0$. In the general three-parameters case things are complicated for two reasons:

- the Fisher information matrix is singular as α goes to 0.
- the profile likelihood function for α has a stationary point at $\alpha = 0$, independently of the observed sample.

Azzalini (1985) addresses the first problem by proposing a different parametrization. The second problem is more serious: as stated in Azzalini and Capitanio (1999), " *...there are cases where the likelihood shape and the MLE are problematic. We are not referring here to difficulties with numerical maximization, but to the intrinsic properties of the likelihood function, not removable by change of parameterization. [...] In cases of this sort, the behavior of the MLE appears qualitatively unsatisfactory, and an alternative estimation method is called for.*"

Likelihood estimation methods are not the only frequentist methods which encounter difficulties with the SN model. The method of moments can give even worse results. The fact that difficulties are intrinsically tied with the likelihood shape may suggest calibrating the likelihood with a weight function. To this aim, Liseo and Loperfido (2003b) propose a standard noninformative prior analysis based on Berger and Bernardo (1992). They show that, in the standard case ($\xi = 0, \tau = 1$), the reference (or Jeffreys') prior for α is proper: in fact it has tails of order $O(\alpha^{-3/2})$. This fact automatically provides a calibration to the odd behavior of the likelihood function and allows for reliable estimation procedures. In the three-parameters case, the joint reference prior for (ξ, τ, α), say $\pi_R(\xi, \tau, \alpha)$, "factorizes" into $\pi_R(\alpha)$ and the usual default prior for location-scale parameters, namely, $\pi_R(\xi, \tau) \propto 1/\tau$; again, $\pi_R(\alpha)$ is, with an abuse of language, marginally proper with tails of order $O(\alpha^{-3/2})$.

Consider first the standard case. The Jeffreys' prior associated with this model is

$$\pi_J(\alpha) \propto \sqrt{I(\alpha)}, \tag{9.6}$$

where

$$I(\alpha) = \mathrm{E}_\alpha\left(\frac{\partial}{\partial\alpha}\log f(Z;\alpha)\right)^2 = \int 2z^2\phi(z)\frac{\phi^2(\alpha z)}{\Phi(\alpha z)}dz,$$

represents the Fisher expected information. Note that $I(\alpha)$ cannot be written in a closed-form (Azzalini, 1985). The following theorem holds.

Theorem 9.4.1 The Jeffreys' prior for the skewness parameter α is such that

(i) $\pi_J(\alpha)$ is symmetric about $\alpha = 0$ and it is decreasing in $|\alpha|$;

(ii) the tails of $\pi_J(\alpha)$ are of order $O(\alpha^{-3/2})$.

Proof. See the Appendix. □

Point (ii) of the above theorem says that $\pi_J(\alpha)$ is proper, even though α is an unbounded parameter. This implies that the posterior mode of α is finite with probability 1, and this indicates a strong disagreement between frequentist and default Bayes results, at least for finite sample sizes. However, the behavior of the posterior mean is not satisfactory, as the following theorem shows.

Theorem 9.4.2 Let $X_1, \ldots, X_n$ be a sample of iid random variables with density $SN(0,1,\alpha)$ and let $\pi_J(\alpha)$ be the prior distribution defined in (9.6). Then the following three events are equivalent:

a) All the observations have the same sign.

b) The maximum likelihood estimate of α is infinite.

c) The posterior mean of α is infinite.

Proof. See the Appendix. □

The above theorem says that, using Jeffreys' prior, the posterior mean behaves like the maximum likelihood estimator. Consequently, it is more reasonable to use other Bayesian point estimates, such as the posterior median or mode.

In the general three-parameters case, it can be shown that the conditional reference prior for the nuisance parameters (ξ, τ) is the usual $\pi_R(\xi,\tau) \propto 1/\tau$. It is then possible to obtain a closed-form of the integrated likelihood for α. The following result is a corollary of Theorem 2 in Liseo and Loperfido (2003b).

Corollary 9.4.1 The reference-integrated likelihood is given by

$$\widetilde{L}_R(\alpha) \propto F_{\Omega_R}\left(\alpha\frac{\mathbf{x} - \bar{x}\mathbf{1}_n}{s}\right), \tag{9.7}$$

where F_{Ω_R} is the cdf of a centered n-dimensional t distribution with n degrees of freedom and scale matrix $\Omega_R(\alpha) = I_n + \frac{\alpha^2}{n}\mathbf{1}_n\mathbf{1}_n^T$, $\mathbf{x}$ is the n-vector of observations, $\bar{x}$ and s are the sample mean and standard deviation.

Liseo and Loperfido (2003b) show that the integrated likelihood may not vanish for large values of $|\alpha|$, and that it makes it difficult to obtain likelihood intervals for α although this problem is not serious for large sample sizes.

A "correct" default analysis for this problem should be performed by deriving Jeffreys' prior for the marginal model whose likelihood function is given by (9.7). Alternatively one can use the reference algorithm approach to show that the actual joint reference prior for the three-parameters case is

$$\pi_R(\xi, \tau, \alpha) = \pi_R(\xi, \tau)\pi_R(\alpha) \propto \frac{1}{\tau} g^{1/2}(\alpha), \qquad (9.8)$$

where $g(\alpha)$ is a complicated function of the parameter of interest α. The use of the exact reference prior (9.8) is cumbersome. However, as in the standard case, it can be shown that the tails of $\pi_R(\alpha)$ are of order $\alpha^{-3/2}$. Again, the marginal prior is proper and it can be used together with (9.7) to produce a proper posterior distribution for α. Alternatively, one can approximate the actual posterior by substituting the correct prior (9.8) with the Jeffreys' prior (9.6) obtained in the standard case.

The fact that the default prior for α is marginally "proper" allows the use of the standard Bayes factor for checking the presence of skewness in the data. More precisely, if we are interested in comparing

$$H_0 : \alpha = 0 \text{ vs. } H_1 : \alpha \neq 0,$$

we can use the above reference prior for the common nuisance parameter and the prior (9.6) to obtain the following Bayes factor

$$B = \frac{\widetilde{L}_R(0)}{\int_\alpha \widetilde{L}_R(\alpha)\pi_J(\alpha)d\alpha}.$$

This can be calculated in several different ways: either from the output of a Metropolis Hastings algorithm (Chib and Jeliazkov, 2001) or with a one dimensional numerical integration.

9.4.2 Some Multivariate Results

In this section, we assume that our sample is a data matrix denoted by $X = (\mathbf{x}_1, \ldots, \mathbf{x}_n)^T$, where the $\mathbf{x}_i$'s are iid multivariate random vectors, $\mathbf{x}_i \sim SN_p(\boldsymbol{\xi}, \Omega, \boldsymbol{\alpha})$, according to the Azzalini and Dalla Valle (1996) definition. Not all the results of the previous section can be easily generalized to the multidimensional case. Liseo and Loperfido (2003c) show that it is possible to obtain a closed-form expression for the integrated likelihood of the shape vector $\boldsymbol{\alpha}$. In detail, let us assume the following conjugate conditional prior for the nuisance parameters $(\boldsymbol{\xi}, \Omega)$:

$$\boldsymbol{\xi} \mid \tau, \Omega \sim N_p(\boldsymbol{\xi}_0, \frac{1}{\tau}\Omega), \qquad \Omega^{-1} \sim W_p(\Omega, h).$$

For computational reasons, we reparametrize the SN_p family by setting

$$\boldsymbol{\alpha} = \boldsymbol{\alpha}(\boldsymbol{\beta}) = \frac{\sqrt{\boldsymbol{\beta}^T\boldsymbol{\beta}}}{\sqrt{\boldsymbol{\beta}^T\Omega\boldsymbol{\beta}}}\boldsymbol{\beta}.$$

Therefore, our parameter of interest is now $\boldsymbol{\beta}$. Then the following result, generalizing (9.7), holds.

Theorem 9.4.3 Let $\bar{\mathbf{x}}$ and S denote the sample mean and covariance matrix respectively. Under the previous assumptions, the integrated likelihood of $\boldsymbol{\beta}$ is given by

$$\widetilde{L}(\boldsymbol{\beta}) \propto \mathrm{P}\left(T_{n+h-p+1} \le \frac{(X - \mathbf{1}_n\mathbf{m}^T)\sqrt{\boldsymbol{\beta}^T\boldsymbol{\beta}}}{\sqrt{\boldsymbol{\beta}^T\Omega_{\mathbf{x}}\boldsymbol{\beta}}}\sqrt{n+h-p+1}\boldsymbol{\beta}\right),$$

where T has an n-dimensional t distribution with $n+h-p+1$ degrees of freedom, centered at the origin and with scale matrix

$$I_n + \mathbf{1}_n\mathbf{1}_n^T\frac{\boldsymbol{\beta}^T\boldsymbol{\beta}}{\tau+n}, \tag{9.9}$$

and $\quad \mathbf{m} = \frac{n\bar{\mathbf{x}}+\tau\boldsymbol{\xi}_0}{\tau+n}, \qquad \Omega_{\mathbf{x}} = \Omega + nS + \frac{n\tau}{\tau+n}\left(\bar{\mathbf{x}} - \boldsymbol{\xi}_0\right)\left(\bar{\mathbf{x}} - \boldsymbol{\xi}_0\right)^T.$

Proof. Liseo and Loperfido (2003c). □

An immediate consequence of Theorem 9.4.3 is the *noninformative* reference expression of the integrated likelihood, when the hyperparameters of the normal-Wishart prior, namely τ, Ω, and h, are chosen to be zero.

Corollary 9.4.2 The noninformative integrated likelihood of $\boldsymbol{\beta}$ is

$$\widetilde{L}_R(\boldsymbol{\beta}) \propto \mathrm{P}\left(T_{n-p+1} \le \frac{(X - \mathbf{1}_n\bar{\mathbf{x}}^T)\sqrt{\boldsymbol{\beta}^T\boldsymbol{\beta}}}{\sqrt{\boldsymbol{\beta}^T nS\boldsymbol{\beta}}}\sqrt{n-p+1}\boldsymbol{\beta}\right), \tag{9.10}$$

where the scale matrix of the multivariate t distribution is now given by (9.9) with $\tau = 0$.

The behavior of (9.10) can be compared with those of other, more classical likelihood functions, such as the profile likelihood and its ramifications, which are all based on the elimination of the nuisance parameters $\boldsymbol{\xi}$ and Ω via conditional maximization. As an example, consider again the data set already discussed in Section 9.3.1. We use the expression (9.10) to compute the maximum integrated likelihood of $\boldsymbol{\beta}$. Figure 9.1 shows the contour levels of $\widetilde{L}_R(\boldsymbol{\beta})$. One can see that in this case the maximum is attained at a finite

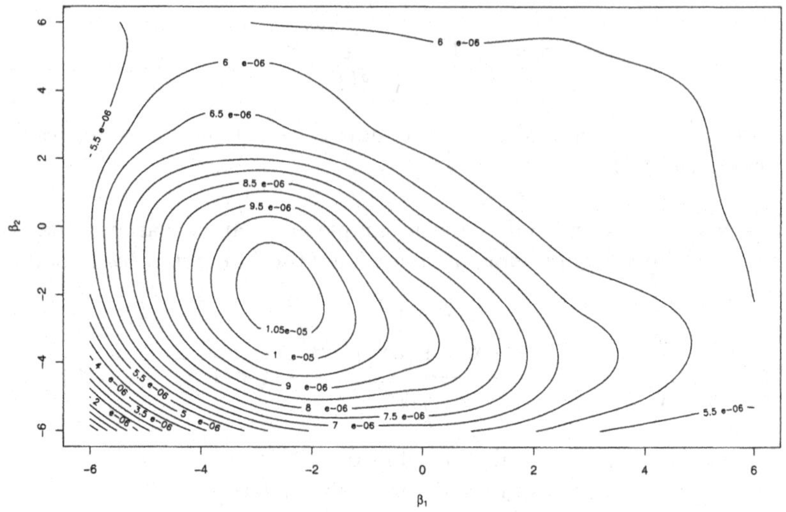

Figure 9.1 *Contour plot of the integrated likelihood for* $\widetilde{L}_R(\boldsymbol{\beta})$.

value, namely $\hat{\boldsymbol{\beta}} = (-2.6, -2.0)^T$. An unsettled problem with $\widetilde{L}_R(\cdot)$ is how to easily compute a standard error of the point estimate.

It is not easy to perform a complete objective Bayesian analysis of the multivariate skew-normal distribution. A manageable expression of the reference prior for the vector of parameters $(\boldsymbol{\xi}, \Omega, \boldsymbol{\alpha})$ seems impossible to obtain. The following theorem enumerates some partial results.

Theorem 9.4.4 Let $\mathbf{x} \sim SN_p(\boldsymbol{\xi}_0, \Omega, \boldsymbol{\alpha})$, where $\boldsymbol{\xi}_0$ is a known vector. Then:

1) when Ω is known, the Jeffreys' prior for $\boldsymbol{\alpha}$ is a proper p-dimensional distribution.
2) when Ω is unknown, the conditional reference prior $\pi_J(\boldsymbol{\alpha} \mid \Omega)$ of $\boldsymbol{\alpha} \mid \Omega$ is proper for all values of Ω.

Proof. Liseo and Loperfido (2003c). □

Theorem 9.4.4 basically says that a reference analysis can be performed when the location parameter $\boldsymbol{\xi}_0$ is known. In fact, in this case, the likelihood function is given by (9.2). Notice that it factorizes into a term which depends only on $\boldsymbol{\alpha}$ and another factor, depending only on Ω, which is the kernel of a multivariate normal density. Then, a separate reference prior for $\boldsymbol{\alpha}$ and Ω can be calculated. Theorem 9.4.4 says that the reference prior for $\boldsymbol{\alpha}$ is proper; the reference prior for Ω can be chosen among those obtained by Yang and Berger (1994). Then, a possible inferential strategy consists of using empirical Bayes ideas. First, obtain a simple and robust estimate of the location parameter $\boldsymbol{\xi}$, maybe assuming a flat prior on it. Second, use the *profile* likelihood $L(\hat{\boldsymbol{\xi}}_{\boldsymbol{\alpha},\Omega}, \Omega, \boldsymbol{\alpha})$ combined with the reference prior $\pi_R(\boldsymbol{\alpha})$ and the reference prior on Ω to obtain a posterior distribution for $\boldsymbol{\alpha}$. This route is currently under investigation.

9.5 Appendix

Proof of Theorem 9.2.1
Since $C\boldsymbol{\theta}_0 + \mathbf{d} \sim N_k\left(C\boldsymbol{\mu} + \mathbf{d}, C\Omega C^T\right)$, the probability of the linear constraints can be denoted as

$$\mathrm{P}\left(C\boldsymbol{\theta}_0 + \mathbf{d} \leq \mathbf{0}\right) = \Phi_k\left(\mathbf{0}; C\boldsymbol{\mu} + \mathbf{d}, C\Omega C^T\right).$$

The marginal density of $\boldsymbol{\theta}$ can be then written as

$$\begin{aligned} f(\boldsymbol{\theta}) &= \frac{1}{\Phi_k\left(\mathbf{0}; C\boldsymbol{\mu} + \mathbf{d}, C\Omega C^T\right)} \int_{C\boldsymbol{\theta}_0 + \mathbf{d} \leq \mathbf{0}} \phi_p\left(\boldsymbol{\theta}; \boldsymbol{\theta}_0, \Sigma\right) \phi_p\left(\boldsymbol{\theta}_0; \boldsymbol{\mu}, \Omega\right) d\boldsymbol{\theta}_0 \\ &= \frac{1}{\Phi_k\left(\mathbf{0}; C\boldsymbol{\mu} + \mathbf{d}, C\Omega C^T\right)} \frac{1}{(2\pi)^{p/2}} \frac{1}{(\mid \Sigma \mid \cdot \mid \Omega \mid)^{1/2}} \\ &\times \exp\left\{-\frac{1}{2}\left[(\boldsymbol{\theta} - \boldsymbol{\mu})^T (\Sigma + \Omega)^{-1} (\boldsymbol{\theta} - \boldsymbol{\mu})\right]\right\} \\ &\times \int_{C\boldsymbol{\theta}_0 + \mathbf{d} \leq \mathbf{0}} \frac{1}{(2\pi)^{p/2}} \exp\left\{-\frac{1}{2}\left[(\boldsymbol{\theta}_0 - \mathbf{z})^T \Delta^{-1} (\boldsymbol{\theta}_0 - \mathbf{z})\right]\right\} d\boldsymbol{\theta}_0, \end{aligned}$$

where

$$\mathbf{z} = \left(\Sigma^{-1} + \Omega^{-1}\right)^{-1} \left(\Sigma^{-1}\boldsymbol{\theta} + \Omega^{-1}\boldsymbol{\mu}\right), \text{ and } \Delta^{-1} = \Sigma^{-1} + \Omega^{-1}.$$

Then $\Delta = \Sigma\left(\Sigma + \Omega\right)^{-1} \Omega$ and

$$\mid \Delta \mid = \frac{\mid \Sigma \mid \cdot \mid \Omega \mid}{\mid \Sigma + \Omega \mid}.$$

With the usual change of variables $\boldsymbol{\theta}_0 = \Delta^{1/2}\mathbf{u} + \mathbf{z}$, one obtains that $C(\Delta^{1/2}\mathbf{u} + \mathbf{z}) + \mathbf{d} \sim N_k\left(C\mathbf{z} + \mathbf{d}, C\Delta C^T\right)$. Then,

$$f(\boldsymbol{\theta}) = \frac{\Phi_k\left(\mathbf{0}; C\mathbf{z} + \mathbf{d}, C\Delta C^T\right)}{\Phi_k\left(\mathbf{0}; C\boldsymbol{\mu} + \mathbf{d}, C\Omega C^T\right)} \frac{\exp\left\{-\frac{1}{2}\left[(\boldsymbol{\theta} - \boldsymbol{\mu})^T (\Sigma + \Omega)^{-1} (\boldsymbol{\theta} - \boldsymbol{\mu})\right]\right\}}{(2\pi)^{p/2} \mid \Sigma + \Omega \mid^{1/2}},$$

and the proof is completed. □

Proof of Theorem 9.4.1
Since the square root is a monotonic transformation, it suffices to prove the results for $I(\alpha)$.

(i) The symmetry of $I(\alpha)$ is proved by splitting the integral in two parts.

$$\begin{aligned} I(\alpha) &= 2\int_0^\infty z^2\phi(z)\frac{\phi^2(\alpha z)}{\Phi(\alpha z)}dz + 2\int_0^\infty z^2\phi(-z)\frac{\phi^2(-\alpha z)}{1-\Phi(\alpha z)}dz \\ &= 2\int_0^\infty z^2\phi(z)\phi^2(-\alpha z)\frac{1}{\Phi(\alpha z)(1-\Phi(\alpha z))} = I(-\alpha). \end{aligned}$$

The monotonicity of $\pi_J(\cdot)$ for $\alpha > 0$ is proved calculating the first derivative:

$$\begin{aligned} \frac{\partial}{\partial\alpha}I(\alpha) &= -\int_0^\infty z^3\frac{\phi(z)\phi^2(\alpha z)}{\Phi^2(\alpha z)(1-\Phi^2(\alpha z))} \\ &\times\ [2\alpha z\Phi(\alpha z)(1-\Phi(\alpha z)) + \phi(\alpha z)(1-2\Phi(\alpha z))]\,d\alpha. \end{aligned} \tag{9.11}$$

To prove that $\partial I(\alpha)/\partial\alpha$ is negative it suffices to show that, for all $s > 0$, $2s\Phi(s)\Phi(-s) + \phi(s)\,(1-2\Phi(s)) > 0$. Since for all $s > 0$, $s\Phi(-s) < \phi(s)$ the result follows immediately.

(ii) From the symmetry of $I(\alpha)$ it is enough to study the right tail only. Let

$$I(\alpha) = A(\alpha) + B(\alpha) = 2\int_0^\infty z^2\phi(z)\frac{\phi^2(\alpha z)}{\Phi(\alpha z)}dz + 2\int_0^\infty z^2\phi(z)\frac{\phi^2(\alpha z)}{1-\Phi(\alpha z)}dz.$$

For all $\alpha > 0$ and $z > 0$, $1 < 1/\Phi(\alpha z) < 2$. Then $A^*(\alpha) < A(\alpha) < 2A^*(\alpha)$, where

$$A^*(\alpha) = \int_0^\infty z^2\phi(z)\phi^2(\alpha z)dz = \frac{1}{2\pi(1+2\alpha^2)^{3/2}};$$

then $A(\alpha) = O(1/\alpha^3)$. Also,

$$\begin{aligned} B(\alpha) &= 2\int_0^{1/\alpha} z^2\phi(z)\frac{\phi^2(\alpha z)}{1-\Phi(\alpha z)}dz \\ &+\ 2\int_{1/\alpha}^\infty z^2\phi(z)\frac{\phi^2(\alpha z)}{1-\Phi(\alpha z)}dz = B_1(\alpha) + B_2(\alpha). \end{aligned}$$

When $0 < z < \alpha^{-1}$, $1/2 < \Phi(\alpha z) < \Phi(1)$; then $2 < (1-\Phi(\alpha z))^{-1} < 6.30 = c$. It follows that $B_1^*(\alpha) < B_1(\alpha) < c/2B_1^*(\alpha)$, where

$$B_1^*(\alpha) = 4\int_0^{1/\alpha} z^2\phi(z)\phi^2(\alpha z)dz = \frac{4}{\pi^{3/2}}\frac{1}{(1+2\alpha^2)^{3/2}}\,\tilde{\Gamma}\left(\frac{3}{2}, 1+\frac{1}{2\alpha^2}\right),$$

where $\tilde{\Gamma}(a, x)$ denotes the incomplete Gamma function

$$\tilde{\Gamma}(a, x) = \int_0^x t^{a-1}\exp(-t)\,dt.$$

It follows that $B_1(\alpha) = O(1/\alpha^3)$. Finally, for fixed positive α and $z > \alpha^{-1}$ we use the following inequality (Feller, 1971)

$$\frac{1}{1-\Phi(\alpha z)} \le \frac{\alpha^3 z^3}{(\alpha^2 z^2 - 1)\phi(\alpha z)}.$$

Then

$$\begin{aligned} B_2(\alpha) &\leq 2\alpha^3 \int_{1/\alpha}^{\infty} \frac{z^5}{\alpha^2 z^2 - 1} \phi(z)\phi(\alpha z) dz \\ &\leq 2\alpha^3 \int_{1/\alpha}^{\infty} z^5 \phi(z)\phi(\alpha z) dz < \frac{2}{\sqrt{2\pi}} \frac{\alpha^3}{(1+\alpha^2)^3} \int_0^{\infty} w^5 \phi(w) dw \\ &= \frac{16}{\sqrt{2\pi}} \frac{\alpha^3}{(1+\alpha^2)^3}. \end{aligned}$$

It follows that $I(\alpha)$ is the sum of three functions of order $O(1/\alpha^3)$, and, consequently, the tails of $\pi_J(\alpha)$ are of order $-3/2$. □

Proof of Theorem 9.4.2
Without loss of generality we consider only the case where all the x_i's are positive and, consequently, the likelihood function will be increasing and the existence of the posterior mean will depend on the order of the right tail.
[a) ⇒ b)] This is obvious.
[b) ⇒ c)] The contribution of each x_i's to the likelihood $L(\alpha)$ is $\Phi(\alpha x_i)$, which is either an increasing or decreasing function. Then, if b) is true, all the observations are positive and the likelihood is a strictly increasing function which tends to 1. It follows that, for large α, the posterior distribution is proportional to the prior

$$\pi_J(\alpha \mid \mathbf{x}) \propto \pi_J(\alpha) L(\alpha) \propto \pi_J(\alpha).$$

The posterior expectation integral can then be written for large k, as

$$\mathrm{E}(\alpha \mid \mathbf{x}) \approx \int_{-\infty}^{k} \alpha \pi_J(\alpha \mid \mathbf{x}) d\alpha + \int_k^{\infty} \alpha \pi_J(\alpha) d\alpha = \text{const} + \infty.$$

[c) ⇒ a)] This can be proved by showing that [not a)] ⇒ [not c)]. Suppose, without loss of generality, there is just one negative observation, say $x_1 = -c$, $c > 0$. Then the right tail of $L(\alpha)$ is, for large k, proportional to $\Phi(-\alpha c) = \phi(\alpha c)(\alpha c) + o(1/\alpha)$. Recalling results (ii) in Theorem 9.4.1, the posterior expectation integral is, for a large k, and some constant a_1 and a_2,

$$\mathrm{E}(\alpha \mid \mathbf{x}) = a_1 + a_2 \int_k^{\infty} \alpha \frac{\phi(\alpha c)}{\alpha c} \left(\frac{1}{\alpha^{3/2}} + o(1/\alpha^2) \right) d\alpha < +\infty,$$

and the proof is complete. □

PART II

Applications and Case Studies

CHAPTER 10

Bayesian Multivariate Skewed Regression Modeling with an Application to Firm Size

José T. A. S. Ferreira and Mark F. J. Steel

10.1 Introduction

In this chapter, we study the application of Bayesian multivariate linear regression models, where the errors have skewed distributions belonging to one of two parametric classes. In particular, we apply the multivariate skewed distributions as defined by Ferreira and Steel (2003), henceforth denoted by FS, and by Sahu, Dey, and Branco (2003), henceforth denoted by SDB. The resulting regression models are employed in a study on firm size.

In FS we introduced a novel method for the generation of multivariate skewed distributions. A p-dimensional skewed distribution is defined via an affine linear transformation of independent univariate variables, each with a possibly skewed distribution. As is shown in FS, this method generates a very general class of distributions. For the distribution of the univariate components involved in the transformation, we suggested the method defined in Fernández and Steel (1998). We note however that other choices could have been made. The multivariate skewed distributions defined in this fashion share a number of interesting characteristics, including: direct analytical form of the probability density function (pdf), ease of moment calculation, analytical form of Mardia's (1970) measure of skewness in most cases, absence of restrictions on mean and covariance structure due to skewness, freedom from conditioning arguments, thus not involving cumulative distribution functions (cdf's) and freedom from the particular choice of coordinate axes. The main disadvantage of this class of distributions is that, in general, it is not closed under marginalization or conditioning.

The second class of multivariate skewed distributions that will be studied in this chapter is the one developed in SDB. Using a hidden truncation model (Arnold and Beaver, 2000a) and conditioning on as many unobserved quantities as variables, SDB extend the work of Azzalini and Dalla Valle (1996), and Branco and Dey (2001), and develop a more general class of skewed distributions. This class of distributions is among the most general within the hidden truncation modeling framework. It is closely related to the class of elliptically contoured distributions and it shares some of the properties of the latter, such as closeness under marginalization and conditioning. The evaluation of the pdf of these multivariate skewed distributions requires the calculation of a p-dimensional cdf, which can be problematic for high dimensions and/or for certain distributions. Further, it imposes that the skewness of the distribution is introduced along the coordinate axes, consequently restricting the flexibility of the class.

We apply both classes of multivariate skewed distributions in a Bayesian linear regression setup. We compare the methodologies using skewed and fat-tailed distributions, namely the skew-t distribution as defined by each method. We also analyze the models without fat tails, using skew-normal distributions. Finally, we compare these alternatives with the symmetric ones: t and normal. The prior distribution is always chosen to be proper and, for the common parameters, equal for both classes of distributions. Formal model comparison is carried out using Bayes factors.

We apply the Bayesian regression models to a study of the distribution of firm size. Using data for three hundred publicly traded companies we evaluate the validity of common economic hypotheses, such as the suitability of the law of proportionate effects (Gibrat, 1931). For all companies, data is available at two points in time: 1980 and 1990, permitting the study not only of the size distribution of the companies, but also of growth in the 1980's. We also examine the influence of research and development effort and investment on the distribution of the quantities of interest.

The remainder of this chapter is organized into four sections. In Section 10.2 we outline the two classes of multivariate skewed distributions. We introduce the Bayesian multivariate regression models in Section 10.3. Section 10.4 is devoted to the analysis of the firm size application. Finally, we offer a brief discussion in Section 10.5.

10.2 Multivariate Skewed Distributions

This section provides a brief review of the two classes of multivariate skewed distributions that are going to be studied in this chapter. Further details are available from the respective references. A number of other classes of multivariate skewed distributions is available in the literature and we refer the interested reader to the first part of the current edition. Our particular choice is inspired by the facts that both classes are quite general and have been analyzed separately in a Bayesian framework, similar to the one here.

In the sequel, we will apply the notation FS and SDB as prefixes to the skewed distributions, dropping the term "skew" (e.g., the skew-normal distribution as defined in Ferreira and Steel (2003) will be denoted as FS-normal).

10.2.1 FS Skewed Distributions

In FS, the authors introduce a general method for the construction of multivariate skewed distributions, based on affine linear transformations of univariate variables with skewed distribution. Let p, a positive integer, be the dimension of the random vector $\boldsymbol{\epsilon} = (\epsilon_1, \ldots, \epsilon_p)^T \in \mathbb{R}^p$ and $\boldsymbol{\gamma} = (\gamma_1, \ldots, \gamma_p)^T \in \mathbb{R}_+^p$. Also, let $\mathbf{f} = (f_1(\cdot), \ldots, f_p(\cdot))^T$ denote a vector of p unimodal and symmetric univariate pdf's. The distribution of $\boldsymbol{\epsilon}$ is a multivariate skewed distribution with independent components where, for $j = 1, \ldots, p$, the pdf of ϵ_j is $g(\epsilon_j|\gamma_j, f_j)$, which corresponds to a skewed version of the distribution with pdf $f_j(\cdot)$.

Following an affine transformation, given a vector $\boldsymbol{\mu} = (\mu_1, \ldots, \mu_p)^T$ and a non-singular matrix $A \in \mathbb{R}^{p \times p}$, the variable $\boldsymbol{\eta} = (\eta_1, \ldots, \eta_p)^T \in \mathbb{R}^p$, defined as

$$\boldsymbol{\eta} = A^T \boldsymbol{\epsilon} + \boldsymbol{\mu}, \tag{10.1}$$

has a general multivariate skewed distribution, with parameters $\boldsymbol{\mu}, A, \boldsymbol{\gamma}$, and $\mathbf{f}$, denoted by $FS(\boldsymbol{\mu}, A, \boldsymbol{\gamma}, \mathbf{f})$. The pdf for $\boldsymbol{\eta}$ is then simply given by

$$||A||^{-1} \prod_{j=1}^{p} g[(\boldsymbol{\eta} - \boldsymbol{\mu})^T A_{\cdot j}^{-1} | \gamma_j, f_j], \tag{10.2}$$

where $A_{\cdot j}^{-1}$ denotes the j-th column of A^{-1}, $||A||$ denotes the absolute value of the determinant of A, and $g(\cdot|\gamma_j, f_j)$ is the pdf corresponding to the univariate skewed distribution.

FS show that, in contrast to the elliptically contoured distribution case, knowledge of $A^T A$ is not sufficient. Further, a decomposition of the non-singular matrix $A = \tilde{O} U$ is applied, where $\tilde{O}$ is a $p \times p$ orthogonal matrix and U is a $p \times p$ upper triangular matrix with strictly positive diagonal elements. Straightforward manipulations show that $A^T A = U^T U$.

The skewed version of the symmetric pdf $f_j(\cdot)$ can be obtained using a number of different methods. In FS, the authors generate univariate skewed distributions using the method proposed by Fernández and Steel (1998). If $f_j(\cdot)$ is a univariate pdf that is symmetric around zero, decreasing in the absolute value of its argument, and if $\gamma_j \in (0, \infty)$, then the latter method defines

$$g(\epsilon_j|\gamma_j, f_j) = \frac{2}{\gamma_j + \frac{1}{\gamma_j}} f_j\left(\epsilon_j \gamma_j^{-\text{sign}(\epsilon_j)}\right), \tag{10.3}$$

to be the pdf of a univariate skewed distribution, where sign$(\cdot)$ is the usual sign function.

The multivariate skewed distributions generated by (10.1)-(10.3) have a number of interesting characteristics of which we highlight: its validity for any vector of univariate distributions $\mathbf{f}$, the dependence of the existence of moments on the existence of moments of the univariate distributions alone, and the possibility of unrestricted modeling of mean, variance, and skewness.

10.2.2 SDB Skewed Distributions

In SDB, the authors introduce a novel method for the introduction of skewness into elliptically contoured distributions.

Let Σ denote a $p \times p$ covariance matrix and $\boldsymbol{\mu} \in \mathbb{R}^p$. Then, a continuous elliptically contoured distribution of the p-dimensional vector $\boldsymbol{\epsilon}$ can be defined by the pdf

$$f(\boldsymbol{\epsilon}|\boldsymbol{\mu}, \Sigma, g^{(p)}) = |\Sigma|^{-\frac{1}{2}} g^{(p)} \left[(\boldsymbol{\epsilon} - \boldsymbol{\mu})^T \Sigma^{-1} (\boldsymbol{\epsilon} - \boldsymbol{\mu})\right], \tag{10.4}$$

where $g^{(p)}(\cdot)$ is a function from $\mathbb{R}_+$ to $\mathbb{R}_+$ given by

$$g^{(p)}(x) = \frac{\Gamma\left(\frac{p}{2}\right)}{\pi^{\frac{p}{2}}} \frac{g(x;p)}{\int_0^\infty r^{\frac{p}{2}-1} g(r;p) dr}, \tag{10.5}$$

with $g(x;p)$ a non-increasing function from $\mathbb{R}_+$ to $\mathbb{R}_+$ such that the integral in (10.5) exists. Now let

$$\boldsymbol{\varphi} = (\boldsymbol{\psi}^T, \boldsymbol{\epsilon}^T)^T, \ \boldsymbol{\mu} = (\boldsymbol{\mu}^{*T}, \mathbf{0}^T)^T, \text{ and } \Sigma = \begin{pmatrix} \Sigma^* & O \\ O & I_p \end{pmatrix},$$

where $\boldsymbol{\psi}, \boldsymbol{\epsilon}$ and $\boldsymbol{\mu}^*$ are in $\mathbb{R}^p$, $\mathbf{0}$ is the p-dimensional zero vector, Σ^* is a $p \times p$ covariance matrix, I_p denotes the p-dimensional identity matrix, and O is a matrix of zeros. Further, let $\boldsymbol{\varphi}$ have pdf $f(\boldsymbol{\varphi}|\boldsymbol{\mu}, \Sigma, g^{(2p)})$ as in (10.4). By defining

$$\boldsymbol{\eta} = D\boldsymbol{\epsilon} + \boldsymbol{\psi},$$

where $D = \text{diag}(\boldsymbol{\delta})$, $\boldsymbol{\delta} \in \mathbb{R}^p$, the random variable $\boldsymbol{\eta}|\boldsymbol{\epsilon} > \mathbf{0}$ has a multivariate skewed distribution as defined by SDB, denoted by $SDB(\boldsymbol{\mu}, \Sigma, \boldsymbol{\delta}, g^{(p)})$. The conditional pdf of $\boldsymbol{\eta}|\boldsymbol{\epsilon} > \mathbf{0}$ is given by

$$2^p f(\boldsymbol{\eta}|\boldsymbol{\mu}, \Sigma + D^2, g^{(p)}) \times$$
$$\mathrm{P}\left[\mathbf{v} > \mathbf{0} \middle| D(\Sigma + D^2)^{-1}(\boldsymbol{\eta} - \boldsymbol{\mu}), I_p - D(\Sigma + D^2)^{-1} D, g_\alpha^{(m)}\right], \tag{10.6}$$

with $\mathbf{v} \in \mathbb{R}^p$ and $\mathrm{P}(\cdot|\boldsymbol{\mu}, \Sigma, g^{(p)})$ the probability function corresponding to the pdf (10.4),

$$g_a^{(p)}(x) = \frac{\Gamma\left(\frac{p}{2}\right)}{\pi^{\frac{p}{2}}} \frac{g(a + x; 2p)}{\int_0^\infty r^{\frac{p}{2}-1} g(a + r; 2p) dr}, \tag{10.7}$$

and finally $\alpha = (\boldsymbol{\eta} - \boldsymbol{\mu})^T (\Sigma + D^2)^{-1} (\boldsymbol{\eta} - \boldsymbol{\mu})$.

The class of distributions described by (10.6) has the property of being closed under marginalization and conditioning. The main practical problem in using these distributions is that the calculation of the cdf required in (10.6) can be complicated, especially in higher dimensions and for certain distributions. SBD remark that an MCMC sampler does not require evaluating this cdf, but does necessitate drawings from truncated multivariate distributions which can be computationally difficult.

10.3 Regression Models

In the sequel, we assume that n observations from an unknown underlying process are available, each of which is given as a pair $(\mathbf{y}_i, \mathbf{x}_i)$, $i = 1, \ldots, n$. For each i, $\mathbf{y}_i \in \mathbb{R}^p$ represents the variable of interest and $\mathbf{x}_i \in \mathbb{R}^k$ is a vector of covariates. Throughout, we condition on $\mathbf{x}_i$ without explicit mention in the text.

We assume that the process generating the variable of interest can be described by independent sampling for $i = 1, \ldots, n$ from the linear regression model

$$\mathbf{y}_i = \lambda_i^{-1/2} \boldsymbol{\eta}_i + B^T \mathbf{x}_i, \tag{10.8}$$

where $\lambda_i \in \mathbb{R}_+$ has some distribution parameterized by ν, B is a $k \times p$ matrix of real coefficients, and $\boldsymbol{\eta}_i \in \mathbb{R}^p$ has a distribution with a specific form.

In this chapter we are going to assume that $\boldsymbol{\eta}_i$ follows one of two alternatives: $FS(\mathbf{0}, A, \boldsymbol{\gamma}, \mathbf{f})$ or $SDB(\mathbf{0}, \Sigma, \boldsymbol{\delta}, g^{(p)})$, where $\mathbf{f}$ is the p-dimensional vector with all components equal to $\phi(\cdot)$, the standard normal pdf and

$$g^{(p)}(u) = \frac{\exp\{-u/2\}}{(2\pi)^{p/2}},$$

corresponding to the skew-normal distributions for both classes.

In a similar manner as when dealing with mixtures of normals, imposing a specific (mixture) distribution on λ_i extends (10.8) to a substantially larger class of distributions. As examples, imposing a Dirac prior on $\lambda_i = 1$ retrieves the skew-normal distributions, while if λ_i has a gamma distribution with both precision and shape parameters equal to $\nu/2$, then $\mathbf{y}_i$ has a skew-t distribution with ν degrees of freedom (df). The present chapter will assume one of the two alternatives above. Assuming that $\boldsymbol{\gamma} = \mathbf{1}_p$ or that $\boldsymbol{\delta} = \mathbf{0}$ generates the symmetric special cases of the distributions, obviously coinciding for both classes. In summary, in the present chapter we will consider one pair of models that can model both skewness and fat tails (FS-t and SDB-t), one pair of models that can model skewness alone (FS-normal and SDB-normal), one model than can model only fat tails (Student-t) and, finally, one model that can neither model skewness nor fat tails (normal).

10.3.1 Prior Distributions

The definition of the Bayesian models is completed with the specification of the prior distribution of the unknown parameters. In order to compare

FS and SDB skewed distributions, we specify common prior distributions whenever possible. For FS, we assume that the prior is of the form

$$\mathrm{P}_{B,A,\gamma,\nu} = \mathrm{P}_{B|A}\mathrm{P}_A\mathrm{P}_\gamma\mathrm{P}_\nu.$$

In a similar manner, we assume that for the SDB models, the prior structure is

$$\mathrm{P}_{B,\Sigma,\boldsymbol{\delta},\nu} = \mathrm{P}_{B|\Sigma}\mathrm{P}_\Sigma\mathrm{P}_{\boldsymbol{\delta}}\mathrm{P}_\nu.$$

Using the decomposition of a non-singular matrix $A = \tilde{O}U$, the decomposition of a covariance matrix $\Sigma = U^T U$, and the fact that $A^T A = U^T U = \Sigma$, we note that $\tilde{O}$ in the FS parameterization has no counterpart in the SDB model. We impose a prior on U for both models through an inverted Wishart distribution on Σ. The latter has as parameters $q > p - 1$ and Q, a $p \times p$ covariance matrix, and its pdf is proportional to

$$|Q|^{\frac{q}{2}}|\Sigma|^{-\frac{p+q+1}{2}} \exp\left(-\tfrac{1}{2}\ \mathrm{tr}\ (\Sigma^{-1}Q)\right),$$

with tr denoting the trace operation.

As in FS, $\tilde{O}$ has a distribution on $\mathcal{O}^p$ that is invariant to linear orthogonal transformations, where $\mathcal{O}^p$ is a set of orthogonal matrices that ensures identifiability (see Appendix B of FS for details).

The prior on B is set conditional on Σ, and is taken to be a matrix-variate normal with parameters B_0, Σ, and M, with B_0 a $k \times p$ matrix of real components and M a $k \times k$ covariance matrix. Then, the pdf of the prior distribution on B is proportional to

$$|M|^{-\frac{p}{2}}|\Sigma|^{-\frac{k}{2}} \exp\left[-\tfrac{1}{2}\ \mathrm{tr}\ (\Sigma^{-1}(B - B_0)^T M^{-1}(B - B_0))\right].$$

For the models with heavy tails, the parameter ν controls the df. An exponential prior on ν, with hyperparameter d and restricted to $(3, \infty)$, is imposed. This prior does not allow for extremely heavy tails, as it imposes the existence of the first three moments of the distributions. This was not seen to be too restrictive for the applications in Section 10.4.

The priors on $\boldsymbol{\gamma}$ and $\boldsymbol{\delta}$ used here are the ones suggested by FS and SDB. We assume that $\boldsymbol{\gamma} \in \mathbb{R}^p_+$ has prior distribution with pdf

$$\prod_{j=1}^p (2\pi s)^{-1/2}\Big\{\gamma_j^{-2} \exp\left[-\tfrac{(1-\gamma_j^{-1})^2}{2s}\right] I_{(0,1)}(\gamma_j) + \exp\left[-\tfrac{(\gamma_j-1)^2}{2s}\right] I_{[1,\infty)}(\gamma_j)\Big\},$$

imposing that for any two constants such that $1 < \gamma_a < \gamma_b$, we have that $\mathrm{P}[\gamma_j \in (\gamma_a, \gamma_b)] = \mathrm{P}\left[\gamma_j \in \left(\frac{1}{\gamma_b}, \frac{1}{\gamma_a}\right)\right]$, which is inspired by symmetry considerations in the FS model. The vector $\boldsymbol{\delta}$ is assumed to have a normal distribution with zero mean and covariance Γ.

10.3.2 Numerical Implementation

In order to conduct inference, numerical methods have to be adopted. In particular, we construct Markov chain Monte Carlo (MCMC) methods. For both classes of models, we used hybrid samplers composed of Metropolis-Hastings steps for all parameters except ν for both classes and for λ_i, $i = 1, \ldots, n$ for the FS models, where Gibbs steps are easy to implement. For the FS models, the sampler is close to the one in FS. For the SDB models, we use a sampler similar to a corrected version of the one suggested in SDB. The correction to SDB's sampler can be found on the webpage of the first author.

Details of the MCMC samplers are omitted here on grounds of brevity, but a description of the samplers, as well as our Matlab implementation of the Bayesian models, can be obtained from the authors upon request.

10.4 Application to Firm Size

The relative sizes of firms, their dynamics, and their relation to the firms' particular characteristics, is an important problem in economics and the focus of substantial research effort. A large review of studies on firm size can be found in Ahn (2001). In this section, we perform a study of the size distribution of a cohort of three hundred companies, in the manufacturing sector, in the years 1980 and 1990.

The set of firms that we study here originates from a larger cohort, created and maintained by Bronwyn H. Hall, containing information on about 3,000 publicly traded companies in the US manufacturing sector. The original data set contains the records of an unbalanced panel from 1951–1991. Further information about the data set and panel can be found in Hall (1993a,b).

From the complete data set we randomly selected a cohort of three hundred firms for which data is available for both 1980 and 1990. We are interested in studying the overall distribution of three measures of firm size (market value, tangible assets, and sales), under the influence of two cofactors: research and development (R&D) effort, and investment. The measures of firm size are expressed as the logarithm of the original values in millions of dollars. The covariates R&D and Investment are measured as the ratio between quantity spent and total assets, both standardized to have mean zero and unit variance. In addition to our predictors, we also included a constant term.

We stress that the cohort of firms that we study is not a random sample from the set of all of US manufacturing firms. All firms in this study were publicly traded during the period 1980–1990, implying that small firms, less likely to be quoted in the stock market, are underrepresented in the cohort. Further, we impose that the firms have survived the period of study, implying that our cohort of firms does not contain failing firms. See Geroski, Lazarova, Urga, and Walters (2003, p. 51) for a discussion of how this could affect the results.

In this chapter, we analyze the joint distribution of the firm size variables for 1980 and 1990 and their growth between 1980 and 1990. We are especially interested in testing for the presence of skewness and fat tails in the distributions. The existence of skewed distributions for measures of firm size is not a novel hypothesis in economic theory. In fact, it has been suggested in many previous studies, a good review of which is provided by Sutton (1997). Gibrat (1931) introduced the law of proportionate effect, also known as Gibrat's law, where the firm size variables are assumed to have a lognormal distribution.

In order to investigate the presence of skewness, Figure 10.1 presents pairwise scatterplots for the firm size measures at 1980 and 1990. Particularly from the plots on the two left-most columns of the figure, some skewness is apparent. The presence of skewness is also suggested by marginal skewness measures. For the 1980 data the sample skewness is 0.34, 0.10 and -0.16 for Market value, Tangible assets, and Sales, respectively; for the 1990 data the sample skewness is, in the same order, 0.24, 0.17 and -0.03.

The full definition of the Bayesian models introduced in Section 10.3 requires the setting of the hyperparameters. In what follows, B_0 and M, in the matrix-variate prior for B, are set to the $k \times p$ zero matrix and $100I_k$, respectively. For the prior on Σ, Q is set to I_p and q is set to $p+2$ (ensuring the existence of a prior mean). The remaining hyperparameters are set as

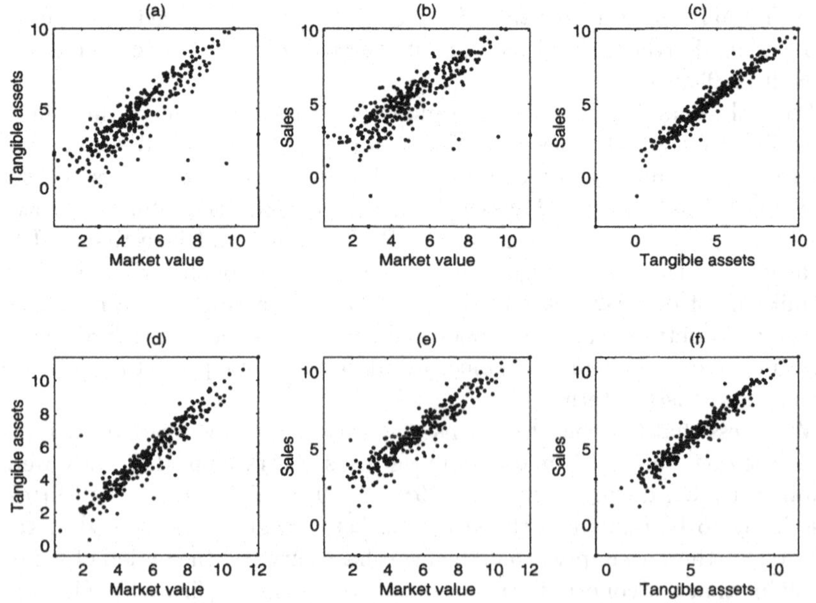

Figure 10.1 *Pairwise scatterplots of the firm size measures at 1980 (top row) and 1990 (bottom row).*

in FS and SDB. In particular, s is set to unity, $\Gamma = 100I_p$, and $d = 0.1$. These settings generate a rather vague prior on the model parameters.

Inference for the models was always conducted using every tenth realization from a chain of 50,000 iterations. A burn-in period of 10,000 samples preceded the collection phase. For all models, convergence was achieved early in the burn-in period.

Formal model comparison will be conducted by comparing marginal likelihoods. The p_4 estimator in Newton and Raftery (1994) is used to provide estimates of marginal likelihoods, with their δ set to 0.1.

We now provide posterior and predictive inference divided into two sections, the first for the cross-sectional studies on the distribution of firm size at 1980 and 1990, and the second for the analysis of firm growth for that period.

10.4.1 Distribution of Firm Size

The presence of skewness in the distribution of firm size can easily be determined by examining the marginal posterior distributions of $\boldsymbol{\gamma}$ and $\boldsymbol{\delta}$. For the FS models, a $\boldsymbol{\gamma}$ different from the p-dimensional vector of ones indicates skewness. For the SDB models, the same holds if $\boldsymbol{\delta}$ differs from the p-dimensional zero vector. Figure 10.2 presents estimates of the posterior density of the components of $\boldsymbol{\gamma}$ (left column) and of $\boldsymbol{\delta}$ (right column), for the data relating to 1980 (upper row) and 1990 (lower row) for the skew-t models. In the plots, the posterior densities are represented by lines, one for each component of $\boldsymbol{\gamma}$ or $\boldsymbol{\delta}$. In addition, in every plot the prior distribution for the parameter is also presented. For both points in time, for the FS models evidence on the presence of skewness is rather strong. The $\boldsymbol{\gamma}$ parameter of the FS-t estimated for the 1980 data has two components with distributions markedly centered away from unity. For 1990, the same model reveals that all components are different from unity. The presence of skewness in the SDB models is less evident, with only one component of $\boldsymbol{\delta}$ substantially different from zero.

We now examine the effect of R&D and Investment in the distribution of firm size. Table 10.1 presents summaries of the marginal posterior distributions of the regression coefficients, estimated using the FS-t model. The magnitude of the coefficients provides evidence that the influence of covariates is restricted. Nevertheless, some conclusions can be drawn. Research and development seem to have a mostly negative effect on the firm size variables in 1980. A possible explanation is the fact that if a firm assigns a substantial amount to R&D, then its immediate sales and turnover are reduced and, as a consequence, so is its value. In contrast, Investment is seen to have a mostly positive effect on Market value and Tangible assets. Both covariates appear to have been more influential in 1980 than in 1990. The SDB model leads to similar inference on the regression coefficients.

So far, we have noted the presence of skewness in the distributions. Now, we study the tails of the distributions. Figure 10.3 presents the posterior density for ν for the heavy tailed models, for 1980 (left) and 1990 (right). As is evident, heavy tails are strongly supported by the data. All models have similar estimates for the pdf of ν, with the symmetric Student-t model focusing on slightly lower values. This may be a consequence of modeling with an elliptically contoured distribution which does not capture the skewness present in the data. Heavier tails are then required to account for the inadequacy of the elliptical model.

We conclude the analysis of the distribution of firm size with a formal

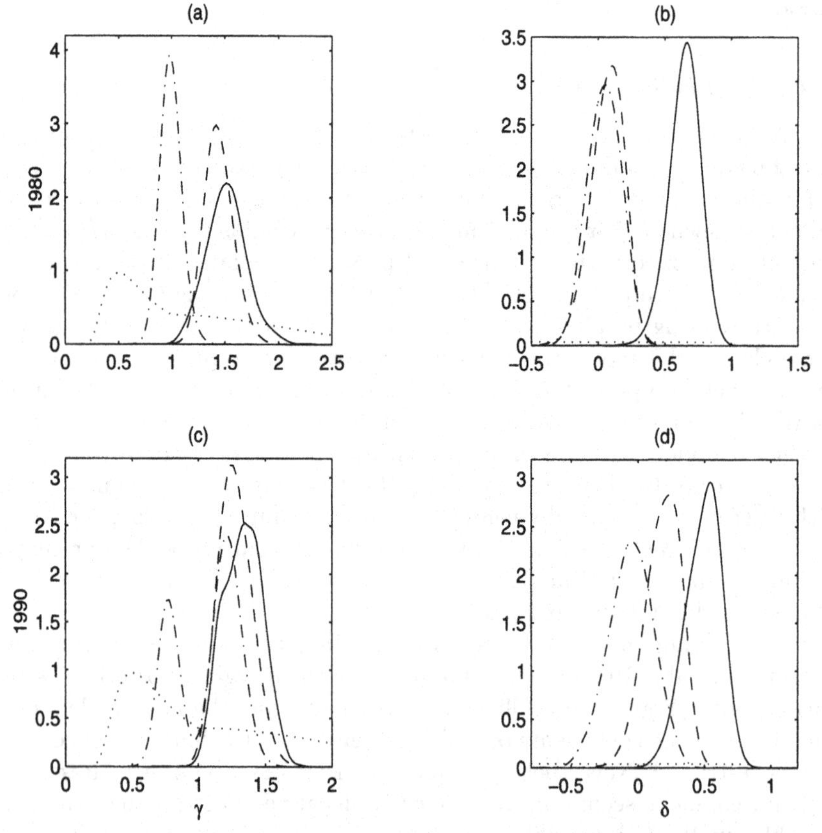

Figure 10.2 *Marginal posterior pdf of the components of* γ *(left column) and* $\boldsymbol{\delta}$ *(right column), for the data relating to 1980 (upper row) and 1990 (lower row), for the skew-t models. The prior distribution of the parameter is shown by the dotted line, with the other lines standing for the different components of vectors* γ *or* $\boldsymbol{\delta}$.

Table 10.1 *Summaries of the marginal posterior distributions of the regression coefficients, estimated using the FS-t model*

Year	Covariate	Size var.	Mean	Std	5% quant.	95% quant.
1980	R&D	M. value	0.02	0.13	−0.18	0.24
		T. assets	−0.36	0.13	−0.57	−0.14
		Sales	−0.30	0.12	−0.50	−0.10
	Investment	M. value	0.32	0.10	0.15	0.50
		T. assets	0.13	0.11	−0.05	0.31
		Sales	0.04	0.10	−0.12	0.22
1990	R&D	M. value	−0.03	0.12	−0.22	0.16
		T. assets	−0.05	0.11	−0.22	0.14
		Sales	0.01	0.10	−0.16	0.18
	Investment	M. value	0.28	0.27	−0.18	0.73
		T. assets	0.15	0.25	−0.26	0.58
		Sales	0.19	0.21	−0.14	0.57

comparison of the different models. Table 10.2 shows the difference in log marginal likelihood between the FS-t model and all other models. A negative value denotes an advantage for the FS-t model. If prior model proba-

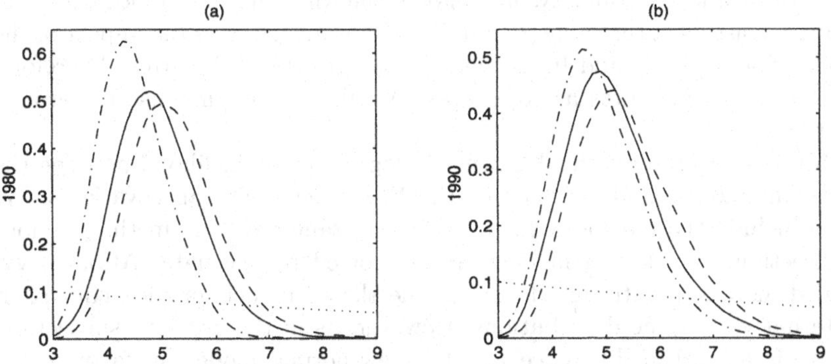

Figure 10.3 *Marginal posterior pdf of ν for the FS-t (solid line), the SDB-t (dashed line) and the Student-t (dot-dashed line) models. Panel (a) plots the estimates for 1980 and (b) the estimates for 1990. The prior distribution of the parameter is shown by the dotted line.*

Table 10.2 *Distribution of firm size – difference in log marginal likelihood between the FS-t model and all other models. Negative values denote advantage for the FS-t model.*

	Student-t			Normal		
	FS	SDB	Elliptical	FS	SDB	Elliptical
1980	0	−6.2	−12.7	−56.0	−95.2	−117.3
1990	0	−3.3	−11.0	−46.5	−49.7	−58.4

bilities are assumed equal for all models, the exponentials of the values in Table 10.2 are posterior odds versus the FS-t model. For both years, models that allow for heavier tails are shown to be more adequate. Also, skewed models are preferred, with the FS models getting more support than their SDB counterparts. In addition, the FS-normal models dominate the SDB-normal ones. In summary, the results from Table 10.2 suggest strong data support for skewness and fat tails. Thus, the symmetry and normal tail behavior assumptions (the law of proportionate effects) do not hold for the set of companies that we analyze in this chapter.

10.4.2 Analysis of Firm Growth

The second part of our application is devoted to the study of firm growth. We analyze the growth of Market value and of Tangible assets between 1980 and 1990. We define $\mathbf{y}_i = (y_i^1, y_i^2)^T,\ i = 1, \ldots, n$, where for observation i, y_i^1 represents the difference in the logarithm of Market value between 1990 and 1980, and equivalently, y_i^2 denotes the difference in the logarithm of Tangible assets in the same period. We also want to assess the impact of the 1980 effort in R&D and Investment on the growth of the firms. Therefore, we use these measures as covariates. We also include a constant term in our analysis.

Table 10.3 presents the difference in log marginal likelihood between the different models and the one for the FS-t model with all covariates. We also include the models with only the constant term. As in the previous subsection, the skewed models are supported by the data. Also, heavy-tailed models are strongly favored. The skew-t models provide equivalent alternatives for the distributions. However, as in the previous subsection, there is a marked difference for the skew-normal models in favor of the FS ones, especially when covariate information is used. This ranking of the models is not affected by the inclusion of the R&D and Investment cofactors, even though the covariates are assessed to be important. For all models, the covariates prove to be relevant.

The influence of the covariates on firm growth can be assessed by the

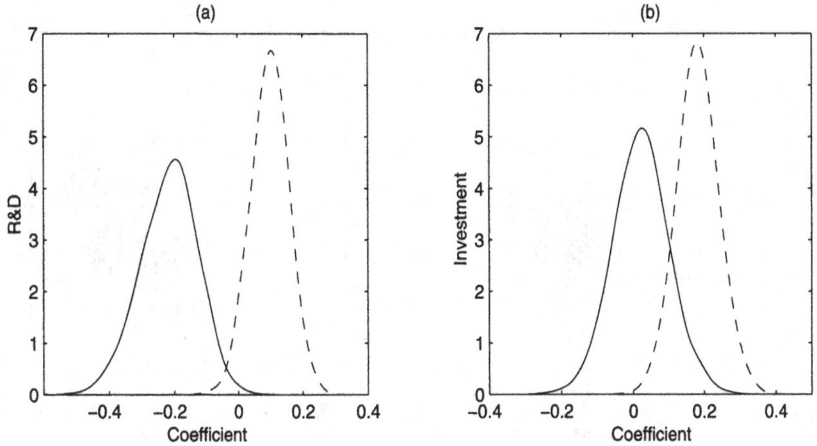

Figure 10.4 *FS-t estimate of the marginal posterior densities of the coefficients of covariates R&D (a) and Investment (b). The solid lines represent the estimates of the coefficient for Market value growth and the dashed lines the ones for Tangible asset growth.*

plots in Figure 10.4, where posterior pdf's for the coefficients of the covariates are presented. These estimates were obtained from the FS-t models. Figure 10.4(a) shows the distinct effect of R&D on the two measures of firm growth. R&D effort in 1980 has a positive effect on Tangible asset growth and a negative effect on the growth of Market value. From Figure 10.4(b) we realize that Investment has almost no effect on Market value and has a strong positive effect on Tangible assets. In summary, growth in Tangible assets is positively affected by both R&D and Investment, while growth in Market value is negatively affected by R&D and is not affected by Investment.

As evidenced in Table 10.3 heavy tails are supported by the data. The posterior median value of ν for the FS-t model is 3.80 when covariates are

Table 10.3 *Distribution of firm growth – difference in log marginal likelihood between the FS-t model with all covariates and all other models. Negative values denote advantage for the FS-t model with covariates.*

	Student-t			Normal		
	FS	SDB	Elliptical	FS	SDB	Elliptical
With covariates	0	0.4	−25.6	−49.3	−68.4	−87.8
Without	−32.2	−33.0	−60.6	−75.6	−77.2	−111.5

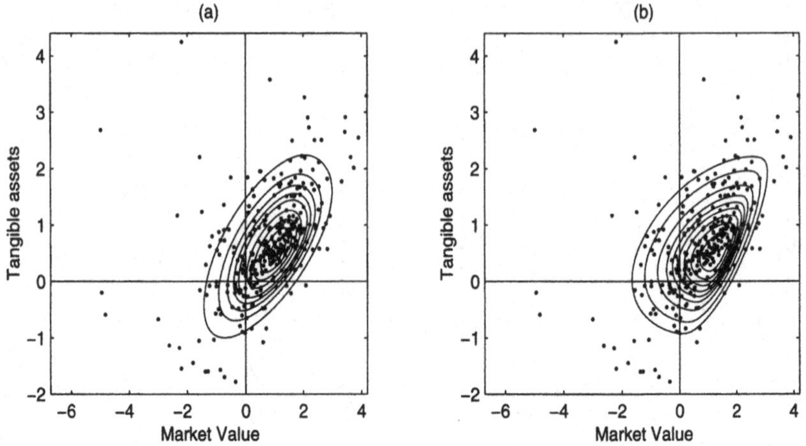

Figure 10.5 *Data (dots) and contours of the posterior predictive density of the FS-t (a) and FS-normal (b) models with only the constant term as regressor.*

used and 3.70 when only the constant term is used. The median values for the remaining heavy-tailed models are similar. The presence of heavy tails does not support the common assumption in economics that firm growth has a normal distribution.

Until now, we have not graphically presented any skewed distribution. The application in Section 10.4.1, with its three covariates, is not directly suitable for visualization. However, here we analyze a bidimensional problem and we can provide some contour plots of the posterior predictive pdf's. Figure 10.5 presents contour plots of the posterior predictive densities, for the FS-t model and FS-normal with only the constant term, overplotted with the data represented by the dots. The presence of skewness in the distribution is obvious, especially for the FS-normal model. By fixing tail behavior to be the one of the normal distribution, the distribution has to be substantially more skewed than if heavier tails are allowed. This can explain the reason for the near equivalence of the FS and SDB models for the Student-t alternatives, and for the advantage of the FS models when only normal tails are allowed. The majority of firms experienced growth during the ten-years period up to 1990. However, a considerable number of firms exhibited a decrease, sometimes considerable, in one or both measures. In particular, after inspecting the contours of the pdf's, we can conclude that a firm having its Market value diminishing while increasing Tangible assets is slightly more likely than the opposite. This can be generated by a poor opinion of the firm by the market, even when Tangible assets are increasing. As these are manufacturing firms, this could be linked with the general decline in the manufacturing sector in that period.

Contour plots of the posterior predictive pdf's for the SDB models are not included due to the fact that their computation is much more demanding. Due to the pdf form in (10.6)-(10.7), the computation of the posterior predictive pdf contours would require substantial computing effort.

10.5 Discussion

In this chapter we review and compare two different alternatives for multivariate skewed distributions. The first methodology, introduced in Ferreira and Steel (2003), generates multivariate skewed distributions by using linear transformations of univariate skewed distributions. The second, proposed by Sahu *et al.* (2003), uses a hidden truncation framework, where one unobserved component is required for each dimension.

We provide general Bayesian linear regression models that can allow for both skewness and heavy tails. FS have derived sufficient conditions for inference under intuitive improper priors. However, as similar conditions are not available for the SDB models, we conduct inference here under a proper prior, when possible common to both models. We also analyze restricted versions of the most general models, by excluding either skewness, heavy tails, or both.

The regression models are used in two econometric applications: cross-sectional studies of the distribution of a cohort of manufacturing firms, in two time periods, and the analysis of the growth of the same firms between the two periods. We also assess the relevance of R&D and Investment on the distributions. Fat tails and skewness found support in both applications.

The FS models were seen to be more suited to the applications at hand. The preference was substantially stronger for the normal models, indicating the flexibility of the FS models. One methodological reason for this flexibility is that the FS class of distributions can model mean, covariance, and skewness separately. The same is not true for the SDB class, where fixing Σ and increasing $\boldsymbol{\delta}$ in absolute value leads to a decrease in the correlation between the variables. Thus, in the SDB class of distributions, it is not possible to model simultaneously highly correlated and heavily skewed data. We feel that this is an advantage of the FS models, and according to the model comparison we performed in this article, one that can have strong practical relevance.

In addition, the models proposed in FS are much more computationally efficient, both in terms of model fitting and model comparison. The numerical methods necessary to conduct inference under the SDB models rely on a data augmentation procedure requiring one truncated multivariate variable for each observation. The updating of these variables can be quite demanding. Model comparison based on marginal posterior probabilities is also an expensive computational issue. Estimation of these marginal probabilities can be done, at least, in two different ways. The first involves calculating multivariate cdf values, which is difficult for most common distributions, especially in high dimensions. The second is to apply Monte Carlo integration using the data augmentation procedure used in model fitting, with the augmented variables sampled from the prior. None of these procedures is required for the FS models, where the pdf of the skewed distribution has a much more explicit form which is straightforward to evaluate for most cases.

Acknowledgments

The work of the first author was supported by grant SFRH BD 1399 2000 from Fundação para a Ciência e Tecnologia, Ministério para a Ciência e Tecnologia, Portugal.

CHAPTER 11

Capital Asset Pricing for UK Stocks under the Multivariate Skew-Normal Distribution

Chris Adcock

11.1 Introduction

The standard theories of modern finance are based on the assumption that returns on risky financial assets follow a multivariate normal distribution. The best known example of the explicit assumption of normality is perhaps the option pricing formula developed by Black and Scholes (1973), and Merton (1973). In other areas of finance theory, the assumption is made implicitly. For example, the theory of mean-variance portfolio selection developed by Markowitz (1952) does not require normality of returns per se, but does assume that asset returns are characterized by their first and second moments and that, furthermore, an investor's utility function will be a function only of the mean and variance of portfolio returns. Implicit normality is also a feature of the CAPM, the capital asset pricing model, of Sharpe (1964), Lintner (1965), and Mossin (1966). The CAPM, which is derived using Markowitz' efficient frontier in conjunction with various assumptions about investor behavior, is a linear model that relates the expected return on an asset, i say, over a given investment period to the expected return on the market portfolio, m say, over the same investment period. This model states that

$$\mu_i = R_f + \beta_i(\mu_m - R_f), \tag{11.1}$$

where R_f is the return on a risk-free instrument, such as a T-bill or LIBOR in the UK. The parameters μ_i and μ_m are, respectively, the expected return on asset i and the market portfolio, and β_i is the covariance between returns on asset i and the market portfolio, divided by the variance of returns on the market portfolio. In the derivation of the CAPM, the market portfolio is a portfolio of "all assets." For the purposes of this article, it may be thought of as a stock market index, such as the FTSE100 or the S&P500.

The assumption of normality is implicit in the market model. This is, in effect, an operational version of the CAPM in which expected returns are replaced by observations. Derived from the CAPM, the standard form of the market model is a linear relationship between the return on an asset and the return on the market portfolio:

$$R_i = R_f + \beta_i(R_m - R_f) + \epsilon_i, \tag{11.2}$$

where R_i and R_m are the observed returns over the investment period for asset i and the market portfolio respectively, and ϵ_i is an unobserved residual component of return with zero expected value and $\text{Var}(\epsilon_i) = \sigma_i^2$. The correspondence between the two equations is clear and the CAPM gives an obvious intuition to the market model above. The parameters of the market model are conventionally estimated using ordinary least squares (OLS). The estimated regression coefficient is universally known in finance as beta. It is probably still the most common measure of the risk of a security relative to the market. Making the conventional assumption that the risk-free return is known, and ignoring ϵ_i for the present, it is clear that according to this model, the variance of returns is $\beta_i^2\sigma_m^2$, where σ_m^2 is the variance of the market portfolio.

The standard regression assumption that the residuals are distributed independently of market returns has an important consequence in finance. This permits the interpretation of the ϵ_i as stock specific returns which are unrelated to market movements and hence to think of the residual standard deviation, σ_i for asset i, as a component of the risk of the security that is unrelated to market risk. It is also customary to assume that the error terms are cross-sectionally independent. This means that the covariance matrix of assets returns conditional on market returns is diagonal with elements σ_i^2. Another important aspect of the market model is that the excess return on a security over the risk-free rate, $R_i - R_f$, is proportional to the excess return on the market: that is, there is no intercept. From an empirical perspective, it is obvious that a security with a positive intercept might, ceteris paribus, be preferred to one which complied with the market model above. Not surprisingly, the "search for alpha," as it is called, is an important activity for investment practitioners.

Formally, the market model is derived by considering the conditional multivariate probability distribution of the return vector $\mathbf{r}$ given R_m and then assuming in addition that the CAPM holds. When returns are not normal, and depending on the probability distribution under consideration, this approach can lead to different insights into the relationship between asset returns and the return on the market portfolio.

It is well-known that in practice returns on risky financial assets are not normally distributed. Most financial assets exhibit excess kurtosis: their return distributions are fat-tailed. Many securities also exhibit skewness. There are many papers in the finance and related literature that examine the implications of both types of non-normality for modeling the distribution of returns and for the task of portfolio selection. Broadly speaking, there are three issues to address. The first is to specify and then to fit a suitable model that describes the non-normal distribution of returns. There are numerous univariate non-normal probability distributions, which are applied in finance. Since the pioneering work of Engle (1982), and Bollerslev (1986), it is also common practice to deal with non-normality using econometric methods. However, the task of portfolio selection really requires the use of a multivariate probability distribution. This is needed to describe the dependencies between returns on different securities, as well as the non-normality that can readily be observed in the marginal distributions of returns on each security. Nonetheless, this is easier said than done and it remains common practice, even for portfolio selection, to use models for which the residual terms are cross-sectionally independent even though the marginal probability distributions are non-normal.

The second issue is to employ a suitable utility function that will include both skewness and kurtosis and also reflect the investor's attitude towards them. To date, the most common general approach is to extend Markowitz' quadratic utility function by including portfolio skewness and kurtosis. The third issue is the construction of the market model and the estimation of its parameters. This is an important area of investigation because it is obvious to all that the return on the market has a major influence on the returns of most securities. As long as the CAPM holds, it is acceptable to posit a linear market model of the general type (11.2) above. However, even though the expected value of ϵ_i will be zero, the appropriate method of estimation will depend on its probability distribution.

The effect of kurtosis in returns has been addressed widely in the finance literature. The literature includes numerous papers that document the use of the t distribution as a model for returns. Examples include the well-known studies of Blattberg and Gonedes (1974), and Praetz (1972), as well as more recent works by Aparicio Acosta and Estrada (2000). Application in finance of the multivariate t distribution, the most commonly used member of the elliptically symmetric family after the normal, is reported in Chamberlain (1983), Zhou (1993), and Adcock and Shutes (2000).

The aim of the study reported here is to study skewness. There are several models which have been used to model skewness for individual financial assets. Examples include several papers by MacDonald and his co-workers, MacDonald and Newey (1988), MacDonald and Nelson (1989), and MacDonald and Xu (1995), as well as a recent empirical study by Theodossiou (1998). However, these are univariate models. The specific aim of this chapter is to apply a version of the increasingly well-known multivariate skew-normal distribution to the task of estimating the market model, which relates asset returns to the return on the market portfolio. The multivariate skew-normal distribution, henceforth SN distribution, offers a coherent model for the returns of any number of assets. This is an advantage over many other non-normal probability distributions that can deal effectively with non-normality in one dimension, but cannot easily be generalized to the multivariate case. As shown below, the SN distribution also leads to an explicit form for the market model which may be correctly estimated using the method of maximum likelihood (ML). The SN-based market model also offers some new insights into the relationship between returns on a security and returns on the market portfolio. Although not studied in detail in this chapter, the SN model also offers a tractable and simple platform for portfolio selection based on the maximization of expected utility.

The structure of this chapter is as follows. Section 11.2 summarizes the form of the multivariate skew-normal distribution that is used. Section 11.3 gives the market model and its properties. Section 11.4 describes the estimation methodology and the data. This is used in an empirical study into some of the constituent stocks of the FTSE100 index and is described in Section 11.5. Section 11.6 concludes.

The notation throughout is that in common use. The letter $\mathbf{r}$ denotes a column vector of random variables, $R_1, \ldots, R_p$ say. An observation on the vector $\mathbf{r}$ at time t is denoted $\mathbf{r}_t$. The use of the subscripts a and m in this article have a specific meaning: R_a is the return on an arbitrary portfolio and R_m is the return on the market portfolio. The notation $\mathbf{1}_p$ denotes a p-vector of ones, thus $\mathbf{1}_p^T\mathbf{x}$ is an alternative notation for $\sum_{i=1}^{p} x_i$ for any vector $\mathbf{x}$. The computations reported in this chapter were carried out using the software Splus.

11.2 The Multivariate Skew-Normal Model

The multivariate skew-normal distribution was introduced by Azzalini and Dalla Valle (1996). It is a natural extension of the univariate skew-normal distribution which was originally due to O'Hagan and Leonard (1976) and was developed in various articles by Azzalini (1985, 1986). A closely related class of distributions is described in Arnold and Beaver (2000a).

The standard form of the multivariate skew-normal distribution may be obtained by considering the distribution of a random vector, $\mathbf{r}$ say, which is defined as

$$\mathbf{r} = \mathbf{y} + \boldsymbol{\delta} U. \tag{11.3}$$

The random vector $\mathbf{y}$ has a full rank multivariate normal distribution with mean vector $\boldsymbol{\mu}$ and covariance matrix Σ. The random variable U which is independent of $\mathbf{y}$, has a standard normal distribution that is truncated below at zero. The elements of the vector $\boldsymbol{\delta}$ measure the sensitivity of the return on each asset to the truncated variable U.

For applications in finance, a modification of this distribution is employed, as reported in Adcock and Shutes (2001), henceforth A&S. The vectors $\mathbf{r}, \mathbf{y}$, and $\boldsymbol{\delta}$ are defined as above. The random variable U has a normal distribution with mean τ and variance 1, truncated below at zero. This modification generates a richer family of probability distributions. In addition, the empirical evidence described below suggests that the inclusion of τ as a parameter to be estimated is a useful feature. From the perspective of applications in finance, the random variable U may be interpreted as a non-negative shock, which is unobserved, but which affects all securities. The $\boldsymbol{\delta}$ parameters measure sensitivity of the return on each asset to this shock, whatever it may be. It may be noted that the idea of adding a skewness shock to a multivariate normally distributed vector of asset returns may be found in Simaan (1993), which predates A&S. Simaan's paper is, however, mainly concerned with the effect of the skewness shock on portfolio selection.

The probability distribution of $\mathbf{r}$ is SN with parameters $\boldsymbol{\mu}, \Sigma, \boldsymbol{\delta}$, and τ, denoted as $\mathbf{r} \sim SN_p(\boldsymbol{\mu}, \Sigma, \boldsymbol{\delta}, \tau)$. The probability density function of this distribution is

$$\phi_p(\mathbf{r}; \boldsymbol{\mu} + \boldsymbol{\delta}\tau, \Sigma + \boldsymbol{\delta}\boldsymbol{\delta}^T) \frac{\Phi\left(\dfrac{\tau + \boldsymbol{\delta}^T \Sigma^{-1}(\mathbf{r} - \boldsymbol{\mu})}{\sqrt{1 + \boldsymbol{\delta}^T \Sigma^{-1} \boldsymbol{\delta}}}\right)}{\Phi(\tau)},$$

where $\Phi(x)$ is the standard normal cumulative distribution function evaluated at x. The notation $\phi_p(\mathbf{x}; \boldsymbol{\omega}, W)$ denotes the probability density function, evaluated at $\mathbf{x}$, of a multivariate normal distribution with mean vector $\boldsymbol{\omega}$ and covariance matrix W. This result, which is reported in A&S, is essentially Azzalini and Dalla Valle's (1996) result with a change of notation and generalization to accommodate a non-zero value of τ; see also Arnold and Beaver (2002). The distribution of any sub-vector of $\mathbf{r}$, including the random variable R_i, is of the same form, based upon the corresponding sub-vectors of $\boldsymbol{\mu}$ and $\boldsymbol{\delta}$ and sub-matrix of Σ.

As reported in A&S, the moment generating function of this distribution, with $\mathbf{t}$ denoting a p-vector, is

$$M_{\mathbf{r}}(\mathbf{t}) = \exp\left[(\boldsymbol{\mu} + \boldsymbol{\delta}\tau)^T \mathbf{t} + \frac{1}{2}\mathbf{t}^T(\Sigma + \boldsymbol{\delta}\boldsymbol{\delta}^T)\mathbf{t}\right] \frac{\Phi(\boldsymbol{\delta}^T \mathbf{t} + \tau)}{\Phi(\tau)}.$$

The first two (multivariate) moments are given by

$$\mathrm{E}(\mathbf{r}) = \boldsymbol{\mu} + \boldsymbol{\delta}\{\tau + \xi_1(\tau)\},$$
$$\mathrm{Var}(\mathbf{r}) = \Sigma + \boldsymbol{\delta}\boldsymbol{\delta}^T\{1 + \xi_2(\tau)\},$$

where

$$\xi_k(x) = \frac{\partial^k \ln \Phi(x)}{\partial x^k}, \quad k = 1, 2, \ldots.$$

The distribution of a linear function of the elements of $\mathbf{r}$, $\mathbf{w}^T\mathbf{r}$ say, is also of the above SN form, albeit in one dimension, with scalar parameters $\mathbf{w}^T\boldsymbol{\mu}$, $\mathbf{w}^T\Sigma\mathbf{w}$, $\mathbf{w}^T\boldsymbol{\delta}$, and τ, respectively. Specifically, this applies to an arbitrary portfolio, a say, for which the return R_a is defined in terms of the elements of $\mathbf{r}$ as

$$R_a = \mathbf{w}^T\mathbf{r} = \sum_{i=1}^{p} w_i R_i,$$

where w_i represents the weight of asset i in portfolio a. If the vector $\mathbf{t}$ is now redefined as

$$\mathbf{t} = -\frac{1}{\theta}\mathbf{w}, \ \theta > 0,$$

the moment generating function is also the expectation of a utility function. Specifically,

$$M_{\mathbf{r}}\left(-\frac{1}{\theta}\mathbf{w}\right) = \mathrm{E}\big(e^{-\frac{1}{\theta}\mathbf{w}^T\mathbf{r}}\big) = \mathrm{E}\big(e^{-\frac{1}{\theta}R_a}\big).$$

Since the argument of the right-hand side is a proper utility function, this expression provides a convenient mechanism for portfolio selection, with increasing values of the parameter θ representing increasing investor appetite for risk. Briefly, this is done by choosing the weights w_i to maximize the expected utility, generally subject to some restrictions.

11.3 The Market Model

As stated in the introduction, the market model (11.2) is correctly obtained by considering the probability distribution of asset returns $\mathbf{r}$ given the return R_m on the market portfolio. When returns are multivariate normal with mean vector $\boldsymbol{\mu}$ and covariance matrix Σ, the conditional distribution of the vector of returns $\mathbf{r}$ given R_m is also multivariate normal:

$$\mathbf{r} \mid R_m \sim N_p(\boldsymbol{\mu}_c, \Sigma_c),$$

where

$$\Sigma_c = \Sigma - \sigma_m^2\boldsymbol{\beta}\boldsymbol{\beta}^T,$$
$$\boldsymbol{\mu}_c = \boldsymbol{\mu} + \boldsymbol{\beta}(R_m - \mu_m),$$
$$\mu_m = \mathbf{w}^T\boldsymbol{\mu}, \ \sigma_m^2 = \mathbf{w}^T\Sigma\mathbf{w}, \ \boldsymbol{\beta} = \frac{\Sigma\mathbf{w}}{\sigma_m^2}.$$

The conditional expected return of asset i given R_m is

$$\mathrm{E}(R_i \mid R_m) = \mu_i + \beta_i(R_m - \mu_m).$$

The elements of the vector $\boldsymbol{\beta}$ are correctly estimated by the fitted coefficient in the regression of R_i, the return of stock i, on R_m, the return on the market. As noted in the introduction, it is commonly assumed that Σ_c is diagonal. However, this is theoretically impossible even though it is a convenient simplification. Adcock and Clark (1999) provide details.

If the CAPM holds, the elements of beta are correctly estimated using excess returns, that is from the regressions of $R_i - R_f$ on $R_m - R_f$. For the rest of this article, the notation $\boldsymbol{\beta}$ is used to denote the vector of covariances of each asset with the market portfolio divided by market variance. This definition is used, regardless of the underlying distribution of returns, assuming that the relevant moments exist.

For the SN distribution, it is well-known (see, e.g., Azzalini and Dalla Valle, 1996) that the conditional distributions are also of the "extended" SN form. As shown in A&S, in the current notation the conditional distribution of returns $\mathbf{r}$ given the market return R_m is

$$\mathbf{r} \mid R_m \sim SN_p(\boldsymbol{\mu}_c, \Sigma_c, \boldsymbol{\delta}_c, \tau_c),$$

with the parameters defined as follows:

$$\tau_c = \frac{\tau + (R_m - \mu_m)\alpha_m/\sigma_m^2}{\sqrt{1+\alpha_m^2/\sigma_m^2}},$$

$$\boldsymbol{\delta}_c = \frac{\boldsymbol{\delta} - \alpha_m\tilde{\boldsymbol{\beta}}}{\sqrt{1+\alpha_m^2/\sigma_m^2}},$$

$$\Sigma_c = \Sigma - \sigma_m^2\tilde{\boldsymbol{\beta}}\tilde{\boldsymbol{\beta}}^T,$$

$$\boldsymbol{\mu}_c = \boldsymbol{\mu} + \tilde{\boldsymbol{\beta}}(R_m - \mu_m),$$

where

$$\mu_m = \mathbf{w}^T\boldsymbol{\mu},\ \sigma_m^2 = \mathbf{w}^T\boldsymbol{\Sigma}\mathbf{w},\ \tilde{\boldsymbol{\beta}} = \frac{\boldsymbol{\Sigma}\mathbf{w}}{\boldsymbol{\sigma}_m^2},\ \alpha_m = \mathbf{w}^T\boldsymbol{\delta}.$$

In this notation, the return on the market portfolio is $SN(\mu_m, \sigma_m^2, \alpha_m, \tau)$. The conditional mean vector of $\mathbf{r}$ given R_m is

$$\mathrm{E}(\mathbf{r} \mid R_m) = \boldsymbol{\mu}_c + \boldsymbol{\delta}_c\{\tau_c + \xi_1(\tau_c)\}.$$

As A&S show, rearrangement gives

$$\mathrm{E}(\mathbf{r} \mid R_m) = \boldsymbol{\mu} + \boldsymbol{\delta}\tau + \boldsymbol{\lambda}(R_m - \mu_m - \alpha_m\tau) + \boldsymbol{\delta}_c\xi_1(\tau_c),$$

where the vector $\boldsymbol{\lambda}$ is defined as

$$\boldsymbol{\lambda} = \frac{1}{\sigma_m^2 + \alpha_m^2}(\Sigma + \boldsymbol{\delta}\boldsymbol{\delta}^T)\mathbf{w} = \frac{\sigma_m^2\tilde{\boldsymbol{\beta}} + \alpha_m\boldsymbol{\delta}}{\sigma_m^2 + \alpha_m^2}.$$

The conditional expected value of $\mathbf{r}$ is in general a non-linear function of R_m through the dependence of τ_c on R_m.

These equations have implications for the market model. There are two cases to consider, depending upon whether or not the market skewness parameter is equal to zero. When $\alpha_m = 0$, it follows that

$$\boldsymbol{\delta}_c = \boldsymbol{\delta},\ \tau_c = \tau, \boldsymbol{\lambda} = \tilde{\boldsymbol{\beta}},$$

and that

$$\tilde{\boldsymbol{\beta}} = \boldsymbol{\beta}.$$

Hence,

$$\mathrm{E}(\mathbf{r} \mid R_m) = \boldsymbol{\mu} + \boldsymbol{\delta}\tau + \boldsymbol{\delta}\xi_1(\tau) + \boldsymbol{\beta}(R_m - \mu_m) = \mathrm{E}(\mathbf{r}) + \boldsymbol{\beta}(R_m - \mathrm{E}(R_m)).$$

That is, when the market portfolio has no skewness, the non-linearity disappears and beta, defined in the traditional way, continues to measure the sensitivity of asset returns to return on the market. Further, if the CAPM holds, the above conditional expected value becomes

$$\mathrm{E}(\mathbf{r} \mid R_m) = R_f\mathbf{1}_p + \boldsymbol{\beta}(R_m - R_f).$$

For the second case when $\alpha_m \neq 0$, however, no such reduction occurs. The general equation for the conditional expected value of $\mathbf{r}$ given R_m is

$$\mathrm{E}(\mathbf{r} \mid R_m) = \mathrm{E}(\mathbf{r}) + \boldsymbol{\lambda}(R_m - \mathrm{E}(R_m)) + \boldsymbol{\delta}_c\Delta,$$

where the scalar Δ is given by

$$\Delta = \left\{\xi_1(\tau_c) - \frac{\sigma_m}{\sqrt{\sigma_m^2 + \alpha_m^2}}\xi_1(\tau)\right\}.$$

The first key point to note is that the vector of measures of risk is now $\boldsymbol{\lambda}$ rather than $\boldsymbol{\beta}$. When the CAPM holds, the vector of conditional expected values is

$$\mathrm{E}(\mathbf{r} \mid R_m) = R_f\mathbf{1}_p + \boldsymbol{\lambda}(R_m - R_f) + (\boldsymbol{\beta} - \boldsymbol{\lambda})(\mathrm{E}(R_m) - R_f) + \boldsymbol{\delta}_c\Delta.$$

Thus the SN market model does have an "alpha." The non-linear component is affected by the elements of $\boldsymbol{\delta}_c$, which varies from stock to stock, and by Δ, which is time varying through the dependence on the time series of returns on R_m.

The main purpose of this chapter is to estimate the parameters of the SN distribution for asset returns and hence to derive the estimated market model using the results above. However, it is appropriate to note that the univariate skew-normal distribution may also be used to estimate a different version of the market model. This does not have a formal theoretical basis but, as noted in the introduction, is nonetheless common practice. We assume the regression model (11.2) holds for each security. The error terms are both temporally and cross-sectionally independent and excess return of the market and the risk-free rate are treated as given. It is then assumed that each ϵ_i has a univariate skew-normal distribution with parameters 0, σ_i^2, α_i and τ_i. This model loses the interrelationships between the stock specific components of return, but does have the ability to specify a different estimated truncation parameter for each security.

11.4 Estimation Methodology and Data

For the multivariate skew-normal model, it is assumed that a time series of observed vectors $\mathbf{r}_t$ is available for $t = 1, \ldots, T$ and that each element is independently distributed as SN. This vector includes observations on each individual asset and on the market portfolio. Using the notation already defined, the log-likelihood function is proportional to

$$-\frac{T}{2}\ln|\Sigma+\boldsymbol{\delta}\boldsymbol{\delta}^T|-\frac{1}{2}\sum_{t=1}^{T}(\mathbf{r}_t-\boldsymbol{\mu}-\boldsymbol{\delta}\tau)^T(\Sigma+\boldsymbol{\delta}\boldsymbol{\delta}^T)^{-1}(\mathbf{r}_t-\boldsymbol{\mu}-\boldsymbol{\delta}\tau)$$
$$+\sum_{t=1}^{T}\ln\Phi\left(\frac{\tau+\boldsymbol{\delta}^T\Sigma^{-1}(\mathbf{r}_t-\boldsymbol{\mu})}{\sqrt{1+\boldsymbol{\delta}^T\Sigma^{-1}\boldsymbol{\delta}}}\right)-T\ln\Phi(\tau).$$

For the univariate SN model, the log-likelihood function for asset i is proportional to

$$-\frac{T}{2}\ln(\sigma_i^2+\alpha_i^2)-\frac{1}{2\sigma_i^2}\sum_{t=1}^{T}(\epsilon_{it}-\alpha_i\tau_i)^2+\sum_{t=1}^{T}\ln\Phi\left(\frac{\tau_i+\frac{\alpha_i\epsilon_{it}}{\sigma_i^2}}{\sqrt{1+\frac{\alpha_i^2}{\sigma_i^2}}}\right)-T\ln\Phi(\tau_i),$$

where ϵ_{it} is defined as

$$\epsilon_{it}=R_{it}-R_{ft}-\beta_i(R_{mt}-R_{ft}).$$

Maximum likelihood (ML) estimation for both models is carried out using the EM algorithm. Details are omitted to conserve space, but are available on request.

The data used in this study are based on weekly prices from Datastream for the securities that were constituents of the FTSE100 index as of October 20, 1995 and that had available price data since January 1978. This gave a total of 67 securities. Also used were the FTSE100 index itself and returns on the risk-free rate (taken as 7 day LIBOR). Weekly returns were calculated in the usual way by taking logarithms. Data for the 250 weeks up to and including October 20, 1995, i.e., approximately 5 years, were used in the investigations.

11.5 Empirical Study

The SN model described in Section 11.2 was applied to the UK FTSE100 based data set as described in Section 11.4 and was estimated using the method of ML. The standard multivariate normal model was also estimated for comparison purposes. The effect of the SN model was tested using a standard likelihood ratio test. The restriction in the test is that the vector $\boldsymbol{\delta}$ equals zero. For the model in p dimensions, the critical values for the test are taken from the chi-squared distribution with p degrees of freedom. The results of this test, as well as summary statistics for the market portfolio, are reported in Table 11.1.

According to the results in the table, the null hypothesis that all values of the skewness parameter vector $\boldsymbol{\delta}$ equal zero is rejected. The SN model is preferred to the standard multivariate normal model. If the formulae

Table 11.1 *Summary of Maximum Likelihood Estimation for the SN Model. Based on 250 observations of 67 FTSE100 stocks and the index, as described in Section 11.4.*

A	**ML estimates for the market portfolio**	
	Normal Distribution	
	μ	0.0022
	σ	0.0184
	Skew-Normal Distribution	
	μ	−0.0004
	σ	0.0172
	λ	0.0864
	τ	−32.8858
B	**Likelihood ratio test**	
	l-normal	38287.54
	l-SN	38374.73
	Lr	87.19
	Pr	0.00

Legend

In panel A – Normal Distribution, μ and σ are the estimated values of the mean and standard deviation respectively for the market portfolio. In panel B– SN, μ, σ, λ, and τ are the estimated values of the parameters for the return on the market portfolio computed from the multivariate skew-normal distribution.

In panel B, l-normal and l-SN are the computed values of the log-likelihood function evaluated at the respective MLEs. Lr denotes the value of the likelihood ratio test statistic and Pr the corresponding probability.

for the first two moments are applied to the estimated parameters for the market portfolio, the computed estimates of expected return and volatility are numerically close to the corresponding estimated parameters based on the normal distribution.

Examination of estimated values of the vector $\boldsymbol{\lambda}$, which measures the sensitivity of asset returns to returns on the market portfolio under the SN model, as well as the estimated values of $\tilde{\boldsymbol{\beta}}$ leads to different insights. Table 11.2 shows summary statistics for the estimated values $\boldsymbol{\lambda}$ and $\tilde{\boldsymbol{\beta}}$. Also shown in the table for comparison purposes are the estimated values of $\boldsymbol{\beta}$ from MLE under the normal model, which are the same as the conventional

Table 11.2 *Comparison of the Values of "Beta." Estimates based on 250 observations of 67 FTSE100 stocks and the index, as described in Section 11.4.*

	Method/Model			
	OLS	uni-sn	msn-beta	msn-lambda
Average	1.0649	1.0616	1.0299	1.2951
St.dev	0.2657	0.2671	0.2574	0.7174
Min	0.4444	0.4272	0.4879	−0.0145
Max	1.6541	1.6546	1.6195	3.2451

Legend

Average	Average value of coefficient of market return.
St.dev	Standard deviation of coefficient of market return.
Min	Minimum value of coefficient.
Max	Maximum value of coefficient.
OLS	Ordinary least squares, equivalently MLE based on normal distribution.
uni-sn	Univariate skew-normal regression model.
msn-beta	Estimated values of $\tilde{\boldsymbol{\beta}}$ based on SN model.
msn-lambda	Estimated value of $\boldsymbol{\lambda}$ based on SN model.

OLS betas, and the estimated regression coefficients from the univariate market model with univariate skew-normal errors.

To summarize the values of 67 parameters, the table shows the average value of each of the four sets of coefficients. Thus, in the first column of the table, the average estimated OLS beta for all 67 stocks is 1.0649. The other three rows in the table give an indication of the variability of each set of estimated coefficients. First, the results in Table 11.2 suggest that the univariate skew-normal regressions have little to add to MLE based on the normal distribution. This is confirmed by an examination of the individual likelihood ratio tests for the regressions: less than 10 securities have likelihood ratio tests that would reject normality in favor of the univariate skew-normal market model. Details of this additional analysis are omitted, but are available on request.

Secondly, at first sight the estimated values of $\tilde{\boldsymbol{\beta}}$ appear also to be numerically similar to the OLS betas. The correlation between the two sets of estimates is over 90%, but the values of $\tilde{\boldsymbol{\beta}}$ are slightly but systemati-

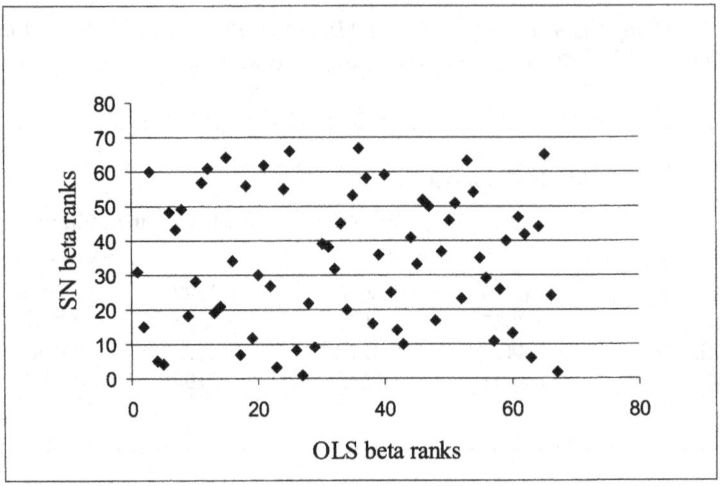

Figure 11.1 *Scatter plot of SN betas ranks vs. OLS betas ranks. Estimates based on 250 observations of 67 FTSE100 stocks and the index, as described in Section 11.4.*

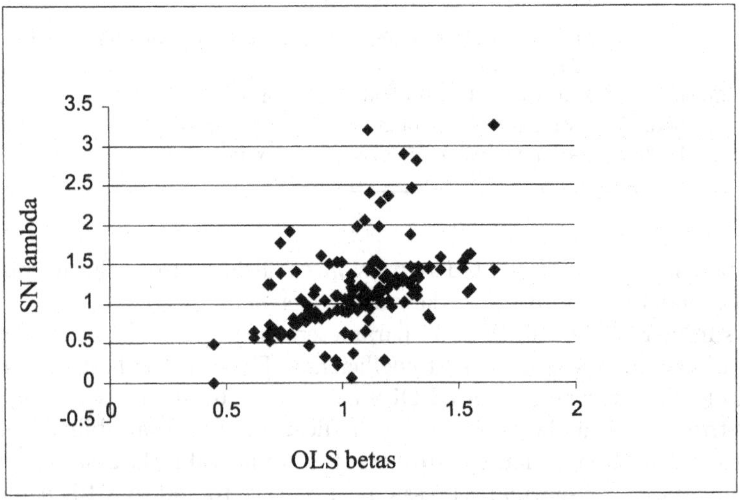

Figure 11.2 *Scatter plot of SN lambdas vs. OLS betas. Estimates based on 250 observations of 67 FTSE100 stocks and the index, as described in Section 11.4.*

cally smaller. The coefficient in the regression of the elements of $\tilde{\beta}$ against the OLS betas is about 0.97. However, Figure 11.1 shows a scatter plot of the corresponding ranks. This makes it clear that there are differences

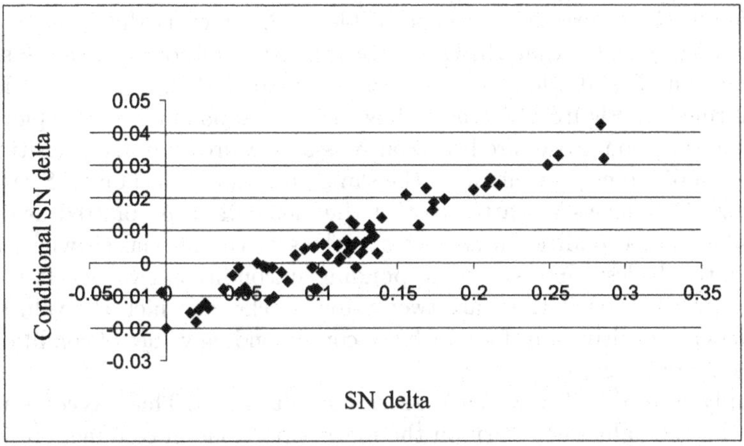

Figure 11.3 *Scatter plot of conditional SN deltas vs. SN deltas. Estimates based on 250 observations of 67 FTSE100 stocks and the index, as described in Section 11.4.*

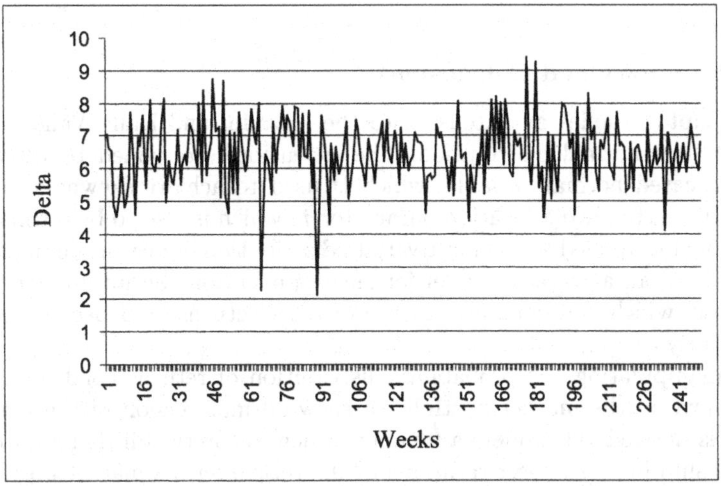

Figure 11.4 *Time series of* Δ. *Estimates based on 250 observations of 67 FTSE100 stocks and the index, as described in Section 11.4.*

and that, in particular, a stock selection process that is based on order will generate different portfolios.

Thirdly, the values of $\boldsymbol{\lambda}$ are clearly different. The correlation between the two sets of values is only about 42%. A regression of the elements of $\boldsymbol{\lambda}$ against the OLS betas gives a regression coefficient of about 1.2, thus confirming the impression given in Table 11.2, which is that the λs are systematically higher than the OLS betas. However, although the regression is significant, F statistic of 13.8 with a p-value 0.0004, the R^2 is low. This is confirmed by Figure 11.2 which shows the corresponding scatter plot. A portfolio selection procedure based on λ as a measure of market sensitivity will give a different portfolio from the same procedure based on OLS betas.

Figure 11.3 shows a scatter plot of the elements of $\boldsymbol{\delta}_c$ plotted against the values of unconditional skewness. Values of conditional skewness are systematically less than the corresponding unconditional values. It is interesting to note that there are twenty-five stocks for which the values of skewness are positive, but for which the corresponding values of conditional skewness are negative.

Finally, Figure 11.4 shows a time series of values of Δ. This was computed using the 250 values of return on the market portfolio in conjunction with the relevant estimated parameter values (this in-sample computation is solely for illustration). The aim of this graph is to demonstrate that the non-linearity is volatile. However, it may be noted that the computed values of Δ based on the estimates and the time series of market returns are all positive.

11.6 Summary and Conclusions

This chapter applies an extension to the Azzalini and Dalla Valle multivariate skew-normal distribution. The extension is obtained by allowing the truncated normal variable, which is used to achieve skewness in the distribution, to have a location parameter τ, which is also to be estimated. The model is applied to the multivariate distribution of returns on financial assets. It is an attractive model for such applications because it includes skewness, which is a common feature of asset returns, and because of its tractability.

By manipulating the probability distribution of returns conditional on the return on the market portfolio, a skew-normal version of the market model is derived. This offers a number of new theoretical insights into the relationship between asset returns and the return on the market portfolio. Most notable of these is that the sensitivity of asset returns to returns on the market portfolio is not measured by the conventional beta and that the estimated values of the sensitivity parameter are substantially different from conventional measures based on OLS. Another interesting feature is that the skew-normal based market model contains a component that is a non-linear function of market return.

Planned developments of this work include use of the model for portfolio selection and application to a number of other established and emerging stock markets as well as to the UK.

CHAPTER 12

A Skew-in-Mean *GARCH* Model

Giovanni De Luca and Nicola M. R. Loperfido

12.1 Introduction

Market operators, policy makers and economists apply statistical models of financial data for option pricing, analyzing market microstructures, and predicting future gains.

Financial data sets record trades, interest rates, or foreign exchange rates. Attention focuses on $R_t = 100(\log P_t - \log P_{t-1})$, that is the log-return at time t (henceforth return) of the stock price process P_t over the interval $(t-1;t]$. This definition is used regardless of the frequency of the data (i.e., intra-daily, daily, or weekly). Rydberg (2000) defines as "stylized facts" some features often observed in returns: they can be either time-independent or time-dependent. Time-independent features deal with the data as a whole, while time-dependent features deal with relationships between a return and the following ones. The most important time-independent features are: negligible mean (or zero mean), negative skewness (or contemporaneous asymmetry) and excess kurtosis (or fat tails). The most important time-dependent features are: negligible autocorrelation (linear prediction of future returns is virtually useless), volatility clustering (large squared returns tend to be followed by other large squared returns) and predictive asymmetry (negative correlation between present returns and future squared returns).

There are several economic theories that try explaining the above stylized facts. According to arbitrage theory, negligible means and correlation of returns are due to the markets' ability to incorporate news into prices. Leverage theory (Christie, 1982) claims that a reduction of the equity value would raise the debt-to-equity ratio, hence raising the riskiness of the firm as shown by an increase in future volatility. As a result, the future volatility will be negatively related to the current return of the stock. However, the leverage effect might be too small to fully account for predictive asymmetry (Schwert, 1989).

Volatility feedback theory (see French, Schwert, and Stambaugh, 1987; Brown, Harlow, and Tinic, 1988; Campbell and Hentschel, 1992) states that stock price reactions to unfavorable events tend to be larger than reactions to favorable ones. More precisely, bad news decreases stock price and increases volatility, therefore determining a further decrease of the price. On the other hand, good news increases stock price and increases volatility, thus mitigating the initial effect on the price. The consequence is a certain amount of negative skewness in the distribution of stock returns and negative correlation between present returns and future volatility. These and other theories of economic behavior form the basis for much of the modern finance and monetary theories (Bollerslev and Engle, 1993).

Qualitative relationships implied in economic theories must be quantified and formalized in statistical models. Models assumptions should have economic interpretations and their implications should be consistent with the stylized facts. Moreover, models estimation and diagnostic checking should be inferentially sound as well as computationally feasible.

A number of authors proposed to model volatility in different ways. Stochastic volatility models describe more stylized facts (i.e., kurtosis and volatility clustering) but inference on their parameters is very difficult (the corresponding likelihood has a high-dimensional integral form). Ghysels, Harvey, and Renault (1996) provide an extensive survey of stochastic volatility models. *GARCH*-type models can describe all the above stylized facts and pose less inferential problems. However, their flexibility often depends on parameters without any real economic meaning. Bollerslev, Chou, and Kroner (1992) extensively analyze many *GARCH*-type models and the corresponding limitations.

Skewness of financial returns can be modeled in several ways, too. McDonald and Newey (1988) apply the generalized T distribution. Harvey and Siddique (1999) assume that the conditional distribution of the returns is a non-central T distribution, indexed by a parameter that controls its skewness. Peiró (1999) models skewness through discrete mixtures of normal distributions. McDonald (1996) gives a complete survey of earlier works on skewed distributions appropriate for returns distributions, such as the generalized Beta of the second type, which includes many important distributions as special cases.

This chapter introduces a skew-in-mean *GARCH* model (henceforth denoted *SGARCH*) which formalizes the volatility feedback theory and quantifies the feedback effect through a single parameter (the feedback parameter). The model describes all the above stylized facts and shows how the feedback effect determines skewness, kurtosis, and predictive asymmetry.

The chapter is structured as follows. Section 12.2 states the assumptions of the model and motivates them through economic considerations. *SGARCH* modeling of market reactions to good and bad news is shown in Section 12.3. Section 12.4 analyzes modeling of the above stylized facts. Section 12.5 introduces data from small capitalized markets and comments on some descriptive statistics. Section 12.6 deals with model estimation and checking. Section 12.7 concludes. The appendix collects all propositions' proofs.

12.2 Assumptions

US returns influence foreign markets without being influenced by any single foreign market (Eun and Shim, 1989; Ling, Engle, and Ito, 1994). Influence is typically short term, that is no longer than one day (Eun and Shim, 1989; Karolyi, 1995) and is thought to be more effective for small capitalized markets. Moreover, markets pay special attention to the sign of US returns, distinguishing bear days (negative signs) from bull days (positive signs).

The above facts motivate the use of negative (positive) signs of previous day US return as proxy for bad (good) news. Influence on returns of small capitalized markets of foreign countries (henceforth foreign returns) is modeled through the following assumptions:

1. *Endogenous and past exogenous shocks are independent and normally distributed.*

   A shock is a random variable quantifying the news impact on the market. We shall denote by ξ_t and ϵ_t endogenous (i.e., originated in the foreign country itself) and exogenous (i.e., originated in the US) shocks at time t, respectively. Both ξ_t and ϵ_t are white noises. Moreover, we shall assume that ξ_t and ϵ_{t-1} are independent, normal (the normal distribution maximizes entropy, that is lack of information, among all continuous unbounded distributions with given expectation and variance), centered at the origin (bad news and good news balance each other) with unit variances (it is not a restrictive assumption, since in the *SGARCH* model ξ_t and ϵ_{t-1} are either multiplied by unknown constants or by random variables with unknown variance). Formally:

$$\begin{pmatrix} \xi_t \\ \epsilon_{t-1} \end{pmatrix} \sim N_2\left(\begin{pmatrix} 0 \\ 0 \end{pmatrix}, \begin{pmatrix} 1 & 0 \\ 0 & 1 \end{pmatrix} \right).$$

2. *Returns have a GARCH structure.*

   News impact on returns is either mitigated or amplified by market volatility. We shall assume that volatility multiplies shocks' effect on returns and that volatilities have a *GARCH* structure. Formally:

$$Y_t = \eta_t \epsilon_t, \qquad X_t = \sigma_t \left(f\left(\epsilon_{t-1}\right) + \xi_t \right),$$

$$\eta_t^2 = \delta_0 + \sum_{i=1}^{q} \delta_i (\eta_{t-i}\epsilon_{t-i})^2 + \sum_{j=q+1}^{p+q} \delta_j \eta_{t+q-j}^2, \qquad (12.1)$$

$$\sigma_t^2 = \omega_0 + \sum_{i=1}^{q} \omega_i (\sigma_{t-i}\xi_{t-i})^2 + \sum_{j=q+1}^{p+q} \omega_j \sigma_{t+q-j}^2, \qquad (12.2)$$

   where Y_t and X_t denote the daily US and foreign returns at time t, respectively. The function $f\left(\epsilon_{t-1}\right)$ models the effect of the lagged exogenous shock ϵ_{t-1} on X_t. In order to ensure $\eta_t^2 > 0$ and $\sigma_t^2 > 0$, δ_0 and ω_0 have to be positive and the remaining parameters in Equations (12.1) and (12.2) non-negative.

3. *Foreign markets' returns linearly depend on shocks and feedback.*

   Volatility feedback theory implies that news increases volatility, which in turn lowers returns. Hence the direct effect of good (bad) news is mitigated (strengthened) by the feedback effect. The *SGARCH* model formalizes the direct and feedback effects through the parameters α and β as follows:

$$\frac{X_t}{\sigma_t} = \sqrt{\frac{2}{\pi}}\beta + \alpha\epsilon_{t-1} - \beta|\epsilon_{t-1}| + \xi_t, \qquad 0 \le \beta \le \alpha. \tag{12.3}$$

   This model for returns is essentially the same as that in Nelson (1991) for volatility.

The above assumptions lead to the following considerations:

- Feedback effect makes returns more reactive to bad news than to good news. Effectively, when $\epsilon_{t-1} = -1$, the conditional expectation of X_t is $\sigma_t\left(\sqrt{\frac{2}{\pi}}\beta - \alpha - \beta\right)$, if $\epsilon_{t-1} = 1$, the expectation is $\sigma_t\left(\sqrt{\frac{2}{\pi}}\beta + \alpha - \beta\right)$, which is smaller in modulus.
- The *SGARCH* model is a threshold model:

$$\frac{X_t}{\sigma_t} = \sqrt{\frac{2}{\pi}}\frac{\theta^- - \theta^+}{2} + \theta^+\epsilon_{t-1}^+ + \theta^-\epsilon_{t-1}^- + \xi_t, \qquad \theta^- \ge \theta^+ \ge 0,$$

  where $\theta^+ = \alpha - \beta$, $\theta^- = \alpha + \beta$, $\epsilon_{t-1}^+ = \max(0, \epsilon_{t-1})$, $\epsilon_{t-1}^- = \min(0, \epsilon_{t-1})$. Other examples of threshold models describing predictive asymmetry are the *EGARCH* (Nelson, 1991) and the *TGARCH* (Zakoian, 1994) models. However, they do not describe either negative skewness or conditional kurtosis.
- The *SGARCH* model generalizes the bivariate *GARCH* model under the constant correlation assumption (Bollerslev, 1990), since they coincide when the feedback parameter β is zero. More formally, assumptions 1, 2, 3, and the constraint $\beta = 0$ imply that

$$\begin{pmatrix} X_t/\psi_t \\ Y_{t-1}/\eta_{t-1} \end{pmatrix} \sim N_2\left(\begin{pmatrix} 0 \\ 0 \end{pmatrix}, \begin{pmatrix} 1 & \rho \\ \rho & 1 \end{pmatrix}\right),$$

  where $\psi_t = \sigma_t\sqrt{1+\alpha^2}$ and $\rho = \alpha/\sqrt{1+\alpha^2}$.

12.3 News

Assumptions in the previous section imply that foreign returns, conditionally on the sign of previous time US return and volatility, are skew-normally distributed (the proof is in the Appendix).

Table 12.1 *Parameters, moments, and correlation between foreign returns and one-lagged US returns under bad and good news.*

	Bad news	Good news
location	$\sqrt{\frac{2}{\pi}}\beta\sigma_t$	$\sqrt{\frac{2}{\pi}}\beta\sigma_t$
scale	$\sigma_t\sqrt{1+(\alpha+\beta)^2}$	$\sigma_t\sqrt{1+(\alpha-\beta)^2}$
shape	$\frac{-(\alpha+\beta)}{\sqrt{1+(\alpha+\beta)^2}}$	$\frac{\alpha-\beta}{\sqrt{1+(\alpha-\beta)^2}}$
expectation	$-\sqrt{\frac{2}{\pi}}\alpha\sigma_t$	$+\sqrt{\frac{2}{\pi}}\alpha\sigma_t$
variance	$\sigma_t^2\left[1+\frac{\pi-2}{\pi}(\alpha+\beta)^2\right]$	$\sigma_t^2\left[1+\frac{\pi-2}{\pi}(\alpha-\beta)^2\right]$
skewness	$\frac{\pi-4}{2}\left\{\frac{2(\alpha+\beta)^2}{\pi+(\pi-2)(\alpha+\beta)^2}\right\}^{1.5}$	$\frac{4-\pi}{2}\left\{\frac{2(\alpha-\beta)^2}{\pi+(\pi-2)(\alpha-\beta)^2}\right\}^{1.5}$
kurtosis	$2(\pi-3)\left\{\frac{2(\alpha+\beta)^2}{\pi+(\pi-2)(\alpha+\beta)^2}\right\}^2$	$2(\pi-3)\left\{\frac{2(\alpha-\beta)^2}{\pi+(\pi-2)(\alpha-\beta)^2}\right\}^2$
correlation	$\frac{(\alpha+\beta)\sqrt{\pi-2}}{\sqrt{\pi+(\pi-2)(\alpha+\beta)^2}}$	$\frac{(\alpha-\beta)\sqrt{\pi-2}}{\sqrt{\pi+(\pi-2)(\alpha-\beta)^2}}$

Econometric applications of the skew-normal distribution appeared in Aigner, Lovell, and Schmidt (1977), and Heckman (1979). Engle and Ng (1993) introduced a model where innovations in the volatility equations are skew-normal (however, they do not mention the skew-normal distribution). Bartolucci, De Luca, and Loperfido (2000) applied skew-normal distributions to financial returns but they do not motivate their choice through economic considerations. In $SGARCH$ models, on the contrary, the skew-normal distribution naturally arises from simple assumptions on the market behavior. More precisely, the distributions of $X_t|\sigma_t, Y_{t-1} < 0$ (bad news) and $X_t|\sigma_t, Y_{t-1} > 0$ (good news) are skew-normal. The corresponding parameters, moments, and correlations between foreign returns and US returns conditionally on volatility and the type of news are reported in Table 12.1. The corresponding proofs are provided in the Appendix.

Mathematical features of $X_t|\sigma_t, Y_{t-1} < 0$ and $X_t|\sigma_t, Y_{t-1} > 0$ are consistent with empirical findings, as well as with economic theory:

- Expected returns are negative (positive) in presence of bad (good) news. The absolute value does not change in the two situations.
- Volatility is higher in presence of bad news.
- Skewness is negative and more remarkable in presence of bad news.

- Bad news involves a stronger correlation between present foreign returns and one-lagged US returns.

When the parameter β in (12.3) equals zero, good news increases stock prices exactly as much as bad news decreases them. More precisely, the distributions of X_t conditionally on $(\sigma_t, Y_{t-1} > 0)$ and $(\sigma_t, Y_{t-1} < 0)$ are skew-normal centered at the origin (the location parameters are both zero). Hence:

- The type of news (good or bad) does not influence either variance or correlations between X_t and Y_{t-1}.
- The type of news (good or bad) influences the sign (but not the absolute value) of skewness.

12.4 Returns

The *SGARCH* model implies that the unconditional distribution of $X_t|\sigma_t$ is the mixture, with equal weights, of two skew-normal distributions (both $X_t|\sigma_t, Y_{t-1} > 0$ and $X_t|\sigma_t, Y_{t-1} < 0$ are skew-normal and Y_{t-1} is symmetric about the origin). It also conforms with some of the well-known stylized facts of asset returns. The proofs of what follows are provided in the Appendix.

(A) *TIME-INDEPENDENT FEATURES*

(A1) *Zero expectation*

$$\mathrm{E}(X_t|\sigma_t) = \mathrm{E}(X_t) = 0,$$

which easily follows from $\mathrm{E}(X_t|\sigma_t, Y_{t-1} > 0) = -\mathrm{E}\left(X_t|\sigma_t, Y_{t-1} < 0\right) = \alpha\sigma_t\sqrt{2/\pi}$ and $\mathrm{P}\left(Y_{t-1} < 0\right) = \mathrm{P}\left(Y_{t-1} > 0\right) = 1/2$.

(A2) *Negative skewness*

$$\mathrm{S}(X_t) < \mathrm{S}(X_t|\sigma_t) = \frac{\sqrt{2}\beta\left(2\pi\beta^2 - 3\pi\alpha^2 - 4\beta^2\right)}{\left(\pi + (\pi - 2)\,\beta^2 + \alpha^2\right)^{3/2}} \leq 0,$$

where the skewness $\mathrm{S}(Z)$ of the random variable Z is the third moment of $\left(Z - \mathrm{E}\left(Z\right)\right)/\sqrt{\mathrm{Var}\left(Z\right)}$. The first inequality follows from unconditional skewness in a *GARCH* type model being smaller than the conditional one when the latter is negative (Hsieh, 1989). The second inequality follows from the assumption $\beta \leq \alpha$. Note that a few securities display positive skewness. If one wishes to take it into account, the constraint to be considered is $0 \leq |\beta| \leq \alpha$.

(A3) *Kurtosis*

$$\mathrm{K}\left(X_t\right) > \mathrm{K}\left(X_t|\sigma_t\right) = 3 + \frac{8\beta^2\left(\pi - 3\right)}{\left(\pi + (\pi - 2)\,\beta^2 + \alpha^2\right)^2} \geq 3,$$

where the kurtosis $\mathrm{K}(Z)$ of the random variable Z is the fourth moment of $(Z - \mathrm{E}(Z)) / \sqrt{\mathrm{Var}(Z)}$. The first inequality follows from unconditional kurtosis being higher than the conditional one, in *GARCH*-type models (Hsieh, 1989). The second inequality follows from β being non-negative by assumption. The *GARCH*-type structure explains part of the kurtosis of X_t, but kurtosis is also found in the conditional distribution of $X_t|\sigma_t$ (Bollerslev, 1987; Hong, 1988; Gallant, Hsieh, and Tauchen, 1989; Engle and Gonzalez-Rivera, 1991). According to the *SGARCH* model, the kurtosis depends on the volatility structure as well as on the volatility feedback effect (kurtosis of $X_t|\sigma_t$ is an increasing function of β).

(B) *TIME-DEPENDENT FEATURES*

(B1) *Zero correlation*

$$\mathrm{Corr}(X_{t+1}, X_t|\sigma_t) = \mathrm{Corr}(X_{t+1}, X_t) = 0.$$

The above equality easily follows from X_{t+1} being independent from X_t conditionally on σ_{t+1} and $\mathrm{E}(X_t|\sigma_t) = \mathrm{E}(X_t) = \mathrm{E}(X_{t+1}|\sigma_{t+1}) = \mathrm{E}(X_{t+1}) = 0$.

(B2) *Volatility clustering*

$$\mathrm{Corr}\left(X_{t+1}^2, X_t^2\right) > 0.$$

The above equation easily follows from X_{t+1} being independent from X_t conditionally on σ_{t+1} and $\mathrm{Var}(X_t|\sigma_t)$ being proportional to σ_t.

(B3) *Predictive asymmetry*

$$\mathrm{Corr}\left(\sigma_{t+1}^2, X_t\right) \leq 0.$$

Note that σ_{t+1}^2 is a random variable, since by definition it is a linear combination of past squared returns. The equality sign holds if and only if $\beta = 0$. Hence we can say that in the *SGARCH* model predictive asymmetry depends on the feedback effect.

12.5 Data

We analyze the returns of three European financial markets: the Swiss, the Italian, and the Dutch market. In order to make this choice clear, that is why the markets chosen are examplarily *small*, we remind readers of the difference in size between the US market and the three European exchanges. The weight of the capitalization of the former on the world capitalization is over 40%, while the weights of each of the latter is approximately between 2% and 3%.

We focus on the returns of representative market indices. For the Swiss market we choose the SMI index, for the Italian market we use the MIB30 index, and for the Dutch market we use the AEX index. The three indices

Table 12.2 *Descriptive statistics for the three stock indices.*

	SMI	MIB	AEX
Average	0.035	0.027	0.033
Standard deviation	1.292	1.559	1.495
Skewness	−0.183	−0.076	−0.232
Kurtosis	6.986	4.938	6.398
Correlation	0.266	0.211	0.321

are analyzed in the period from January 18, 1995 to November 29, 2002. The US market is represented by the Standard & Poor 500 (S&P500) which is certainly the most popular index for the New York Stock Exchange.

In Table 12.2, the most relevant statistics are reported for the indices. It is possible to observe average returns very close to zero and negative skewness coefficients. As expected, a high degree of kurtosis is a common feature. Moreover, in the last rows we report the correlation with the one-lagged S&P500 returns. These correlations are quite effective (they would be slightly higher if we deleted the observation of September 11, 2001).

Then we divide the entire sample into two subsamples, according to the sign of the one-lagged S&P500 return. In order to take into account the differences in closing days of the stock exchanges, some hypotheses have to be made. We assume that if the foreign exchange is open at time t and the US exchange is closed at time $t-1$, then ϵ_{t-1} is set to zero (there is no information from the US market). If the foreign exchange is closed at time t, the US exchange information at time $t-1$ is useless; next the foreign exchange return (at time $t+1$) is related to ϵ_t. We compute the same statistics summarized in Table 12.3.

The resulting statistics seem to be very interesting. They show a different behavior of the foreign market indices according to the sign of the last trading day in the American stock exchange.

In the three markets the average return is positive (negative) when the one-lagged return of the S&P500 is positive (negative). The standard deviation of the returns of the foreign market indices is always greater in the presence of a negative sign coming from the US market. The skewness coefficient is negative and stronger when the American stock exchange return is negative; it is positive in the opposite case. Finally the relationship between present foreign returns and past US returns is clearly stronger when the S&P500 return is negative. All these empirical findings match the theoretical features of the *SGARCH* model. Only the observed kurtosis is always smaller in the presence of a negative US return, which contradicts the model.

Table 12.3 *Descriptive statistics for the three stock indices according to the sign of one-lagged S&P500 return.*

		SMI	MIB	AEX
Average	SP>0	0.245	0.259	0.381
	SP<0	−0.200	−0.220	−0.350
Standard deviation	SP>0	1.178	1.488	1.373
	SP<0	1.358	1.588	1.511
Skewness	SP>0	0.157	0.015	0.219
	SP<0	−0.287	−0.037	−0.436
Kurtosis	SP>0	7.876	5.294	7.419
	SP<0	6.496	4.575	5.395
Correlation	SP>0	0.125	0.059	0.173
	SP<0	0.291	0.228	0.263

12.6 Estimation

In order to fit the $SGARCH$ model (12.3) for the three stock indices, we need the time series of ϵ_{t-1} which enters the mean equation of the model. We obtain it after estimating a $GARCH(1,1)$ for S&P500 returns. The estimates are listed in Table 12.4.

The autocorrelation functions of the squared foreign returns steadily decrease, hinting at the orders $p = q = 1$ in the variance Equation (12.2).

Maximum likelihood estimates of the $SGARCH$ model are listed in Table 12.5. All parameter estimates are significant. In particular, α is always clearly greater than β. The highest value of the quantity $\alpha - \beta$ refers to the AEX returns involving the major distance between the behaviors of the index conditionally on the signals coming from the US market.

On the whole, the economic interpretation is straightforward: it is relevant to distinguish between bad and good news from the US market. The

Table 12.4 *Parameter estimation of the $GARCH$ model for the S&P500 returns. * means significant at 5% level, ** means significant at 1% level.*

Parameter	Estimate
δ_0	0.011*
δ_1	0.095**
δ_2	0.904**

Table 12.5 *Parameter estimation of the SGARCH model for the three indices returns. * means significant at 5% level, ** means significant at 1% level.*

Parameter	SMI	MIB	AEX
ω_0	0.040**	0.091**	0.016**
ω_1	0.130**	0.106**	0.082**
ω_2	0.842**	0.844**	0.889**
α	0.263**	0.186**	0.343**
β	0.134**	0.114**	0.072*

inclusion of the effect of exogenous news significantly improves the predictive performance.

Model diagnostic is based on the estimated standardized squares of the random variable $X_t|\sigma_t, \text{sgn}(Y_{t-1})$, defined as follows:

$$\widehat{R}_t^2 = \begin{cases} \dfrac{\left(X_t - \widehat{\sigma}_t\widehat{\beta}\sqrt{2/\pi}\right)^2}{\widehat{\sigma}_t^2\left[1+\left(\widehat{\alpha}+\widehat{\beta}\right)^2\right]} & \text{if} \quad Y_{t-1} < 0, \\[2ex] \dfrac{\left(X_t - \widehat{\sigma}_t\widehat{\beta}\sqrt{2/\pi}\right)^2}{\widehat{\sigma}_t^2\left[1+\left(\widehat{\alpha}-\widehat{\beta}\right)^2\right]} & \text{if} \quad Y_{t-1} > 0. \end{cases}$$

Since the sample size is very large, the distribution of $\widehat{R}_t^2$ approximates the distribution of R_t^2; that is,

$$R_t^2 = \begin{cases} \dfrac{\left(X_t - \sigma_t\beta\sqrt{2/\pi}\right)^2}{\sigma_t^2\left[1+(\alpha+\beta)^2\right]} & \text{if} \quad Y_{t-1} < 0, \\[2ex] \dfrac{\left(X_t - \sigma_t\beta\sqrt{2/\pi}\right)^2}{\sigma_t^2\left[1+(\alpha-\beta)^2\right]} & \text{if} \quad Y_{t-1} > 0. \end{cases}$$

The random variables R_t are mutually independent and their distribution is a mixture of skew-normals with location parameter zero and scale parameter 1. Azzalini (1985) shows that $Z^2 \sim \chi_1^2$ when $Z \sim SN(0, 1, \alpha)$. Hence R_t^2 are approximately independent and identically distributed according to a χ_1^2 distribution.

Testing for skew-normality can be done using squared $\widehat{R}_t$ and the statistic

$$\sqrt{\frac{n}{2}}\left(\frac{1}{n}\sum_{t=1}^{n}\widehat{R}_t^2 - 1\right) \overset{a}{\sim} N(0,1),$$

where n is the size of the time series. The observed values are close to their zero expectation with satisfactory p-values as shown in Table 12.6.

In order to obtain a visual check for the hypothesis, we plot the quantiles of $\widehat{R}_1^2, \ldots, \widehat{R}_n^2$ against the corresponding quantiles of a χ_1^2 distribution in

Table 12.6 *Skew-normality test and corresponding p-values.*

	SMI	MIB	AEX
Statistics	0.272	0.287	0.561
p-value	0.785	0.774	0.575

Figure 12.1. Almost all points lie on the bisector, hinting that the model is correctly specified and ensures a good fit. A very few points are far away from the bisector and correspond to the highest quantiles, but they are not enough to state the inadequacy of the tails of the model.

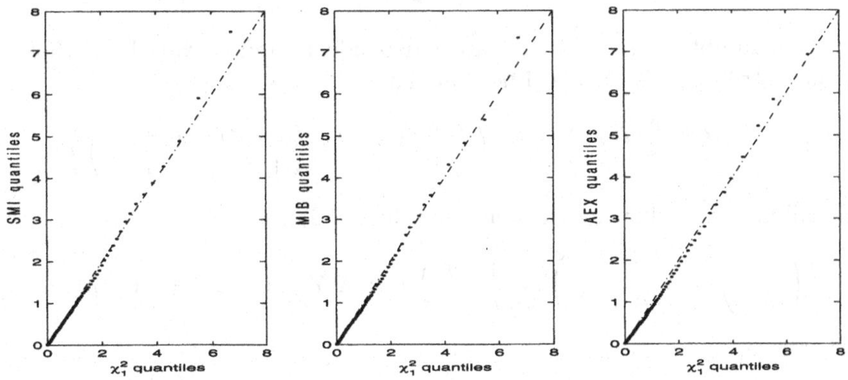

Figure 12.1 *QQ plot for the standardized squared* $\hat{R}_t$.

12.7 Conclusions

This chapter introduces a $GARCH$-type model to describe the main features of small capitalized markets characterized by the short term influence exerted by the US stock exchange, as stressed by a number of empirical studies. The model takes into account well-known stylized facts of financial returns, such as zero mean, time-varying volatility, and fat tails, as well as the contemporaneous and predictive asymmetry. It is denoted by skew-in-mean $GARCH$, or $SGARCH$. The mean equation includes endogenous (i.e., country specific) and past exogenous (i.e., originated in the US market) shocks. As a result, the distribution of the returns, conditionally on the sign of the one-day lagged US market, is a skew-normal. The model was successfully applied to three European stock indices.

12.8 Appendix

Proofs in Section 12.2

We shall prove how to get the distribution of X_t given $Y_{t-1} < 0$ only. The proof for $Y_{t-1} > 0$ is similar.

- *Distribution:* from the definition of X_t we have:

$$(X_t|\sigma_t, Y_{t-1} < 0) \equiv \sigma_t \left(\sqrt{\frac{2}{\pi}}\beta + \alpha\epsilon_{t-1} - \beta|\epsilon_{t-1}| + \xi_t\right)\Bigg| \; Y_{t-1} < 0.$$

Ordinary properties of the absolute value lead to the following:

$$(X_t|\sigma_t, Y_{t-1} < 0) \equiv \sigma_t \left(\sqrt{\frac{2}{\pi}}\beta + (\alpha+\beta)\,\epsilon_{t-1} + \xi_t\right).$$

By assumption, ξ_t and ϵ_{t-1} are independent random variables whose joint distribution is standard normal. Then

$$\begin{pmatrix} \xi_t + (\alpha+\beta)\,\epsilon_{t-1} \\ \epsilon_{t-1} \end{pmatrix} \sim N_2\left(\begin{pmatrix} 0 \\ 0 \end{pmatrix}, \begin{pmatrix} 1+(\alpha+\beta)^2 & \alpha+\beta \\ \alpha+\beta & 1 \end{pmatrix}\right).$$

Recall now the following theorem (Azzalini, 1985):

$$\begin{pmatrix} X \\ Y \end{pmatrix} \sim N_2\left(\begin{pmatrix} 0 \\ 0 \end{pmatrix}, \begin{pmatrix} 1 & \rho \\ \rho & 1 \end{pmatrix}\right) \Rightarrow X|Y < 0 \sim SN\,(0, 1, -\rho)\,.$$

Therefore,

$$\xi_t + (\alpha+\beta)\,\epsilon_{t-1}|\; Y_{t-1} < 0 \sim SN\left(0, \sqrt{1+(\alpha+\beta)^2}, \frac{-(\alpha+\beta)}{\sqrt{1+(\alpha+\beta)^2}}\right).$$

Azzalini (1985) also shows that

$$Z \sim SN(\xi, \psi, \lambda) \Rightarrow aZ{+}b \sim SN(a\xi{+}b, |a|\,\psi, \lambda a^+), \qquad a^+ = \begin{cases} +1 & a \geq 0, \\ -1 & a < 0. \end{cases}$$

When $Y_{t-1}<0$, X_t is a linear transformation of $\xi_t + (\alpha+\beta)\,\epsilon_{t-1}$, so that the above theorem implies

$$(X_t|\sigma_t, Y_{t-1} < 0) \sim SN\left(\sqrt{\frac{2}{\pi}}\beta\sigma_t, \sigma_t\sqrt{1+(\alpha+\beta)^2}, \frac{-(\alpha+\beta)}{\sqrt{1+(\alpha+\beta)^2}}\right),$$

and this part of the proof is complete.

- *Expected value:* if $Z \sim SN(\xi, \psi, \delta)$, then $\mathrm{E}(Z) = \xi + \psi\sqrt{2/\pi}\delta$ (Azzalini, 1985). Since the distribution of X_t is skew-normal when $Y_{t-1} < 0$ we can apply the above proposition and write:

$$\begin{aligned}\mathrm{E}\left(X_t|\sigma_t, Y_{t-1}<0\right) &= \sqrt{\frac{2}{\pi}}\beta\sigma_t + \sigma_t\sqrt{\frac{1+(\alpha_t+\beta)^2}{2/\pi}}\frac{-(\alpha+\beta)}{\sqrt{1+(\alpha+\beta)^2}} \\ &= \sqrt{\frac{2}{\pi}}\sigma_t\left(\beta-(\alpha+\beta)\right) = -\sqrt{\frac{2}{\pi}}\alpha\sigma_t.\end{aligned}$$

- *Variance:* if $Z \sim SN(\xi, \psi, \delta)$, then $\mathrm{Var}(Z) = \psi^2\left(1-(2/\pi)\,\delta^2\right)$ (Azzalini, 1985). Since the distribution of X_t is skew-normal when $Y_{t-1}<0$ we can apply the above proposition and write:

$$\begin{aligned}\mathrm{Var}\left(X_t|\sigma_t, Y_{t-1}<0\right) &= \sigma_t^2\left(1+(\alpha+\beta)^2\right)\left(1-\frac{2(\alpha+\beta)^2}{\pi(1+(\alpha+\beta)^2)}\right) \\ &= \sigma_t^2\left(1+\frac{\pi-2}{\pi}(\alpha+\beta)^2\right).\end{aligned}$$

- *Skewness:* we shall now apply the following property of skew-normal distributions (Azzalini, 1985):

$$Z \sim SN(\xi, \psi, \delta) \Rightarrow \mathrm{S}\left(Z\right) = \frac{4-\pi}{2}\delta^+\left(\frac{2\delta^2}{\pi-2\delta^2}\right)^{3/2}, \ \delta^+ = \begin{cases}+1 & \delta \geq 0, \\ -1 & \delta < 0.\end{cases}$$

Since the distribution of X_t is skew-normal when $Y_{t-1}<0$ we can apply the above proposition and write:

$$\begin{aligned}\mathrm{S}\left(X_t|\sigma_t, Y_{t-1}<0\right) &= \frac{\pi-4}{2}\left(\frac{2(\alpha+\beta)^2/\left(1+(\alpha+\beta)^2\right)}{\pi-2(\alpha+\beta)^2/\left(1+(\alpha+\beta)^2\right)}\right)^{3/2} \\ &= \frac{\pi-4}{2}\left(\frac{2(\alpha+\beta)^2}{\pi+(\pi-2)(\alpha+\beta)^2}\right)^{3/2}.\end{aligned}$$

- *Correlation:* in order to keep formulas simple, let

$$Z = \frac{X_t - \beta\sqrt{2/\pi}}{\sigma_t\sqrt{1+(\alpha+\beta)^2}}.$$

Then the joint density of $(Z, Y_{t-1})\,|Y_{t-1}<0$ is

$$g\left(z, y_{t-1}|Y_{t-1}<0\right) = \frac{1}{\pi\sqrt{1-\lambda^2}}\exp\left(-\frac{z^2-2\lambda z y_{t-1}+y_{t-1}^2}{2(1-\lambda^2)}\right),$$

where $\lambda = (\alpha+\beta)/\sqrt{1+(\alpha+\beta)^2}$. The expected value of the random variable $Z\,Y_{t-1}|Y_{t-1}<0$ is

$$\int_{-\infty}^{+\infty}\int_{-\infty}^{0}\frac{1}{\pi\sqrt{1-\lambda^2}}\exp\left(-\frac{z^2-2\lambda z y_{t-1}+y_{t-1}^2}{2\left(1-\lambda^2\right)}\right)dzdy_{t-1}.$$

By symmetry we have:

$$\begin{aligned}&\mathrm{E}\left(Z\,Y_{t-1}|Y_{t-1}<0\right)\\&=\int_{-\infty}^{+\infty}\int_{0}^{+\infty}\exp\left(-\frac{z^2-2\lambda z y_{t-1}+y_{t-1}^2}{2\left(1-\lambda^2\right)}\right)dzdy_{t-1}.\end{aligned}$$

Then we can write

$$\mathrm{E}\left(Z\,Y_{t-1}|Y_{t-1}<0\right)=\mathrm{E}\left(Z\,Y_{t-1}\right)=\frac{\alpha+\beta}{\sqrt{1+\left(\alpha+\beta\right)^2}}.$$

Apply now some basic properties of the skew-normal distribution:

$$\begin{aligned}\mathrm{E}\left(X_t-\sqrt{\frac{2}{\pi}}\sigma_t\beta\middle|Y_{t-1}<0\right)&=-\sqrt{\frac{2}{\pi}}\sigma_t\left(\alpha+\beta\right),\\\mathrm{Var}\left(X_t-\sqrt{\frac{2}{\pi}}\sigma_t\beta\middle|Y_{t-1}<0\right)&=\sigma_t^2\left(1+\frac{\pi-2}{\pi}\left(\alpha+\beta\right)^2\right).\end{aligned}$$

From the definition of covariance we get:

$$\begin{aligned}\mathrm{Cov}(X_t,Y_{t-1}|\sigma_t,Y_{t-1}<0)&=\sigma_t\left(\alpha+\beta\right)-\frac{2}{\pi}\sigma_t\left(\alpha+\beta\right)\\&=\sigma_t\left(1-\frac{2}{\pi}\right)\left(\alpha+\beta\right).\end{aligned}$$

Then the correlation between $X_t, Y_{t-1}|Y_{t-1}<0$ is:

$$\begin{aligned}\mathrm{Corr}(X_t,Y_{t-1}|Y_{t-1}<0)&=\frac{\sigma_t\left(\alpha+\beta\right)\left(1-\frac{2}{\pi}\right)}{\sqrt{\left(1-\frac{2}{\pi}\right)\sigma_t^2\left(1+\frac{\pi-2}{\pi}t\left(\alpha+\beta\right)^2\right)}}\\&=\frac{\left(\alpha+\beta\right)\sqrt{\pi-2}}{\sqrt{\pi+\left(\pi-2\right)\left(\alpha+\beta\right)^2}},\end{aligned}$$

and the proof is complete.

□

Proofs in Section 12.4
The proof that $\mathrm{E}(X_t)=0$ is trivial.
We shall now evaluate the skewness of X_t. The third moment from the origin of a $SN\left(0,1,\delta\right)$ distribution is $\mathrm{E}\left(X^3\right)=\sqrt{2/\pi}\delta\left(3-\delta^2\right)$ (Chiogna,

1998). Hence the third moment from the origin of the random variable $\epsilon_t + (\alpha+\beta)\, Y_{t-1} |\, \sigma_t, Y_{t-1} < 0$ is

$$\mathrm{E}\left(\left(\epsilon_t + (\alpha+\beta)\, Y_{t-1}\right)^3 \middle|\, \sigma_t, Y_{t-1} < 0\right) = -\sqrt{\frac{2}{\pi}}\,(\alpha+\beta)\left(3 + 2\,(\alpha+\beta)^2\right).$$

Let $Y = \epsilon_t + (\alpha+\beta)\, Y_{t-1}$. From the definition of X_t we get

$$\mathrm{E}\left(\frac{X_t^3}{\sigma_t^3} \middle|\, \sigma_t, Y_{t-1} < 0\right) = \mathrm{E}\left(\left(\beta\sqrt{\frac{2}{\pi}} + Y\right)^3 \middle|\, \sigma_t, Y_{t-1} < 0\right),$$

the distribution of $Y\,|Y_{t-1} < 0$ is skew-normal, then

$$\begin{aligned}
\mathrm{E}\left(\frac{X_t^3}{\sigma_t^3} \middle|\, \sigma_t, Y_{t-1} < 0\right) &= \mathrm{E}\left(Y^3 | Y_{t-1}\right) + 3\beta\sqrt{\frac{2}{\pi}}\mathrm{E}\left(Y^2\,| Y_{t-1} < 0\right) \\
&\quad + \beta^2 \frac{6}{\pi}\mathrm{E}\left(Y\,| Y_{t-1} < 0\right) + \beta^3 \frac{2}{\pi}\sqrt{\frac{2}{\pi}} \\
&= -\sqrt{\frac{2}{\pi}}\,(\alpha+\beta)\left(3 + 2\,(\alpha+\beta)^2\right) + \beta^3 \frac{2}{\pi}\sqrt{\frac{2}{\pi}} \\
&\quad + 3\beta\sqrt{\frac{2}{\pi}}\left(1 + (\alpha+\beta)^2\right) - \beta^2\frac{6}{\pi}\sqrt{\frac{2}{\pi}}\,(\alpha+\beta).
\end{aligned}$$

In a similar way it can be shown that

$$\begin{aligned}
\mathrm{E}\left(\frac{X_t^3}{\sigma_t^3} \middle|\, \sigma_t, Y_{t-1} < 0\right) &= \sqrt{\frac{2}{\pi}}\,(\alpha-\beta)\left(3 + 2\,(\alpha-\beta)^2\right) + \beta^3 \frac{2}{\pi}\sqrt{\frac{2}{\pi}} \\
&\quad + 3\beta\sqrt{\frac{2}{\pi}}\left(1 + (\alpha-\beta)^2\right) + \beta^2\frac{6}{\pi}\sqrt{\frac{2}{\pi}}\,(\alpha-\beta).
\end{aligned}$$

The third moment of $X_t^3|\sigma_t$ from the origin is the average of the conditional expectations $\mathrm{E}\left(X_t^3 \middle|\, Y_{t-1} < 0\right)$ and $\mathrm{E}\left(X_t^3 \middle|\, Y_{t-1} > 0\right)$. Then

$$\begin{aligned}
\mathrm{E}\left(\frac{X_t^3}{\sigma_t^3} \middle|\, \sigma_t\right) &= \frac{1}{2}\sqrt{\frac{2}{\pi}}\left((\alpha-\beta)\left(3 + 2\,(\alpha-\beta)^2\right)\right. \\
&\quad \left. - (\alpha+\beta)\left(3 + 2\,(\alpha+\beta)^2\right)\right) \\
&\quad + \frac{3}{2}\beta\sqrt{\frac{2}{\pi}}\left(\left(1 + (\alpha-\beta)^2\right) + \left(1 + (\alpha+\beta)^2\right)\right) \\
&\quad + \beta^2\frac{3}{\pi}\sqrt{\frac{2}{\pi}}\left((\alpha-\beta) - (\alpha+\beta)\right) + \beta^3\frac{2}{\pi}\sqrt{\frac{2}{\pi}}.
\end{aligned}$$

A little algebra shows the first term of the sum in the above equation is

$-2\sqrt{2/\pi}\beta\left(6\alpha^2+3+2\beta^2\right)$. We can then write:

$$\begin{aligned}
\mathrm{E}\left(\frac{X_t^3}{\sigma_t^3}\middle|\sigma_t\right) &= -\sqrt{\frac{2}{\pi}}\beta\left(6\alpha^2+3+2\beta^2\right)+3\beta\sqrt{\frac{2}{\pi}}\left(1+\alpha^2+\beta^2\right) \\
&\quad -\beta^3\frac{4}{\pi}\sqrt{\frac{2}{\pi}} \\
&= \sqrt{\frac{2}{\pi}}\beta\left(2\beta^2-3\alpha^2-\beta^2\frac{4}{\pi}\right),
\end{aligned}$$

and the final result follows.

We shall now prove predictive asymmetry. By assumption the random variables $\epsilon_t, \xi_t, \epsilon_{t-1}, \xi_{t-1}, \ldots$ introduced in Section 12.3 belong to a white noise process, so that they are mutually independent. Let $W_t = \beta\sqrt{2/\pi} + \alpha\epsilon_{t-1} - \beta|\epsilon_{t-1}| + \xi_t$. Hence

$$W_t \perp (W_{t-1}, \sigma_t, W_{t-2}, \sigma_{t-1}, \ldots)$$

We shall now prove that $\mathrm{E}\left(\sigma_{t-i+1}^2 X_t\right) = \mathrm{E}\left(X_{t-i}^2 X_t\right) = 0$ if $i > 0$. The definitions of W_t and X_t lead to

$$\mathrm{E}\left(\sigma_{t-i+1}^2 X_t\right) = \mathrm{E}\left(\sigma_{t-i+1}^2 \sigma_t W_t\right), \quad \mathrm{E}\left(X_{t-i}^2 X_{t,}\right) = \mathrm{E}\left(\sigma_{t-i}^2 W_{t-i}^2 \sigma_t W_t\right).$$

We already know that W_t and $W_{t-i}, \sigma_t, \sigma_{t-i}, \sigma_{t-i+1}$ are independent when $i > 0$. We can therefore apply a well-known property of the expected value:

$$\begin{aligned}
\mathrm{E}\left(\sigma_{t-i+1}^2 X_t\right) &= \mathrm{E}\left(\sigma_{t-i+1}^2 \sigma_t\right)\mathrm{E}\left(W_t\right), \\
\mathrm{E}\left(X_{t-i}^2 X_{t,}\right) &= \mathrm{E}\left(\sigma_{t-i}^2 W_{t-i}^2 \sigma_t\right)\mathrm{E}\left(W_t\right).
\end{aligned}$$

Since $\mathrm{E}(W_t) = 0$ (Nelson, 1991) the above equations imply that

$$\mathrm{E}\left(\sigma_{t-i+1}^2 X_t\right) = \mathrm{E}\left(X_{t-i}^2 X_t\right) = 0, \qquad i > 0,$$

and this part of the proof is complete.

We shall now complete the proof by showing that $\mathrm{Corr}(\sigma_{t+1}^2, X_t) \leq 0$. From the definition of σ_t the covariance between σ_{t+1}^2 and X_t can be written as follows:

$$\mathrm{Cov}\left(\sigma_{t+1}^2, X_t\right) = \mathrm{Cov}\left(\omega_0 + \sum_{i=1}^{k}\omega_i X_{t-i+1}^2 + \sum_{i=1}^{h}\omega_i \sigma_{t-i+1}^2\,,\ X_t\right).$$

Apply now linear properties of the covariance operator:

$$\mathrm{Cov}\left(\sigma_{t+1}^2, X_t\right) = \sum_{i=1}^{k}\omega_i \mathrm{Cov}\left(X_t, X_{t-i+1}^2\right) + \sum_{i=1}^{h}\omega_i \mathrm{Cov}\left(\sigma_{t-i+1}^2, X_t\right).$$

Since $\mathrm{E}(X_t) = 0$ the above equation can be written as follows:

$$\mathrm{Cov}\left(\sigma_{t+1}^2, X_t\right) = \sum_{i=1}^{k} \omega_i \mathrm{E}\left(X_t X_{t-i+1}^2\right) + \sum_{i=1}^{h} \omega_i \mathrm{E}\left(\sigma_{t-i+1}^2 X_t\right).$$

Recall now that $\mathrm{E}(\sigma_{t-i+1}^2 X_t) = \mathrm{E}\left(X_{t-i}^2 X_t\right) = 0$ if $i > 0$, then

$$\mathrm{Cov}\left(\sigma_{t+1}^2, X_t\right) = \omega_1 \mathrm{E}\left(X_t X_t^2\right) = \omega_1 \mathrm{E}\left(X_t^3\right).$$

By assumption, $\omega_1 \geq 0$ and $\mathrm{E}(X_t^3) = \mathrm{S}(X_t^3)\mathrm{Var}^{3/2}\left(X_t^3\right)$. Since X_t is negatively skewed it follows that $\mathrm{Corr}(\sigma_{t+1}^2, X_t) \leq 0$ and the proof is complete.

We shall now prove kurtosis. Let $W_t = \xi_t + \alpha\epsilon_{t-1} - \beta|\epsilon_{t-1}|$.

- We first prove that $\mathrm{E}(W_t^4) = 3\left(\beta^4 + 6\alpha^2\beta^2 + 2\beta^2 + \alpha^4 + 1 + 2\alpha^2\right)$. From the Total Probability theorem we get that $\mathrm{E}\left(W_t^4\right)$ equals

$$\mathrm{E}\left(W_t^4 | Y_{t-1} < 0\right) \mathrm{P}\left(Y_{t-1} < 0\right) + \mathrm{E}\left(W_t^4 | Y_{t-1} > 0\right) \mathrm{P}\left(Y_{t-1} > 0\right).$$

By assumption the distribution of Y_{t-1} is normal centered at the origin, so that $\mathrm{P}(Y_{t-1} < 0) = \mathrm{P}\left(Y_{t-1} > 0\right) = 1/2$. The conditional distributions of $W_t|Y_{t-1} < 0$ and $W_t|Y_{t-1} > 0$ are skew-normal with location parameter equal to zero. Therefore the corresponding even moments do not depend on the skewness parameters:

$$\begin{aligned} \mathrm{E}\left(W_t^4 | Y_{t-1} < 0\right) &= 3\left(1 + (\alpha+\beta)^2\right)^2 \quad \mathrm{E}\left(W_t^4 | Y_{t-1} > 0\right) \\ &= 3\left(1 + (\alpha-\beta)^2\right)^2. \end{aligned}$$

We can then write:

$$\begin{aligned} \mathrm{E}\left(W_t^4\right) &= \frac{3}{2}\left(\left(1 + (\alpha+\beta)^2\right)^2 + \left(1 + (\alpha-\beta)^2\right)^2\right) \\ &= 3\left(\beta^4 + 6\alpha^2\beta^2 + 2\beta^2 + \alpha^4 + 1 + 2\alpha^2\right). \end{aligned}$$

- We shall now prove that $\mathrm{E}(X_t^4/\sigma_t^4) = 3\left(1 + \frac{\pi-2}{\pi}\beta^2 + \alpha^2\right)^2 + \frac{8}{\pi}\left(1 - \frac{3}{\pi}\right)$. From the definitions of X_t and W_t we get

$$\mathrm{E}\left(\frac{X_t^4}{\sigma_t^4}\right) = \mathrm{E}\left(\left(W_t + \beta\sqrt{\frac{2}{\pi}}\right)^4\right).$$

Apply now the Binomial Theorem on the right-hand side of the above equation:

$$\mathrm{E}\left(\frac{X_t^4}{\sigma_t^4}\right) = \mathrm{E}\left(W_t^4\right) + 4\beta\sqrt{\frac{2}{\pi}}\mathrm{E}\left(W_t^3\right) + \frac{12\beta^2}{\pi}\mathrm{E}\left(W_t^2\right) + \frac{8\beta^3}{\pi}\sqrt{\frac{2}{\pi}}\mathrm{E}\left(W_t\right) + \frac{4\beta^4}{\pi^2}.$$

It is known that $\mathrm{E}(W_t^3) = -\beta\sqrt{2/\pi}\,(6\alpha^2+3+2\beta^2)$, $\mathrm{E}(W_t^2) = 1+\alpha^2+\beta^2$, $\mathrm{E}(W_t) = -\beta\sqrt{2/\pi}$. We can therefore write

$$\begin{aligned}
\mathrm{E}\left(\frac{X_t^4}{\sigma_t^4}\middle|\,\sigma_t\right) &= 3\beta^4+18\alpha^2\beta^2+6\beta^2+3\alpha^4+3+6\alpha^2 \\
&\quad -\frac{8\beta^2}{\pi}\left(6\alpha^2+3+2\beta^2\right) \\
&\quad +\frac{12\beta^2}{\pi}\left(1+\alpha^2+\beta^2\right)-\frac{16\beta^4}{\pi^2}+\frac{4\beta^4}{\pi^2}.
\end{aligned}$$

Simple algebra leads to the following:

$$\begin{aligned}
\mathrm{E}\left(\frac{X_t^4}{\sigma_t^4}\middle|\,\sigma_t\right) &= \beta^4\left(3-\frac{4}{\pi}-\frac{12}{\pi^2}\right)+\beta^2\left(6-\frac{12}{\pi}+18\alpha^2-\frac{36\alpha^2}{\pi}\right) \\
&\quad +3\alpha^4+3+6\alpha^2 \\
&= 3\left(1+\frac{\pi-2}{\pi}\beta^2+\alpha^2\right)^2+\frac{8}{\pi}\left(1-\frac{3}{\pi}\right).
\end{aligned}$$

- Recall now the definition of kurtosis:

$$\mathrm{K}\left(X_t\right)=\frac{\mathrm{E}\left(\left(X_t-\mathrm{E}\left(X_t\right)\right)^4\right)}{\mathrm{Var}^2\left(X_t\right)}.$$

We know that $\mathrm{E}(X_t)=0$ and it can be shown that the variance is equal to $\mathrm{Var}(X_t)=\sigma_t^2\left(1+(\pi-2)\left(\beta^2/\pi\right)+\alpha^2\right)$. Therefore

$$\begin{aligned}
\mathrm{K}\left(X_t|\,\sigma_t\right) &= \frac{3\left(1+\frac{\pi-2}{\pi}\beta^2+\alpha^2\right)^2+\frac{8}{\pi}\left(1-\frac{3}{\pi}\right)}{\left(1+\frac{\pi-2}{\pi}\beta^2+\alpha^2\right)^2} \\
&= 3+\frac{8\beta^2\left(\pi-3\right)}{\left(\pi+\left(\pi-2\right)\beta^2+\alpha^2\right)^2},
\end{aligned}$$

and the proof is complete.

□

CHAPTER 13

Skew-Normality in Stochastic Frontier Analysis

J. Armando Domínguez-Molina, Graciela González-Farías, and Rogelio Ramos-Quiroga

13.1 Introduction

A problem of interest to econometricians is the specification and estimation of a frontier production function. The original formulation of the stochastic frontier model is due to Aigner, Lovell, and Schmidt (1977) and Meeusen and Van den Broeck (1977):

$$Y = f(\mathbf{x}; \boldsymbol{\beta}) + \varepsilon, \tag{13.1}$$

where the error term $\varepsilon = V - U$, is composed by a symmetric component, V, representing measurement error, and by the non-negative technical inefficiency component U. The stochastic frontier $f(\mathbf{x}; \boldsymbol{\beta}) + V$ allows for firms with same inputs, $\mathbf{x}$, to have different frontiers due to unobservable shocks; indeed, (13.1) models the inefficiency of a company to attain its production frontier; see Parsons (2002) for a treatment of stochastic frontier analysis (SFA) in marketing science.

Some issues of interest in stochastic frontier analysis are the estimation of the common frontier $f(\mathbf{x}; \boldsymbol{\beta})$, the estimation of the technical efficiency, typically related to $\tau^2 = \text{Var}(U)$, and the estimation of the measurement error $\sigma^2 = \text{Var}(V)$. Before we discuss these objectives, we will take a closer look at the structure of the disturbance term ε, and its relation to Azzalini's (1985) proposal of a skew-normal distribution. The distributional properties of Y will be inherited from those of ε by a translation of the location parameter via $f(\mathbf{x}; \boldsymbol{\beta})$.

Aigner *et al.* (1977) discuss the model when U has a positive half-normal distribution and is independent of V which is assumed to be normally distributed. Stevenson (1980) generalizes this model by considering U to have a normal distribution with mean μ and variance σ^2 truncated below at zero. From Aigner *et al.* (1977) we have that the density of $\varepsilon = V - U$, with U and V independent, $V \sim N(0, \sigma^2)$, and U positive half-normal, is given by

$$g(\varepsilon) = 2\frac{1}{\eta}\phi\left(\frac{\varepsilon}{\eta}\right)\Phi\left(-\frac{\lambda}{\eta}\varepsilon\right), \tag{13.2}$$

where $\eta = \sqrt{\tau^2 + \sigma^2}$, $\lambda = \tau/\sigma$, $\phi(\cdot)$ and $\Phi(\cdot)$ denote the probability density function (pdf) and the cumulative distribution function (cdf) of a standard normal random variable, respectively. A recent proposal is given in Tancredi (2003) where a skew-t distribution is used to model the composed error.

Comparing this function with the skew-normal density introduced by Azzalini (1985)

$$g(x) = 2\phi(x)\Phi(\alpha x), \ \alpha \in \mathbb{R}, \tag{13.3}$$

we see that the density of the disturbance ε is just the density of a scaled skew-normal distribution; that is, $\varepsilon = \eta X$ where X is skew-normal with $\alpha = -\lambda$.

Densities related to Azzalini's can be traced back to the work of Birnbaum (1950). For a historical note see Remark 2.2 in Arnold and Beaver (2002). Azzalini's work, however, was the first to fully study the properties of the skew-normal distribution and to give it its name. Azzalini's (1985) work was followed by several developments related to the skew-normal, among which we can mention Henze (1986). He showed how to obtain it as the sum of two random variables, one of them sign free, and the other one positive. Azzalini (1985) also generalizes the skew-normal density as

$$g(x) = \Phi^{-1}\left(\nu/\sqrt{1+\alpha^2}\right)\phi(x)\Phi(\alpha x + \nu). \tag{13.4}$$

Arnold and Beaver (2000a, 2002, 2003) describe several procedures that lead to skewed densities, of which the skew-normal is a special case. One of these procedures considers the distribution of random variables of the form

$$Y = U + \alpha V(c),$$

where U and V can have arbitrary distributions and $V(c)$ is distributed as V truncated below at c; they give expressions for the cases U and V arbitrary, U and V symmetric, U and V non-independent.

The usual setting in SFA is when we consider models for optimal production functions, thus the disturbance term is constructed as $\varepsilon = V - U$. If the setting under consideration involves minimal cost frontiers, then the usual device is to switch the sign of U: $\varepsilon = V + U$. By doing this we have that the cost observations would be above the minimum cost frontier. One direct generalization is to consider

$$\varepsilon = V + \alpha U(c),$$

where α is fixed, and $U(c)$ is a random variable truncated below at a positive constant c. The sign of α would indicate the direction of the asymmetry.

The corresponding density is given by

$$g(\varepsilon) = \int_{-\infty}^{\infty} f(\varepsilon - \alpha u)\, h(u)\, du.$$

If $V \sim N(\mu, \sigma^2)$ and $U \sim N^c(\nu, \tau^2)$, i.e., U has a $N(\nu, \tau^2)$ distribution truncated below at $c \in \mathbb{R}$, then the density is

$$g(\varepsilon) = \frac{\Phi^{-1}\left(\frac{\nu - c}{\tau}\right)}{\sqrt{\sigma^2 + \delta^2\tau^2}} \phi\left(\frac{\varepsilon - \mu - \alpha\nu}{\sqrt{\sigma^2 + \alpha^2\tau^2}}\right) \times \Phi\left[\frac{\alpha\tau(\varepsilon - \mu - \alpha\nu)}{\sigma\sqrt{\sigma^2 + \alpha^2\tau^2}} + \frac{(\nu - c)\sqrt{\alpha^2\tau^2 + \sigma^2}}{\sigma\tau}\right]. \tag{13.5}$$

This density is derived in the example presented in the Appendix.

If $\varepsilon = U - V$, $U \sim N(0, \sigma^2)$, $V \sim N^c(\nu, \tau^2)$, U independent of V, then if in (13.5) $\mu = c = 0$, and $\alpha = -1$, we obtain

$$g(\varepsilon) = \frac{\Phi^{-1}\left(\frac{\nu}{\tau}\right)}{\eta} \phi\left(\frac{\varepsilon + \nu}{\eta}\right) \Phi\left[\frac{-\tau(\varepsilon + \nu)}{\sigma\eta} + \frac{\nu\eta}{\sigma\tau}\right],$$

where $\eta = \sqrt{\tau^2 + \sigma^2}$. It is the same as in Kumbhakar and Lovell (2000, Equation 3.2.46) after simplification, i.e.,

$$g(\varepsilon) = \frac{1}{\eta}\phi\left(\frac{\varepsilon + \nu}{\eta}\right) \Phi\left(\frac{\nu}{\eta\lambda} - \frac{\varepsilon\lambda}{\eta}\right) / \Phi\left(\frac{\nu}{\sigma}\right),$$

where $\lambda = \tau/\sigma$.

Also, observe that if we define the random variable $W = (\varepsilon - \mu - \alpha\nu) / \sqrt{\sigma^2 + \alpha^2\tau^2}$ and $\alpha^* = \alpha\tau/\sigma$, $\nu^* = (\nu - c)\sqrt{\alpha^2\tau^2 + \sigma^2}/\sigma\tau$ then the density of W is the same as (13.4).

The error structure, $\varepsilon = V + \alpha U$, has been examined under several distributional assumptions for V and U; see Arnold and Beaver (2002, 2003), Kumbhakar and Lovell (2000).

This chapter is structured as follows. Section 13.2 discusses estimation issues in the SFA model. Section 13.3 presents a correlated structure for the compound error. SFA with skew-elliptical components is discussed in Section 13.4. We conclude in Section 13.5, and some theoretical results are described in the Appendix.

13.2 Estimation

Let us assume a data structure of cross-sectional type, composed of independent observations on n firms. We have their output production levels and the corresponding values of exogenous variables collected at a fixed period of time. We assume a model of the form

$$Y_i = f(\mathbf{x}_i; \boldsymbol{\beta}) + \varepsilon_i, \quad \text{where} \quad \varepsilon_i = V_i + \alpha_i U_i,\ \alpha_i \in \mathbb{R}, \quad i = 1, \ldots, n.$$

In the next subsection we will state a series of distributional assumptions on the error vector $\boldsymbol{\varepsilon}= (\varepsilon_1, \ldots, \varepsilon_n)^T$. We are interested in the estimation of the production technology parameter $\boldsymbol{\beta}$ in $f(\mathbf{x}; \boldsymbol{\beta})$, and also in the prediction of the technical efficiency of each firm. This last problem implies that we need to separate the statistical noise from the technical inefficiency.

13.2.1 Model Assumptions

A standard set of assumptions in stochastic frontier analysis would include: a measurement error V_i, distributed as independent random shocks $N(0, \sigma^2)$; efficiencies U_i, distributed as independent truncated normal variables, $N^c(\nu, \tau^2)$; and an independence assumption on U_i and V_j for all i and j. Finally, inputs $\mathbf{x}_i$ are known as non-stochastic variables.

We will use the notation $\mathbf{u} \sim N_m^{\mathbf{c}}(\boldsymbol{\nu}, \Lambda)$ to denote a truncated $N_m(\boldsymbol{\nu}, \Lambda)$ random vector below at $\mathbf{c}$, that is the truncation is of the type $\mathbf{w} \geq \mathbf{c}$, where $\mathbf{w} \geq \mathbf{c}$ means $W_j \geq c_j$, $j = 1, \ldots, m$; see Kotz, Balakrishnan, and Johnson (2000, Section 45.10).

In this work we propose a general model that contains submodels of the stochastic frontier model with normal and truncated normal errors, that is:

$$\mathbf{y} = \mathbf{f}(X; \boldsymbol{\beta}) + \mathbf{v} + D\mathbf{u}, \tag{13.6}$$

where $\mathbf{v} = (V_1, \ldots, V_n)^T \sim N_n(\mathbf{0}, \Sigma)$, $\mathbf{u} = (U_1, \ldots, U_m)^T \sim N_m^{\mathbf{c}}(\boldsymbol{\nu}, \Lambda)$, $m \geq n$, D is an arbitrary $n \times m$ matrix, $\mathbf{v}$ is independent of $\mathbf{u}$, $\mathbf{f}(X; \boldsymbol{\beta}) = (f(\mathbf{x}_1; \boldsymbol{\beta}), \ldots, f(\mathbf{x}_n; \boldsymbol{\beta}))^T$ and $X = (\mathbf{x}_1, \ldots, \mathbf{x}_n)^T$ a known matrix of covariables, $\mathbf{x}_i = (x_{i1}, \ldots, x_{ip})^T$ and $\boldsymbol{\beta} = (\beta_0, \beta_1, \ldots, \beta_p)^T$ is unknown.

The matrix D gives flexibility to the model. If we leave it unspecified, we can estimate it and use this estimate to help validate the model assumptions. On the other hand, we can set it to $D = I_n$ or $D = -I_n$ to evaluate efficiencies or inefficiencies, respectively. Finally, we can induce particular correlation structures in the errors by allowing non-zero off-diagonal parameters.

Next, we define the closed skew-normal (CSN) distribution as given in González-Farías, Domínguez-Molina, and Gupta (2003). We say that a random vector $\mathbf{y}$ has a CSN distribution if its probability density function (pdf) is given by

$$g_{p,q}(\mathbf{y}) = C\phi_p(\mathbf{y}; \boldsymbol{\mu}, \Sigma)\, \Phi_q[D(\mathbf{y} - \boldsymbol{\mu}); \boldsymbol{\nu}, \Delta], \quad \mathbf{y} \in \mathbb{R}^p, \tag{13.7}$$

where

$$C^{-1} = \Phi_q\left(\mathbf{0}; \boldsymbol{\nu}, \Delta + D\Sigma D^T\right), \tag{13.8}$$

and $\phi_p(\cdot\,; \boldsymbol{\mu}, \Sigma)$ and $\Phi_p(\cdot\,; \boldsymbol{\mu}, \Sigma)$ denote the pdf and the cdf of a p-variate normal distribution with mean $\boldsymbol{\mu}$ and covariance matrix Σ, respectively. We denote this as $\mathbf{y} \sim CSN_{p,q}(\boldsymbol{\mu}, \Sigma, D, \boldsymbol{\nu}, \Delta)$. If $\mathbf{y} \sim CSN_{p,q}(\boldsymbol{\mu}, \Sigma, D, \boldsymbol{\nu}, \Delta)$, the moment generating function of $\mathbf{y}$ is

$$M_{\mathbf{y}}(\mathbf{t}) = \frac{\Phi_q\left(D\Sigma\mathbf{t};\boldsymbol{\nu},\Delta + D\Sigma D^T\right)}{\Phi_q\left(\mathbf{0};\boldsymbol{\nu},\Delta + D\Sigma D^T\right)} e^{\mathbf{t}^T\boldsymbol{\mu}+\frac{1}{2}\mathbf{t}^T\Sigma\mathbf{t}}, \quad \mathbf{t} \in \mathbb{R}^p. \tag{13.9}$$

From Proposition 13.6.1 in the Appendix, we have that the density of the compound error $\boldsymbol{\varepsilon} = \mathbf{v} + D\mathbf{u}$ is

$$g(\boldsymbol{\varepsilon}) = \Phi_m^{-1}(\mathbf{0};\mathbf{c}-\boldsymbol{\nu},\Lambda)\,\phi_n(\boldsymbol{\varepsilon};D\boldsymbol{\nu},\Theta)\,\Phi_m\left[\Lambda D^T\Theta^{-1}(\boldsymbol{\varepsilon}-D\boldsymbol{\nu});\mathbf{c}-\boldsymbol{\nu},\Upsilon\right], \tag{13.10}$$

where:

$$\Theta = \Sigma + D\Lambda D^T \quad \text{and } \Upsilon = \left(D^T\Sigma^{-1}D + \Lambda^{-1}\right)^{-1}, \tag{13.11}$$

using the particular case of the Sherman-Morrison-Woodbury formula:

$$(A+BCD)^{-1} = A^{-1} - A^{-1}B\left(C^{-1} + DA^{-1}B\right)^{-1}DA^{-1}, \tag{13.12}$$

where A and B are non-singular $m \times m$ and $n \times n$ matrices, respectively; see Equation (A.2.4f) of Mardia, Kent, and Bibby (1979). We can write Υ as $\Upsilon = \Lambda - \Lambda D^T\Theta^{-1}D\Lambda$. Thus $\boldsymbol{\varepsilon}$ has a closed skew-normal distribution, i.e.,

$$\boldsymbol{\varepsilon} \sim CSN_{n,m}\left(D\boldsymbol{\nu},\Theta,\Lambda D^T\Theta^{-1},\mathbf{c}-\boldsymbol{\nu},\Upsilon\right), \tag{13.13}$$

see Equation (13.7) and González-Farías *et al.* (2003).

The model with error structure (13.13) includes the following cases as submodels:

Model I: Homoscedastic and Uncorrelated Errors

If we set in the model (13.6) $D = \alpha I_n, \Sigma = \sigma^2 I_n, \Lambda = \tau^2 I_n$, where I_n is the $n \times n$ identity matrix, we have the case of homoscedastic and uncorrelated observations: $\text{Var}(Y_i) = \textit{constant}$, $i = 1,\ldots,n$ and $\text{Cov}(Y_i,Y_j) = \text{Cov}(\varepsilon_i,\varepsilon_j) = 0$, $i \neq j$, $i,j = 1,\ldots,n$.

Model II: Heteroscedastic and Uncorrelated Errors

If the matrices D, Σ and Λ are diagonal and if any of them is of the form

$$D = \text{diag}(\alpha_1,\ldots,\alpha_n), \quad \Sigma = \text{diag}\left(\sigma_1^2,\ldots,\sigma_n^2\right), \quad \Lambda = \text{diag}\left(\tau_1^2,\ldots,\tau_n^2\right),$$

then we have the case of heteroscedastic but uncorrelated observations: $\text{Var}(Y_i) = k_i$, $i = 1,\ldots,n$ and $k_i \neq k_j$ for some $i \neq j$, and $\text{Cov}(Y_i,Y_j) = \text{Cov}(\varepsilon_i,\varepsilon_j) = 0$, $i \neq j$, $i,j = 1,\cdots,n$.

Model III: Correlated Errors

If any of the matrices D, Σ or Λ is non-diagonal we have the case of correlated errors $\text{Cov}(Y_i,Y_j) = \text{Cov}(\varepsilon_i,\varepsilon_j) \neq 0$ for some $i \neq j$. In the case of

$$\Sigma > 0, \; D = \text{diag}(\alpha_1,\ldots,\alpha_n)\;, \; \Lambda = \text{diag}\left(\tau_1^2,\ldots,\tau_n^2\right),$$

we can use the properties of (13.13) in order to evaluate the marginal distribution of the error, since $V_i \sim N\left(0, \sigma_i^2\right)$ and $U_i \sim N^c\left(\nu_i, \tau_i^2\right)$ are independent. Thus we can use (13.5) in order to evaluate the marginal distributions of the errors.

Using Remark 1 of González-Farías *et al.* (2003), we have that the marginal distribution of the error, in model (13.6) is

$$\varepsilon_i \sim CSN_{1,m}\left(0, \Theta_{\mathbf{a}_i}, D_{\mathbf{a}_i}, \boldsymbol{\nu}, \Upsilon_{\mathbf{a}_i}\right), \tag{13.14}$$

where

$$\begin{aligned}
\Theta_{\mathbf{a}_i} &= \mathbf{a}_i^T \Theta \mathbf{a}_i, \\
D_{\mathbf{a}_i} &= \Lambda D^T \mathbf{a}_i \Theta_{\mathbf{a}_i}^{-1}, \\
\Upsilon_{\mathbf{a}_i} &= \Upsilon + \Lambda D^T \Theta^{-1} D \Lambda - \Lambda D^T \mathbf{a}_i \mathbf{a}_i^T D \Lambda \Theta_{\mathbf{a}_i}^{-1},
\end{aligned}$$

with $\mathbf{a}_i$ the i-th unit vector in $\mathbb{R}^n$ and Θ, Υ given in (13.11).

13.2.2 Likelihood

Given a vector of observations $\mathbf{y} = (Y_1, \ldots, Y_n)^T$ and the corresponding set of inputs $X = (\mathbf{x}_1, \ldots, \mathbf{x}_n)^T$, then under the assumptions of model (13.6) we can write the likelihood function and base our inferences on it.

From (13.10) and (13.6) we have that the likelihood function of the parameters $\boldsymbol{\beta}, \Sigma, D, \boldsymbol{\nu}, \mathbf{c}, \Lambda$ is

$$\begin{aligned}
L\left(\boldsymbol{\beta}, \Sigma, D, \boldsymbol{\nu}, \mathbf{c}, \Lambda\right) =& \Phi_m^{-1}\left(\mathbf{0}; \mathbf{c} - \boldsymbol{\nu}, \Lambda\right) \phi_n\left(\mathbf{y} - \mathbf{f}\left(X; \boldsymbol{\beta}\right); D\boldsymbol{\nu}, \Theta\right) \\
& \times \Phi_m\left\{\Lambda D^T \Theta^{-1}\left[\mathbf{y} - \mathbf{f}\left(X; \boldsymbol{\beta}\right) - D\boldsymbol{\nu}\right]; \mathbf{c} - \boldsymbol{\nu}, \Upsilon\right\}.
\end{aligned} \tag{13.15}$$

In the case of Model II (heteroscedastic and uncorrelated errors), the likelihood reduces to

$$\begin{aligned}
& L\left(\boldsymbol{\beta}, \Sigma, D, \boldsymbol{\nu}, \mathbf{c}, \Lambda\right) \\
&= \prod_{i=1}^{n}\left\{\Phi^{-1}\left(\frac{\nu_i - c_i}{\tau_i}\right) \phi\left(Y_i - f\left(\mathbf{x}_i; \boldsymbol{\beta}\right); \alpha_i \nu_i, \sigma_i^2 + \alpha_i^2 \tau_i^2\right)\right. \\
& \times \Phi\left[\tau_i^2 \alpha_i \left(\sigma_i^2 + \alpha_i^2 \tau_i^2\right)^{-1}\left(Y_i - f\left(\mathbf{x}_i; \boldsymbol{\beta}\right) - \alpha_i \nu_i\right);\right. \\
& \left.\left. c_i - \nu_i, \tau_i^2 - \tau_i^4 \alpha_i^2 \left(\sigma_i^2 + \alpha_i^2 \tau_i^2\right)^{-1}\right]\right\} \\
&= \prod_{i=1}^{n}\left\{\frac{\Phi^{-1}\left(\frac{\nu_i - c_i}{\tau_i}\right)}{\sqrt{\sigma_i^2 + \alpha_i^2 \tau_i^2}} \phi\left(\frac{Y_i - f\left(\mathbf{x}_i; \boldsymbol{\beta}\right) - \alpha_i \nu_i}{\sqrt{\sigma_i^2 + \alpha_i^2 \tau_i^2}}\right)\right. \\
& \left. \times \Phi\left(\frac{\tau_i \alpha_i \left(Y_i - f\left(\mathbf{x}_i; \boldsymbol{\beta}\right) - \alpha_i \nu_i\right)}{\sigma_i \sqrt{\sigma_i^2 + \alpha_i^2 \tau_i^2}} + \left(\nu - c_i\right) \frac{\sqrt{\sigma_i^2 + \alpha_i^2 \tau_i^2}}{\tau_i \sigma_i}\right)\right\}.
\end{aligned}$$

13.2.3 Estimation of Inefficiencies/Efficiencies

In the univariate model of SFA, the error is of the form $\varepsilon = V + \alpha U$. The main interest in SFA is to assess, for each $\mathbf{x}$, efficiencies or inefficiencies of the firm, through the sign and the value of the estimated parameter α. After the estimation of $\boldsymbol{\beta}$ by maximum likelihood we can evaluate the residuals:

$$\hat{\varepsilon}_i = Y_i - f\left(\mathbf{x}_i; \widehat{\boldsymbol{\beta}}\right).$$

To estimate efficiencies, a straightforward approach would try to predict u_i for a common set of inputs and then take its average or some other measure of central tendency. However, we cannot separate u_i from the compound error. The most common proposals in the stochastic frontier literature are based on the conditional distribution of $\mathbf{u}|\boldsymbol{\varepsilon}$ which has proved to be tractable. In particular we consider the following estimate of efficiency based on the mean of $\mathbf{u}|\boldsymbol{\varepsilon}$:

$$\begin{aligned} \text{eff}\left(\boldsymbol{\beta}, \sigma, \alpha, \nu, \tau, c, \varepsilon; \mathbf{x}\right) &\equiv \text{efficiency}\left(\boldsymbol{\beta}, \sigma, \alpha, \nu, \tau, c, \varepsilon; \mathbf{x}\right) \\ &= \alpha \text{E}\left(\mathbf{u}|\boldsymbol{\varepsilon}\right). \end{aligned}$$

Thus we estimate the efficiency of firm i by

$$\text{eff}_i^* = \text{eff}\left(\widehat{\boldsymbol{\beta}}, \hat{\sigma}, \hat{\alpha}, \hat{\nu}, \hat{\tau}, \hat{c}, \hat{\varepsilon}_i; \mathbf{x}_i\right).$$

Let us assume $\mathbf{v} \sim N_n\left(\mathbf{0}, \Sigma\right)$ and $\mathbf{u} \sim N_m^{\mathbf{c}}\left(\boldsymbol{\nu}, \Lambda\right)$, $\mathbf{v}$ and $\mathbf{u}$ independent. If

$$\boldsymbol{\varepsilon} = \mathbf{v} + D\mathbf{u},$$

then the joint density of $\mathbf{u}$ and $\boldsymbol{\varepsilon}$ is given by

$$g\left(\mathbf{u}, \boldsymbol{\varepsilon}\right) = \Phi_n^{-1}\left(-\mathbf{c}; -\boldsymbol{\nu}, \Lambda\right) \phi_n\left(\boldsymbol{\varepsilon} - D\mathbf{u}; \mathbf{0}, \Sigma\right) \phi_n\left(\mathbf{u}; \boldsymbol{\nu}, \Lambda\right), \tag{13.16}$$

which, after some algebra and the aid of Equation (13.12), can be reduced to

$$\begin{aligned} g\left(\mathbf{u}, \boldsymbol{\varepsilon}\right) = {} & \Phi_n^{-1}\left(-\mathbf{c}; -\boldsymbol{\nu}, \Lambda\right) \phi_n\left(\mathbf{u}; \boldsymbol{\nu} + \Lambda D^T \Theta^{-1}\left(\boldsymbol{\varepsilon} - D\boldsymbol{\nu}\right), \Upsilon\right) \\ & \times \phi_n\left(\boldsymbol{\varepsilon}; D\boldsymbol{\nu}, \Theta\right), \end{aligned} \tag{13.17}$$

where Θ and Υ are defined in (13.11). From (13.17) and (13.26) we have that the conditional density of $\mathbf{u}|\boldsymbol{\varepsilon}$ is

$$\begin{aligned} g\left(\mathbf{u}|\boldsymbol{\varepsilon}\right) &= g\left(\mathbf{u}, \boldsymbol{\varepsilon}\right) / g\left(\boldsymbol{\varepsilon}\right) \\ &= K \phi_n\left(\mathbf{u}; \boldsymbol{\nu} + \Lambda D^T \Theta^{-1}\left(\boldsymbol{\varepsilon} - D\boldsymbol{\nu}\right), \Upsilon\right), \end{aligned}$$

where

$$K^{-1} = \Phi_n\left(\boldsymbol{\nu} + \Lambda D^T \Theta^{-1}\left(\boldsymbol{\varepsilon} - D\boldsymbol{\nu}\right); \mathbf{c}, \Upsilon\right),$$

that is, the conditional distribution of $\mathbf{u}|\boldsymbol{\varepsilon}$ is a truncated multivariate normal

$$\mathbf{u}|\boldsymbol{\varepsilon} \sim N_n^{\mathbf{c}}\left(\boldsymbol{\psi}, \Upsilon\right), \tag{13.18}$$

where $\boldsymbol{\psi} = \boldsymbol{\nu} + \Lambda D^T \Theta^{-1} (\boldsymbol{\varepsilon} - D\boldsymbol{\nu})$. Now from Lemma 13.6.2 in the Appendix, we have a closed form expression for the mean of $\mathbf{u}|\boldsymbol{\varepsilon}$,

$$\mathrm{E}(\mathbf{u}|\boldsymbol{\varepsilon}) = \boldsymbol{\psi} + \Upsilon \frac{\Phi_n^*(\boldsymbol{\psi}; \mathbf{0}, \Upsilon)}{\Phi_n(\boldsymbol{\psi}; \mathbf{0}, \Upsilon)}, \tag{13.19}$$

where $\Phi_n^*(\mathbf{s}; \boldsymbol{\nu}, \Lambda)$ is defined in (13.27) and (13.28). This formula would be the basis for an estimate of the efficiency in the multivariate case. Thus we obtain an estimation of the inefficiencies/efficiencies for the model (13.6), by substituting the corresponding estimates in:

$$\mathrm{eff}(\boldsymbol{\beta}, \Sigma, D, \boldsymbol{\nu}, \Lambda, \mathbf{c}, \boldsymbol{\varepsilon}; \mathbf{x}) = D\mathrm{E}(\mathbf{u}|\boldsymbol{\varepsilon}).$$

From (13.19), we obtain the efficiency for $\mathbf{x}_i$, that is $\mathrm{E}(U_i|\boldsymbol{\varepsilon})$ is the i-th element of the n-vector of efficiencies (13.19). The efficiency for $\mathbf{x}_i$, given ε_j, is $\mathrm{E}(U_i|\varepsilon_j)$. Now, since the ε_j's are correlated, U_i is not independent of ε_j. Thus the efficiency in general will depend on the performance of all other firms.

13.3 A Correlated Structure for the Compound Error

Consider the particular case of Model III defined in Section 13.2.1. Let $U_i \sim N^0(\nu, \tau^2)$ be iid random variables representing inefficiencies and consider the particular case when the measurement error V_i follows a first order autoregressive process:

$$V_i = \rho V_{i-1} + T_i, \tag{13.20}$$

where $|\rho| < 1$, $V_0 \sim N(0, \sigma^2/(1-\rho^2))$ and for all $i = 1, \ldots, n$, T_i's are iid with mean 0 and variance σ^2 and independent of V_0. This correlation structure would be reasonable for the case in which the stochastic frontiers for contemporaneous companies vary in time in a correlated fashion.

The autoregressive model (13.20) is one of the simplest and most useful models that captures temporal dependencies. It is easy to check that $\mathbf{v} = (V_1, \ldots, V_n)^T$ has joint distribution

$$\mathbf{v} \sim N_n\left(\mathbf{0}, \frac{\sigma^2}{1-\rho^2} R\right),$$

where $R = (r_{ij})$, with $r_{ij} = \rho^{|i-j|}$.

The compound error $\varepsilon_i = V_i + \alpha U_i$, with V_i and U_i as before, can be written as $\boldsymbol{\varepsilon} = \mathbf{v} + D\mathbf{u}$ as in (13.10) with $\Sigma = (\sigma^2/(1-\rho^2)) R$, $\Lambda = \tau^2 I_n$, $D = \alpha I_n$, $\mathbf{c} = \mathbf{0}$, and $\boldsymbol{\nu} = \nu \mathbf{1}_n$.

Maximum likelihood estimation is based on maximizing

$$\begin{aligned}\tilde{L}(\boldsymbol{\beta}, \sigma^2, \rho, \nu, \tau^2) &= L(\boldsymbol{\beta}, \Sigma, \alpha I_n, \nu \mathbf{1}_n, \mathbf{0}, \tau^2 I_n) \\ &= \Phi^{-n}(\nu/\tau)\, \phi_n(\mathbf{y} - \mathbf{x}^T\boldsymbol{\beta}; \alpha\nu\mathbf{1}_n, \Theta) \\ &\quad \times \Phi_n\left[\alpha\tau^2\Theta^{-1}(\mathbf{y} - \mathbf{x}^T\boldsymbol{\beta} - \alpha\nu\mathbf{1}_n); -\nu\mathbf{1}_n, \Upsilon\right].\end{aligned}$$

This expression can be derived from (13.15), where $\Upsilon = \left(\alpha^2\Sigma^{-1} + \tau^{-2}I_n\right)^{-1} = \alpha^{-2}\Sigma\left(I_n - \Theta^{-1}\Sigma\right)$, with $\Theta = \Sigma + \alpha^2\tau^2 I_n$. The corresponding log-likelihood function is

$$\begin{aligned}\ell\left(\boldsymbol{\beta},\sigma^2,\rho,\nu,\tau^2\right) &= \ln\tilde{L}\left(\boldsymbol{\beta},\sigma^2,\rho,\nu,\tau^2\right) \\ &= -n\ln\Phi\left(\nu/\tau\right) - \tfrac{1}{2}\ln|\Theta| \\ &\quad - \tfrac{1}{2}\left(\mathbf{y} - \mathbf{x}^T\boldsymbol{\beta} - \alpha\nu\mathbf{1}_n\right)^T\Theta^{-1}\left(\mathbf{y} - \mathbf{x}^T\boldsymbol{\beta} - \alpha\nu\mathbf{1}_n\right) \\ &\quad + \ln\left\{\Phi_n\left[\alpha\tau^2\Theta^{-1}\left(\mathbf{y} - \mathbf{x}^T\boldsymbol{\beta} - \alpha\nu\mathbf{1}_n\right); -\nu\mathbf{1}_n, \Upsilon\right]\right\}.\end{aligned} \tag{13.21}$$

From relation (13.18) it follows that

$$\mathbf{u}|\boldsymbol{\varepsilon} \sim N_n^{\mathbf{0}}\left(\boldsymbol{\psi},\Upsilon\right),$$

where

$$\boldsymbol{\psi} = \Sigma\Theta^{-1}\boldsymbol{\nu} + \alpha^{-1}\left(I_n - \Sigma\Theta^{-1}\right)\boldsymbol{\varepsilon}.$$

Now from Lemma 13.6.2 we get that the conditional mean of $\mathbf{u}|\boldsymbol{\varepsilon}$ is given by

$$\mathrm{E}\left(\mathbf{u}|\boldsymbol{\varepsilon}\right) = \boldsymbol{\psi} + \Upsilon\frac{\Phi_n^*\left(\boldsymbol{\psi};\mathbf{0},\Upsilon\right)}{\Phi_n\left(\boldsymbol{\psi};\mathbf{0},\Upsilon\right)}, \tag{13.22}$$

where $\Phi_n^*\left(\mathbf{s};\boldsymbol{\nu},\Lambda\right)$ is defined in (13.27) and (13.28).

13.3.1 Simulated Example with Correlated Compound Errors

We simulate 50 values from the model

$$Y_i = \beta_0 + \beta_1 x_i + V_i - U_i, \tag{13.23}$$

Table 13.1 *Estimated values for the model (13.23). ℓ is the evaluation of the log-likelihood (13.21) on its corresponding MLE's. $\hat{\ell} = -4.744$, is the evaluation of (13.21) in the global MLE of model (13.23).*

	OLS	COLS	$\mathrm{ML}_{\alpha=0}$	Aigner	Stevenson	(13.23)
$\hat{\beta}_0$	−0.009	1.447	0.005	0.684	0.733	0.751
$\hat{\beta}_1$	1.979	1.979	1.958	2.344	2.323	2.191
$\hat{\sigma}$	0.727	0.727	0.682	0.329	0.307	0.332
$\hat{\tau}$				1.089	0.942	0.703
$\hat{\nu}$					0.477	0.658
$\hat{\rho}$			0.312			0.626
ℓ	−8.525	−107.735	−5.918	−7.272	−7.193	−4.744
$e^{(\ell-\hat{\ell})}$	0.023	$\cong 0$	0.309	0.080	0.086	1

where $\mathbf{v} = (V_1, \ldots, V_{50})^T$ follows model (13.20) and $U_1, \ldots, U_{50}$ are iid observations from $N^c(\nu, \tau^2)$. The values of the parameters used to generate the sample are $\sigma = 0.1$, $\tau = \sqrt{2}$, $\rho = 0.7$, $\nu = -1/2$, $c = 0$, $\alpha = -1$, $\beta_0 = 1$, $\beta_1 = 2$, $v_0 = -0.209$. The computations were made with Splus 6.0 using an adaptation of the algorithm of Genz (1992) for computing the multivariate normal distribution function.

The maximization of (13.21) gives maximum likelihood estimators (MLE) for the model (13.23). These values are shown in Table 13.1. Given that $f(\mathbf{x}; \boldsymbol{\beta}) = \beta_0 + \beta_1 x$ we must take $c = 0$, because $\beta_0 + u_i \sim N^{c+\beta_0}(\nu + \beta_0, \tau)$ and then it would not be possible to separate c, ν and β_0 from $c + \beta_0$ and $\nu + \beta_0$. Table 13.1 also shows the estimated values for the following particular cases of (13.23):

1) Aigner *et al.* (1977) model: $\rho = \nu = c = 0$.
2) Stevenson (1980) model: $\rho = 0$, $c = 0$.
3) Maximum likelihood estimation with $\alpha = 0$, ($\mathrm{ML}_{\alpha=0}$). The parameters $\boldsymbol{\beta}$, σ, and ρ are estimated by maximum likelihood assuming $\alpha = 0$.
4) Ordinary least square (OLS): $\rho = 0$, $\alpha = 0$. The parameter $\boldsymbol{\beta}$ is

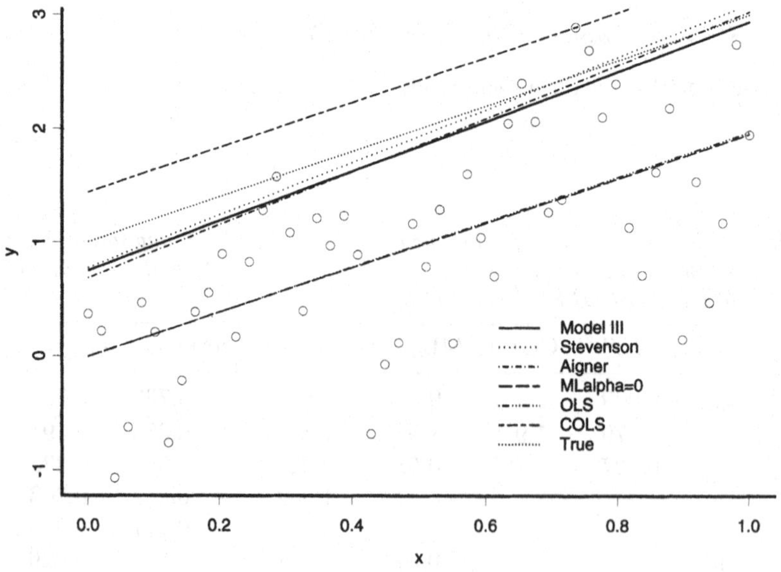

Figure 13.1 *Scatterplot of* $\mathbf{y}$ *vs* $\mathbf{x}$ *with the estimated line* $y = \hat{\beta}_0 + \hat{\beta}_1 x$ *for different models.*

estimated by OLS, and σ is estimated by the standard deviation of the OLS residuals.

5) Corrected ordinary least squares (COLS): $\rho = 0$, $\alpha = 0$. The parameter $\boldsymbol{\beta}$ is estimated by OLS with the estimated production function shifted upwards until all the residuals are non-negative, with at least one being zero (Winsten, 1957).

In models 3) to 5) the parameter $\alpha = 0$. In this case ν, c, τ are not identifiable by $\boldsymbol{\varepsilon}$, because the density function (13.10) of $\boldsymbol{\varepsilon}$ for the set of parameters $\Sigma, D = O, \boldsymbol{\nu}, \mathbf{c}, \Lambda$ is the same as the density of $\boldsymbol{\varepsilon}$, for the set parameters $\Sigma, D = O, \boldsymbol{\nu}^*, \mathbf{c}^*, \Lambda^*$, for any values of $\boldsymbol{\nu}^*, \mathbf{c}^*, \Lambda^*$. In fact when $D = O$, the parameters $\boldsymbol{\nu}, \mathbf{c}, \Lambda$ disappear from the density function of $\boldsymbol{\varepsilon}$.

The Euclidean norm of the gradient of ℓ evaluated on $\hat{\boldsymbol{\beta}}, \hat{\sigma}^2, \hat{\rho}, \hat{\nu}, \hat{\tau}^2$ for model (13.23) is not 0, in fact it is 2183.7, which means that the MLE of $\boldsymbol{\beta}, \sigma^2, \rho, \nu, \tau^2$ are not regular solutions. The estimators of $\boldsymbol{\beta}$ and σ by means of COLS are not based on a related likelihood procedure. For the other estimators the Euclidean norm of their corresponding gradients is zero.

Let $\hat{\ell} = \ell(\hat{\boldsymbol{\beta}}, \hat{\sigma}^2, \hat{\rho}, \hat{\nu}, \hat{\tau}^2)$. The set of values of $(\boldsymbol{\beta}, \sigma^2, \rho, \nu, \tau^2)$ for which $R = \exp(\ell(\boldsymbol{\beta}, \sigma^2, \rho, \nu, \tau^2) - \hat{\ell}) \geq p$ is called a $100p\%$ likelihood region (LR) for $(\boldsymbol{\beta}, \sigma^2, \rho, \nu, \tau^2)$. It is common to consider 50%, 10% and 1% likelihood regions. Values inside the 10% LR will be referred to as *plausible*, and values outside this region as *implausible*. Similarly we will refer to values inside the 50% LR *as very plausible,* and values outside the 1% LR as *very implausible*; see Kalbfleish (1985, Sec. 9.3, 10.3, and 10.8).

Table 13.2 *Estimated efficiencies using Equation (13.22) for model (13.23).*

i	$E(U_i \mid \varepsilon)$	i	$E(U_i \mid \varepsilon)$	i	$E(U_i \mid \varepsilon)$	i	$E(U_i \mid \varepsilon)$
1	0.365	14	0.345	27	0.426	40	0.222
2	0.444	15	0.185	28	1.553	41	1.210
3	1.561	16	0.555	29	0.277	42	1.513
4	1.188	17	1.127	30	0.878	43	0.710
5	0.281	18	0.383	31	1.308	44	0.208
6	0.547	19	0.610	32	0.269	45	1.992
7	1.439	20	0.326	33	0.121	46	0.701
8	1.017	21	0.515	34	0.409	47	1.753
9	0.601	22	1.831	35	1.132	48	1.310
10	0.580	23	1.264	36	1.137	49	0.143
11	0.351	24	1.207	37	0.000	50	0.927
12	1.091	25	0.390	38	0.219		
13	0.608	26	0.839	39	0.632		

Table 13.3 *Estimated efficiencies using Equation (13.32) for model (13.23).*

i	$\mathrm{E}(U_i\|\varepsilon)$	i	$\mathrm{E}(U_i\|\varepsilon)$	i	$\mathrm{E}(U_i\|\varepsilon)$	i	$\mathrm{E}(U_i\|\varepsilon)$
1	0.479	14	0.312	27	0.654	40	0.332
2	0.612	15	0.229	28	1.732	41	1.337
3	1.793	16	0.453	29	0.492	42	1.759
4	1.431	17	1.031	30	0.974	43	0.977
5	0.531	18	0.432	31	1.320	44	0.553
6	0.764	19	0.620	32	0.326	45	2.392
7	1.676	20	0.472	33	0.224	46	1.175
8	1.223	21	0.757	34	0.358	47	2.176
9	0.727	22	2.211	35	0.976	48	1.584
10	0.631	23	1.700	36	0.916	49	0.356
11	0.432	24	1.568	37	0.161	50	0.965
12	1.034	25	0.679	38	0.208		
13	0.531	26	1.041	39	0.459		

From Table 13.1, we observe that the estimator for the model COLS is very implausible given that for this model $R \cong 0$ and it is outside the 1% LR. The MLE with $\alpha = 0$ is a plausible estimator because $R = 0.309$ and it is inside the 10% LR. The estimators from OLS, Aigner, and Stevenson are implausible because they are outside the 10% LR.

Figure 13.1 shows the scatter plot of $\mathbf{y}$ vs $\mathbf{x}$ for the simulated data with the estimated line for the corresponding model. Let us notice that the estimated lines with OLS, COLS, and $\mathrm{ML}_{\alpha=0}$ are far from the lines estimated considering $\alpha \neq 0$. The estimated line for model (13.23) (solid line in Figure 13.1) is closer to the corresponding estimated lines for the true Aigner and Stevenson models.

The estimated efficiencies given in Tables 13.2 and 13.3 were obtained from Equations (13.22) and (13.32) evaluated on the MLE's for model (13.23), respectively. Observe that most of the values of the estimations of $\mathrm{E}(U_i|\varepsilon)$ are smaller than the estimations of $\mathrm{E}(U_i|\varepsilon_i)$.

13.4 SFA with Skew-Elliptical Components

Recall that a p-vector $\mathbf{w}$ has a spherical distribution if and only if for its characteristic function $\psi(\mathbf{t})$ there exists a function $\varphi(\cdot)$ of scalar variable such that $\varphi(\mathbf{t}) = \psi\left(\mathbf{t}^T\mathbf{t}\right)$. The function $\psi(\cdot)$ is called the *characteristic generator.* A $p \times 1$ random vector $\mathbf{v}$ is said to have an *elliptically contoured* distribution or *elliptically symmetric* distribution with parameters $\boldsymbol{\mu} \in \mathbb{R}^p$ and $\Sigma\,(p \times p)$ if

$$\mathbf{v} = \boldsymbol{\mu} + A^T\mathbf{w},$$

where $\mathbf{w}$ has a spherical distribution with characteristic generator ψ, and $A\,(k \times p)$, $A^T A = \Sigma$ with rank$(\Sigma) = k$. We will write $\mathbf{v} \sim EC_p\,(\boldsymbol{\mu}, \Sigma, \psi)$.

A random vector $\mathbf{v} \sim EC_p\,(\boldsymbol{\mu}, \Sigma, \psi)$, in general, does not necessarily possess a density. It is however possible to show that the density of $\mathbf{v}$, if it exists, must be of the form $h\left(\mathbf{v}^T \mathbf{v}\right)$ for some non-negative function $h\,(\cdot)$ of a scalar variable such that

$$\int_0^\infty y^{\frac{p}{2}-1} h\,(y)\,dy < \infty. \tag{13.24}$$

In this case we will write $\mathbf{v} \sim EC_p\,(\mu, \Sigma, h)$; see Fang, Kotz, and Ng (1990, p. 35).

Consider the stochastic frontier model

$$\mathbf{y} = \mathbf{f}\,(X; \boldsymbol{\beta}) + \mathbf{v} + D\mathbf{u},$$

where $\mathbf{v} \sim EC_n\,(\mathbf{0}, \Sigma, h_1)$, $\mathbf{u} \sim EC_m^{\mathbf{c}}\,(\boldsymbol{\nu}, \Lambda, h_2)$, $m \geq n$, $\mathbf{v}$ independent of $\mathbf{u}$, D is an arbitrary matrix, $\mathbf{f}\,(X; \boldsymbol{\beta}) = \left(f\,(\mathbf{x}_1; \boldsymbol{\beta}), \ldots, f\,(\mathbf{x}_n; \boldsymbol{\beta})\right)^T$, and $X = (\mathbf{x}_1, \ldots, \mathbf{x}_n)^T$, $\mathbf{x}_i$ is a known vector of $\mathbb{R}^{d_1}, i = 1, \ldots, n$, and $\boldsymbol{\beta} \in \mathbb{R}^{d_2}$.

The density function of $\boldsymbol{\varepsilon} = \mathbf{v} + D\mathbf{u}$ is of the form

$$g\,(\boldsymbol{\varepsilon}) = K \int_{c_1}^\infty \cdots \int_{c_n}^\infty h_1 \left[(\boldsymbol{\varepsilon} - D\mathbf{u})^T (\boldsymbol{\varepsilon} - D\mathbf{u})\right] h_2\,(\mathbf{u})\,d\mathbf{u},$$

where K is a normalizing constant. This expression, in general, is difficult to evaluate.

The moment generating function (mgf) could be useful to get the density function of $\boldsymbol{\varepsilon}$ or, alternatively, we could use it to obtain the moments of $\boldsymbol{\varepsilon}$ and use the method of moments to estimate the parameters of the model. If the moment generating functions of $\mathbf{u}$ and $\mathbf{v}$ are available, then the mgf of $\boldsymbol{\varepsilon}$ is of the form

$$M_{\boldsymbol{\varepsilon}}\,(\mathbf{t}) = M_{\mathbf{v}}\,(\mathbf{t})\, M_{\mathbf{u}}\left(D^T \mathbf{t}\right). \tag{13.25}$$

13.5 Conclusions

Stochastic frontier analysis is a natural setting for applications of skewed distributions. Its compound error structure has been present since the pioneering works of Aigner *et al.* (1977), and Meeusen and Van den Broeck (1977). In this chapter, we have considered a general model for the disturbances based on the closed skew-normal distribution that accommodates from homoscedastic and uncorrelated errors (Model I) to general structures for correlated errors (Model III). The computational aspects of maximum likelihood estimation are always an issue when using skew-normal distributions. However, for parsimonious models, it can be done as we showed in Section 13.3 on estimation for correlated structures on the disturbances. We ended the exposition with a brief treatment of an extension of the stochastic frontier model to the case of skew-elliptical errors.

Acknowledgments

We thank Alonso Nuñez-Páez for the adaptation to Splus of the Genz (1992) program, needed in the example. This study was partially supported by CONACYT research grant 39017E, DGYCT Grant SEC01-0890, and CONCYTEG research grant 03-16-K118-027, Anexo 01.

13.6 Appendix

13.6.1 Distributional Properties of Multivariate Compound Errors

Let $\mathbf{w} = (W_1, \ldots, W_p)^T$. We consider truncations of the type $\mathbf{w} \geq \mathbf{c}$, where $\mathbf{w} \geq \mathbf{c}$ means $W_j \geq c_j$, $j = 1, \ldots, p$, that is, values of W_j less than c_j are excluded, see Kotz *et al.* (2000, Sec. 45.10) for a discussion on multivariate truncated distributions.

For an arbitrary vector $\mathbf{c} \in \mathbb{R}^p$ define the function:

$$I(\mathbf{x}; \mathbf{c}) = \begin{cases} \mathbf{x}, & \text{if } \mathbf{x} \geq \mathbf{c}, \\ \mathbf{0}, & \text{if } \mathbf{x} < \mathbf{c}. \end{cases}$$

Let $\mathbf{w}$ be a random vector, we denote by $\mathbf{w}(\mathbf{c})$ a random vector with truncation of the type $\mathbf{w} \geq \mathbf{c}$, for $\mathbf{c} \in \mathbb{R}^p$. Thus, if $\mathbf{w}$ has density f, and we let $\mathbf{u} \stackrel{D}{=} \mathbf{w}(\mathbf{c})$, where $\stackrel{D}{=}$ means equality in distribution, the density function of $\mathbf{u}$ is given by

$$g(\mathbf{u}) = \frac{f(\mathbf{u})}{\mathrm{P}(\mathbf{u} \geq \mathbf{c})} I(\mathbf{u}; \mathbf{c}).$$

If $\mathbf{w} \sim EC_p(\boldsymbol{\mu}, \Sigma, h)$ with h satisfying (13.24), and $\mathbf{u} \stackrel{D}{=} \mathbf{w}(\mathbf{c})$, then the density function of $\mathbf{u}$ is

$$\begin{aligned} g(\mathbf{u}) &= \frac{f(\mathbf{u}; \boldsymbol{\mu}, \Sigma)}{\mathrm{P}(\mathbf{u} \geq \mathbf{c})} I(\mathbf{u}; \mathbf{c}) \\ &= \frac{f(\mathbf{u}; \boldsymbol{\mu}, \Sigma)}{\mathrm{P}(-\mathbf{u} \leq -\mathbf{c})} I(\mathbf{u}; \mathbf{c}) \\ &= \frac{f(\mathbf{u}; \boldsymbol{\mu}, \Sigma)}{F(-\mathbf{c}; -\boldsymbol{\mu}, \Sigma)} I(\mathbf{u}; \mathbf{c}), \end{aligned}$$

where f and F are the density and the cumulative distribution function of $\mathbf{u}$, respectively.

If $\mathbf{w} \sim EC_p(\boldsymbol{\mu}, \Sigma, h)$, and $h(w) = e^{-w/2}$, that is $\mathbf{w} \sim N_p(\boldsymbol{\mu}, \Sigma)$, and $\mathbf{u} \stackrel{D}{=} \mathbf{w}(\mathbf{c})$, then

$$\begin{aligned} g(\mathbf{u}) &= [\mathrm{P}(\mathbf{x} \geq \mathbf{c})]^{-1} \phi_p(\mathbf{u}; \boldsymbol{\mu}, \Sigma) I(\mathbf{u}; \mathbf{c}) \\ &= [\mathrm{P}(-\mathbf{x} \leq -\mathbf{c})]^{-1} \phi_p(\mathbf{u}; \boldsymbol{\mu}, \Sigma) I(\mathbf{u}; \mathbf{c}) \\ &= \Phi_p^{-1}(-\mathbf{c}; -\boldsymbol{\mu}, \Sigma) \phi_p(\mathbf{u}; \boldsymbol{\mu}, \Sigma) I(\mathbf{u}; \mathbf{c}). \end{aligned}$$

We will denote by $\mathbf{u} \sim EC_p^{\mathbf{c}}(\boldsymbol{\mu}, \Sigma, h)$ if $\mathbf{u} \stackrel{D}{=} \mathbf{w}(\mathbf{c})$ and $\mathbf{w} \sim EC_p(\boldsymbol{\mu}, \Sigma, h)$, similarly $\mathbf{u} \sim N_p^{\mathbf{c}}(\boldsymbol{\mu}, \Sigma)$ if $\mathbf{u} \stackrel{D}{=} \mathbf{w}(\mathbf{c})$ and $\mathbf{w} \sim N_p(\boldsymbol{\mu}, \Sigma)$.

Proposition 13.6.1 If $\mathbf{v} \sim N_p(\boldsymbol{\mu}, \Sigma)$ and $\mathbf{u} \sim N_q^{\mathbf{c}}(\boldsymbol{\nu}, \Lambda)$, and $\mathbf{v}$ is independent of $\mathbf{u}$ and

$$\boldsymbol{\varepsilon} = \mathbf{v} + D\mathbf{u},$$

where D is a $p \times q$ matrix, then

$$\boldsymbol{\varepsilon} \sim CSN_{p,q}\left(\boldsymbol{\mu} + D\boldsymbol{\nu}, \Theta, \Lambda D^T \Theta^{-1}, \mathbf{c}-\boldsymbol{\nu}, \Upsilon\right),$$

where Θ and Υ are defined in (13.11). That is, the density function of $\boldsymbol{\varepsilon}$ is

$$\begin{aligned} g(\boldsymbol{\varepsilon}) = {} & \Phi_p^{-1}(\mathbf{0}; \mathbf{c} - \boldsymbol{\nu}, \Lambda)\, \phi_p(\boldsymbol{\varepsilon}; \boldsymbol{\mu} + D\boldsymbol{\nu}, \Theta) \\ & \times \Phi_p\left(\Lambda D^T \Theta^{-1}(\boldsymbol{\varepsilon} - \boldsymbol{\mu} - D\boldsymbol{\nu}); \mathbf{c} - \boldsymbol{\nu}, \Upsilon\right). \end{aligned} \tag{13.26}$$

Before proving Proposition 13.6.1, we present a Lemma related to the mgf of a truncated multivariate normal distribution.

Lemma 13.6.1 If $\mathbf{u} \sim N_q^{\mathbf{c}}(\boldsymbol{\nu}, \Lambda)$ then its mgf is given by

$$M_{\mathbf{u}}(\mathbf{t}) = \Phi_q^{-1}(\mathbf{0}; \mathbf{c} - \boldsymbol{\nu}, \Lambda)\, e^{\mathbf{t}^T\boldsymbol{\nu} + \frac{1}{2}\mathbf{t}^T\Lambda\mathbf{t}}\, \Phi_q(\Lambda\mathbf{t}; \mathbf{c} - \boldsymbol{\nu}, \Lambda).$$

Proof.

$$\begin{aligned} M_{\mathbf{u}}(\mathbf{t}) &= \mathrm{E}(e^{\mathbf{t}^T\mathbf{u}}) \\ &= \Phi_q^{-1}(-\mathbf{c}; -\boldsymbol{\nu}, \Lambda) \int_{c_1}^{\infty} \cdots \int_{c_q}^{\infty} e^{\mathbf{t}^T\mathbf{u}} \phi_q(\mathbf{u}; \boldsymbol{\nu}, \Lambda)\, dw_1 \cdots dw_q. \end{aligned}$$

Given that

$$e^{\mathbf{t}^T\mathbf{u}} \phi_p(\mathbf{u}; \boldsymbol{\nu}, \Lambda) = e^{\mathbf{t}^T\boldsymbol{\nu} + \frac{1}{2}\mathbf{t}^T\Lambda\mathbf{t}} \phi_p(\mathbf{u}; \boldsymbol{\nu} + \Lambda\mathbf{t}, \Lambda),$$

we get

$$\begin{aligned} M_{\mathbf{u}}(\mathbf{t}) = {} & \Phi_q^{-1}(-\mathbf{c}; -\boldsymbol{\nu}, \Lambda)\, e^{\mathbf{t}^T\boldsymbol{\nu} + \frac{1}{2}\mathbf{t}^T\Lambda\mathbf{t}} \\ & \times \int_{c_1}^{\infty} \cdots \int_{c_q}^{\infty} \phi_q(\mathbf{u}; \boldsymbol{\nu} + \Lambda\mathbf{t}, \Lambda)\, dw_1 \cdots dw_q \\ = {} & \Phi_q^{-1}(-\mathbf{c}; -\boldsymbol{\nu}, \Lambda)\, e^{\mathbf{t}^T\boldsymbol{\nu} + \frac{1}{2}\mathbf{t}^T\Lambda\mathbf{t}} \Phi_q(-\mathbf{c}; -\boldsymbol{\nu} - \Lambda\mathbf{t}, \Lambda). \end{aligned}$$

□

Proof of Proposition 13.6.1. From Lemma 13.6.1 and the fact that the mgf satisfies $M_{D\mathbf{u}}(\mathbf{t}) = M_{\mathbf{u}}\left(D^T\mathbf{t}\right)$, we obtain:

$$M_{D\mathbf{u}}(\mathbf{t}) = \Phi_q^{-1}(\mathbf{0}; \mathbf{c} - \boldsymbol{\nu}, \Lambda)\, e^{\mathbf{t}^T D\boldsymbol{\nu} + \frac{1}{2}\mathbf{t}^T D\Lambda D^T \mathbf{t}} \Phi_q\left(\Lambda D^T\mathbf{t}; \mathbf{c} - \boldsymbol{\nu}, \Lambda\right),$$

and given that $\mathbf{w} \sim N_p(\boldsymbol{\mu}, \Sigma)$, the mgf of $\mathbf{w}$ is

$$M_{\mathbf{w}}(\mathbf{t}) = e^{\mathbf{t}^T\boldsymbol{\mu} + \frac{1}{2}\mathbf{t}^T\Sigma\mathbf{t}}.$$

By independence of $\mathbf{v}$ and $\mathbf{u}$ the mgf of $\mathbf{v}+D\mathbf{u}$ is

$$\begin{aligned} M_{\boldsymbol{\varepsilon}}(\mathbf{t}) &= M_{\mathbf{v}+D\mathbf{u}}(\mathbf{t}) \\ &= M_{\mathbf{v}}(\mathbf{t})\, M_{\mathbf{u}}\left(D^T\mathbf{t}\right) \\ &= \Phi_q^{-1}\left(\mathbf{0};\mathbf{c}-\boldsymbol{\nu},\Lambda\right)\Phi_q\left[\Lambda D^T\Theta^{-1}\Theta\mathbf{t};\mathbf{c}-\boldsymbol{\nu},\Lambda\right]e^{\mathbf{t}^T(\boldsymbol{\mu}+D\boldsymbol{\nu})+\frac{1}{2}\mathbf{t}^T\Theta\mathbf{t}}. \end{aligned}$$

Comparing with the mgf (13.9) of the CSN, we get that the density function of $\boldsymbol{\varepsilon}$ is given by formula (13.26). This density could have been obtained by integrating (13.17) with respect to $\mathbf{u}$. □

Example: (*The stochastic frontier error model*). If $V \sim N\left(\mu,\sigma^2\right)$, $W \sim N^c\left(\nu,\tau^2\right)$, V and W are independent, then the density function of

$$\varepsilon = V + \alpha W,$$

is given by

$$\begin{aligned} g(\varepsilon) &= \Phi^{-1}\left(\frac{\nu-c}{\tau}\right)\phi\left(\varepsilon;\mu+\alpha\nu,\sigma^2+\alpha^2\tau^2\right) \\ &\times \Phi\left[\frac{\alpha\tau^2(\varepsilon-\mu-\alpha\nu)}{\sigma\tau\sqrt{\sigma^2+\alpha^2\tau^2}}+\frac{(\nu-c)\sqrt{\alpha^2\tau^2+\sigma^2}}{\sigma\tau}\right]. \end{aligned}$$

Proof. From Proposition 13.6.1 with $\Sigma=\sigma^2, D=\alpha, \Lambda=\tau^2$ and the identities

$$\Theta=\sigma^2+\alpha^2\tau^2,\ \Lambda D^T\Theta^{-1}=\alpha\tau^2\left(\sigma^2+\alpha^2\tau^2\right)^{-1},$$

and

$$\Upsilon=\frac{\sigma^2\tau^2}{\alpha^2\tau^2+\sigma^2},$$

we obtain that the density function of ε is

$$\begin{aligned} g(\varepsilon) &= \Phi^{-1}\left(\frac{\nu-c}{\tau}\right)\phi\left(\varepsilon;\mu+\alpha\nu,\sigma^2+\alpha^2\tau^2\right) \\ &\times \Phi\left[\alpha\tau^2\left(\sigma^2+\alpha^2\tau^2\right)^{-1}(\varepsilon-\mu-\alpha\nu);c-\nu,\frac{\sigma^2\tau^2}{\alpha^2\tau^2+\sigma^2}\right]. \end{aligned}$$

Simplifying the former expression we get the stated density. □

13.6.2 Expectation of the Truncated Multivariate Normal Distribution

Lemma 13.6.2 If $\mathbf{w} \sim N_n^{\mathbf{c}}(\boldsymbol{\nu},\Lambda)$, then

$$\mathrm{E}(\mathbf{w})=\boldsymbol{\nu}+\Lambda\frac{\Phi_n^*(\boldsymbol{\nu};\mathbf{c},\Lambda)}{\Phi_n(\boldsymbol{\nu};\mathbf{c},\Lambda)},$$

with

$$\Phi_n^*(\mathbf{s};\boldsymbol{\nu},\Lambda)=\frac{\partial}{\partial\mathbf{s}}\Phi_n(\mathbf{s};\boldsymbol{\nu},\Lambda), \qquad (13.27)$$

and

$$\frac{\partial}{\partial s_i}\Phi_n\left(\mathbf{s};\boldsymbol{\nu},\Lambda\right)=\phi\left(s_i;\mu_i,\lambda_{ii}\right)\times\Phi_{n-1}\left(\mathbf{s}_{\neg i};\boldsymbol{\nu}_{\neg i}+\Lambda_{i\neg i}\lambda_{ii}^{-1}\left(s_i-\nu_i\right),\Lambda_{\neg i\neg i}\right),\tag{13.28}$$

where, $\mathbf{s}_{\neg i}=\left(s_1,\ldots,s_{i-1},s_{i+1},\ldots,s_p\right)^T$, $\Lambda_{\neg i\neg i}$ is the $(n-1)\times(n-1)$ matrix derived from $\Lambda=(\lambda_{ij})$ by eliminating its i-th row and its i-th column and $\Lambda_{i\neg i}$ is the $(n-1)$ vector derived from the i-th column of Λ by removing the i-th row term.

Proof. Given that

$$M_{\mathbf{u}}\left(\mathbf{t}\right)=\Phi_n^{-1}\left(\boldsymbol{\nu};\mathbf{c},\Lambda\right)e^{\mathbf{t}^T\boldsymbol{\nu}+\frac{1}{2}\mathbf{t}^T\Lambda\mathbf{t}}\Phi_n\left(\Lambda\mathbf{t}+\boldsymbol{\nu};\mathbf{c},\Lambda\right),$$

and

$$\frac{\partial}{\partial\mathbf{t}}e^{\mathbf{t}^T\boldsymbol{\nu}+\frac{1}{2}\mathbf{t}^T\Lambda\mathbf{t}}=\left(\boldsymbol{\nu}+\Lambda\mathbf{t}\right)e^{\mathbf{t}^T\boldsymbol{\nu}+\frac{1}{2}\mathbf{t}^T\Lambda\mathbf{t}},$$

$$\frac{\partial}{\partial\mathbf{t}}\Phi_n\left(\Lambda\mathbf{t}+\boldsymbol{\nu};\mathbf{c},\Lambda\right)=\Lambda\left.\frac{\partial}{\partial\mathbf{s}}\Phi_n\left(\mathbf{s},\Lambda\right)\right|_{\mathbf{s}=\Lambda\mathbf{t}+\boldsymbol{\nu}}=\Lambda\Phi_n^*\left(\Lambda\mathbf{t}+\boldsymbol{\nu};\mathbf{c},\Lambda\right),$$

it follows that

$$\mathrm{E}\left(\mathbf{w}\right)=\left.\frac{\partial}{\partial\mathbf{t}}M_{\mathbf{u}}\left(\mathbf{t}\right)\right|_{\mathbf{t}=\mathbf{0}}=\boldsymbol{\nu}+\Lambda\frac{\Phi_n^*\left(\boldsymbol{\nu};\mathbf{c},\Lambda\right)}{\Phi_n\left(\boldsymbol{\nu};\mathbf{c},\Lambda\right)}.$$

The expression for (13.28) is obtained from the Appendix of Domínguez-Molina, González-Farías, and Gupta (2003). □

In particular, if $n=1$, $\Lambda=\lambda^2$, then

$$\frac{\partial}{\partial s}\Phi\left(s;\nu,\lambda^2\right)=\phi\left(s;\nu,\lambda^2\right),$$

and thus

$$\mathrm{E}(W)=\nu+\lambda^2\frac{\phi\left(\nu;c,\lambda^2\right)}{\Phi\left(\nu;c,\lambda^2\right)}.$$

Notice that if $\mathbf{s}\sim N_n\left(\boldsymbol{\nu},\Lambda\right)$, then $\mathrm{Var}\left(\mathbf{s}_{\neg i}\right)=\Lambda_{\neg i\neg i}$, $\mathrm{Cov}\left(\mathbf{s}_{\neg i},s_i\right)=\Lambda_{\neg i,i}$, $\mathrm{Var}\left(s_i\right)=\lambda_{ii}$, and the joint distribution of $\left(\mathbf{s}_{\neg i}^T,s_i\right)^T$ is $N_p\left(\boldsymbol{\nu}^\dagger,\Lambda^\dagger\right)$, where $\boldsymbol{\nu}^\dagger=\left(\boldsymbol{\nu}_{\neg i}^T,\nu_i\right)^T$ and

$$\Lambda^\dagger=\begin{pmatrix}\Lambda_{\neg i\neg i} & \Lambda_{i\neg i}\\ \Lambda_{i\neg i}^T & \lambda_{ii}\end{pmatrix}.$$

13.6.3 Efficiencies for Individual Errors of Model III

Formula (13.19) gives the efficiency of $\mathbf{u}$ given the vector $\boldsymbol{\varepsilon}$, in particular an element of $\mathbf{u}|\boldsymbol{\varepsilon}$ is given by $U_i|\boldsymbol{\varepsilon}$. It could be of interest to know also the efficiency for a particular ε_j. Integrating out $\boldsymbol{\varepsilon}_{\neg j}$ in (13.16) we deduce that

$$g(\mathbf{u},\varepsilon_j) = \Phi_n^{-1}(-\mathbf{c};-\boldsymbol{\nu},\Lambda)\,\phi(\varepsilon_j - D_j\mathbf{u};0,\sigma_{jj})\,\phi_n(\mathbf{u};\boldsymbol{\nu},\Lambda), \quad (13.29)$$

where D_j is the j-th row of D. After similar manipulations as those to obtain (13.17), we rewrite (13.29) as

$$g(\mathbf{u},\varepsilon_j) = \Phi_n^{-1}(-\mathbf{c};-\boldsymbol{\nu},\Lambda)\,\phi_n\left(\mathbf{u};\boldsymbol{\nu}+B_j^{\dagger},B_j\right)\phi\left(\varepsilon_j;D_j\boldsymbol{\nu},\sigma_{jj}+D_j\Lambda D_j^T\right), \quad (13.30)$$

where $B_j = \left(\frac{1}{\sigma_{jj}}D_j^T D_j + \Lambda^{-1}\right)^{-1}$, $B_j^{\dagger} = B_j D_j^T \frac{(\varepsilon_j - D_j\boldsymbol{\nu})}{\sigma_{jj}}$. Now, integrating out $\mathbf{u}_{\neg i}$ in (13.30) we obtain that the joint density of U_i and ε_j is

$$\begin{aligned} &g(u_i,\varepsilon_j)\\ &= \Phi_n^{-1}(-\mathbf{c};-\boldsymbol{\nu},\Lambda)\,\phi\left[u_i;\nu_i+\left(B_j^{\dagger}\right)_i,(B_j)_{ii}\right]\phi\left(\varepsilon_j;D_j\boldsymbol{\nu},\sigma_{jj}+D_j\Lambda D_j^T\right)\\ &\times \Phi_{n-1}\left\{-\mathbf{c}_{\neg i};\boldsymbol{\nu}_{\neg i}+\left(B_j^{\dagger}\right)_{\neg i}+(B_j)_{i\neg i}(B_j)_{ii}^{-1}(u_i-\nu_i-c_i),(B_j)_{\neg i\neg i}\right\}.\end{aligned}$$

Taking $D = \alpha I_n$ and $\Lambda = \operatorname{diag}(\tau_i)$, and observing that $D_j = \alpha\mathbf{a}_j^T$, where $\mathbf{a}_j$ is the j-th unit vector in $\mathbb{R}^n$, we see that

$$\begin{aligned} B_j &= \left(\frac{\alpha^2}{\sigma_{jj}}\mathbf{a}_j\mathbf{a}_j^T + \Lambda^{-1}\right)^{-1}\\ &= \operatorname{diag}\left(\tau_{11},\ldots,\tau_{j-1,j-1},\frac{\sigma_{jj}\tau_j}{\sigma_{jj}+\alpha^2\tau_j},\tau_{j+1,j+1}\right),\end{aligned}$$

$$B_j^{\dagger} = B_j\alpha\mathbf{a}_j\frac{(\varepsilon_j-\nu_j)}{\sigma_{jj}} = \frac{\alpha\tau_j(\varepsilon_j-\nu_j)}{\sigma_{jj}+\tau_j}\mathbf{a}_j,$$

which implies that

$$\begin{aligned} g(u_i,\varepsilon_j) =&\Phi_n^{-1}(-\mathbf{c};-\boldsymbol{\nu},\Lambda)\,\phi\left(\varepsilon_j;\nu_j,\sigma_{jj}+\alpha^2\tau_j\right) \quad (13.31)\\ &\times\phi\left(u_i;\nu_i+\left(B_j^{\dagger}\right)_i,(B_j)_{ii}\right)\\ &\times\Phi_{n-1}\left\{-\mathbf{c}_{\neg i};\boldsymbol{\nu}_{\neg i}+\left(B_j^{\dagger}\right)_{\neg i},(B_j)_{\neg i\neg i}\right\}.\end{aligned}$$

The conditional density of $U_i|\varepsilon_j$ is obtained with (13.31) and the marginal density of ε_j given in (13.14). When $i = j$ Equation (13.31) reduces to

$$\begin{aligned} g(u_i,\varepsilon_i) &= \Phi^{-1}(-c_i;-\nu_i,\tau_i)\,\phi\left(\varepsilon_i;\nu_i,\sigma_{ii}+\alpha^2\tau_i\right)\\ &\times\phi\left(u_i;\nu_i+\frac{\alpha\tau_i(\varepsilon_i-\nu_i)}{\sigma_{ii}+\tau_i},\frac{\sigma_{ii}\tau_i}{\sigma_{ii}+\alpha^2\tau_i}\right)\\ &\Phi^{-1}\left(\frac{\nu_i-c_i}{\tau_i}\right)\phi\left(\frac{\varepsilon_i-\nu_i}{\sqrt{\sigma_{ii}+\alpha^2\tau_i}}\right)\\ &\times\phi\left(u_i;\frac{\nu_i\sigma_{ii}+\alpha\tau_i\varepsilon_i-\alpha\tau_i\nu_i+\nu_i\tau_i}{\sigma_{ii}+\tau_i},\frac{\sigma_{ii}\tau_i}{\sigma_{ii}+\alpha^2\tau_i}\right).\end{aligned}$$

Since $\varepsilon_i \sim V_i + \alpha U_i$ has a density given in (13.5) it follows:

$$U_i|\varepsilon_i \sim N^c\left[\frac{\nu\sigma^2 + (\varepsilon_i - \mu)\,\tau^2\alpha}{\sigma^2 + \tau^2\alpha^2}, \frac{\sigma^2\tau^2}{\sigma^2 + \tau^2\alpha^2}\right].$$

Letting $\nu^* = \nu + \frac{\sigma^2\alpha}{\alpha^2\tau^2+\sigma^2}\left(\varepsilon - \mu - \alpha\nu\right),\ \sigma^{2*} = \frac{\sigma^2\tau^2}{\alpha^2\tau^2+\sigma^2}$ we get that

$$\mathrm{E}\left(U_i|\varepsilon_i\right) = \nu^* + \sigma^* \frac{\phi\left(\frac{\nu^*-c}{\sigma^*}\right)}{\Phi\left(\frac{\nu^*-c}{\sigma^*}\right)}. \tag{13.32}$$

□

CHAPTER 14

Coastal Flooding and the Multivariate Skew-*t* Distribution

Keith R. Thompson and Yingshuo Shen

14.1 Introduction

Globally averaged sea level rose by about 15 cm over the course of the 20th century. Superimposed on this global rise are regional variations associated with vertical land movement. In some regions the land is rising so quickly that sea level is sinking with respect to the land (e.g., Fennoscandia). In other regions the land is sinking and this can result in local rates of sea level rise that greatly exceed the globally averaged value of 1.5 mm per year.

Projections of global sea level rise over the next century range from 1 to 9 mm per year with a most likely increase of about 5 mm per year (IPCC, 2001). The major concern over an acceleration in the rate of rise of global sea level is its effect on the frequency and severity of coastal flooding. During the last century there were many disastrous floods that caused significant loss of life. For example, 6,000 lives were lost in Galveston Texas in 1900 and 500,000 in Bangladesh in 1970 (Pugh, 1987). An acceleration in the rate of global sea level rise will only exacerbate the problem, particularly in low lying areas and in regions of pronounced land subsidence. For a full discussion of how and why sea level has changed, and may change in the future, see IPCC (2001).

Extensive sea level records from coastal tide gauges have been used to estimate global sea level rise over the last two centuries. Sea level is however a physical variable of broader interest to both oceanographers and geophysicists. It is used, for example, to estimate ocean currents, changes in ocean heat content, melting of ice caps, and vertical crustal movement. In this study we will show how the skew-*t* distribution can be used to project flooding risk over the next century, and also to quality control sea level observations leading to a more reliable database for a wide range of applications.

The first step in the present analysis is to split the hourly sea level into a tidal component and a part that is left over, the so-called "residual." The tides are due to variations in the gravitational pull of the sun and moon and their frequencies are known with great accuracy. Over the last century effective statistical schemes have been developed to estimate tidal amplitudes and phases from observed sea level records and thereby predict the tide. The result is that, given an hourly sea level record for a year or so in length, extremely accurate tidal predictions can be made with lead times of days to centuries. From a statistical standpoint the tides can be treated as deterministic. The residual is usually dominated by wind and air pressure effects in mid-latitudes and we will treat the residual as stochastic.

The skew-t distribution is used in this study to model the seasonally changing probability density of the tidal residuals. This will lead to a parsimonious description of the residual distribution and an effective way of identifying suspect observations. It will also allow us to estimate the return periods of extreme sea levels from short records, an important practical consideration in many regions, and also project how flooding risk might change with accelerating global sea level rise and other climate change possibilities such as increased storminess. This type of "what if" calculation of flooding risk is not possible using classical extremal analysis. The reason is that classical analysis is based only on annual maxima and cannot separate, or project, the contribution from different physical phenomena such as global sea level rise or storms.

This study complements a recent analysis by Genton and Thompson (2003) in which the bivariate skew-t distribution was used to assess flooding risk for Charlottetown, a low lying city on Prince Edward Island in the Gulf of St. Lawrence. This study examines a long record from another eastcoast Canadian city and also extends the earlier analysis by (i) introducing a skew-t distribution with seasonally varying parameters, and (ii) showing how the skew-t distribution can be used to quality control sea level observations.

The structure of the chapter is as follows. A brief review of the salient features of the multivariate skew-t distribution, and the estimation of its parameters, are given in Section 14.2. This is followed in Section 14.3 by a discussion of the 80-year sea level record for Halifax, Nova Scotia, Canada. The fitting of the skew-t distribution to the residuals is described in Section 14.4 and some applications, including quality control and flooding risk estimation, are described in Section 14.5. The results are summarized and discussed in the final section.

14.2 A Seasonally Varying Skew-t Distribution

A multivariate skew-t distribution is defined in this section and some of its properties are reviewed. All of these properties are discussed in detail by Azzalini and Capitanio (2003) and Capitanio, Azzalini, and Stanghellini (2003). This is followed by a description of the seasonally varying, bivariate skew-t distribution used later in this study. The section concludes with comments on estimation of the skew-t parameters and assessment of model fit.

Let $\mathbf{x}$ denote a p-dimensional random vector. It is taken to have a skew-t distribution if its density is of the form

$$g(\mathbf{x}) = 2\, t_p\,(\mathbf{x}; \nu)\, T_1 \left(\boldsymbol{\alpha}^T W^{-1}(\mathbf{x} - \boldsymbol{\xi}) \left[\frac{\nu + p}{Q + \nu} \right]^{1/2} ; \nu + p \right), \qquad (14.1)$$

where t_p is the density of a p-dimensional t variate with ν degrees of freedom:

$$t_p(\mathbf{x}; \nu) = \frac{\Gamma((\nu + p)/2)}{|\Omega|^{1/2}(\pi\nu)^{p/2}\Gamma(\nu/2)}(1 + Q/\nu)^{-(\nu+p)/2}, \qquad (14.2)$$

where

$$Q = (\mathbf{x} - \boldsymbol{\xi})^T \Omega^{-1}(\mathbf{x} - \boldsymbol{\xi}). \qquad (14.3)$$

The scalar parameter ν denotes the degrees of freedom of the multivariate t distribution. The p-dimensional vector $\boldsymbol{\xi}$ is a location parameter. Ω is a $p \times p$ covariance matrix and its diagonal part is denoted by W^2. The associated correlation matrix is then $\overline{\Omega} = W^{-1}\Omega W^{-1}$. The skewing function, $T_1(\cdot; \nu + p)$, is a univariate t distribution function with $\nu + p$ degrees of freedom. The p-dimensional vector $\boldsymbol{\alpha}$ controls the skewness.

The expected value of $\mathbf{x}$ is given by

$$\mathrm{E}(\mathbf{x}) = \boldsymbol{\xi} + W\boldsymbol{\mu}, \qquad (14.4)$$

and its variance by

$$\mathrm{Var}(\mathbf{x}) = \frac{\nu}{\nu - 2}\Omega - W\boldsymbol{\mu}\boldsymbol{\mu}^T W, \qquad (14.5)$$

where

$$\boldsymbol{\mu} = \left[\frac{\nu/\pi}{1 + \boldsymbol{\alpha}^T \overline{\Omega} \boldsymbol{\alpha}} \right]^{1/2} \frac{\Gamma((\nu - 1)/2)}{\Gamma(\nu/2)} \overline{\Omega} \boldsymbol{\alpha}. \qquad (14.6)$$

In the present study we will take $\mathbf{x}$ to be a pair of consecutive, hourly sea level residuals, i.e., $p = 2$. We will also assume the residuals have zero mean. This is justifiable because the residual is defined to be the variation about the tide which includes a seasonally varying mean. Substituting $\mathrm{E}(\mathbf{x}) = \mathbf{0}$ in (14.4) gives an expression for $\boldsymbol{\xi}$ in terms of the elements of Ω, ν, and $\boldsymbol{\alpha}$. To simplify the model further, we will assume the variance of the residuals does not change significantly over one hour and assume

$$\Omega = \sigma^2 \begin{pmatrix} 1 & \rho \\ \rho & 1 \end{pmatrix},$$

and

$$\boldsymbol{\alpha} = \alpha_s\,(1,\,1)^T,$$

where α_s is a scalar. The result is a bivariate skew-t with 4 scalar parameters: σ, ρ, ν, and α_s.

Following the discussion of observed sea level variability in the next section, it will be clear that some of the skew-t parameters must vary with season. We have modeled the seasonal variation of σ by

$$\sigma(t) = \theta_1 \left[1 + \theta_2 \cos(2\pi t/T - \theta_3)\right], \tag{14.7}$$

where t denotes time and T is the annual period. The parameter θ_1 determines the mean value of σ over the course of a year, and θ_2 controls the amplitude of its seasonal variation. The parameter θ_3 determines the time of year at which σ is a maximum. To ensure that the scale factor is positive throughout the year, we assume $\theta_1 > 0$ and $0 < \theta_2 < 1$. Similar models can be constructed for the seasonal variation of the other parameters. For example we experimented with

$$\rho(t) = \tanh(\theta_4 + \theta_5 \cos(2\pi t/T - \theta_6)),$$

and

$$\nu(t) = \frac{1}{\theta_7 + \theta_8 \cos(2\pi t/T - \theta_9)},$$

and a similar type of non-linear equation for α_s with 3 additional parameters. The result is a seasonally varying, bivariate skew-t distribution determined by 12 constant parameters $\{\theta_1, \ldots, \theta_{12}\}$ subject to various constraints.

Given a time series of hourly sea levels, it is straightforward to remove the tide and obtain a corresponding time series of residuals $\{x_1, x_2, \ldots, x_n\}$. These residuals can then be grouped into consecutive pairs of residuals: $(x_1, x_2), (x_2, x_3), \ldots, (x_{n-1}, x_n)$. The method of maximum likelihood can then be used to estimate the θ_i by maximizing numerically the log-likelihood function that depends on the θ_i through σ, ρ, ν and α_s.

The difficulty with maximum likelihood estimation in the present context is that the residual pairs are dependent. This means the log-likelihood function is not simply the sum of the log-likelihoods of the individual bivariate observations. To solve this problem, Genton and Thompson (2003) sub-sampled the residual pairs every τ time steps where τ was sufficiently large for the assumption of independence to be reasonable. Sub-sampling every τ hours gives τ parameter estimates because the start time of the sub-sampling can begin on any one of τ hours. Arguing that each estimate is asymptotically unbiased, Genton and Thompson (2003) averaged their τ estimates to obtain a more reliable estimate of the skew-t parameters than that provided by any one sub-sample. A similar approach was used in the present study.

Having fitted the skew-t to the residuals, the quality of the fit can be assessed using standard diagnostic plots, e.g., quantile and probability plots. For example, Azzalini and Capitanio (2003) show that

$$Q/2 \sim F_{2,\nu} \,. \tag{14.8}$$

Thus to assess the fit of the skew-t, we can calculate $Q/2$ using (14.3) and check if its empirical distribution differs significantly from an F distribution with 2 and ν degrees of freedom using a probability plot. Azzalini and Capitanio (2003) give several examples of this approach. Another set of diagnostic plots can be generated by noting that if $\mathbf{x}$ has a skew-t distribution, so does any linear transformation of $\mathbf{x}$. Thus the marginal distribution of the residual is also skew-t, as is the change in residual from one hour to the next. Azzalini and Capitanio (2003) and Capitanio *et al.* (2003) give expressions for the skew-t density of linear transformations of $\mathbf{x}$ and so it is straightforward to find quantiles of, for example, the residuals for a given time of year, and compare them against the quantiles from the marginal distribution of the fitted bivariate skew-t. Some quantile plots are given in Section 14.4.

14.3 Observations of Coastal Sea Level

Hourly sea levels for Halifax, from 1920 to 2001, were extracted from data disks produced and distributed by the WOCE Data Products Committee (2002). When plotting the time series it was obvious that some of the raw observations were highly suspect, particularly in the early part of the record. These values were deleted prior to analysis.

Halifax annual mean sea level rose at an average rate of 3.3 mm per year over the last 80 years; see Figure 14.1. This is faster than the generally accepted rate of global sea level rise and the difference is thought to be due to subsidence of the Earth's crust in this region. The annual maxima (the largest hourly values recorded in each calendar year) also exhibit a clear increase through time; see Figure 14.1. This is presumably due in part to the increase in the annual means. One way to remove this effect is to subtract the annual means from the annual maxima and thereby define the maxima with respect to the slowly rising mean sea level. This has been done in the analysis that follows.

The return period plot shown in Figure 14.2 is based on Halifax annual maxima, corrected for mean sea level rise. This plot suggests that, to a first approximation, the distribution of Halifax annual maxima is close to that of a Gumbel distribution. Note however that there is the hint of upward curvature, suggesting that a generalized extreme value distribution with a small, positive shape parameter may fit better. We generalized the Gumbel distribution to allow its location parameter to change with time ($\mu = \beta_0 + \beta_1 t$) and then checked if the addition of the $\beta_1 t$ term led to a significant improvement in model fit. Based on the likelihood ratio test, we conclude that the removal of the annual means has effectively eliminated the trend in the annual maxima. Thus the trend in the extremes appears due primarily to the slow rise of sea level and not changes in the intensity of higher frequency phenomena such as storms.

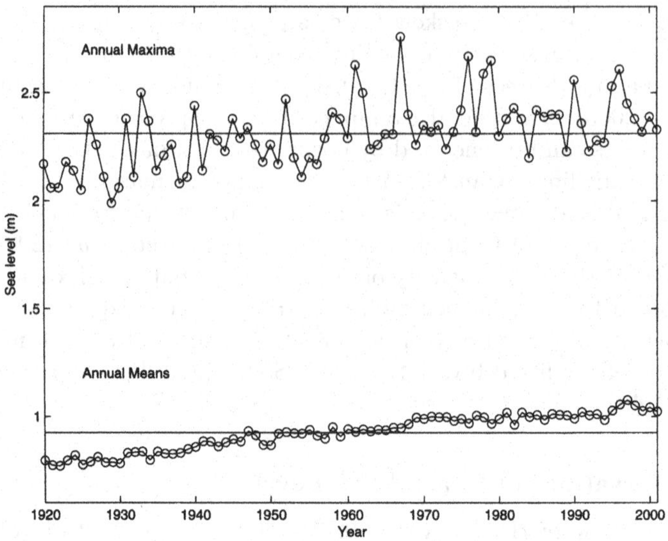

Figure 14.1 *Annual means and annual maxima of Halifax hourly sea level, 1920–2001. Both means and maxima are referenced to Chart Datum.*

The highest sea level recorded at Halifax between 1920 and 2001 was 2.76 m; see Figure 14.1. This record was reached at 7 pm on February 23, 1967 when it caused flooding of the Halifax ferry terminal and an important coastal road. The hourly sea levels for this event are plotted in Figure 14.3.

Comparing the top and bottom panels of Figure 14.3 it is clear that the record sea level was due to the coincidence of a high tide and a large residual (in fact a "storm surge" generated by gale force winds and low air pressure). If the storm surge had occurred 6 hours earlier it would have coincided with low tide and no flooding would have occurred. It is also clear from this figure that if the surge had occurred at high tide a few days later, closer to spring tides, the flooding would have been worse. We will show later that by combining the distribution of the residuals with the predicted tide, it is possible to get more reliable estimates of flooding risk than can be achieved by analyzing just the observed levels. The idea of decomposing sea level into tide and surge in order to estimate return periods goes back to Pugh and Vassie (1980). For modifications of this early work, see Tawn and Vassie (1989). Middleton and Thompson (1986) used a similar decomposition in their development of a method for estimating return periods from short sea level records.

Residuals for the period 1920 to 2001 were calculated by first predicting the tide for each year and then subtracting it from the observed hourly sea levels. The tidal analysis package of Pawlowicz, Beardsley, and Lentz (2002)

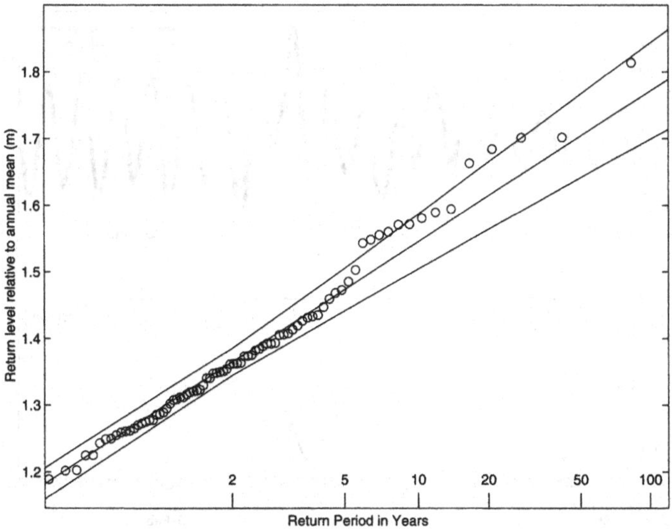

Figure 14.2 *Return level plot for Halifax, 1920–2001. Annual maxima, corrected for their corresponding annual mean, and the Gumbel distribution were used. The center line was fit using maximum likelihood. The bounding lines are 95% confidence intervals calculated using the delta method (e.g., Coles, 2001).*

was used. Each yearly analysis included the standard set of 68 tidal constituents.

Time series plots of the residuals had a number of spikes that looked suspiciously like recording errors. There was also variability at the Nyquist frequency that may have been due to aliasing of a natural resonance of Halifax harbor. To attenuate such spikes, and remove aliased signals, the residual record was low-pass filtered to suppress all variations with periods shorter than 4 hours. Another feature of the residual plots was the occasional appearance of tidal energy with a period close to that of the dominant tidal constituent (M_2). It is possible that this tidal energy is the result of slight timing errors: a timing error of 10 minutes can cause residuals exceeding 10 cm at Halifax. In order to remove the occasional tidal signals appearing in the Halifax residuals, a band-pass filter was used to suppress variations with periods between 11 and 14 hours. Henceforth, we will refer to the filtered hourly residuals simply as residuals.

The standard deviation of the January residuals is about twice that of the July residuals; see Figure 14.4. This is to be expected because the residual variability is due primarily to storm surges and storms are strongest in winter. There is also a pronounced seasonal variation in the autocorrelation of the Halifax residuals at a lag of one hour: the highest sample autocorrelation was found for March (0.982) and the lowest for July (0.962). Thus

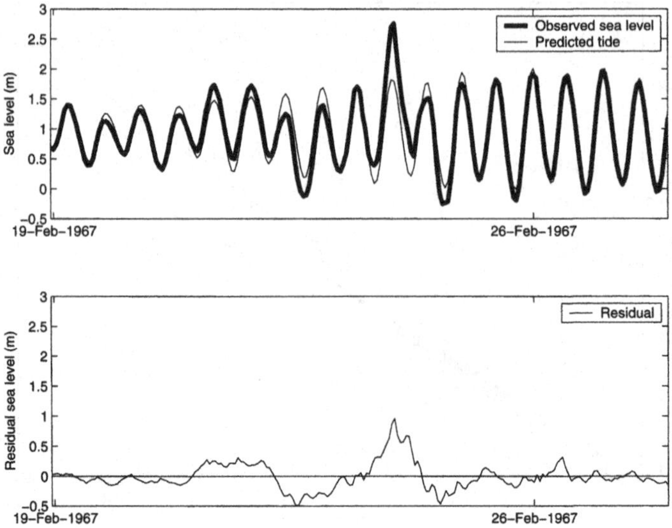

Figure 14.3 *Top panel: Observed hourly sea level and the predicted tide for a 10-day period that includes 7 pm, February 23, 1967. This is the time at which the highest sea level in the 1920-2001 record was reached. Bottom panel: Difference between the observed hourly sea level and the predicted tide, i.e., the residual.*

the "memory" of the residual process is longer in winter and this is another aspect of its seasonality. Both these features were found by Genton and Thompson (2003) in an earlier analysis of Charlottetown sea level.

The skewness and kurtosis of the complete residual record are 0.47 and 5.2, respectively. There is no clear seasonal variation in these statistics but it is important to note that in every month the skewness was found to be positive and the kurtosis greater than 3. This leads us to conclude that the assumption of normally distributed residuals is not tenable. Further, the positive skewness evident in all seasons cannot be modeled by a standard t distribution. For these reasons we will attempt to fit a skew-t distribution to the residuals in the next section.

14.4 Fitting the Skew-t Distribution

The bivariate skew-t was fit to pairs of adjacent residuals using the likelihood based approach, with sub-sampling, described in Section 14.2. We experimented with various forms of seasonal variation in the skew-t parameters. We restrict our attention in this chapter to the following simple case

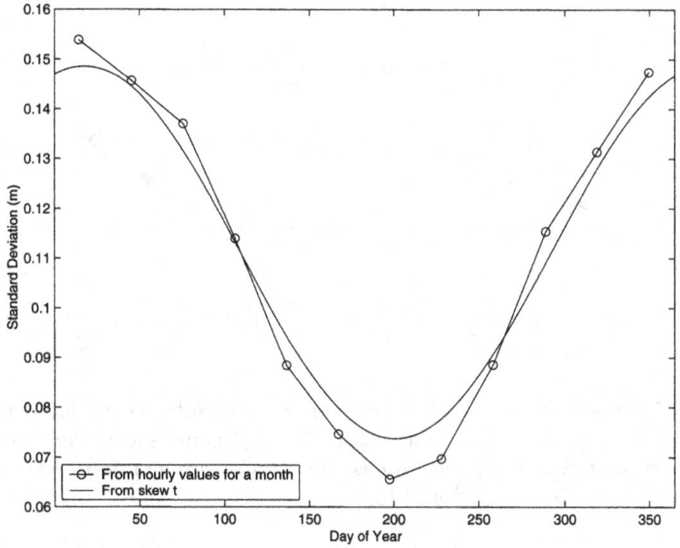

Figure 14.4 *Seasonal variation of the standard deviation of the Halifax hourly residuals. Circles show the standard deviation for each month, estimated by binning residuals from 1920–2001 by month, and then calculating the sample standard deviation in the standard way. The smooth line shows the standard deviation calculated from the skew-t distribution. A seasonally varying, bivariate skew-t was fit to all residual pairs, (x_t, x_{t+1}), as explained in the text. The standard deviation was then calculated from the marginal distribution of x_t for the middle of each month.*

in which seasonality is expressed only through σ:

$$\begin{aligned} \sigma &= \theta_1 \left[1 + \theta_2 \cos(2\pi t/T - \theta_3)\right], \\ \rho &= \tanh(\theta_4), \\ \nu &= 1/\theta_5, \\ \delta &= \theta_6. \end{aligned} \tag{14.9}$$

The skewing parameter α_s is obtained from δ using Equation (4) of Azzalini and Capitanio (2003). Upper and lower bounds were placed on the θ_i to ensure the skew-t parameters were reasonable, e.g., $\sigma > 0$.

Figure 14.5 shows the probability density of the seasonally varying skew-t for January and July. This figure clearly illustrates the increased variance in winter, the high correlation between consecutive residuals, and the skewness of the fitted distribution. Note the symmetry of the distributions about the $y = x$ line. This results from the simplified forms assumed for Ω and $\boldsymbol{\alpha}$.

To assess the overall fit of the bivariate skew-t we made a probability

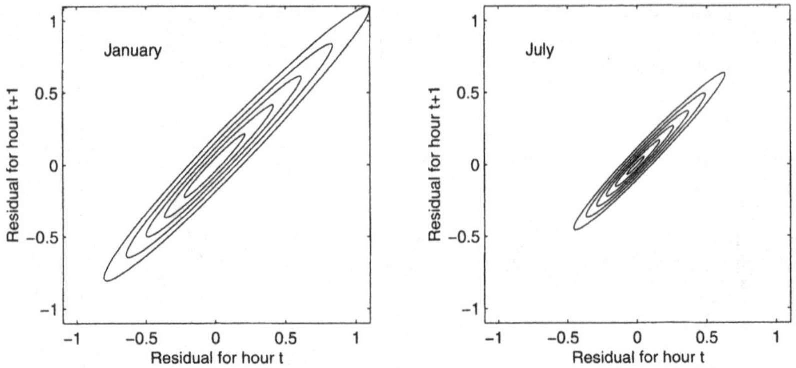

Figure 14.5 *Probability density contours of the bivariate skew-t fit to pairs of consecutive Halifax residuals, 1920–2001. The left panel shows the density for mid-January and the right panel shows the density for mid-July. The contour values are* 10^2, 10, 10^{-1}, 10^{-2} *and* 10^{-3}.

plot of Q based on (14.8). The result (not shown) was a set of points that was almost indistinguishable from the 1:1 line.

The marginal density of the residuals can be readily determined from the bivariate skew-t parameters using results provided by Azzalini and Capitanio (2003), and Capitanio *et al.* (2003). From the marginal density it is straightforward to calculate the standard deviation, skewness, and kurtosis of the residuals and also to make quantile plots. The latter provide a much more critical assessment of the fit of the skew-t in the tails than that provided by probability plots.

The standard deviation of the residuals according to the skew-t distribution undergoes a similar seasonal variation to that observed; see Figure 14.4. The skewness and kurtosis are constant through time for the present model and they are 0.43 and 4.4, respectively, compared to 0.47 and 5.2 calculated directly from the residuals in the standard way. The quantile plots of the residuals, stratified by month, are shown in Figure 14.6. Our interpretation of this plot is that the skew-t is doing a reasonable job of describing the overall distribution of the residuals, particularly in winter when the largest storm surges occur. The skew-t does not however capture all the features in the empirical distribution. For example the quantiles based on the skew-t are too large in some winter months at high levels, resulting in small, upward "kinks" in some of the QQ-plots. Although the reason for this is not known, it is important to note that each kink is defined by less that 10 hourly values (out of a total of about 60,000 hourly values used for each QQ-plot).

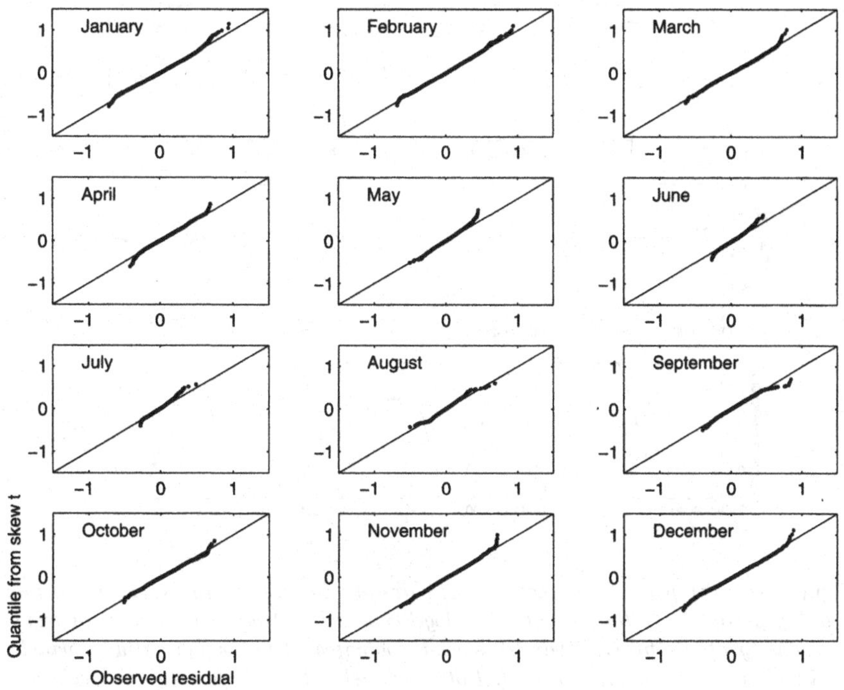

Figure 14.6 *Quantile plots of the residuals, stratified by month. The x-axis shows the empirical quantiles obtained directly from the ordered residuals for 1920–2001. The y-axis shows the quantiles according the fitted skew-t distribution. These quantiles are from the marginal distribution of the fitted bivariate skew-t, evaluated at the middle of each month.*

14.5 Applications of the Skew-t Distribution

Based on the diagnostic plots presented in the previous section, we conclude that the bivariate skew-t gives a reasonable description of the residuals. We will now show how the fitted skew-t density can be used to tackle some practical problems.

14.5.1 Quality Control of Sea Level Observations

Given a pair of residuals, and the estimated skew-t parameters, it is straightforward to calculate a log-likelihood using Equation (14.1). A sequence of log-likelihoods for October 2000 is plotted in the top panel of Figure 14.7. Clearly some of the log-likelihoods point to observations that may be considered outliers. The middle panel shows the residuals and the bottom panel shows the observed sea levels. Note the highly unusual, and aphysical, feature in the observed sea levels corresponding to the anomalous

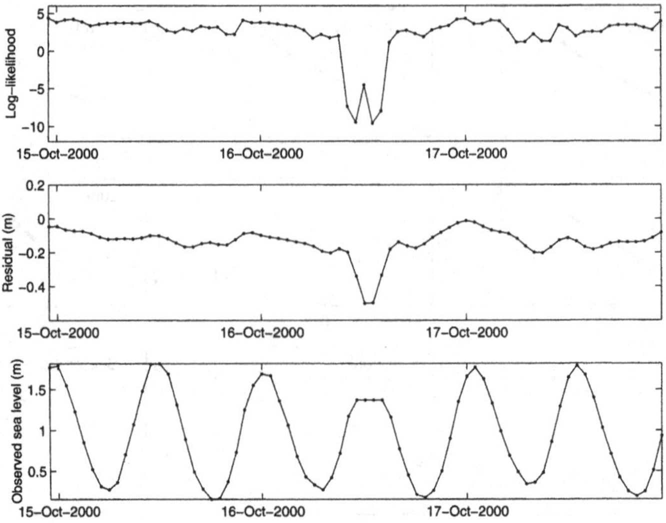

Figure 14.7 *Top panel: Log-likelihood of pairs of adjacent residuals,* (x_t, x_{t+1}) *for a 6-day period in October 2000. The log-likelihood at time t was obtained from the density of the fitted, bivariate skew-t evaluated at* (x_t, x_{t+1}). *Middle panel: Residual,* x_t, *for the same period. Bottom panel: Observed hourly sea levels for the same period.*

log-likelihoods: a tide that has been "chopped off" and replaced by 4 constant values. This is probably the result of interpolating the raw sea levels over a gap. Clearly these data must be treated as suspect. The main point of this plot is not to show that the data for this day are strange, but rather to demonstrate that the skew-t, through the associated log-likelihood function, is potentially useful in checking the quality of sea level observations.

14.5.2 Detecting Secular Changes in the Sea Level Distribution

As speculation about climate change grows, it is important to check if changes have already taken place. One attraction of the above skew-t density is that it has only 6 parameters and these can be readily tested for trends either by (i) adding terms of the form θt to (14.9), or (ii) fitting the density to blocks of data and fitting a trend line through the resulting sequence of parameter estimates. We used the second approach on four consecutive 20-year blocks starting in 1920 and ending in 1999. Overall the skew-t parameters are quite stable (not shown) and so there is no strong evidence for secular changes in the residual distribution at Halifax over the last 80 years. This is consistent with our earlier remark that the trends in the Halifax annual maxima are due primarily to changes in mean level and not higher frequency phenomena such as storms.

14.5.3 Estimating Flooding Risk

Following Genton and Thompson (2003), let x_c denote some specified critical level and let $p_{12\ldots n}$ denote the probability that all hourly sea levels between hours 1 and n are below it. Another way of describing this event is by saying that the maximum sea level over the period is below x_c.

The asymptotic distribution theory of the maxima of stationary processes is quite complete (e.g., Leadbetter, Lindgren, and Rootzén, 1983). For example, the limiting distribution of the maximum of an independent and identically distributed (iid) sequence is known to be one of three extreme value types. The limiting distribution of the maximum of a dependent sequence is the same as that of the corresponding iid sequence if stationarity is assumed and conditions are imposed to limit long-range dependence at high levels and clustering of exceedances. These conditions are made precise by the conditions $D(u_n)$ and $D'(u_n)$ of Leadbetter *et al.* (1983).

The probability that the maximum sea level between times 1 and n is below x_c can be written

$$p_{12\ldots n} = p_1 \, p_{2|1} \, p_{3|2,1} \, \cdots \, p_{n|n-1,\ldots,1} \tag{14.10}$$

where $p_{t|t-1,t-2,\ldots,1}$ is the probability that sea level is below x_c at time t, given it was also below this level for earlier times. In this study, (14.10) has been approximated as follows:

$$p_{12\ldots n} \approx p_{M+1}^{(M)} \, p_{M+2}^{(M)} \, p_{M+3}^{(M)} \, \cdots \, p_{n-1}^{(M)} \, p_n^{(M)}, \tag{14.11}$$

where $p_t^{(M)} = p_{t|t-1,t-2,\ldots,t-M}$. Based on the asymptotic theory referenced above, $M = 0$ is expected to give a reasonable approximation of $p_{12\ldots n}$ at high levels if the residual process is stationary and conditions $D(u_n)$ and $D'(u_n)$ apply. The sensitivity of flooding risk to changing M from 1 to 0 is explored below.

To evaluate $p_t^{(M)}$, we decompose sea level into a tide (x_T) and residual (x) as before. Sea level below the critical value is equivalent to $x < x_c - x_T$, i.e., the residual is not large enough to "jump the gap" between the predicted tide and the critical level. Through this decomposition, $p_t^{(M)}$ can be expressed in terms of x_c, the predicted tide and the residual density which, in turn, is given by an $M + 1$ dimensional skew-t distribution fit to the residuals.

The result of a typical calculation of $p_{12\ldots n}$ is shown by the trace labeled "0 mm per year" in Figure 14.8. The trend in sea level was assumed zero and the critical level was taken to be 1.81 m, the maximum level observed at Halifax between 1920 and 2001 relative to its annual mean. Overall the trace

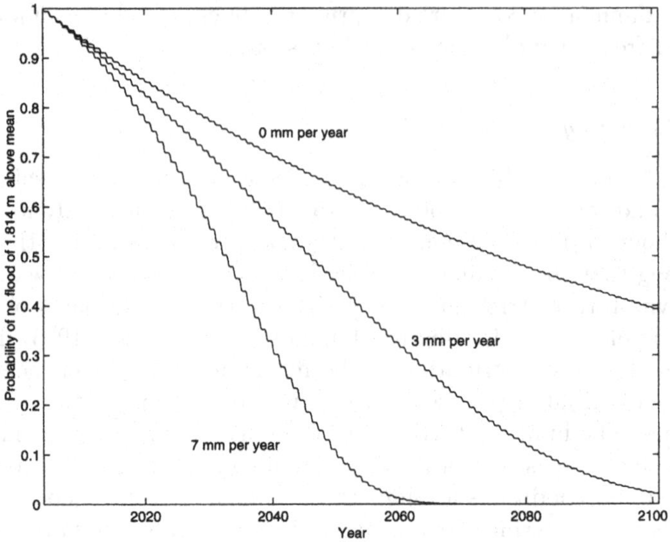

Figure 14.8 *Flooding risk at Halifax over the next century. The x-axis shows time. The y-axis shows the probability of not exceeding 1.81 m, the maximum level observed between 1920 and 2001. The line labeled 0 mm per year assumes no increase in annual mean sea level over the next century. The other two lines are for linear increases of 3 and 7 mm per year.* $M = 1$ *was used for all three cases.*

exhibits an exponential-like decline upon which are superimposed small steps. The steps occur in winter when the residuals tend to be larger because of storm activity, and the probability of an exceedance of x_c greater.

According to Figure 14.8, the probability that sea level will not exceed 1.81 m between 2004 and 2100 is 0.391. This is also the probability that all 97 annual maxima over this period are below x_c. If we assume the annual maxima are independent, and 0.391 is indeed the probability that all annual maxima are below x_c, the probability that a single annual maximum will not exceed x_c is $0.391^{1/97} = 0.990$. This in turn implies a return period for this critical level of $1/(1 - 0.990) = 100$ years. Similar calculations have been performed for a range of x_c and the resulting return periods plotted in Figure 14.9. The agreement between the return periods estimated from the observed annual maxima and the skew-t distribution, fit to all the residuals, is quite encouraging. Also shown on this figure are return periods calculated from 20-year blocks of hourly sea level. Again the results are in reasonable agreement with the annual maxima suggesting that 100-year return levels can indeed be estimated by only 20 years of data if the tide and residual processes are treated separately, as proposed by Pugh and Vassie (1980).

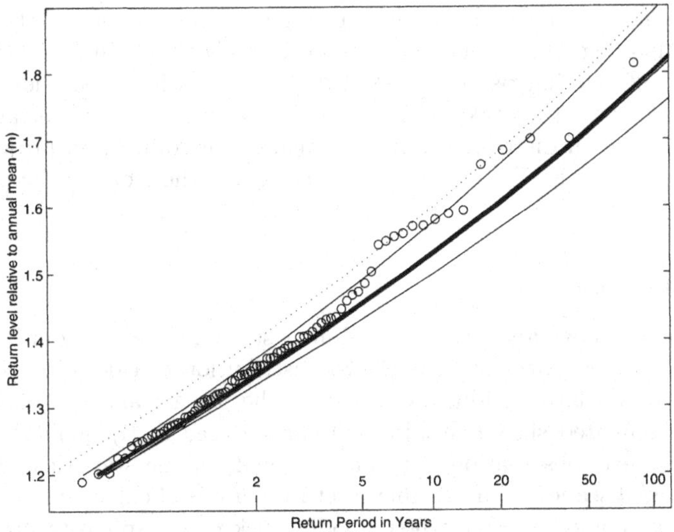

Figure 14.9 *Return period plot for Halifax. The circles are observed annual maxima plotted as in Figure 14.2. The lines are estimates of return periods based on seasonally varying, skew-t distributions fit to the residuals. The thick line is based on residuals for 1920–2001 and $M = 1$. The dotted line is for residuals from the same period and $M = 0$. The four thin lines are estimates based on distinct, 20 year blocks of residuals, all with $M = 1$. The upper and lower thin lines are for 1960–1979 and 1940–1959, respectively; there are two additional thin lines, corresponding to 1920–1939 and 1980–1999, but they are almost completely obscured by the thick line.*

The above calculations were all based on $M = 1$. To explore the sensitivity of flooding risk to changes in M, return periods were also calculated with $M = 0$, i.e., approximating $p_{12...n}$ by $p_1 p_2 \cdots p_n$. The result is shown in Figure 14.9. Comparing return periods estimated with $M = 0$ (dotted line) and $M = 1$ (thick line), it can be seen that the results are not strongly sensitive to the choice of M but that, overall, better estimates are obtained with $M = 1$.

One advantage of using (14.10) to calculate flooding risk is that it can accommodate trends in mean sea level and the residual distribution. The effect of adding a trend of 3 mm per year (the present rate of rise at Halifax) and 7 mm per year (the most likely rate based on IPCC projections) is also shown in Figure 14.8. These results show that if the IPCC projections are indeed correct, there is a 70% chance that Halifax will experience a "1967 flood" at least once by the year 2040. Similar curves could be added to this figure to show the sensitivity of flooding risk to changes in the distribution of the residuals (e.g., increasing storm surge variance). At this stage of analysis, Figure 14.8 should be viewed simply as an indication of the sensitivity of flooding risk to changing sea level trends rather than a definitive prediction of flooding risk over the next century. Such a prediction would require error bars that take into account, for example, the uncertainty in the fitted residual distribution and the many uncertainties in projections of sea level rise (e.g., IPCC, 2001). This is beyond the scope of the present study.

14.6 Discussion

Based on the above analysis, and a similar study by Genton and Thompson (2003), it appears that the skew-t distribution can describe the distribution of sea level residuals quite well. The present analysis illustrates how the estimated skew-t distribution can be used to (i) quality control hourly sea level observations, (ii) detect trends in the statistical properties of coastal sea level, (iii) estimate return periods of extreme levels from short records, and (iv) assess how flooding risk may respond to projected changes in the rate of rise of global sea level.

There are a number of ways in which this study can be improved. Perhaps the most important is the way the skew-t parameters have been estimated. In this study the dependence was dealt with by sub-sampling. This is not ideal and we are presently trying to develop a more rigorous approach that will allow for calculation of standard errors for the skew-t parameters, and formal hypothesis testing.

There are also a number of ways in which this study could be extended. For example we have ignored the possibility that the tide and surge "interact." This is a well-known physical phenomenon that can occur when the ocean offshore of the tide gauge is fairly shallow. It arises from nonlinearities in the governing physical equations. In principle, the skew-t parameterization could be extended to allow the tide to play a role in shaping the density. Another extension would be to use trivariate (and possibly higher order) distributions. It would be interesting to examine the effect on the quality control of the observations. Finally it may be better to model the observation and timing errors rather than attempting to filter them out. This would not be a trivial task and it is not clear at this point if it would outperform the simple bandpass filtering approach used here.

With regard to future work it would appear worthwhile extending the above analysis to include a large number of coastal tide gauges and checking for regional to global scale changes in the distribution of sea level over the last century. It may also be worthwhile using information from recent climate change scenarios on the frequency and severity of storms. Together, these two extensions could provide a better idea of how flooding risk has changed over the last century, and may change over the next.

CHAPTER 15

Time Series Analysis with a Skewed Kalman Filter

Philippe Naveau, Marc G. Genton, and Caspar Ammann

15.1 Introduction

The analysis of time series has always played a central role in statistics. In this chapter, we focus on temporal state-space models for time series that can be decomposed in a sum of three components: a smooth trend, a skewness part, and noise. The first one is supposed to be deterministic and the other two random. Such time series decompositions are common in paleoclimate studies. The central element of the estimation procedure is the Kalman filter that has to be modified to deal with skewness.

The overwhelming assumption of normality in the Kalman filter literature can be understood for many reasons. For example, the Bayesian derivation of the classical Kalman filter operations (Meinhold and Singpurwalla, 1983) needs normality. Note that the very same expressions can be obtained through different means, e.g., by solving a least-squares problem. A major reason to use the normal distribution is that the multivariate distribution is completely characterized by its first two moments. In addition, the stability of the multivariate normal distribution under summation and conditioning offers tractability and simplicity. Therefore, the Kalman filter operations can be performed easily whenever the normality assumption holds. However, this assumption is not satisfied for a large number of applications. For example, some distributions used in a state-space model can be skewed.

There have been many attempts to deal with non-Gaussian state-space models in the past and Monte Carlo based non-linear and non-Gaussian filtering and smoothing have been extensively studied recently; see, e.g., the bibliography in Kitagawa and Gersch (1984), and the large literature survey listed in Schick's (1989) thesis. To name a few, Smith and Miller (1986), Carlin, Polson, and Stoffer (1992), and Meinhold and Singpurwalla (1989), have proposed alternative approaches to the classical Kalman filter. Meinhold and Singpurwalla (1989) assumed a multivariate distribution with Student t marginals. Carlin *et al.* (1992) proposed a methodology based on normal scale mixtures. Smith and Miller (1986) worked with exponential variables conditionally on unobserved variables. A common characteristic between these three studies is that the cost associated with their departures from normality is fairly high, stability under addition and conditioning being lost.

As an alternative to these methods, we recently developed a skewed Kalman filter (Naveau, Genton, and Shen, 2004) by working with closed skew-normal distributions. Besides introducing skewness to the normal distribution, this class of distribution has the advantage of being closed under marginalization and conditioning. It has been introduced by Domínguez-Molina, González-Farías, and Gupta (2003), and is an extension of the multivariate skew-normal distribution first proposed by Azzalini and his coworkers (Azzalini, 1985, 1986; Azzalini and Dalla Valle, 1996; Azzalini and Capitanio, 1999). These distributions are particular types of generalized skew-elliptical distributions recently introduced by Genton and Loperfido (2002), i.e., they are defined as the product of a multivariate elliptically contoured density with a skewing function. At this juncture, we would like to stress that the main advantage of our proposed method is that the skewness in Kalman filter operations can be modeled by a few parameters whose temporal behavior can be explicitly described when the skew-normal distribution is used, see Section 15.3.2.

Besides recalling the basic framework of our skewed Kalman filter, our main goal in this chapter is to present the advantages of introducing skewness when analyzing time series with a Kalman filter. We focus on applications for which a time series can be decomposed into a smooth trend and skewed perturbations.

This chapter is organized as follows. In Section 15.2, we recall the basic steps of the Kalman filter and of the state-space modeling in the Gaussian case. Section 15.3 presents the definition of the closed skew-normal distribution and the Kalman filter procedures for such a distribution is proposed to estimate the temporal evolution of the state vector. In Section 15.4, we illustrate our approach by looking at the climatic impacts of large explosive volcanic eruptions (strong but short-lived perturbations) on temperature time series over a few centuries (smooth trend). In Section 15.5, we summarize our findings and discuss future work.

15.2 The Classical Kalman Filter

15.2.1 State-Space Model

The *state-space model* has been widely studied (e.g., West and Harrison, 1997; Shephard, 1994; Shumway and Stoffer, 1991; Harrison and Stevens, 1971, 1976). This model has become a powerful tool for modeling and forecasting dynamical systems and it has been used in a wide range of disciplines such as biology, economics, engineerings, and statistics (Guo, Wang, and Brown, 1999; Kitagawa and Gersch, 1984). The basic idea of the state-space model is that the d-dimensional vector of observation $\mathbf{y}_t$ at time t is generated by two equations, the *observational* and the *system* equations. The first equation describes how the observations vary in function of the unobserved state vector $\mathbf{x}_t$ of length h:

$$\mathbf{y}_t = F_t\mathbf{x}_t + \boldsymbol{\epsilon}_t, \tag{15.1}$$

where $\boldsymbol{\epsilon}_t$ represent an added noise and F_t is a $d \times h$ matrix of scalars. The essential difference between the state-space model and the conventional linear model is that the state vector $\mathbf{x}_t$ is not assumed to be constant but may change in time. The temporal dynamical structure is incorporated via the system equation:

$$\mathbf{x}_t = G_t\mathbf{x}_{t-1} + \boldsymbol{\eta}_t, \tag{15.2}$$

where $\boldsymbol{\eta}_t$ represents an added noise and G_t is an $h \times h$ matrix of scalars. There exists an extensive literature about the estimation of the parameters for such models. In particular, the Kalman filter provides an optimal way to estimate the model parameters if the assumption of gaussianity holds. Following the definition by Meinhold and Singpurwalla (1983), the term "Kalman filter" used in this work refers to a recursive procedure for inference about the state vector. To simplify the exposition, we assume that the observation errors $\boldsymbol{\epsilon}_t$ are independent of the state errors $\boldsymbol{\eta}_t$ and that the sampling is equally spaced, $t = 1, \ldots, n$. The results shown in this chapter could be easily extended without the latter constraint. But, the loss of clarity in the notations would make this work more difficult to read without bringing any new important concepts.

15.2.2 The Kalman Filter Procedure in the Gaussian Case

In this section, we recall the basic steps of the Gaussian Kalman filter. Following the work of Meinhold and Singpurwalla (1983), we use a Bayesian formulation to present the different stages of the Kalman filtering. The key notion is that given the data $\vec{\mathbf{y}}_t = (\mathbf{y}_1^T, \ldots, \mathbf{y}_t^T)^T$, inference about the state vector values can be carried out through a direct application of Bayes' theorem. In the Gaussian case, the conditional distribution of $(\mathbf{x}_{t-1}|\vec{\mathbf{y}}_{t-1})$ is assumed to follow a Gaussian distribution at time $t-1$

$$(\mathbf{x}_{t-1}|\vec{\mathbf{y}}_{t-1}) \sim N_h(\hat{\boldsymbol{\psi}}_{t-1}, \hat{\Omega}_{t-1}), \tag{15.3}$$

where $\hat{.}$ represents the posterior mean and variance parameters of the variable $(\mathbf{x}_{t-1}|\vec{\mathbf{y}}_{t-1})$. Then, we look forward in time, but in two stages: prior to observing $\mathbf{y}_t$, and after observing $\mathbf{y}_t$. Suppose that the initial state vector $\mathbf{x}_0$ of the system composed by equations (15.1) and (15.2) follows a Gaussian distribution, $N_h(\boldsymbol{\psi}_0, \Omega_0)$ and that the noise $\boldsymbol{\epsilon}_t$, respectively, $\boldsymbol{\eta}_t$, is an independent and identically distributed (iid) Gaussian vector with mean $\boldsymbol{\mu}_\epsilon$ and covariance Σ_ϵ, respectively, $\boldsymbol{\mu}_\eta$ and Σ_η.

Prior to observing $\mathbf{y}_t$, our best guess for $\mathbf{x}_t$ comes from Equation (15.2) that gives $(\mathbf{x}_t|\mathbf{y}_{t-1}) \sim N_h(\tilde{\psi}_t, \tilde{\Omega}_t)$ with

$$\begin{cases} \tilde{\psi}_t &= G_t\hat{\psi}_{t-1} + \boldsymbol{\mu}_{\boldsymbol{\eta}}, \\ \tilde{\Omega}_t &= G_t\hat{\Omega}_{t-1}G_t^T + \Sigma_{\boldsymbol{\eta}}. \end{cases} \tag{15.4}$$

After observing $\mathbf{y}_t$, the parameters of the posterior distribution of $\mathbf{x}_t$ defined by (15.3) are computed through the next cycle by the following sequential procedure:

$$\begin{cases} \hat{\psi}_t &= \tilde{\psi}_t + \tilde{\Omega}_t F_t^T \Sigma_t^{-1}\mathbf{e}_t, \\ \hat{\Omega}_t &= \tilde{\Omega}_t - \tilde{\Omega}_t F_t^T \Sigma_t^{-1} F_t \tilde{\Omega}_t, \end{cases} \tag{15.5}$$

where $\Sigma_t = \Sigma_{\boldsymbol{\epsilon}} + F_t\tilde{\Omega}_t F_t^T$ and $\mathbf{e}_t$ denotes the error in predicting $\mathbf{y}_t$ from the point $t-1$:

$$\mathbf{e}_t = \mathbf{y}_t - F_t\tilde{\psi}_t - \boldsymbol{\mu}_{\boldsymbol{\epsilon}}.$$

This series of equations, (15.4) and (15.5), constitutes the Kalman filtering steps for the Gaussian state-space model.

15.3 A Skewed Kalman Filter

At this stage, the main question is how to introduce some skewness in the Kalman filter presented in the previous section. In particular, we would like to keep the simplicity of Equations (15.4) and (15.5), while introducing new equations for the skewness parameters. In this section, we present the principal steps to reach such a goal. The proofs of the mathematical results presented therein can be found in Naveau *et al.* (2004).

15.3.1 The Closed Skew-Normal Distribution

The closed skew-normal distribution is a family of distributions including the normal one, but with extra parameters to regulate skewness. It allows for a continuous variation from normality to non-normality, which is useful in many situations.

An n-dimensional random vector $\mathbf{x}$ is said to have a multivariate closed skew-normal distribution (Domínguez-Molina *et al.*, 2003; González-Farías, Domínguez-Molina, and Gupta, 2003), denoted by $CSN_{n,m}(\boldsymbol{\mu}, \Sigma, D, \boldsymbol{\nu}, \Delta)$, if it has a density function of the form

$$\frac{1}{\Phi_m(\mathbf{0}; \boldsymbol{\nu}, \Delta + D\Sigma D^T)}\phi_n(\mathbf{x}; \boldsymbol{\mu}, \Sigma)\Phi_m(D(\mathbf{x}-\boldsymbol{\mu}); \boldsymbol{\nu}, \Delta), \qquad \mathbf{x} \in \mathbb{R}^n, \tag{15.6}$$

where $\boldsymbol{\mu} \in \mathbb{R}^n$, $\boldsymbol{\nu} \in \mathbb{R}^m$, $\Sigma \in \mathbb{R}^{n\times n}$ and $\Delta \in \mathbb{R}^{m\times m}$ are both covariance matrices, $D \in \mathbb{R}^{m\times n}$, $\phi_n(\mathbf{x}; \boldsymbol{\mu}, \Sigma)$ and $\Phi_n(\mathbf{x}; \boldsymbol{\mu}, \Sigma)$ are the n-dimensional normal probability density function (pdf) and cumulative distribution function (cdf) with mean $\boldsymbol{\mu}$ and covariance matrix Σ. When $D = O$, the zero matrix, the density (15.6) reduces to the multivariate normal one, whereas it reduces to Azzalini and Capitanio's (1999) density when $m = 1$ and $\boldsymbol{\nu} = D\boldsymbol{\mu}$. The matrix parameter D is referred to as a "shape parameter." The moment generating function $M_{n,m}(\boldsymbol{t})$ for a CSN distribution is given by

$$M_{n,m}(\boldsymbol{t}) = \frac{\Phi_m(D\Sigma\boldsymbol{t}; \boldsymbol{\nu}, \Delta + D\Sigma D^T)}{\Phi_m(\boldsymbol{0}; \boldsymbol{\nu}, \Delta + D\Sigma D^T)} \exp(\boldsymbol{\mu}^T\boldsymbol{t} + \frac{1}{2}\boldsymbol{t}^T\Sigma\boldsymbol{t}), \tag{15.7}$$

for any $\boldsymbol{t} \in \mathbb{R}^n$. This expression of the moment generating function is important to understanding the closure properties of the CSN distribution for summation. It is straightforward to see that the sum of two CSN of dimension (n, m) is not another CSN of dimension (n, m). Despite this limitation, it is possible to show that the sum of two CSN of dimension (n, m) is a CSN of dimension $(n, 2m)$; see González-Farías *et al.* (2003). Hence the CSN is closed under summation whenever the dimension m is allowed to vary. Although important for specific applications (adding a relatively small number of variables), this closure property is not appropriate when dealing with state-space models. These models are based on sequential transformations from time $t-1$ to time t. Implementing a sum at each time step rapidly increases the dimension m and the sizes of the matrix Δ and D quickly become unmanageable. For this reason, we will propose a new way of introducing skewness without paying this dimensionality cost.

15.3.2 Extension of the Linear Gaussian State-Space Model

Our strategy to derive a model with a more flexible skewness is to directly incorporate a term for skewness into the observation equation

$$\begin{aligned} \mathbf{y}_t &= F_t\mathbf{x}_t + \boldsymbol{\epsilon}_t \qquad (15.8)\\ &= Q_t\mathbf{u}_t + P_t\mathbf{s}_t + \boldsymbol{\epsilon}_t, \text{ with } F_t = (Q_t, P_t) \text{ and } \mathbf{x}_t = (\mathbf{u}_t^T, \mathbf{s}_t^T)^T, \\ &= \text{ Linear Part } + \text{ Skewed Part } + \text{ Gaussian noise}, \end{aligned}$$

where the random vector $\mathbf{u}_t$ of length k and the $d \times k$ matrix of scalar Q_t represent the linear part of the observation equation. In comparison, the random vector $\mathbf{s}_t$ of length l and the $d \times l$ matrix of scalar P_t correspond to the additional skewness. We have $h = k + l$. Hence, the observational time series has a temporal structure driven by two random components, a linear Gaussian part and a skewed-Gaussian component. An iid Gaussian random noise is added to this structure. The most difficult task in this construction is to propose a simple dynamical structure of the skewness vector $\mathbf{s}_t$ and the "linear" vector $\mathbf{u}_t$ while keeping the independence between these two vectors (the last condition is not theoretically necessary but it is useful when interpreting the parameters). To reach this goal, we suppose that the random vector $(\mathbf{u}_t^T, \mathbf{v}_t^T)^T$ is generated from a linear system

$$\begin{cases} \mathbf{u}_t &= K_t \mathbf{u}_{t-1} + \boldsymbol{\eta}_t^*, \\ \mathbf{v}_t &= -L_t \mathbf{v}_{t-1} + \boldsymbol{\eta}_t^+, \end{cases} \tag{15.9}$$

where the Gaussian noise $\boldsymbol{\eta}_t^* \sim N_k(\boldsymbol{\mu}_\eta^*, \Sigma_\eta^*)$ is independent of the Gaussian noise $\boldsymbol{\eta}_t^+ \sim N_l(\boldsymbol{\mu}_\eta^+, \Sigma_\eta^+)$, and where K_t , respectively L_t, represents a $k \times k$ matrix of scalars, respectively an $l \times l$ matrix of scalars. In this chapter, we take $l = 1$ in order to simplify the notations and this special case suffices for our applications in Section 15.4.

To continue our construction of the system, a few notations are needed. The multivariate normal distribution of the vector $(\mathbf{u}_t^T, \mathbf{v}_t^T)^T$ is denoted by

$$\begin{pmatrix} \mathbf{u}_t \\ \mathbf{v}_t \end{pmatrix} \sim N_{k+l}\left(\begin{pmatrix} \boldsymbol{\psi}_t^* \\ \boldsymbol{\psi}_t^+ \end{pmatrix}, \begin{pmatrix} \Omega_t^* & O \\ O & \Omega_t^+ \end{pmatrix}\right). \tag{15.10}$$

The parameters of such vectors can be sequentially derived from an initial vector $(\mathbf{u}_0^T, \mathbf{v}_0^T)^T$ with a normal distribution.

Let $D_t^+ = \Omega_{t-1}^+ L_t^T (\Omega_t^+)^{-1}$, $\boldsymbol{\psi}_t^+ = -L_t \boldsymbol{\psi}_{t-1}^+ + \boldsymbol{\mu}_\eta^+$, and define $\Phi_{(t)}(\cdot) = \Phi_1(\cdot, \boldsymbol{\psi}_t^+, \Omega_t^+)$. The skewness part $\mathbf{s}_t$ of the state vector $\mathbf{x}_t = (\mathbf{u}_t^T, \mathbf{s}_t^T)^T$ is defined as

$$\mathbf{s}_t = \boldsymbol{\eta}_t^+ - L_t \mathbf{w}_t, \tag{15.11}$$

where the vector $\mathbf{w}_t$ is defined in function of $\mathbf{v}_{t-1}$ as follows. If $D_t^+ \boldsymbol{\psi}_t^+ \leq \boldsymbol{\psi}_{t-1}^+$, then

$$\mathbf{w}_t = \begin{cases} \mathbf{v}_{t-1}, & \text{if } \mathbf{v}_{t-1} \leq D_t^+ \boldsymbol{\psi}_t^+, \\ 2\boldsymbol{\psi}_{t-1}^+ - \mathbf{v}_{t-1}, & \text{if } \mathbf{v}_{t-1} > 2\boldsymbol{\psi}_{t-1}^+ - D_t^+ \boldsymbol{\psi}_t^+, \quad \text{otherwise} \\ \Phi_{(t-1)}^{-1}\left(\Phi_{(t-1)}(D_t^+ \boldsymbol{\psi}_t^+) \frac{\Phi_{(t-1)}(\mathbf{v}_{t-1}) - \Phi_{(t-1)}(D_t^+ \boldsymbol{\psi}_t^+)}{\Phi_{(t-1)}(2\boldsymbol{\psi}_{t-1}^+ - D_t^+ \boldsymbol{\psi}_t^+) - \Phi_{(t-1)}(D_t^+ \boldsymbol{\psi}_t^+)}\right), \end{cases} \tag{15.12}$$

otherwise,

$$\mathbf{w}_t = \begin{cases} \mathbf{v}_{t-1}, & \text{if } \mathbf{v}_{t-1} \leq D_t^+ \boldsymbol{\psi}_t^+, \quad \text{otherwise} \\ \Phi_{(t-1)}^{-1}\left(\Phi_{(t-1)}(D_t^+ \boldsymbol{\psi}_t^+) \frac{\Phi_{(t-1)}(\mathbf{v}_{t-1}) - \Phi_{(t-1)}(D_t^+ \boldsymbol{\psi}_t^+)}{1 - \Phi_{(t-1)}(D_t^+ \boldsymbol{\psi}_t^+)}\right). \end{cases} \tag{15.13}$$

With these definitions, the variable $\mathbf{s}_t$ follows a closed skew-normal distribution $\mathbf{s}_t \sim CSN_{l,1}(\boldsymbol{\psi}_t^+, \Omega_t^+, D_t^+, \boldsymbol{\nu}_t^+, \Delta_t^+)$, where we have $\boldsymbol{\nu}_t^+ = \boldsymbol{\psi}_{t-1}^+ - D_t^+ \boldsymbol{\psi}_t^+$, $\Delta_t^+ = \Omega_{t-1}^+ - D_t^+ \Omega_t^+ (D_t^+)^T$, and $\Omega_t^+ = L_t \Omega_{t-1}^+ L_t^T + \Sigma_\eta^+$. Although Equations (15.12) and (15.13) may look complex, it is easy to show that $\mathbf{w}_t$ has the same distribution as $[\mathbf{v}_{t-1} | \mathbf{v}_{t-1} \leq D_t^+ \boldsymbol{\psi}_t^+]$. It follows from (15.9) that the vector $\mathbf{s}_t$ defined from Equation (15.11) has the same distribution as $[\mathbf{v}_t | \mathbf{v}_{t-1} \leq D_t^+ \boldsymbol{\psi}_t^+]$. The former variable is usually used as a more classical definition of a skew-normal vector (Domínguez-Molina *et al.*, 2003) when there is not temporal structure. The advantage of using $\mathbf{w}_t$ over the classical definition is that it is much easier to deal with than the event $E_t = \{\mathbf{v}_{t-1} \leq D_t^+ \boldsymbol{\psi}_t^+\}$ that rarely occurs for all t (except for a very few special cases). Re-drawing $\mathbf{v}_t$ such that E_t is satisfied for all t will imply re-running the whole history of $\mathbf{v}_t$ (with a very high computational cost, even for small t). Consequently, defining the skewness vector $\mathbf{s}_t$ directly from E_t, although possible in theory, is not a manageable solution in practice. In comparison, the vector $\mathbf{w}_t$ allows us to deal easily with $\overline{E}_t = \{\mathbf{v}_{t-1} > D_t^+ \boldsymbol{\psi}_t^+\}$. In this situation, a "folding-type" transformation (see Corcoran and Schneider, 2003) is used to "force" the event E_t to occur, via equations (15.12) and (15.13).

It is possible to show that the state vector also has a closed skew-normal distribution $\mathbf{x}_t = (\mathbf{u}_t^T, \mathbf{s}_t^T)^T \sim CSN_{k+l,k+l}(\boldsymbol{\psi}_t, \Omega_t, D_t, \boldsymbol{\nu}_t, \Delta_t)$ with

$$\boldsymbol{\psi}_t = \begin{pmatrix} \boldsymbol{\psi}_t^* \\ \boldsymbol{\psi}_t^+ \end{pmatrix}, \ \Omega_t = \begin{pmatrix} \Omega_t^* & O \\ O & \Omega_t^+ \end{pmatrix}, \text{ and}$$

$$D_t = \begin{pmatrix} O & O \\ O & D_t^+ \end{pmatrix}, \boldsymbol{\nu}_t = \begin{pmatrix} \mathbf{0} \\ \boldsymbol{\nu}_t^+ \end{pmatrix}, \Delta_t = \begin{pmatrix} I_k & O \\ O & \Delta_t^+ \end{pmatrix}.$$

Hence, the variable $\mathbf{s}_t$ through the matrix L_t introduces at each time step a different skewness (if needed) in the state vector whose temporal structure is defined by $\mathbf{v}_t$ in (15.9). The price for this gain in skewness flexibility is that this state vector (because of (15.12) and (15.13)) does not have any more a linear structure like the one defined by (15.2).

Figure 15.1 illustrates how the temporal evolution of the skewness introduced by L_t (top-left panel) changes the temporal distribution of $\mathbf{s}_t$ (see QQ-plots in bottom panels).

15.3.3 The Steps of our Skewed Kalman Filter

For the skewed Kalman filter, we follow the sequential approach described in Section 15.2.2, but instead of assuming (15.3), we suppose the following conditional distribution at time $t-1$:

$$(\mathbf{x}_t | \vec{\mathbf{y}}_t) \sim CSN_{k+l,k+l}(\hat{\boldsymbol{\psi}}_t, \hat{\Omega}_t, \hat{D}_t, \hat{\boldsymbol{\nu}}_t, \hat{\Delta}_t), \tag{15.14}$$

$$\text{with } \hat{\boldsymbol{\psi}}_t = \begin{pmatrix} \hat{\boldsymbol{\psi}}_t^* \\ \hat{\boldsymbol{\psi}}_t^+ \end{pmatrix}, \ \hat{\Omega}_t = \begin{pmatrix} \hat{\Omega}_t^* & O \\ O & \hat{\Omega}_t^+ \end{pmatrix},$$

$$\hat{D}_t = \begin{pmatrix} O & O \\ O & \hat{D}_t^+ \end{pmatrix}, \hat{\boldsymbol{\nu}}_t = \begin{pmatrix} \mathbf{0} \\ \hat{\boldsymbol{\nu}}_t^+ \end{pmatrix}, \text{ and } \hat{\Delta}_t = \begin{pmatrix} I_k & O \\ O & \hat{\Delta}_t^+ \end{pmatrix}.$$

The main steps explaining the derivation of the skewed Kalman filter steps described in this section can be found in the Appendix. As presented in Section 15.2.2, we look at the evolution of the posterior distribution in two stages: prior to observing $\mathbf{y}_t$, and after observing $\mathbf{y}_t$. For the former,

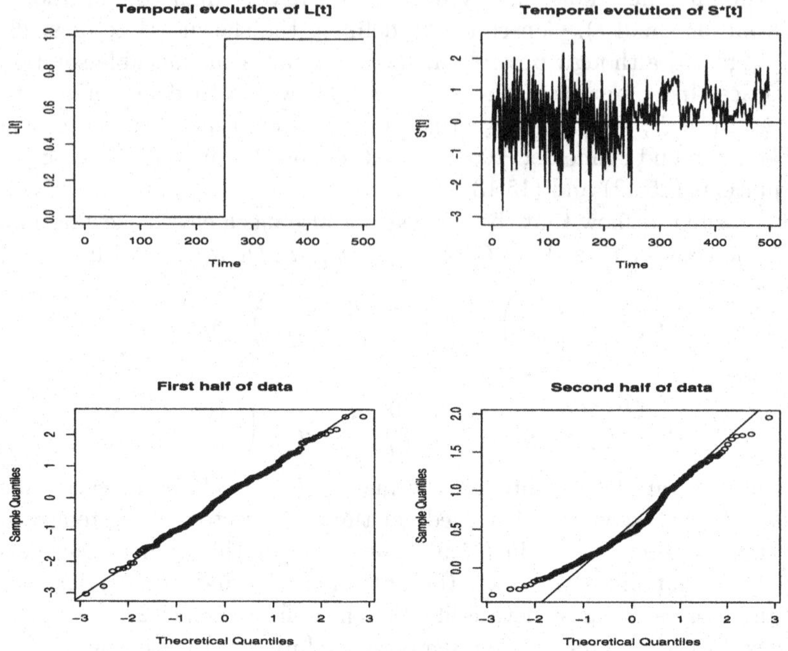

Figure 15.1 *Temporal evolution of* $\mathbf{s}_t$ *in function of* L_t *(top-left panel) and QQ-plot of the data after and before* $t = 250$*, bottom left and right panels, respectively.*

the step described by (15.4) for the first and second moments becomes

$$\begin{array}{ll} \text{Linear part} & \left\{ \begin{array}{ll} \tilde{\boldsymbol{\psi}}_t^* & = K_t \hat{\boldsymbol{\psi}}_{t-1}^* + \boldsymbol{\mu}_{\boldsymbol{\eta}}^*, \\ \tilde{\Omega}_t^* & = K_t \hat{\Omega}_{t-1}^* K_t^T + \Sigma_{\boldsymbol{\eta}}^*, \end{array} \right. \\ \text{Skewed part} & \left\{ \begin{array}{ll} \tilde{\boldsymbol{\psi}}_t^+ & = -L_t \hat{\boldsymbol{\psi}}_{t-1}^+ + \boldsymbol{\mu}_{\boldsymbol{\eta}}^+, \\ \tilde{\Omega}_t^+ & = L_t \hat{\Omega}_{t-1}^+ L_t^T + \Sigma_{\boldsymbol{\eta}}^+. \end{array} \right. \end{array} \tag{15.15}$$

In addition, the skewness parameters are equal to

$$\left\{ \begin{array}{ll} \tilde{D}_t^+ & = \hat{\Omega}_{t-1}^+ L_t^T (\tilde{\Omega}_t^+)^{-1}, \\ \tilde{\boldsymbol{\nu}}_t^+ & = \hat{\boldsymbol{\psi}}_{t-1}^+ - \tilde{D}_t^+ \tilde{\boldsymbol{\psi}}_t^+, \\ \tilde{\Delta}_t^+ & = \hat{\Omega}_{t-1}^+ - \tilde{D}_t^+ \tilde{\Omega}_t^+ (\tilde{D}_t^+)^T. \end{array} \right. \tag{15.16}$$

After observing $\mathbf{y}_t$, the first and second moments instead of following

(15.5) are now defined by

$$
\begin{array}{ll}
\text{Linear part} & \left\{ \begin{array}{ll} \hat{\boldsymbol{\psi}}_t^* & = \tilde{\boldsymbol{\psi}}_t^* + \tilde{\Omega}_t^* Q_t^T \Sigma_t^{-1} \mathbf{e}_t, \\ \hat{\Omega}_t^* & = \tilde{\Omega}_t^* - \tilde{\Omega}_t^* Q_t^T \Sigma_t^{-1} Q_t \tilde{\Omega}_t^*, \end{array} \right. \\
\text{Skewed part} & \left\{ \begin{array}{ll} \hat{\boldsymbol{\psi}}_t^+ & = \tilde{\boldsymbol{\psi}}_t^+ + C_t P_t^T \Sigma_t^{-1} \mathbf{e}_t, \\ \hat{\Omega}_t^+ & = \tilde{\Omega}_t^+ - C_t P_t^T \Sigma_t^{-1} P_t C_t, \end{array} \right.
\end{array}
\tag{15.17}
$$

where $\Sigma_t = Q_t \tilde{\Omega}_t^* Q_t^T + P_t(\tilde{\Omega}_t^+ + \tau_t^{(2)}) P_t^T + \Sigma_\epsilon$, and the error in predicting $\mathbf{y}_t$ from time $t-1$ is now

$$
\mathbf{e}_t = \mathbf{y}_t - Q_t \tilde{\boldsymbol{\psi}}_t^* - P_t[\tilde{\boldsymbol{\psi}}_t^+ + \boldsymbol{\tau}_t^{(1)}] - \boldsymbol{\mu}_\epsilon,
$$

where C_t is the conditional covariance $C_t = \mathrm{Cov}(\mathbf{v}_t, \mathbf{s}_t | \vec{\mathbf{y}}_{t-1})$ with

$$
\boldsymbol{\tau}_t^{(1)} = \frac{\tilde{D}_t^+ \tilde{\Omega}_t^+}{\tilde{\sigma}_t} \tau\left(\frac{\tilde{\boldsymbol{\nu}}_t^+}{\tilde{\sigma}_t}\right) \text{ and } \tau_t^{(2)} = \boldsymbol{\tau}_t^{(1)} \frac{\tilde{D}_t^+ \tilde{\Omega}_t^+}{\tilde{\sigma}_t} \left[\frac{\tilde{\boldsymbol{\nu}}_t^+}{\tilde{\sigma}_t} - \tau\left(\frac{\tilde{\boldsymbol{\nu}}_t^+}{\tilde{\sigma}_t}\right)\right],
\tag{15.18}
$$

with $\tilde{\sigma}_t^2 = \tilde{\Delta}_t^+ + \tilde{D}_t^+ \tilde{\Omega}_t^+ (\tilde{D}_t^+)^T$ and the function $\tau(\cdot)$ being the hazard rate function for the standard Gaussian distribution. In addition, the skewness parameters follow

$$
\left\{ \begin{array}{ll} \hat{D}_t^+ & = \overline{\Omega}_{t-1}^+ \overline{L}_t^T (\hat{\Omega}_t^+)^{-1}, \\ \hat{\boldsymbol{\nu}}_t^+ & = \hat{\boldsymbol{\psi}}_{t-1}^+ - \hat{D}_t^+ \hat{\boldsymbol{\psi}}_t^+, \\ \hat{\Delta}_t^+ & = \overline{\Omega}_{t-1}^+ - \hat{D}_t^+ \hat{\Omega}_t^+ (\hat{D}_t^+)^T, \end{array} \right.
\tag{15.19}
$$

where $\overline{\Omega}_{t-1}^+ = \hat{\Omega}_{t-1}^+ - C_t P_t^T \Sigma_t^{-1} P_t C_t$, $\overline{L}_t = L_t + \Sigma_\nu^+ P_t^T \Sigma_t^{-1} P_t \tilde{C}_t (\overline{\Omega}_{t-1}^+)^{-1}$, and $C_t = \Sigma_\eta^+ - L_t \tilde{C}_t$.

The series of equations (15.15), (15.16), (15.17), and (15.19), constitute the Kalman filtering steps for this skewed state-space model. Although the notations are a little more complex than for the Gaussian Kalman filter, the Kalman filtering steps for this skewed state-space model do not present any computational difficulties.

To illustrate the advantage of using the skewed Kalman filter over the classical one, we show in Figure 15.2 a simple case where some skewness is introduced in time through L_t (see top panel). The bottom panel clearly indicates that the estimated state-space evolution (solid line with circles) departs from the simulated state-space vector (dotted line with crosses) when the injected skewness becomes too strong, i.e., after $L_t \geq 0.5$. In comparison, the skewed Kalman filter (middle panel) is much more capable of following the evolution of the given state vector, even when the skewness is high.

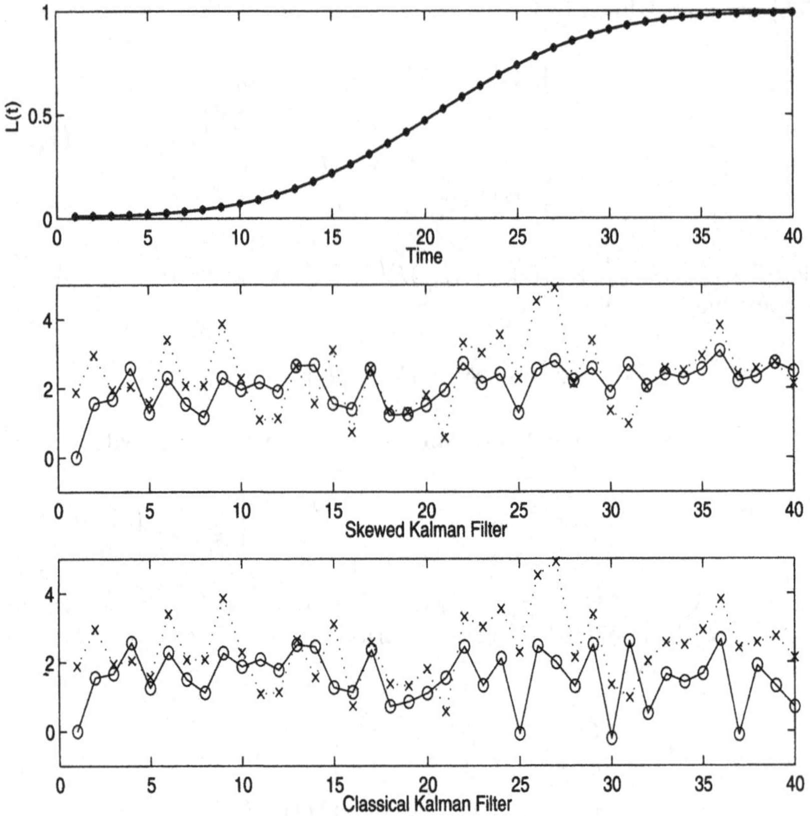

Figure 15.2 *Estimation of the temporal evolution of the state-space variable* $\mathbf{x}_t$ *by using the classical Gaussian Kalman filter (bottom panel) and by implementing the non-linear skewed Kalman filter (middle panel). The crosses represent the simulated values and the circles the estimated values. The top panel shows the introduced skewness through the time evolution of* L_t.

15.4 Applications to Paleoclimate Time Series

Externally forced natural climate variability played an important role over a large part of the 20th century climate. Unfortunately, projections into the future using a host of different scenarios do not include the natural components (solar variations, volcanic eruptions, etc.) but focus on anthropogenic elements only. Although anthropogenic forcing is expected to be dominant, the lack of natural variations is a clear deficiency in regard to uncertainty estimates of future climates.

To understand the full range of natural climate variability, it is important to attribute past climate variations to particular forcing factors. Our main focus is to study the impact of strong but short-lived perturbations from large explosive volcanic eruptions on climate. A statistical method to simultaneously model the slowly changing background climate component and the superposed volcanic pulse-like events is presented. For example, after the recent eruption of Mount Pinatubo in 1991, a climate signal was detected for about two years until the end of 1993, while the aerosol did not reach background conditions until roughly 1996. Our approach based on a statistical multi-state-space model provides an accurate estimator of the climate response to an eruption. Our goal in this application is to simulate large explosive volcanic eruptions and to assess their impacts on climate. The distribution of the derived magnitudes from these largest volcanic coolings will be modeled by a closed skew-normal distribution that gives a more realistic representation of volcanic forcing than the Gaussian distribution. This modeling can be seen as an important step toward adding forcing into future scenario runs in climate models. To learn more about the statistical analysis of volcanic forcing on climate, we refer to Naveau, Ammann, Oh, and Guo (2003).

15.4.1 Multi-Process Linear Models

To illustrate our skewed Kalman filter, we look at the following classical linear model

$$Y_t = h_t + S_t + \epsilon_t, \qquad (15.20)$$

where Y_t is the observed time series, h_t represents a long slowly changing trend, S_t corresponds to a skewed process varying in time, and ϵ_t is a Gaussian white noise. All the different components of (15.20) are assumed to be independent

In our climate application, the time series S_t corresponds to the volcanic signal that is added sporadically to the smooth climatic trend h_t due to human influence or other long term trends. The particularity of S_t is that it does not follow a Gaussian distribution but a skewed one. The next step is to model h_t and S_t into a state-space framework.

15.4.2 The Smoothing Spline Model for Trends

A smoothing spline is a natural choice for the changing baseline because of its flexibility. Wahba (1978) showed that a smoothing spline can be obtained through a signal extraction approach by introducing a stochastic smoothing spline model. More recently, Wecker and Ansley (1983) expressed this smoothing spline model in a state-space form. Following their formulation, we express the trend h_t in function of its derivatives $\mathbf{u}_t = (h_t, h_t^{(1)}, h_t^{(2)}, \ldots, h_t^{(m-1)})^T$ by writing

$$\mathbf{u}_t = K_t \mathbf{u}_{t-1} + \boldsymbol{\eta}_t^*, \qquad (15.21)$$

where $K_t[i,j] = 1/(j-i)!$ for $j \geq i$, or 0 otherwise, and $\boldsymbol{\eta}_t^*$ is an m-dimensional Gaussian random vector with zero mean and covariance elements $\lambda(\sigma_{\eta}^*)^2/[(i+j-1)(i-1)!(j-1)!]$, where λ is the smoothing parameter of the spline. A cubic spline corresponds to $m=2$ and will be used in this work. Equation (15.21) is the first part of the system defined by (15.9). More precisely, we have for a cubic spline

$$\mathbf{u}_t = \begin{pmatrix} h_t \\ h_t^{(1)} \end{pmatrix} = \begin{pmatrix} 1 & 1 \\ 0 & 1 \end{pmatrix} \mathbf{u}_{t-1} + \boldsymbol{\eta}_t^*,$$

where

$$\boldsymbol{\eta}_t^* \sim N_2\left(\begin{pmatrix} 0 \\ 0 \end{pmatrix}, \lambda(\sigma_{\eta}^*)^2 \begin{pmatrix} 1 & 1/2 \\ 1/2 & 1/3 \end{pmatrix} \right).$$

Figure 15.3 shows the estimation of a smooth trend when using the classical Kalman filter. The top panels correspond to the estimation results when a Gaussian noise is added to the trend. In comparison, the bottom panels present the outputs when a closed skew-normal noise is added. From this figure, it is clear that the classical Kalman filter performs well when processing Gaussian noise, but it loses its efficiency when the noise becomes skewed. This indicates that the investigation of a skewed Kalman filter is necessary when dealing with smooth trend estimations.

15.4.3 The Skewed Component

Because we are interested in future climate scenarios and there are no observations available for this time period, we will work with idealized simulations based on the typical outputs from a coupled Ocean-Atmosphere General Circulation Model (GCM); see Ammann, Kiehl, Zender, Otto-Bliesner, and Bradley (2002), and Boville and Gent (1998). In the modeling framework, all the forcings, including greenhouse gases and solar irradiance changes, as well as tropospheric and volcanic aerosol evolution, have to be specified. Consequently, the timing of large volcanic eruptions has to be set, and therefore is known. But the volcanic impact on climate is unknown and has to be estimated. This quantity is modeled by $\mathbf{v}_t$ in our Kalman filter, i.e., the skewed state-space component. As the forcing decreases over several years, at some point the signal can no longer be clearly detected against the background noise. Thus, the extracted volcanic cooling ends before the last volcanic aerosol has left the atmosphere.

From a statistical point of view, we assume that the known volcanic occurrences and aerosols are represented by P_t, which is equal to zero, if there is no volcanic signal, and equal to a pre-determined quantity for the few years following an eruption

$$P_t = aP_{t-1} + o_{t-1}, \text{ with } o_t = \begin{cases} 1, & \text{if an eruption occurs at time } t, \\ 0, & \text{if no eruption occurs at time } t. \end{cases}$$

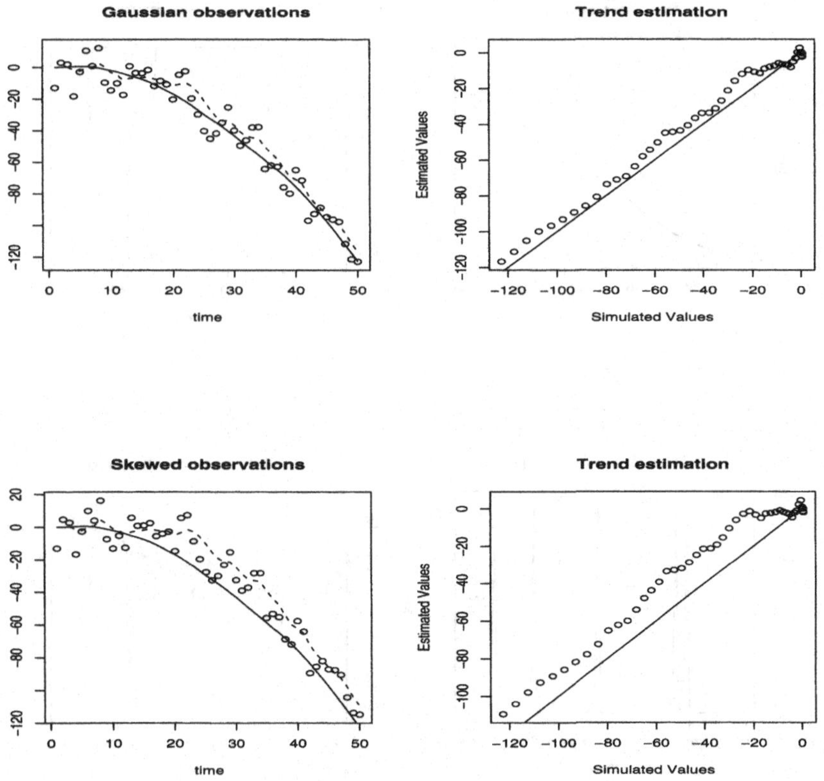

Figure 15.3 *Comparison between the estimation of a smooth trend. Top panels: Classical Kalman filter applied when the added noise is Gaussian. Bottom panels: Classical Kalman filter applied when the added noise is closed skew-normal. In the left panels, the solid lines represent the original smooth trend and the dashed lines correspond to the estimated trend. The points are the noisy observations (trend + noise).*

The bottom-left panel of Figure 15.4 illustrates the introduced aerosol quantity for each volcanic event ($a = 0.2$). The forcing decreases over several years, and at some point, the signal has disappeared.

The impact of each pre-determined eruption is modeled by

$$S_t = \eta_t^+ - L_t W_t, \text{ with } \eta_t^+ \sim N(0, \Sigma_\eta^*),$$

where the random variable W_t is defined from (15.12) and (15.13) and

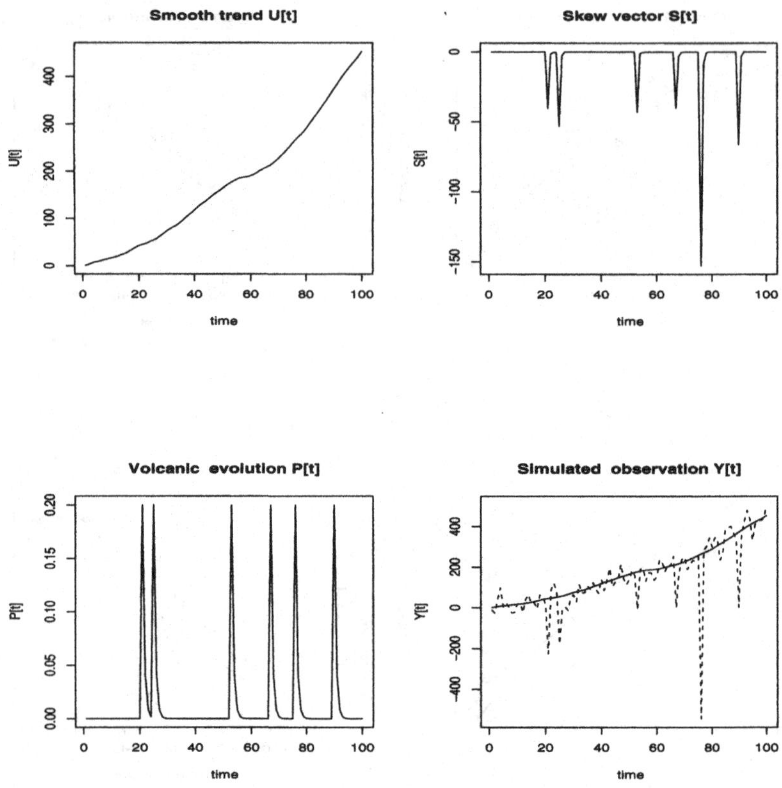

Figure 15.4 *Temporal evolution of each component of the state-space model: the smooth cubic spline U_t (top-left panel), the skewed vector S_t (top-right panel), the function P_t (bottom-left panel), and the simulated observations Y_t (bottom-right panel).*

generated from

$$V_t = -L_t V_{t-1} + \eta_t^+, \text{ with } L_t = \theta P_t,$$

for some constant θ. The top-right panel of Figure 15.4 shows the temporal evolution of S_t.

Figure 15.4 illustrates the different steps to obtain the simulated time series Y_t like the one shown in dotted line in the bottom-right panel. The smooth trend is shown in the top-left panel, the volcanic occurrences are represented in the bottom-left panel, and they are used to derive the volcanic impact (top-right panel). Note that the volcanic forcing has a negative impact on climate. This explains the negative S_t.

15.4.4 The State-Space Model

To combine all components, S_t, h_t, and the associated noise in a state-space model, we write the observational equation in the following way:

$$\begin{aligned} Y_t &= (1,0,1)\begin{pmatrix} h_t \\ h_t^{(1)} \\ S_t \end{pmatrix} + \epsilon_t, \qquad (15.22) \\ &= Q_t \mathbf{u}_t + P_t S_t + \epsilon_t, \\ &= F_t \mathbf{x}_t + \epsilon_t, \end{aligned}$$

with $Q_t = (1,0)^T$, $P_t = 1$, $\mathbf{x}_t = (\mathbf{u}_t^T, S_t)^T$, and $\mathbf{u}_t = (h_t, h_t^{(1)})^T$. The observational Equation (15.22) describes how the different forcings influence the observations and the system equation described in Section 15.4.3 models the temporal dynamics of the forcings. Hence, the state-space model has the advantage of providing a global formulation of our problem.

15.4.5 Simulations and Results

In the bottom-right panel of Figure 15.4, the dashed line corresponds to the simulated observations. To assess the quality of our statistical procedure, we would like to estimate the original trend (the solid line). This type of data set corresponds to an idealized output from a 20th century experiment presented by Ammann *et al.* (2002), employing the National Center for Atmospheric Research Climate System Model (Boville and Gent, 1998), a state-of-the-art coupled Ocean-Atmosphere General Circulation Model.

First, we apply the classical Kalman filter to this data. The results are presented in the top panels of Figure 15.5. As one may expect, the classical Kalman filter has difficulties dealing with the volcanic cooling events (sharp pulse-like events) and it has a tendency to underestimate the original trend (see top-right panel). In comparison, the skewed Kalman filter is more successful at finding the long-term trend and at capturing the intensity and occurrence of almost all simulated volcanic events (see bottom-right panel).

Because in the modeling framework, all the forcings (including greenhouse gases, solar irradiance changes, and tropospheric and volcanic aerosol evolution) had to be specified (and therefore known) in the GCM, it was natural to suppose the parameters of the state-space model were known. For past paleoclimate time series proxies, this is not the case. Current research based on a Bayesian framework is currently undertaken to estimate parameters in such situations.

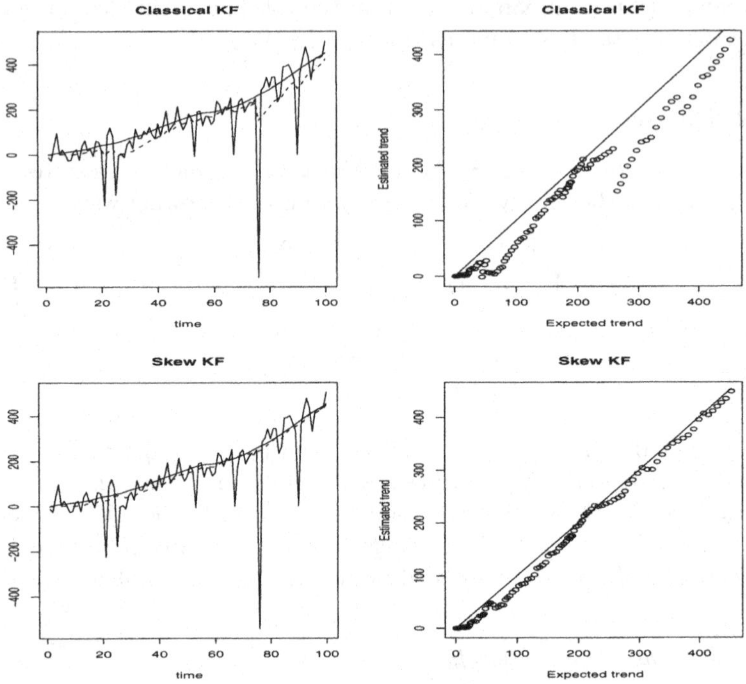

Figure 15.5 *Comparison between the classical Kalman filter and the skewed one. In the left panels, the solid lines represent the original trend and the dashed lines correspond to the estimated trend. The right panels show the difference between the expected and estimated trend for both Kalman filters.*

15.5 Conclusions

In this work, we showed that extending the normal distribution to the closed skew-normal distribution for state-space models reduced neither the flexibility nor the traceability of the operations associated with Kalman filtering. To the contrary, the introduction of a few skewness parameters provides a simple source of asymmetry needed for our applications. The added cost with the skewed filter is mostly due to the introduction of the skewness parameters $(D_t, \Delta_t, \boldsymbol{\nu}_t)$ that have to be computed in both prediction and observation steps. The main numerical difficulty of our method is that the computation of C_t, $\boldsymbol{\tau}^{(1)}$, $\tau^{(2)}$, is not straightforward, those expressions being not linear. Despite this disadvantage, the derivation is still computationally manageable compared to the dimension doubling of the CSN under summation which occurs at every time step.

To validate our research into a larger context, future work is needed. First, a detailed comparison with other non-Gaussian methods has to be undertaken to put into light the limitations and advantages of our approach. To adapt our strategy to a Bayesian paradigm will be of great interest. We believe that taking such an extension should be relatively easy because of the stability under conditioning of the closed skew-normal distribution. Another field of interest is studying the robustness of our method. In terms of geophysics, we plan to apply our approach to a larger variety of paleoclimatic data sets, in particular multivariate time series.

Compared with past approaches based on mixture of normals (Carlin *et al.*, 1992), we believe that the two approaches (closed skew-normal and mixture of normals) are in fact complementary when modeling data and they could be combined when necessary. One is more adapted when dealing with skewness and the other is better for representing multimodality. Finally, we would like to stress that the skewness is clearly identifiable in our parameterization. The interpretation of parameters in mixture models is sometimes not as clear.

Acknowledgment

Part of the research was sponsored by an NFS grant (ATM 0327936).

15.6 Appendix

In this Appendix, we detail the main steps used to derive our skewed Kalman filter operations. To simplify the notations, we assume that the linear Gaussian part does not exist, i.e., $K_t = Q_t = O$ and $\mathbf{u}_t = \mathbf{0}$. To find the full proof without this assumption, we refer to Naveau *et al.* (2004).

First, we need the following two results that could be easily shown by using the basic properties of the moment generating function of the closed skew-normal distribution.

1. If $\mathbf{y} \sim CSN_{n,m}(\boldsymbol{\mu}, \Sigma, D, \boldsymbol{\nu}, \Delta)$ and A is an $r \times n$ matrix such that $A^T A$ is non-singular, then $A\mathbf{y} \sim CSN_{r,m}(A\boldsymbol{\mu}, A\Sigma A^T, DA^{\leftarrow}, \boldsymbol{\nu}, \Delta)$ where $A^{\leftarrow}$ is the left inverse of A.

2. If $\mathbf{x} \sim CSN_{n,m}(\boldsymbol{\psi}, \Omega, D, \boldsymbol{\nu}, \Delta)$ and $\mathbf{z}$ is an n-dimensional Gaussian vector with mean $\boldsymbol{\mu}$ and covariance Σ, and independent of $\mathbf{x}$, then, $\mathbf{x} + \mathbf{z} \sim CSN_{n,m}(\boldsymbol{\mu}_{\mathbf{x}+\mathbf{z}}, \Sigma_{\mathbf{x}+\mathbf{z}}, D_{\mathbf{x}+\mathbf{z}}, \nu_{\mathbf{x}+\mathbf{z}}, \Delta_{\mathbf{x}+\mathbf{z}})$ where $\boldsymbol{\mu}_{\mathbf{x}+\mathbf{z}} = \boldsymbol{\psi} + \boldsymbol{\mu}$, $\Sigma_{\mathbf{x}+\mathbf{z}} = \Omega + \Sigma$, $D_{\mathbf{x}+\mathbf{z}} = D\Omega(\Omega + \Sigma)^{-1}, \boldsymbol{\nu}_{\mathbf{x}+\mathbf{z}} = \boldsymbol{\nu}$, and $\Delta_{\mathbf{x}+\mathbf{z}} = \Delta + (D - D_{\mathbf{x}+\mathbf{z}})\Omega D^T$.

The two stability results, linear transformation and additivity for the closed skew-normal distribution, will be used frequently during the different steps for our skewed Kalman filter operations. The proof is based on the following sequential argument.

For the non-linear state-space model defined by (15.8), (15.9), and (15.11), we assume that we have up to time $t-1$

$$(\mathbf{v}_{t-1}|\vec{\mathbf{y}}_{t-1}) \sim N_l\left(\hat{\boldsymbol{\psi}}^{+}_{t-1}, \hat{\Omega}^{+}_{t-1}\right), \tag{15.23}$$

where $\hat{}$ represents the posterior mean and covariance. From Equation (15.9), we deduce that

$$\left(\begin{array}{c}\mathbf{v}_t\\ \mathbf{v}_{t-1}\end{array}\middle|\vec{\mathbf{y}}_{t-1}\right) = \left(\begin{array}{c}-L_t\mathbf{v}_{t-1}\\ \mathbf{v}_{t-1}\end{array}\middle|\vec{\mathbf{y}}_{t-1}\right) + \left(\begin{array}{c}\boldsymbol{\eta}^{+}_t\\ \mathbf{0}\end{array}\right).$$

Hence, the variable $(\mathbf{v}_t^T, \mathbf{v}_{t-1}^T|\vec{\mathbf{y}}_{t-1})^T$ is Gaussian with the mean and variance equal to

$$N_{2l}\left(\left(\begin{array}{c}-L_t\hat{\boldsymbol{\psi}}^{+}_{t-1} + \boldsymbol{\mu}^{+}_{\boldsymbol{\eta}}\\ \hat{\boldsymbol{\psi}}^{+}_{t-1}\end{array}\right), \left(\begin{array}{cc}\tilde{\Omega}^{+}_t & -L_t\hat{\Omega}^{+}_{t-1},\\ -\hat{\Omega}^{+}_{t-1}L_t^T & \hat{\Omega}^{+}_{t-1}\end{array}\right)\right), \tag{15.24}$$

with $\tilde{\Omega}^{+}_t = L_t\hat{\Omega}^{+}_{t-1}L_t^T + \Sigma^{+}_{\boldsymbol{\eta}}$.

On observing $\mathbf{y}_t$, our objective is to compute the posterior of $(\mathbf{v}_t^T, \mathbf{v}_{t-1}^T)$ first, i.e., $\mathrm{P}(\mathbf{v}_t^T, \mathbf{v}_{t-1}^T|\vec{\mathbf{y}}_t)$ and then to obtain $\mathrm{P}(\mathbf{x}_t|\vec{\mathbf{y}}_t)$. For the former, we introduce $\mathbf{e}_t = \mathbf{y}_t - P_t[\mathrm{E}(\mathbf{s}_t|\vec{\mathbf{y}}_{t-1})] - \boldsymbol{\mu}_{\boldsymbol{\epsilon}}$, where $\mathrm{E}(\mathbf{s}_t|\vec{\mathbf{y}}_{t-1})$ is the conditional expectation of $\mathbf{s}_t$ given $\vec{\mathbf{y}}_{t-1}$. To compute this quantity, it is possible to show from the definition of $\mathbf{s}_t$, see Equation (15.11), that the variable $(\mathbf{s}_t|\vec{\mathbf{y}}_{t-1})$ follows approximately a closed skew-normal distribution given by $CSN_{l,1}(\tilde{\boldsymbol{\psi}}^{+}_t, \tilde{\Omega}^{+}_t, \tilde{D}^{+}_t, \tilde{\boldsymbol{\nu}}^{+}_t, \tilde{\Delta}^{+}_t)$ with the parameters $\tilde{\boldsymbol{\psi}}^{+}_t = -L_t\hat{\boldsymbol{\psi}}^{+}_{t-1} + \boldsymbol{\mu}^{+}_{\boldsymbol{\eta}}$, $\tilde{D}^{+}_t = \hat{\Omega}^{+}_{t-1}L_t^T(\tilde{\Omega}^{+}_t)^{-1}$, $\tilde{\boldsymbol{\nu}}^{+}_t = \hat{\boldsymbol{\psi}}^{+}_{t-1} - \tilde{D}^{+}_t\tilde{\boldsymbol{\psi}}^{+}_t$, and $\tilde{\Delta}^{+}_t = \hat{\Omega}^{+}_{t-1} - \tilde{D}^{+}_t\tilde{\Omega}^{+}_t(\tilde{D}^{+}_t)^T$. The mean and the variance of $(\mathbf{s}_t|\vec{\mathbf{y}}_{t-1})$ can be obtained from the moment generating function $M(\boldsymbol{\theta})$

$$M(\boldsymbol{\theta}) = \frac{\Phi_m(\tilde{D}^{+}_t\tilde{\Omega}^{+}_t\boldsymbol{\theta}; \tilde{\boldsymbol{\nu}}^{+}_t, \tilde{\Delta}^{+}_t + \tilde{D}^{+}_t\tilde{\Omega}^{+}_t(\tilde{D}^{+}_t)^T)}{\Phi_m(\mathbf{0}; \tilde{\boldsymbol{\nu}}^{+}_t, \tilde{\Delta}^{+}_t + \tilde{D}^{+}_t\tilde{\Omega}^{+}_t(\tilde{D}^{+}_t)^T)}\exp\{(\tilde{\boldsymbol{\psi}}^{+}_t)^T\boldsymbol{\theta} + \frac{1}{2}\boldsymbol{\theta}^T\tilde{\Omega}^{+}_t\boldsymbol{\theta}\}.$$

It provides $\mathrm{E}(\mathbf{s}_t|\vec{\mathbf{y}}_{t-1}) = -L_t\hat{\boldsymbol{\psi}}^{+}_{t-1} + \boldsymbol{\mu}^{+}_{\boldsymbol{\eta}} + \boldsymbol{\tau}^{(1)}_t$ and $\mathrm{Var}(\mathbf{s}_t|\vec{\mathbf{y}}_{t-1}) = \tilde{\Omega}^{+}_t + \tau^{(2)}_t$ where $\boldsymbol{\tau}^{(1)}$ and $\tau^{(2)}$ are defined by (15.18). Genton, He, and Liu (2001) also computed these moments in the special case $\boldsymbol{\nu} = \mathbf{0}$.

From the observation Equation (15.8), $(\mathbf{e}_t|\mathbf{v}_t, \mathbf{v}_{t-1}, \vec{\mathbf{y}}_{t-1})$ is equal to

$$(P_t[\mathbf{s}_t - \mathrm{E}(\mathbf{s}_t|\vec{\mathbf{y}}_{t-1})] + \boldsymbol{\epsilon}_t - \boldsymbol{\mu}_{\boldsymbol{\epsilon}}|\mathbf{v}_t, \mathbf{v}_{t-1}, \vec{\mathbf{y}}_{t-1}).$$

This last equality, the normality of $\boldsymbol{\epsilon}_t$, and the fact that the variable $\mathbf{s}_t$ is entirely defined from $(\boldsymbol{\eta}^{+}_t, \mathbf{v}_{t-1})$, imply that the variable $(\mathbf{e}_t|\mathbf{v}_t, \mathbf{v}_{t-1}, \vec{\mathbf{y}}_{t-1})$ follows

$$N_h(P_t[\mathbf{s}_t - \mathrm{E}(\mathbf{s}_t|\vec{\mathbf{y}}_{t-1})], \Sigma_{\boldsymbol{\epsilon}}). \tag{15.25}$$

Applying the classical properties of the multivariate normal distribution to the variables $(\mathbf{e}_t|\mathbf{v}_t, \mathbf{v}_{t-1}, \vec{\mathbf{y}}_{t-1})$ and $(\mathbf{v}_t^T, \mathbf{v}_{t-1}^T|\vec{\mathbf{y}}_{t-1})$, respectively, described by (15.24) and (15.25), allows us to derive that the random vector

$(\mathbf{v}_t^T, \mathbf{v}_{t-1}^T, \mathbf{e}_t^T | \vec{\mathbf{y}}_{t-1})^T$ follows

$$N_{2l+h}\left(\left(\begin{array}{c} -L_t\hat{\boldsymbol{\psi}}_{t-1}^+ + \boldsymbol{\mu}_{\boldsymbol{\eta}}^+ \\ \hat{\boldsymbol{\psi}}_{t-1}^+ \\ \mathbf{0} \end{array}\right), \left(\begin{array}{ccc} \tilde{\Omega}_t^+ & -L_t\hat{\Omega}_{t-1}^+ & C_tP_t^T \\ -\hat{\Omega}_{t-1}^+L_t^T & \hat{\Omega}_{t-1}^+ & \tilde{C}_tP_t^T \\ P_tC_t & P_t\tilde{C}_t & \Sigma_t \end{array}\right)\right),$$

with $C_t = \mathrm{Cov}(\mathbf{v}_t, \mathbf{s}_t|\vec{\mathbf{y}}_{t-1})$, $\tilde{C}_t = \mathrm{Cov}(\mathbf{v}_{t-1}, \mathbf{s}_t|\vec{\mathbf{y}}_{t-1})$ and $\Sigma_t = Q_t\tilde{\Omega}_t^*Q_t^T + P_t\mathrm{Var}(\mathbf{s}_t|\vec{\mathbf{y}}_{t-1})P_t^T + \Sigma_{\boldsymbol{\epsilon}}$. Since we have $(\mathbf{v}_t^T, \mathbf{v}_{t-1}^T|\vec{\mathbf{y}}_t) = (\mathbf{v}_t^T, \mathbf{v}_{t-1}^T|\mathbf{e}_t, \vec{\mathbf{y}}_{t-1})$, the distribution of this vector is multivariate normal with mean

$$\left(\begin{array}{c} -L_t\hat{\boldsymbol{\psi}}_{t-1}^+ + \boldsymbol{\mu}_{\boldsymbol{\eta}}^+ \\ \hat{\boldsymbol{\psi}}_{t-1}^+ \end{array}\right) + H_t\Sigma_t^{-1}\mathbf{e}_t, \text{ where } H_t = \left(\begin{array}{c} C_tP_t^T \\ \tilde{C}_tP_t^T \end{array}\right),$$

and its variance is equal to $\left(\begin{array}{cc} \tilde{\Omega}_t^+ & -L_t\hat{\Omega}_{t-1}^+ \\ -\hat{\Omega}_{t-1}^+L_t^T & \hat{\Omega}_{t-1}^+ \end{array}\right) - H_t\Sigma_t^{-1}H_t^T$. With the same kind of argument, $(\mathbf{v}_t|\vec{\mathbf{y}}_t)$ follows

$$N_l\left(-L_t\hat{\boldsymbol{\psi}}_{t-1}^+ + \boldsymbol{\mu}_{\boldsymbol{\eta}}^+ + J_t\Sigma_t^{-1}\mathbf{e}_t, \tilde{\Omega}_t^+ - J_t\Sigma_t^{-1}J_t^T\right),$$

with $J_t = C_tP_t^T$. This distribution is used to implement the first update of the Kalman filter, i.e., the parameters of (15.23) are now set for a new cycle

$$\hat{\boldsymbol{\psi}}_t^+ = -L_t\hat{\boldsymbol{\psi}}_{t-1}^+ + \boldsymbol{\mu}_{\boldsymbol{\eta}}^+ + C_tP_t^T\Sigma_t^{-1}\mathbf{e}_t, \text{ and } \hat{\Omega}_t^+ = \tilde{\Omega}_t^+ - C_tP_t^T\Sigma_t^{-1}P_tC_t.$$

To get the final part, i.e., $\mathrm{P}(\mathbf{x}_t|\vec{\mathbf{y}}_t)$, we note that $(\mathbf{v}_t^T, \mathbf{v}_{t-1}^T|\vec{\mathbf{y}}_t)$ follows a normal distribution with mean

$$\left(\begin{array}{c} -L_t\hat{\boldsymbol{\psi}}_{t-1}^+ + \boldsymbol{\mu}_{\boldsymbol{\eta}}^+ + C_tP_t^T\Sigma_t^{-1}\mathbf{e}_t \\ \hat{\boldsymbol{\psi}}_{t-1}^+ + \tilde{C}_tP_t^T\Sigma_t^{-1}\mathbf{e}_t \end{array}\right),$$

and variance

$$\left(\begin{array}{cc} \tilde{\Omega}_t^+ & -L_t\hat{\Omega}_{t-1}^+ \\ -\hat{\Omega}_{t-1}^+L_t^T & \hat{\Omega}_{t-1}^+ \end{array}\right) - W_t\Sigma_t^{-1}W_t^T,$$

where

$$W_t = \left(\begin{array}{c} C_tP_t^T \\ \tilde{C}_tP_t^T \end{array}\right), W_t\Sigma_t^{-1}W_t^T = \left(\begin{array}{cc} C_tP_t^T\Sigma_t^{-1}P_tC_t & C_tP_t^T\Sigma_t^{-1}P_t\tilde{C}_t \\ \tilde{C}_tP_t^T\Sigma_t^{-1}P_tC_t & \tilde{C}_tP_t^T\Sigma_t^{-1}P_t\tilde{C}_t \end{array}\right).$$

Next, define the quantities $\overline{\Omega}_{t-1} = \hat{\Omega}_{t-1}^+ - C_tP_t^T\Sigma_t^{-1}P_tC_t$, and $\overline{L}_t = L_t + \Sigma_{\nu}^+P_t^T\Sigma_t^{-1}P_t\tilde{C}_t(\overline{\Omega}_{t-1}^+)^{-1}$. The covariance matrix of the vector $(\mathbf{v}_t^T, \mathbf{v}_{t-1}^T|\vec{\mathbf{y}}_t)$ is then equal to

$$\left(\begin{array}{cc} \hat{\Omega}_t^+ & -\overline{L}_t\overline{\Omega}_{t-1} \\ -\overline{\Omega}_{t-1}\overline{L}_t^T & \overline{\Omega}_{t-1} \end{array}\right).$$

It follows from (15.11) that $(\mathbf{s}_t|\vec{\mathbf{y}}_t) \sim CSN_{l,l}(\hat{\boldsymbol{\psi}}_t^+, \hat{\Omega}_t^+, \hat{D}_t^+, \hat{\boldsymbol{\nu}}_t^+, \hat{\Delta}_t^+)$, with parameters $\hat{D}_t^+ = \overline{\Omega}_{t-1}^+ \overline{L}_t^T (\hat{\Omega}_t^+)^{-1}$, $\hat{\boldsymbol{\nu}}_t^+ = \hat{\boldsymbol{\psi}}_{t-1}^+ - \hat{D}_t^+ \hat{\boldsymbol{\psi}}_t^+$, and $\hat{\Delta}_t^+ = \overline{\Omega}_{t-1}^+ - \hat{D}_t^+ \hat{\Omega}_t^+ (\hat{D}_t^+)^T$.

□

CHAPTER 16

Spatial Prediction of Rainfall Using Skew-Normal Processes

Hyoung-Moon Kim, Eunho Ha, and Bani K. Mallick

16.1 Introduction

Modeling spatial data with Gaussian processes is the common thread of all geostatistical analyses due to their close relationship with kriging (Cressie, 1993). In some instances, the assumption that the distribution is Gaussian is clearly violated. One basic objective of this chapter is to develop spatial models to handle data showing non-Gaussian characteristics, such as skewness. For example, modeling rainfall data from a statistical perspective presents the challenge of dealing with a skewed distribution. Several sophisticated models have been developed to capture the underlying physical dynamics that govern rainfall. Smith and Robinson (1997) presented a likelihood-based approach and also used a Bayesian method to analyze it. A different approach is to use a truncated normal distribution (Stidd, 1973; Sansó and Guenni, 2000) to model the rainfall distribution. De Oliveira, Kedem, and Short (1997) used a Bayesian transformed Gaussian (BTG) model to analyze a rainfall data set.

We propose a richer (flexible) parametric class of multivariate distributions to represent features of the data while reducing unrealistic assumptions. The pioneering work was started by Zellner (1976) who used the regression model with multivariate Student t error terms. To model skewed data, Azzalini and Dalla Valle (1996) and Azzalini and Capitanio (1999) developed a skew-normal distribution which has many properties similar to the normal distribution. Exploiting this idea, we suggest a spatial model based on a skew-normal distribution for rainfall prediction.

The chapter is organized as follows. In Section 16.2, we describe the rainfall data followed by a brief introduction to the skew-normal model. We employ an MCMC-based simulation method to generate samples from the posterior distribution and the predictive distribution. We describe our data analysis in Section 16.3. Finally we offer a brief discussion about generalizing the current model in Section 16.4.

16.2 Data and Model

16.2.1 Automatic Weather Stations and Their Sensors

In Korea, about 400 Automatic Weather Stations (AWS) have been installed since 1998 in order to detect small meteorological phenomena as well as to perform accurate predictions to reduce meteorological disasters. AWS consist of wind speed and temperature measurements, rain detectors, and a data logger on a tower, installed at a height of 10 m or 20 m. There are several sensors in each station and they have to be adequately ventilated. Each sensor has different accuracy levels as described in Table 16.1. Wind speed is observed by an anemometer of three-cup form and direction by a wind vane.

Rain data have been collected using two sensors of rain detector and tipping bucket rain gauges. Rain is first detected by the rain detector using electric current. The rain gauge is mounted in a relatively horizontal level spot at the height of 30 cm. We analyze the data collected from these stations.

Table 16.1 *Standard instruments and accuracy of sensors in an AWS.*

Element	Standard Instrument	Accuracy
Temperature	Ventilated Psychrometer	±0.5°C
Humidity	Ventilated Psychrometer or Humidity Calibrator	±5%
Wind Direction	Protractor	±10°
Wind Speed i) Below 10 m/s ii) Over 10 m/s	Wind Speed Calibrator or Portable Anemometer	i) ±1 m/s ii) ±10%
Rainfall	Rain Gauge	±5 mm (*per* 100 mm)
Pressure	Accuracy Style Aneroid Barometer or Mercury Barometer	±0.7 hPa

16.2.2 Model Description

In this section, we describe the skew-normal model as given in Kim and Mallick (2004) to extend the popular Gaussian process. In spatial problems, each measured value in the data set is located in the domain $D \subseteq \mathbb{R}^d$ and we call it a regionalized value. Considering the regionalized values at all points in a given region, the associated function $Z(\mathbf{x})$ for $\mathbf{x} \in D$ is a regionalized variable. The set of values $\{Z(\mathbf{x}), \mathbf{x} \in D\}, D \subseteq \mathbb{R}^d$, can be viewed as a single draw from an infinite set of random variables (one random variable at each point of the domain). The family of those random variables is called the random field (random function). Let the family of the random variables be the random field of interest, and suppose that we have n observations $\mathbf{z} = (Z(\mathbf{x}_1), \ldots, Z(\mathbf{x}_n))^T$ from a single realization of this field, where $\mathbf{x}_1, \ldots, \mathbf{x}_n$ are known distinct locations in D. Based on $\mathbf{z}$ and on our prior knowledge about the random field, we want to predict the unobserved $Z(\mathbf{x}_0)$, where $\mathbf{x}_0$ is in D. It is assumed that $Z(\mathbf{x}_0)$ comes from the same realization as the data vector $\mathbf{z}$. In the sequel, we will develop a Bayesian spatial prediction model using a skew-normal random field assuming a multivariate skew-normal distribution for Z.

The density function of a centered and scaled multivariate skew-normal distribution can be written as $f(\mathbf{z}|\boldsymbol{\alpha}, \Omega) \propto \phi_n(\mathbf{z}|\Omega)\Phi(\boldsymbol{\alpha}^T\mathbf{z})$, $\mathbf{z} \in \mathbb{R}^n$, where ϕ_n is the n-dimensional normal density, Φ is the normal cumulative distribution function (cdf), Ω is the correlation matrix, and $\boldsymbol{\alpha}$ is the n-dimensional vector of skewness parameters (Azzalini and Capitanio, 1999). When $\boldsymbol{\alpha}$ is a null vector then we get back the Gaussian distribution. Extensions to a more general setup with location parameters $\boldsymbol{\mu} = (\mu_1, \ldots, \mu_n)^T$ and scale parameters $W = \mathrm{diag}(w_1, \ldots, w_n)$ can be done using a transformation $\mathbf{y} = \boldsymbol{\mu} + W\mathbf{z}$ whose density function becomes $f(\mathbf{y}) \propto \phi_n(\mathbf{y} - \boldsymbol{\mu}|\Omega, W)\Phi(\boldsymbol{\alpha}^T W^{-1}(\mathbf{y} - \boldsymbol{\mu}))$.

In our case, we observe $\mathbf{z} = (Z(\mathbf{x}_1), \ldots, Z(\mathbf{x}_n))^T$ and we focus on the prediction of $Z(\mathbf{x}_0)$, which is unknown at the spatial location $\mathbf{x}_0$. We simplify the spatial model by assuming a single scale parameter σ, spatial correlation matrix $K^*_{\boldsymbol{\theta}}$ (parameterized by the parameter $\boldsymbol{\theta}$) and a single skewness parameter α. With the presence of q-dimensional covariates, the mean function is modeled as $\boldsymbol{\mu} = F\boldsymbol{\beta}$, where $\boldsymbol{\beta}$ are the regression parameters with design matrix F. The design matrix contains covariate observations like temperature, humidity, wind direction, pressure, and wind speed, as given in Table 16.1.

To develop the predictive model, we assume the joint distribution of $\mathbf{z}^* = (Z(\mathbf{x}_0), \mathbf{z}^T)^T$ is an $SN_{n+1}(F^*\boldsymbol{\beta}, \sigma^2 K^*_{\boldsymbol{\theta}}, \frac{\alpha}{\sigma}\mathbf{1}_{n+1})$, where $F^* = \begin{pmatrix} g^T(\mathbf{x}_0) \\ F \end{pmatrix}$, $K^*_{\boldsymbol{\theta}} = \begin{pmatrix} k_{\boldsymbol{\theta}}(0) & \mathbf{k}^T_{\boldsymbol{\theta}} \\ \mathbf{k}_{\boldsymbol{\theta}} & K_{\boldsymbol{\theta}} \end{pmatrix}$, $k_{\boldsymbol{\theta}}(0) = K_{\boldsymbol{\theta}}(\mathbf{x}_0, \mathbf{x}_0)$ and $\mathbf{k}_{\boldsymbol{\theta}} = [K_{\boldsymbol{\theta}}(\mathbf{x}_0, \mathbf{x}_i)]_{n\times 1}$. Here g is the covariate values (part of the design matrix) corresponding to $\mathbf{x}_0$. Then $\mathbf{z} = (Z(\mathbf{x}_1), \ldots, Z(\mathbf{x}_n))^T$ is a realization from the real-valued skew-normal random field on $\mathbb{R}^n$. Furthermore $\sigma^2 K_{\boldsymbol{\theta}}$ is a positive definite matrix and $\sigma^2 K_{\boldsymbol{\theta}}(\mathbf{x}, \mathbf{y}) = \sigma^2 K_{\boldsymbol{\theta}}(\|\mathbf{x} - \mathbf{y}\|)$, where $\boldsymbol{\theta} \in \Theta$ is a $p \times 1$ vector of structural parameters, Θ is a subset of $\mathbb{R}^p$, $\sigma^2 \in \mathbb{R}_+$ and $\|\cdot\|$ denotes the Euclidean norm. Structural parameters control the range of correlation and the smoothness of the random field, where for every $\boldsymbol{\theta} \in \Theta$, $K_{\boldsymbol{\theta}}(\cdot)$ is an isotropic correlation function, assumed to be continuous in $\boldsymbol{\theta}$.

In this setup, the marginal distribution again comes out to be a skew-normal distribution, while the conditional distribution becomes a member of an extended skew-normal class of densities. They are, respectively, given by the following expressions (Kim and Mallick, 2004):

$$\begin{aligned} f(\mathbf{z}|\boldsymbol{\eta}) &= 2\phi_n\left(\mathbf{z} - F\boldsymbol{\beta}, \sigma^2 K_{\boldsymbol{\theta}}\right)\Phi\left(\varphi\right), \text{where } \mathbf{z} \in \mathbb{R}^n, \quad (16.1)\\ f(z(\mathbf{x}_0)|\mathbf{z}, \boldsymbol{\eta}) &= \phi_1\left(z(\mathbf{x}_0) - \mu_{1.2}, \sigma^2 K_{1.2}\right)\\ &\quad \times \frac{\Phi\left(\frac{\alpha}{\sigma}(z(\mathbf{x}_0) - \mu_{1.2}) + \varphi\sqrt{1+\alpha^2 K_{1.2}}\right)}{\Phi\left(\varphi\right)}, \quad (16.2) \end{aligned}$$

where $\boldsymbol{\beta} = \left(\boldsymbol{\beta}^T, \sigma^2, \boldsymbol{\theta}^T, \boldsymbol{\alpha}_z^T\right)^T$, $z(\mathbf{x}_0) \in \mathbb{R}$, $\varphi = \frac{\alpha}{\sigma}\boldsymbol{\alpha}_z^T(\mathbf{z} - F\boldsymbol{\beta})$, and $\boldsymbol{\alpha}_z^T = \mathbf{1}_{n+1}^T\left(K_{\boldsymbol{\theta}}^{-1}\mathbf{k}_{\boldsymbol{\theta}} \quad I_n\right)^T / \sqrt{1+\alpha^2 K_{1.2}}$, $\mu_{1.2} = g^T(\mathbf{x}_0)\boldsymbol{\beta} + \mathbf{k}_{\boldsymbol{\theta}}^T K_{\boldsymbol{\theta}}^{-1}(\mathbf{z} - F\boldsymbol{\beta})$, $K_{1.2} = k_{\boldsymbol{\theta}}(0) - \mathbf{k}_{\boldsymbol{\theta}}^T K_{\boldsymbol{\theta}}^{-1}\mathbf{k}_{\boldsymbol{\theta}}$. These distributions will be useful to perform spatial predictions. Even though the results obtained in this chapter are very general and can be applied to any correlation function, we limit our presentation to the popular power exponential family owing to the ease of computation. The mathematical expression of the power exponential family (valid in $\mathbb{R}^l$, $l \geq 1$) is

$$\begin{aligned} K_{\boldsymbol{\theta}}^{PE}(d) &= \exp\left(-\nu d^{\theta_2}\right) = \theta_1^{d^{\theta_2}}; \quad \nu > 0,\\ \theta_1 &= \exp(-\nu) \in (0,1), \quad \theta_2 \in (0,2], \quad (16.3) \end{aligned}$$

and $\boldsymbol{\theta} = (\theta_1, \theta_2)^T$, the vector of correlation parameters. This family contains the exponential ($\theta_2 = 1$) and Gaussian ($\theta_2 = 2$) models, which are often used in applications (De Oliveira *et al.*, 1997). In this family, $\theta_1 > 0$, called a range parameter, controls how fast the correlation decays with distance, and θ_2, called a smoothness or roughness parameter, controls geometrical properties of the random field, such as continuity and differentiability.

16.2.3 Bayesian Analysis

The Bayesian model specification requires prior distributions for all the unknown parameters and we will assume prior independence so the joint prior distribution $\pi(\boldsymbol{\eta}) = \pi(\boldsymbol{\beta}, \sigma^2, \boldsymbol{\theta}, \alpha) = \pi(\boldsymbol{\beta})\pi(\sigma^2)\ \pi(\boldsymbol{\theta})\pi(\alpha)$. With this assumption, the posterior distribution is proportional to $f(\mathbf{z}|\boldsymbol{\eta})\pi(\boldsymbol{\eta}) = f(\mathbf{z}|\boldsymbol{\beta}, \sigma^2, \boldsymbol{\theta}, \alpha)\pi(\boldsymbol{\beta})\pi(\sigma^2)\ \pi(\boldsymbol{\theta})\pi(\alpha)$, where $\boldsymbol{\eta} = (\boldsymbol{\beta}^T, \sigma^2, \boldsymbol{\theta}^T, \alpha)^T$.

Improperness of the posterior distribution with assumption of improper prior distribution is always an important issue in Bayesian analysis and recently Berger, De Oliveira, and Sansó (2001) obtained some valuable results corresponding to the Gaussian process model. In our situation, with the assumption of independent prior distributions, it is simple to show that all the properiety results obtained by Berger *et al.* (2001) are also valid for the skew-normal process as long as we use a proper prior distribution for α.

We consider informative prior distributions as $\boldsymbol{\beta} \sim N_q(\boldsymbol{\beta}_0, \Sigma_0)$, $\sigma^2 \sim IG(\alpha_0, \beta_0)$ an inverse Gaussian distribution, $\pi(\boldsymbol{\theta})$, and $\alpha \sim N(\alpha_0, \tau_0^2)$. Then a posterior distribution $\pi(\boldsymbol{\eta}|\mathbf{z})$ is proportional to

$$IG\left(\sigma^2; \alpha_{\sigma^2}, \beta_{\sigma^2}\right) \Phi\left(\varphi\right) |K_{\boldsymbol{\theta}}|^{-\frac{1}{2}} \pi(\boldsymbol{\beta})\pi(\boldsymbol{\theta})\pi(\alpha), \tag{16.4}$$

where $\alpha_{\sigma^2} = \frac{n}{2} + \alpha_0$, $\beta_{\sigma^2} = \frac{2\beta_0}{(\mathbf{z}-F\boldsymbol{\beta})^T K_{\boldsymbol{\theta}}^{-1}(\mathbf{z}-F\boldsymbol{\beta})\beta_0+2}$. Furthermore, full conditional distributions that will be used in the MCMC method (Gilks, Richardson, and Spiegelhalter, 1996) are given by

$$\pi(\boldsymbol{\beta}|\mathbf{z}, \sigma^2, \boldsymbol{\theta}, \alpha) \quad \propto \quad N_q\left(\mu_{\boldsymbol{\beta}}, \Sigma_{\boldsymbol{\beta}}\right) \Phi\left(\varphi\right), \tag{16.5}$$

where $\mu_{\boldsymbol{\beta}} = \Sigma_{\boldsymbol{\beta}} \left(\frac{F^T K_{\boldsymbol{\theta}}^{-1} F}{\sigma^2} \hat{\boldsymbol{\beta}} + \Sigma_0^{-1} \boldsymbol{\beta}_0 \right)$, $\Sigma_{\boldsymbol{\beta}} = \left(\frac{F^T K_{\boldsymbol{\theta}}^{-1} F}{\sigma^2} + \Sigma_0^{-1} \right)^{-1}$, and $\hat{\boldsymbol{\beta}} = (F^T K_{\boldsymbol{\theta}}^{-1} F)^{-1} F^T K_{\boldsymbol{\theta}}^{-1} \mathbf{z}$, and

$$\begin{aligned}
\pi(\sigma^2|\mathbf{z}, \boldsymbol{\beta}, \boldsymbol{\theta}, \alpha) &\propto IG\left(\alpha_{\sigma^2}, \beta_{\sigma^2}\right) \Phi\left(\varphi\right), & (16.6)\\
\pi(\boldsymbol{\theta}|\mathbf{z}, \boldsymbol{\beta}, \sigma^2, \alpha) &\propto |K_{\boldsymbol{\theta}}|^{-\frac{1}{2}} \exp\left(-\frac{1}{2\sigma^2} (\mathbf{z} - F\boldsymbol{\beta})^T K_{\boldsymbol{\theta}}^{-1} (\mathbf{z} - F\boldsymbol{\beta}) \right) \\
&\quad \times \Phi\left(\varphi\right) \pi(\boldsymbol{\theta}), & (16.7)\\
\pi(\alpha|\mathbf{z}, \boldsymbol{\beta}, \sigma^2, \boldsymbol{\theta}) &\propto \Phi\left(\varphi\right) \pi(\alpha). & (16.8)
\end{aligned}$$

We can modify the prior distribution of α by using mixture distributions. To implement it we have to use additional hyper-priors as γ and p and the hierarchical structure becomes $\pi(\alpha|\gamma, p)\pi(\gamma|p)\pi(p)$ instead of $\pi(\alpha)$. A conditional prior for α is $\pi(\alpha|\gamma, p) = \pi(\alpha|\gamma) = 0$ if $\gamma = 0$ and equal to $\phi_1(\alpha - \alpha_0, \tau_0^2)\Phi\left(\varphi\right)$, otherwise. We assume $\gamma|p \sim Ber(p)$, $p \sim Beta(p_0, p_0)$ with $\mathrm{E}(p) = 1/2$. We will denote a Bernoulli distribution with parameter p and a beta distribution with parameter λ and δ by $Ber(p)$ and $Beta(\lambda, \delta)$ respectively. So the marginal prior distribution for α is $\alpha|p \sim (1-p)I(\alpha = 0) + p\,\phi_1(\alpha - \alpha_0, \tau_0^2)$, where $I(\alpha = 0)$ denotes the indicator function, and hence it is a mixture distribution. Therefore this model does contain a Gaussian process as a special case and $(1-p)$ is the prior probability of it. With the use of this mixture prior, the prediction is done using a mixture of Gaussian and skew-normal processes.

The full conditional distributions of α, γ, p have been changed as follows but the other full conditional distributions remained exactly the same as in (16.5)-(16.7):

$$\begin{aligned}
\pi(\alpha|\mathbf{z}, \gamma, p, \boldsymbol{\beta}, \sigma^2, \boldsymbol{\theta}) &\propto \pi(\alpha|\gamma) f(\mathbf{z}|\boldsymbol{\eta}), & (16.9)\\
\pi(\gamma|\mathbf{z}, \alpha, p, \boldsymbol{\beta}, \sigma^2, \boldsymbol{\theta}) &\propto \pi(\alpha|\gamma)\pi(\gamma|p) \propto Ber\left(\frac{a}{a+b}\right), & (16.10)\\
\pi(p|\mathbf{z}, \gamma, \alpha, \boldsymbol{\beta}\sigma^2, \boldsymbol{\theta}) &\propto \pi(\gamma|p)p(p) \propto Beta(p_0 + \gamma, p_0 + 1 - \gamma), & (16.11)
\end{aligned}$$

where $a = 1 - p$ and $b = p\phi_1(\alpha - \alpha_0, \tau_0^2)$.

Note that φ contains all parameters which are $\boldsymbol{\eta} = (\boldsymbol{\beta}^T, \sigma^2, \boldsymbol{\theta}^T, \alpha)^T$. Finally a Bayesian predictive distribution is given by the following expression: $f(z(\mathbf{x}_0)|\mathbf{z}) \propto \int_{\boldsymbol{\eta}} f(z(\mathbf{x}_0)|\boldsymbol{\eta}, \mathbf{z})\pi(\boldsymbol{\eta}|\mathbf{z})\, d\boldsymbol{\eta}$. The posterior distribution is in a complicated form and we will use a Markov chain Monte Carlo (MCMC) method (Gilks *et al.*, 1996) to generate samples from the posterior distributions. To generate samples from the posterior distribution $p(\boldsymbol{\eta}|\mathbf{z})$ where $\boldsymbol{\eta} = (\boldsymbol{\beta}^T, \sigma^2, \boldsymbol{\theta}^T, \alpha)^T$, we exploit the full conditional distributions in a Gibbs sampling (Gelfand and Smith, 1990) framework. All the five full conditional distributions except $\pi(\boldsymbol{\theta}|\mathbf{z}, \boldsymbol{\beta}, \sigma^2, \alpha)$ can be written as $\pi(\cdot) \propto \psi(\cdot)\Phi(\cdot)$. Here $\psi(\cdot)$ is a density that can be sampled easily and $\Phi(\cdot)$ is a uniformly bounded function, so the probability of a move in the Metropolis-Hastings algorithm requires only the computation of the $\Phi(\cdot)$ function and is given by

$$p(x, y) = \min\left(\frac{\Phi(y)}{\Phi(x)}, 1\right), \tag{16.12}$$

which is known to be an efficient solution when it exists (Chib and Greenberg, 1995). The probability of a move for $\boldsymbol{\theta}$ is

$$p(x, y) = \min\left(\frac{|K_y|^{-\frac{1}{2}} \exp\left(-\frac{1}{2\sigma^2}(\mathbf{z} - F\boldsymbol{\beta})^T K_y^{-1}(\mathbf{z} - F\boldsymbol{\beta})\right) \Phi\left(\varphi_y\right)}{|K_x|^{-\frac{1}{2}} \exp\left(-\frac{1}{2\sigma^2}(\mathbf{z} - F\boldsymbol{\beta})^T K_x^{-1}(\mathbf{z} - F\boldsymbol{\beta})\right) \Phi\left(\varphi_x\right)}, 1\right),$$

where φ_x (φ_y) is the same as φ having x (y) as an argument.

16.3 Data Analyses

We made a univariate plot of the data in Figure 16.1 which shows the presence of skewness, so the use of a skew-normal process is natural here. We used the five covariates from Table 16.1 and the intercept term in F. As our working family of isotropic correlation functions, we use the general (power) exponential correlation function (Yaglom, 1987; De Oliveira *et al.*, 1997). Another flexible family is the Matérn class of correlation functions used by Handcock and Stein (1993). That family covers a broader range of behaviors than the one we use here, but these correlation functions are computationally more demanding and involve the evaluation of Bessel functions.

We develop two classes of prior distributions. The first one (SGP 1) is as follows. We assume $\boldsymbol{\beta} \sim N_6((\bar{z}, 0, 0, 0, 0, 0), 100 I_6))$, $\sigma^2 \sim IG(0.05, 0.05)$, $\theta_1 \sim U(0, 1), \theta_2 \sim U(0, 2]$, and $\alpha \sim N(0, 100)$. We assumed uniform priors for θ_1, θ_2.

Applying the MCMC method of Section 16.2, we obtain samples from the marginal posterior distributions. We found that none of the covariates are significant. After obtaining samples from the marginal posterior distributions, we obtain values of each predictive density function for $Z(\mathbf{x}_0)$

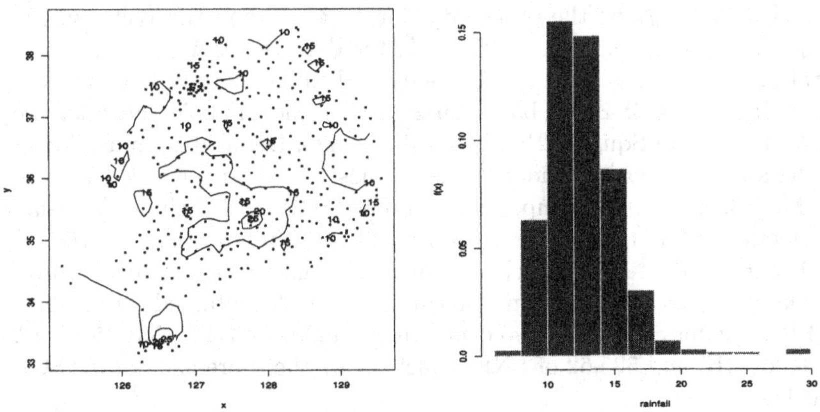

Figure 16.1 *Position of the 365 automatic weather stations with contours of rainfall (left) and histogram of rainfall data (right).*

using the composition method (Tanner, 1996; Gelfand, 1996) for the locations $\mathbf{x}_0 = (126.5, 33.9), (127.5, 35.5), (126.9, 37), (128.1, 35.7), (128.7, 36.5)$ and $(129.4, 36.1)$ covering different sections of the region of interest. The predicted positions with 365 stations are plotted in Figure 16.2. We plotted the predictive distributions corresponding to each location in Figure 16.3.

The second class of priors (SGP 2) is only different for α. That is, $p \sim Beta(1,1)$, $\gamma|p \sim Ber(p)$ and $\alpha|p \sim (1-p)I(\alpha = 0) + p\phi_1(0, 100)$. Notice that the mean of p is $1/2$, which means a priori skewness or not is equally probable. Using similar generating techniques as those used previously, we generated the marginal posterior distributions. The posterior median of p is 0.72, which indicates the presence of skewness in the data.

We have also obtained the predictive distributions for SGP 2 plotted in Figure 16.4. We found that for each of the locations, the two predictive densities of SGP 1 and SGP 2 are visually indistinguishable, and for all practical purposes they provide identical inferences.

Figure 16.5 shows the predicted map with a contour plot obtained by computing $\hat{Z}(\mathbf{x}_0) =$ median of $(Z(\mathbf{x}_0)|\mathbf{z})$ on every $\mathbf{x}_0$ of a 40×40 grid. As a measure of predictive uncertainty at any location $\mathbf{x}_0$ we use the MAD (Median Absolute Deviation) $=$ (median of $|(z(\mathbf{x}_0)|\mathbf{z})_i - \hat{Z}(\mathbf{x}_0)|)/.6745$ which is equal to SIQR (Semi Inter Quartile Range) and $0.6745\ \sigma$ for a normal distribution with mean μ and standard deviation σ, where $(z(\mathbf{x}_0)|\mathbf{z})_i$ denote realizations of $(Z(\mathbf{x}_0)|\mathbf{z})$. Figure 16.6 shows the map of this uncertainty measure with a contour plot.

To check the model adequacy for prediction purposes, we use a cross-validation approach based on single-point-deletion predictive distributions as described by Gelfand, Dey, and Chang (1992). Let $Z(\mathbf{x}_i)$ be the random variable, $z_{i,obs}$ be the observed value of $Z(\mathbf{x}_i)$, and the vector $\mathbf{z}_{(i)} = (z_{1,obs}, \ldots, z_{i-1,obs}, z_{i+1,obs}, \ldots, z_{n,obs})^T$, the data vector with the i-th observation deleted, $i = 1, \ldots, n$. The idea is then to quantify how well the model predicts each $Z(\mathbf{x}_i)$ based on $\mathbf{z}_{(i)}$. As a model choice criterion we use MSPR (Mean Squared Prediction Residuals). Detailed computation of MSPR and other cross-validation techniques have been provided in Kim and Mallick (2004). We compared our model (SGP) with a single Gaussian process model (GP), Bayesian transformed Gaussian process model (BTG) by Oliveira *et al.* (1997), as well as with some kriging methods such as ordinary kriging (OK) and lognormal kriging (LNK). We obtained the smallest MSPR value for SGP as 451.96 compared to 482.32 of BTG, 531.45 of GP, 542.94 of OK, and 525.62 of LNK, which shows the supremacy of the SGP model.

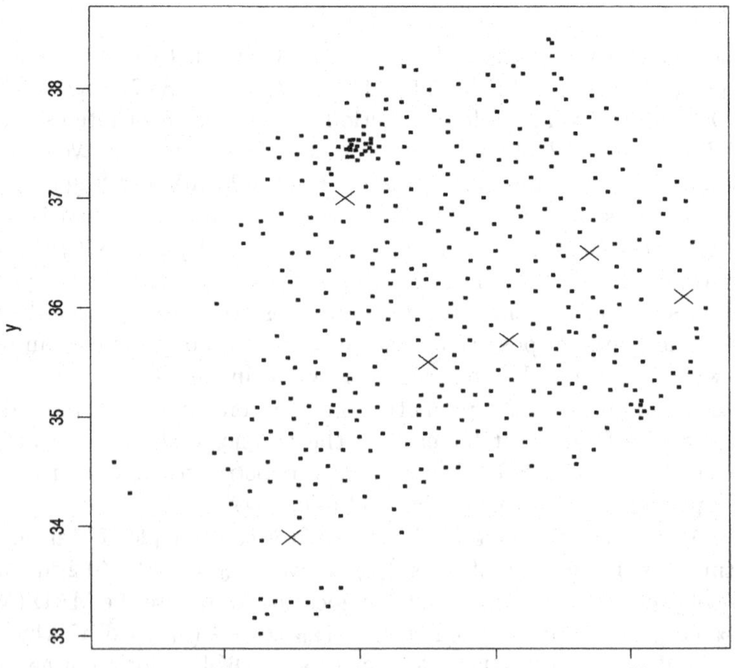

Figure 16.2 *Positions of the 365 stations with 6 new locations for prediction (crosses).*

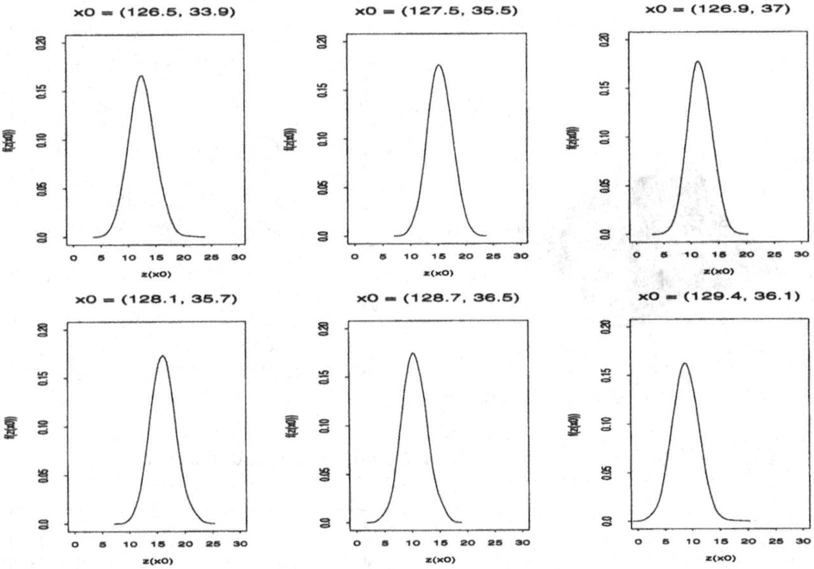

Figure 16.3 *Prediction densities for SGP 1 at 6 locations* $\mathbf{x}_0$.

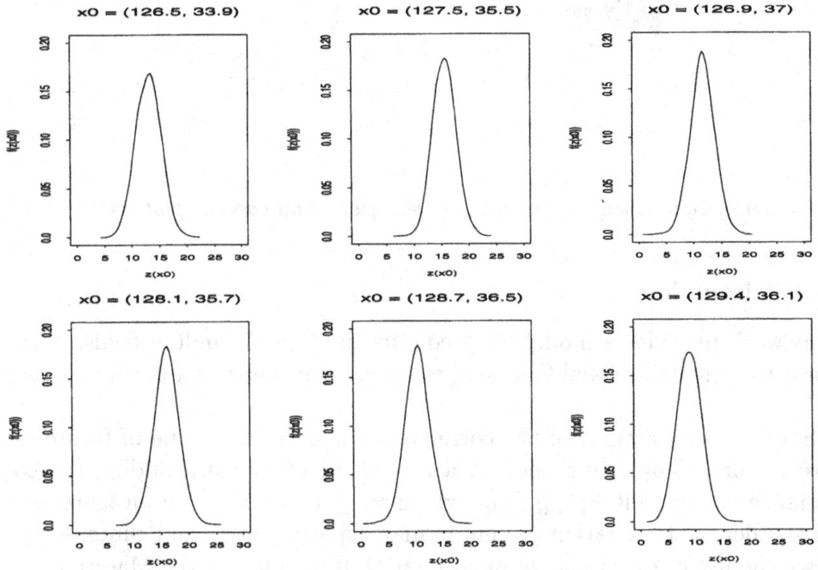

Figure 16.4 *Prediction densities for SGP 2 at 6 locations* $\mathbf{x}_0$.

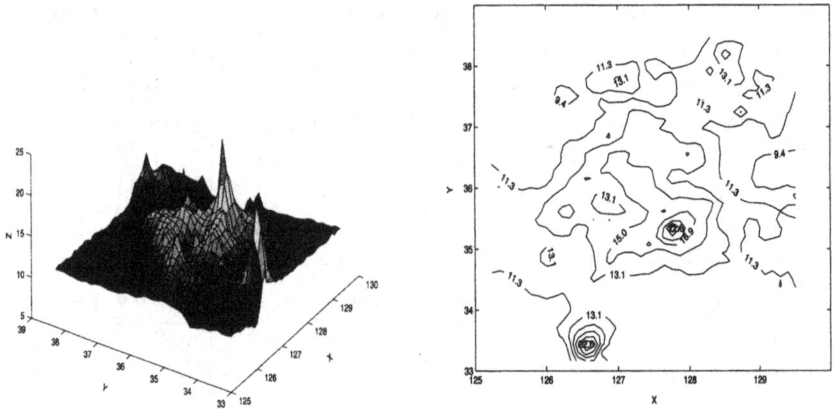

Figure 16.5 *Predicted map: 3D plot (left) and contour plot (right).*

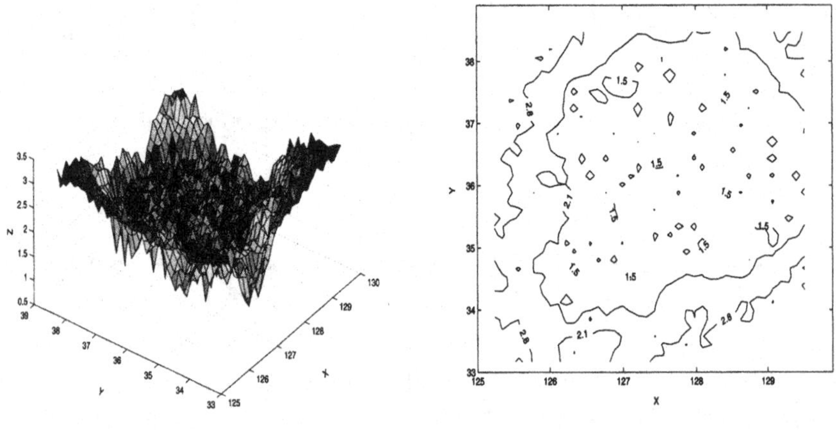

Figure 16.6 *Uncertainty map: 3D plot (left) and contour plot (right).*

16.4 Discussion

This work provides a model for prediction in skewed random fields. It extends the results of usual Gaussian random fields which is a particular case of it.

Some generalizations of the correlation functions might be of future interest. For example, inclusion of a nugget effect, namely, adding the covariance component $\theta_3 \mathbf{1}_{\{dist=0\}}$, where $\theta_3 \geq 0$ controls the measurement error/microscale variation present in many spatial data, will be interesting. Also, the use of richer classes of correlation function, like the Matérn class, could improve the prediction.

In addition, rather than using a linear model for the covariates, we can use non-linear models, which may identify significant covariate effects as well as perform better predictions.

As this chapter only deals with skew-normal processes, an extension to skew-elliptical processes is also desired. Bayesian spatial prediction using the elliptical distribution is proposed by Kim and Mallick (2003), so it is desirable to extend this idea to skew-elliptical distributions.

CHAPTER 17

Shape Representation with Flexible Skew-Symmetric Distributions

Sajjad H. Baloch, Hamid Krim, and Marc G. Genton

17.1 Introduction

The goal of shape modeling is to seek mathematical representations to not only capture the intrinsic morphologies of various shapes, but to also account for their variability. Formally, part of *pattern theory*, whose formalism is to a large extent due to Grenander in the 1970s (Grenander, 1976, 1978), seeks to unravel and to quantify the structure of patterns present in an image. A considerable research activity has yielded a wealth of perspectives and approaches, each with its advantages and limitations. The earliest approach to modeling structure was based on rigid models, and fell short of accounting for the inherent variability, as evidenced by biological and anatomical shapes, e.g., a beating heart or a stomach shape. More flexible and adapted models, namely, deformable templates, were later proposed by Grenander (1996), which addressed the statistical variability of shapes. These models specifically looked upon object boundaries as a set of sites (carefully selected landmarks) joined by arcs (or segments) whose spatial attributes were captured by probability distributions. This mathematically elegant and conceptually sound approach led to inferences on infinite dimensional spaces with an often prohibitive computational demand.

The Kendall school, in an attempt to alleviate this dimensional complexity, proposed a finite dimensional statistical description of shapes by introducing the notion of landmarks of the shape with appropriately embedding spaces normalized with respect to a set of transformation groups (Kendall, 1989; Small, 1996).

Cootes, Taylor, Cooper, and Graham (1995) later capitalized on this formulation and coupled it to the more deterministic and so-called active contour approach (first proposed by Kass, Witkin, and Terzopoulos, 1987), to propose the active appearance model (Cootes, Edwards, and Taylor, 1998). An active contour is nothing but a closed curve which is evolved to approximate a shape via an energy optimization. Details of active shape models, deformable templates, and probabilistic models may be found in Blake and Isard (1998).

An important aspect of shape modeling is the investigation of the accuracy of the learned shape. Ye, Bresler, and Moulin (2000, 2001, 2003) presented Cramér-Rao bounds on the accuracy of parametric shape estimation. Using these bounds, they found the confidence region around the true boundary and the probability with which the estimated boundary lies within the confidence region. These bounds are very important for modeling uncertainty in shapes.

Template learning has been explored from many other perspectives including wavelets by Scott and Nowak (2001). An exhaustive review of this rich literature is clearly beyond the scope of this chapter.

In contrast to all the approaches above, we view the variability of shapes as one that allows realization contours to remain within a certain neighborhood range around a "mean." This, in turn, suggests that for any given angle, as we traverse a shape, a probability density function may be found to capture the potential excursion of the curves at the given angle.

Specifically, we exploit a class of skew-symmetric distributions first introduced by Wang, Boyer, and Genton (2004a) and further generalized by Ma and Genton (2004) to incorporate additional flexibility. The work described in this chapter is built to a large extent on that framework to address a very important applied problem.

The chapter starts with the problem statement in Section 17.2, which also describes the class of shapes we will be considering. In Section 17.3, we describe the probabilistic model we wish to develop. In Section 17.4, we provide illustrating examples and some conclusions in Section 17.5.

17.2 Problem Statement

While the scope of shape analysis may be as broad as one wishes it to be, we will focus our effort, in this chapter, on a constructive methodology of probabilistic models of 2-D planar shapes.

Let a shape S_i be given by a curve $C_{S_i}(t)$:

$$C_{S_i} : I \subseteq \mathbb{R}_+ \to \mathbb{R} \times [0, \pi), \tag{17.1}$$

and for convenience and clarity we take $I \subset \mathbb{N}$ (a sampled curve). Given a set $\{C_{S_i}{}^j\}_{j=1,\dots,n}$, we ask to provide a probabilistic model for S_i in terms of its radius (from the centroid) and angle around it. Note that alternatively we may view $C_{S_i}(t)$ as a parametric representation $(x(t), y(t))$ or $\left(\sqrt{x^2(t) + y^2(t)}, \arctan\left(y(t)/x(t)\right)\right)$.

To better illustrate the overall approach, we choose to learn a specific template of a heart. By learning, we mean to extract from training data the relevant parameters of the underlying distribution of the shape. Given several realizations of a heart, we are asked to learn and to subsequently capture all objects having similar shape realizations. Heart shapes appear-

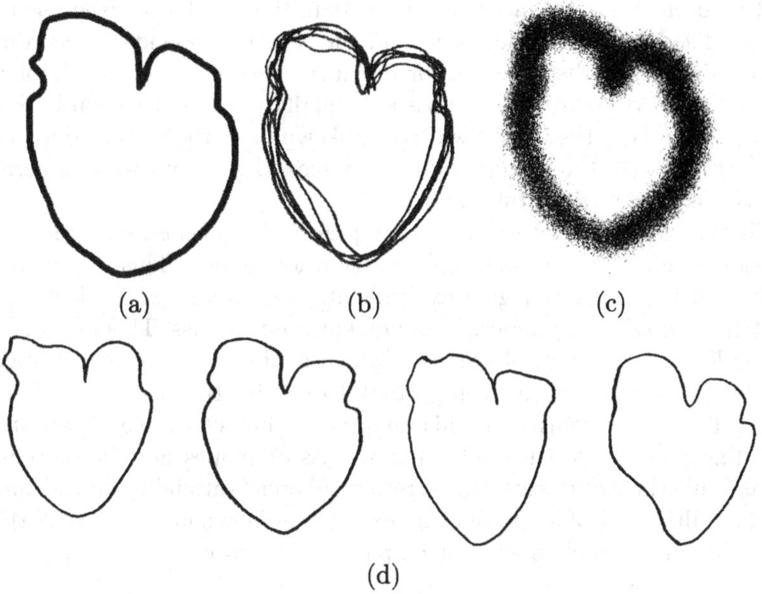

Figure 17.1 *(a) Heart shape; (b) Some realizations superimposed on each other; (c) Sampled superimposed realizations; (d) Constituent realizations.*

ing in two images may differ on any of the following counts. A heart capture may, for instance, be occluded by some other organ and hence be distorted during the imaging process. Noise due to breathing, for instance, or due to a beating heart may also corrupt an image, making edge detection and segmentation tasks difficult. Objects may also appear different from different viewing angles. Despite this variability, it is reasonable to assume that these different realizations share much in common, see Figure 17.1. This means that the realizations, when normalized to a pre-specified area, can deviate from the actual shape only by some multiple of standard deviation σ and lie within a tubular region shown in Figure 17.1(c) with probability close to 1. Hence, we proceed to model all of the deviations and to capture the essence of the shape.

The combination of all such realizations is tantamount to a cloud of points in the neighborhood of a template boundary as shown above in Figure 17.1(c). The cloud of points may be dense and would not likely be statistically independent. By sampling an image at specified angles, the points may be assumed independent and identically distributed or at most may be modeled as a first order Markov process. The boundary of any realization of a given shape, which is modeled by a template, will be a subset of points within a tubular cloud shown in Figure 17.1(c). This tubular region may be interpreted as the permissible shape domain.

Based on these realizations, the points in the cloud at a given angle may be modeled to be distributed around the template boundary according to a certain skewed distribution, for instance, skew-normal, skew-Laplace or skew-t. In particular, if the points are equally likely to occur on both sides of the boundary, then there will be no skewness in the distribution and we will be left with the corresponding non-skewed density that is generalized by the skewed representation.

Skewness of data naturally arises in practical problems and is the striving reason of our emphasis on skewed distribution models. This may be readily observed in medical imaging experiments, e.g., X-Ray or MRI, where the patient's motion and imperfect alignment are the rules. This misalignment often leads to a source of skewness which in the end, is to be captured.

Before we proceed, it is important to specify that the class of shapes we will be investigating in this chapter is that of simple closed shapes, see Figure 17.2. We also note that a class of shapes may be marginal in that while their intrinsic form is simple, inherent variability may ultimately yield multiplicity of loops. Such an example is shown in Figure 17.2(a)4 and 17.2(b)4 and this class will not be part of our discussion herein.

17.3 Shape Analysis

In an image, each pixel on a shape boundary is usually localized by Cartesian coordinates (x, y) with origin at the upper left corner of the image. In this chapter, we will find it simpler and more convenient to use a polar coordinate system, which is equivalent to the Cartesian coordinates. The polar coordinates (r, θ) of a pixel at (x, y) are given below:

$$\begin{aligned} |r| &= \sqrt{x^2 + y^2}, \\ \theta &= \arctan(y/x). \end{aligned} \tag{17.2}$$

There is one difference, however: $\theta \in [0, \pi)$ instead of $[0, 2\pi)$, the reason for which will become clear later on. For tractability and convenience of representation, we choose to translate the origin from the upper left corner to the centroid of the shape (Weisstein, 2002). If the centroid is not a regular admissible center, then take the centroid of the region of regular admissible centers (Poliannikov and Krim, 2003) as the origin.

Following the determination of a centroid for a given shape, we proceed to randomly sample it at angles $\Theta \in [0, \pi)$ according to a distribution $p(\theta)$. For a given fixed $\Theta = \theta_i$, we identify all samples lying at angle θ_i. We exploit the closure of the contour defining the shape to associate two clusters of samples on either side of the centroid at θ_i, with a relative phase difference of π. Combining the two clusters helps reduce the learning space to half the actual space. This also suggests that a bimodal conditional distribution (for fixed θ_i) is a very reasonable choice as illustrated in Figure 17.3. Depending

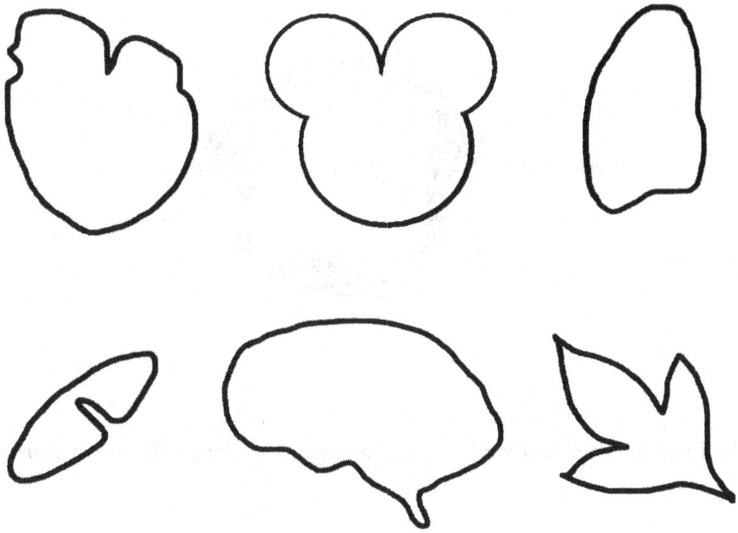

(a)

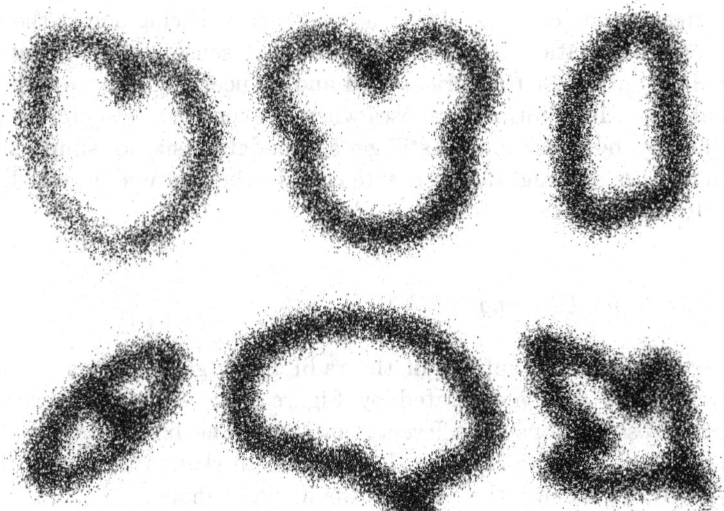

(b)

Figure 17.2 *Class of shapes: (a) Ideal shapes: (1) Heart; (2) Mickey mouse face; (3) Right lung; (4) Ellipse with a slit; (5) Human brain; (6) Sassafras leaf. (b) Corresponding realizations.*

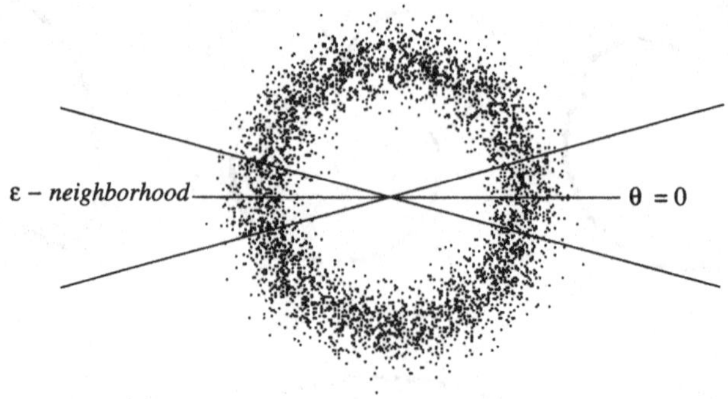

Figure 17.3 *Circular cloud sliced at angle $\theta = 0$ along with ε-neighborhood.*

on the available sample size, note that in the course of parameter learning, it may be necessary to consider an ε-neighborhood around a fixed θ_i and to weight the distance R from the centroid by $+1$ or -1 to distinguish the two corresponding clusters. As mentioned earlier, slicing across the image at a given orientation is advantageous in the sense that it results in a single distribution for that orientation and reduces angle space to half the original space. In addition, for cases when the centroid lies external to the area enclosed by the shape, we still get a bimodal density for simple shapes. It also ensures a straightforward extension to the multimodal distribution for multi-loop shapes.

17.3.1 Posterior Learning

The conditional distributions of the radii R for given angles Θ may be highly variable as demonstrated by Figure 17.4, where these represent different cross sections of different templates. The bimodal normality in Figure 17.4(a) is to be contrasted to a bimodal skewed normality in Figure 17.4(d), each being the result of the intrinsic shape.

In Wang *et al.* (2004a), the following model is proposed to represent such a class of conditional distributions:

$$p(r|\theta) = 2f\left(\frac{r-\xi}{\sigma}\right)\pi\left(\frac{r-\xi}{\sigma}\right), \tag{17.3}$$

where f is any symmetric pdf and π is a skewing function that satisfies the

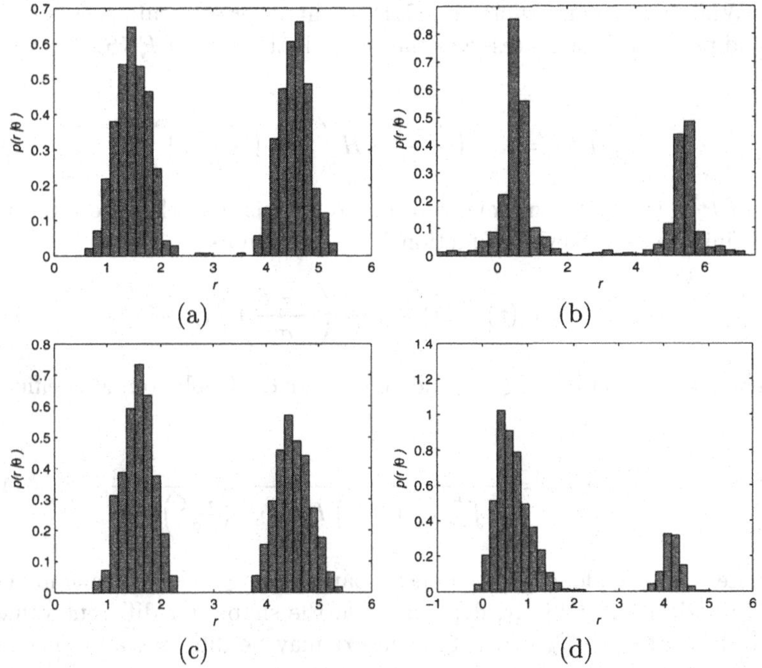

Figure 17.4 *Examples of conditional histograms for distance r given some angle $\Theta = \theta$: Note that the r-axis is normalized. (a) Bimodal normal for circular cloud for $\theta = 0$; (b) Skew-t for $\theta = 0$; (c) Normal and skewed modes for star shape for $\theta = 0$; (d) Normal and skewed modes for star shape for $\theta = \pi/2$.*

following two conditions:

$$0 \leq \pi(r) \leq 1, \qquad (17.4)$$
$$\pi(-r) = 1 - \pi(r).$$

Ma and Genton (2004) suggest a flexible skew-symmetric (FSS) representation by choosing $\pi(r) = H(P_K(r))$, where H is any cumulative distribution function of a continuous random variable that is symmetric around zero and P_K is an odd polynomial of order K. To that end, it was shown that on a compact support, FSS distributions

$$p(r|\theta) = 2f\left(\frac{r-\xi}{\sigma}\right) H\left(P_K\left(\frac{r-\xi}{\sigma}\right)\right), \qquad (17.5)$$

form a dense subclass of skew-symmetric (SS) distributions (17.3).

In the course of this investigation, we have established that an odd polynomial P_K fails to capture the symmetry of two modes of a bimodal density with a simple proof given in Appendix 17.6. To avoid this shortcoming, we propose, as an alternative, that P_K be a polynomial of an arbitrary order

K^*. While this clearly violates (17.4), an appropriate scaling factor yields a valid pdf, which also preserves the general structure of FSS. It is written as

$$p(r|\theta) = 2\omega f\left(\frac{r-\xi}{\sigma}\right) H\left(P_{K^*}\left(\frac{r-\xi}{\sigma}\right)\right), \tag{17.6}$$

where P_{K^*} is any polynomial of order K^*. This may then be interpreted as defining a new skewing function $\pi^*(r)$ written as

$$\pi^*(r) = \omega H\left(P_{K^*}\left(\frac{r-\xi}{\sigma}\right)\right), \tag{17.7}$$

where ω is a function of ξ, σ, and the vector $\boldsymbol{\alpha}$ of polynomial coefficients, given by:

$$\omega(\xi,\sigma,\boldsymbol{\alpha}) = \frac{1}{2\int_{-\infty}^{+\infty} f\left(\frac{r-\xi}{\sigma}\right) H\left(P_{K^*}\left(\frac{r-\xi}{\sigma}\right)\right) dr}. \tag{17.8}$$

Note the dependence of the density parameters ξ, σ and $\boldsymbol{\alpha}$ on θ in (17.3), (17.5) and (17.6). Clearly, depending on the shape, for different values of θ, we have that π, P_K, P_{K^*}, ξ, σ and $\boldsymbol{\alpha}$ may be different and may result in different posterior densities.

Upon specifying the conditional distribution model, and using a data sample of sufficient size, we may proceed to learn the density by standard techniques. We may, for instance, maximize the log-likelihood function:

$$\begin{aligned} L(\xi,\sigma,\boldsymbol{\alpha}) &= m\log\left(2\omega(\xi,\sigma,\boldsymbol{\alpha})\right) + \sum_{i=1}^{m}\log f\left(\frac{r_i-\xi}{\sigma}\right) \\ &+ \sum_{i=1}^{m}\log H\left(\sum_{k=1}^{K^*}\alpha_k\left(\frac{r_i-\xi}{\sigma}\right)^k\right), \end{aligned} \tag{17.9}$$

the negative of which for Gaussian pdf f may be written as

$$\begin{aligned} J(\xi,\sigma,\boldsymbol{\alpha}) &= -L(\xi,\sigma,\boldsymbol{\alpha}) \\ &= m\log\sigma - m\log\omega(\xi,\sigma,\boldsymbol{\alpha}) + \sum_{i=1}^{m}\frac{(r_i-\xi)^2}{2\sigma^2} \\ &\quad - \sum_{i=1}^{m}\log H\left(\sum_{k=1}^{K^*}\alpha_k\left(\frac{r_i-\xi}{\sigma}\right)^k\right), \end{aligned} \tag{17.10}$$

where m is the number of radius samples, $r_1,\ldots,r_m$, while the unknown parameters are the location parameter ξ, polynomial coefficients $\boldsymbol{\alpha}$, and the standard deviation σ.

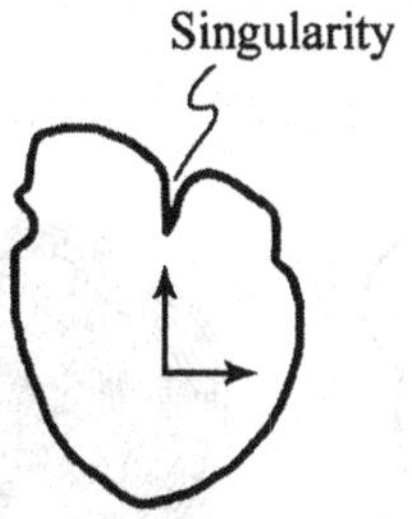

Figure 17.5 *Singularity in a shape at angle $\theta = \pi/2$.*

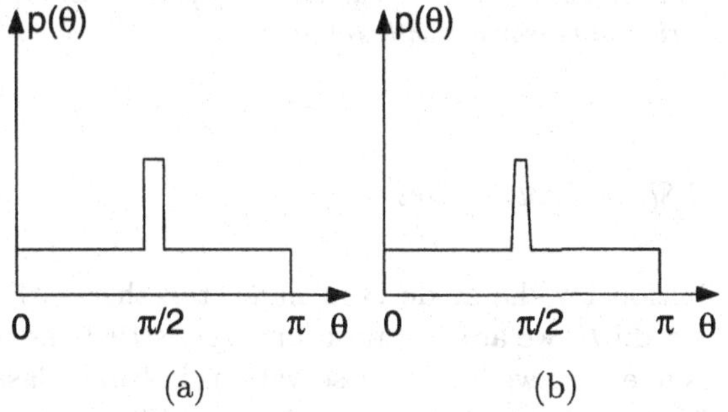

Figure 17.6 *Distribution of the angle: (a) Piecewise uniform; (b) Piecewise uniform tapered.*

17.3.2 Selection of a Distribution for the Angle

Since the points (X, Y) are scattered randomly around the mean shape, keeping in mind the dependence of R and Θ on (X, Y) according to (17.2), we conclude that both R and Θ are random. Hence, in order to completely represent a shape, we need to assign a distribution on Θ.

The challenge in determining a distribution for the angle in a shape descriptor is the presence of singularities such as that shown as a cusp in Figure 17.5. Given the presence of such events and their non-uniform occurrence throughout indicates that a better choice than a uniform distribution is desirable. The aggregation of sample pixels around a singularity heuristically explains the larger mass attributed to the sector in question. We, hence, propose instead a piecewise uniform distribution for the angle Θ, as shown in Figure 17.6(a). We have also found that better results may be achieved if a piecewise uniform tapered (PUT) distribution, shown in Figure 17.6(b), is used to avoid the discontinuities of a piecewise uniform pdf.

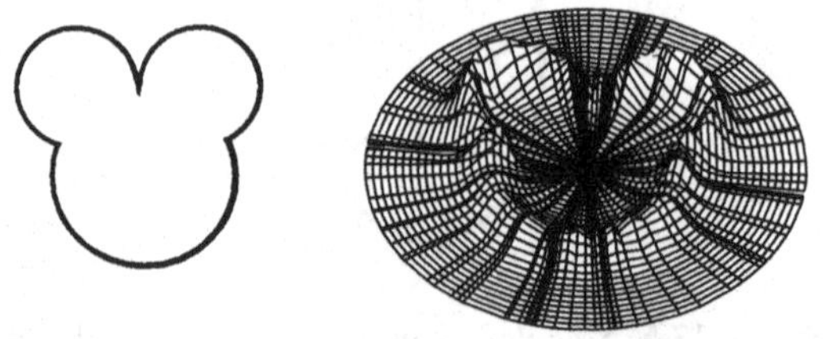

(a) (b)

Figure 17.7 *Illustration of using modes as boundary points: (a) Mickey face shape; (b) Bimodal distributions learned with 100 angles.*

17.3.3 Overall Shape Distribution

With a distribution for the angle Θ in hand, together with a conditional density for the radii R, we are in a position to construct the overall density of a shape. Assume that we have n observations in total. Based on m angle samples, $\boldsymbol{\theta} = (\theta_1, \ldots, \theta_m)^T$, drawn according to $p(\theta)$, the overall shape has a conditional likelihood:

$$p(Z_1, \ldots, Z_n|\boldsymbol{\theta}) = \prod_{i=1}^{m} \left[\prod_{\substack{j=1 \\ Z_j \in N_\varepsilon^{\theta_i}}}^{n} p(r_j|\theta_i) p(\tilde{\theta}_j) \right], \tag{17.11}$$

where $\{Z_j\}_{j=1,\ldots,n}$ is the point with radius r_j and angle $\tilde{\theta}_j$, $N_\varepsilon^{\theta_i}$ is the ε-neighborhood of $\theta = \theta_i$ and $Z_j = (r_j, \tilde{\theta}_j) \in N_\varepsilon^{\theta_i}$.

Comparing two shapes may now be carried out by way of their respective distributions. A maximum likelihood estimation technique is utilized to search for a best matching shape. Hence, corresponding to each angle, we estimate the modes of the posterior, which are assumed to lie on the boundary. In the limit, as the number of angle samples grows to infinity, the set of modes will constitute a closed contour of our template. This is illustrated in Figure 17.7, where conditional distributions $p(r|\theta_i), i = 1, \ldots, m$, given m Θ-samples drawn randomly, have been plotted for a *Mickey* face shape. An examination of the crests in the figure will reveal the presence of the desired shape pattern represented by the modes of the bimodal distributions.

The proposed technique for learning a shape is summarized in Table 17.1.

Table 17.1 *Summary of the proposed technique for learning a shape.*

Preprocessing:

1. Superimpose the realizations on top of each other. The realizations may be the output of edge detection and/or denoising algorithm(s). Sample the superimposed image.
2. Find the centroid of the aggregate of realizations. It will subsequently be the origin.
3. If not known *a priori*, find angles at which singularities are present. Densities of sample points may be used to identify their presence. Select accordingly an appropriate *PUT* angle distribution.

Shape Learning:

4. Choose m samples for $\Theta \in [0, \pi)$ according to the distribution $p(\theta)$. For each $\Theta = \theta$, repeat Step 5 through 7.
5. Slice through the image at angle θ. Identify two clusters of points that lie within some ε-neighborhood of θ.
6. Learn the bimodal conditional distribution $p(r|\theta)$, given by (17.6), for distances $\{r_j\}$ from the origin. For instance, the objective function J, given by (17.10), may be minimized to find the parameters of interest.
7. Compute the modes of the bimodal density.
8. The set of modes for all the angles will constitute the boundary.

17.3.4 Performance Assessment of the Learning Process

In order to assess the quality of a learned template, we may consider evaluating the cumulative deviation between an "*ideal*" shape, the realization of a shape as viewed from a perfect angle in a noise-free environment, and a learned template with a specific angle distribution. In addition, we will also present the relationship between the template and its realizations.

Let us denote the modes at a specified angle θ by $\hat{r}_1(\theta)$ and $\hat{r}_2(\theta)$ (noting the fact that it is a bimodal density), which represent the template boundary. For each realization from the data set, corresponding to the same angle θ, the deviation of the boundary points $r_1(\theta)$ and $r_2(\theta)$ from the modes is given by:

$$dr_i(\theta) = r_i(\theta) - \hat{r}_i(\theta). \tag{17.12}$$

For each i, the deviation in turn, has the same unimodal distribution as the underlying distribution for R with the same variance. Hence, any realization in the training set can be generated from the template by adding the corresponding deviations. In addition, any unobserved data, i.e., any future realization of the learned template, may also be drawn according to the same distribution. This implies that it will also lie in the same permissible domain.

As a performance measure, we may use an L_2-norm of the difference between the two shapes as defined below:

$$D = \sqrt{\int_0^{\pi} (dr_1^2(\theta) + dr_2^2(\theta))\, d\theta}. \tag{17.13}$$

Discretizing the angle space and considering some ε-neighborhood of θ, the departure of the learned template from the ideal shape becomes the sum of squared deviations within ε-neighborhood of all the sampled angles:

$$\tilde{D} = \sqrt{\sum_{\theta \in [0,\pi]} \sum_{\theta_j \in N_\varepsilon^\theta} \left(dr_{1,\theta_j}^2 + dr_{2,\theta_j}^2 \right)}, \tag{17.14}$$

where N_ε^θ is some ε-neighborhood around θ and dr_{i,θ_j} is the deviation for some angle θ_j in N_ε^θ. The inner summation is, therefore, the total deviation within N_ε^θ.

17.4 Experimental Results

In this section, we present some practical applications of the proposed technique. We have tried to give diverse examples related to various fields instead of only focusing on medical applications. We compare the two distributions for angle Θ discussed in Section 17.3.2 and also present the results for very fine sampling of the angle space. These results demonstrate the generality and effectiveness of the proposed method. Instead of only considering synthetic images, we have concentrated more on real images as our test cases. Later in the section, we describe how to simulate different realizations using the model (17.11). Note that in all the examples, we have employed flexible skew-normal distributions with third order polynomial ($K^* = 3$) in (17.6).

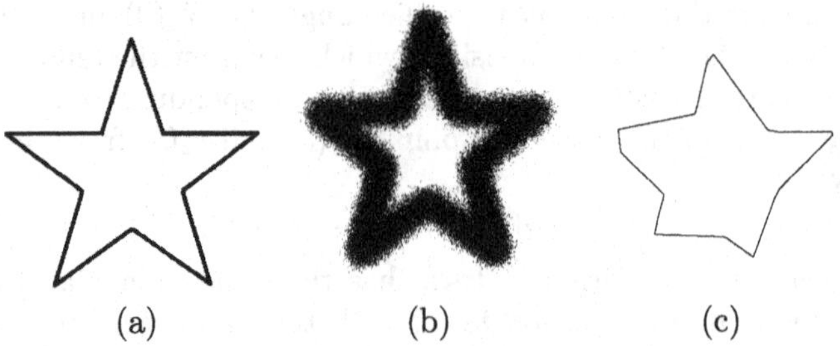

Figure 17.8 *Case study 1: (a) Ideal star shape; (b) Simulated realizations; (c) Template learned with uniform distribution for the angle.*

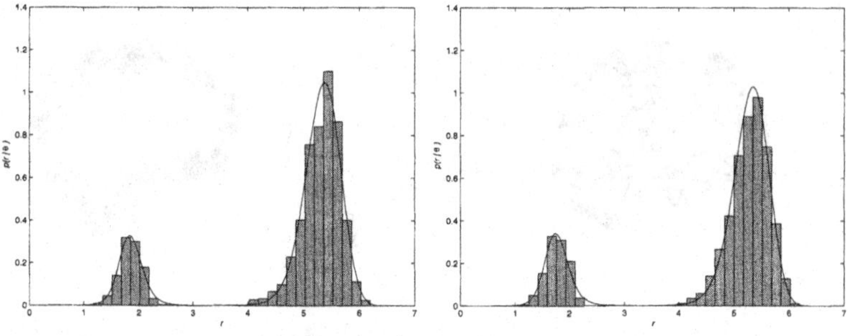

(a) (b)

Figure 17.9 *Case study 1: Two of the learned distributions, $p(r|\theta)$, for: (a) $\theta = 5\pi/17$; (b) $\theta = 20\pi/29$.*

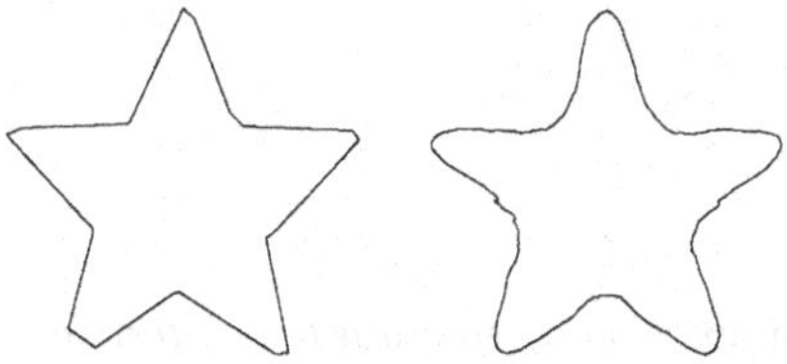

(a) (b)

Figure 17.10 *Case study 1: (a) Star template learned with 10 angle-samples drawn from PUT distribution; (b) Template using 100 angle samples.*

17.4.1 Case Study 1

As the first case study, we consider a star displayed in Figure 17.8(a), which is acquired from the US flag. A standard normal variation is added to the boundary to produce the image shown in Figure 17.8(b). This corresponds to all possible variations in acquired shape within the permissible shape domain that may appear due to the flag waving. The $[0, \pi)$ space is sampled for 10 sample points according to a uniform distribution. The learned template is given in Figure 17.8(c), while two of the learned posterior distributions for $\theta = 5\pi/17$ and $\theta = 20\pi/29$ are shown in Figure 17.9.

Using a *PUT* distribution, similar to the one in Figure 17.6(b), we get the template of Figure 17.10(a). A template learned with 100 angle samples is shown in Figure 17.10(b). Clearly, the template in Figure 17.10(a) closely resembles the ideal shape better than that in Figure 17.8(c). This is also evident by the performance measures for the two distributions that are given in Table 17.2.

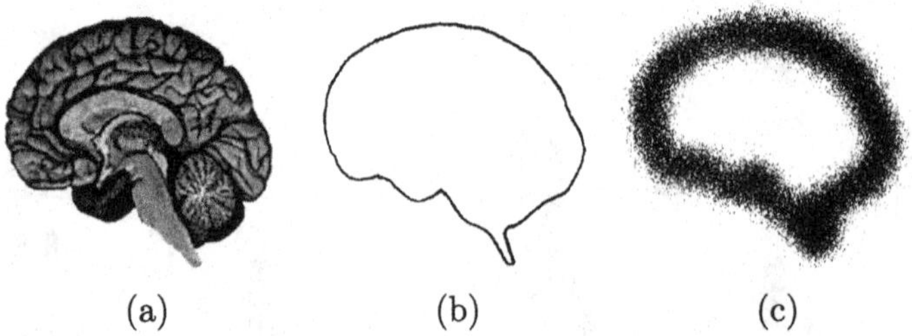

Figure 17.11 *Case study 2: (a) Human brain; (b) Brain contour; (c) Simulated realizations.*

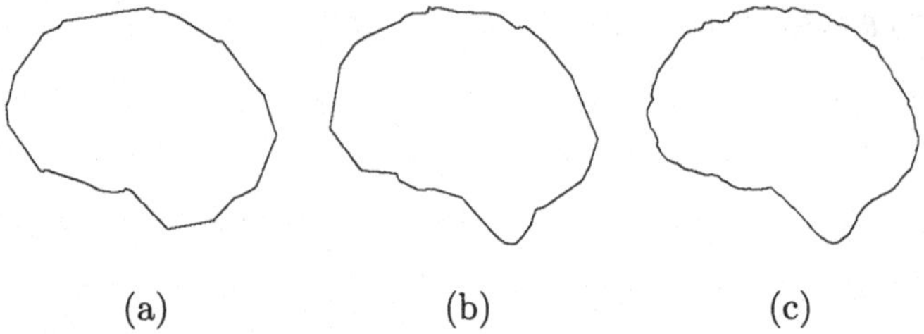

Figure 17.12 *Case study 2: Learned brain templates. (a) Using 20 angle samples drawn from a uniform distribution; (b) Using 20 angle samples drawn from a PUT distribution; (c) Using 100 angle samples drawn from a PUT distribution.*

17.4.2 Case Study 2

In this case study, we try to learn the template of a brain. The ideal shape and realizations are given in Figure 17.11.

Learned templates using 20 angle samples and 100 angle samples are shown in Figure 17.12, clearly demonstrating the advantage of a tapered uniform distribution over the uniform. The performance measures are tabulated in Table 17.2.

17.4.3 Case Study 3

The heart shape shown in Figure 17.5 has a singularity at $\theta = \pi/2$. Hence, a piecewise tapered distribution appears to be a better choice. Although Figure 17.13 visually shows that it captures singularities better than the uniform angle distribution, the performance measure in Table 17.2 is approximately the same for both cases. This is the incurred cost of weighting certain directions more than others while using a small sample size. This, in turn, results in some degradation in the shape in other directions.

With very fine sampling of the angle space, e.g., 100 samples, $\tilde{D}$ is indistinguishable for both distributions.

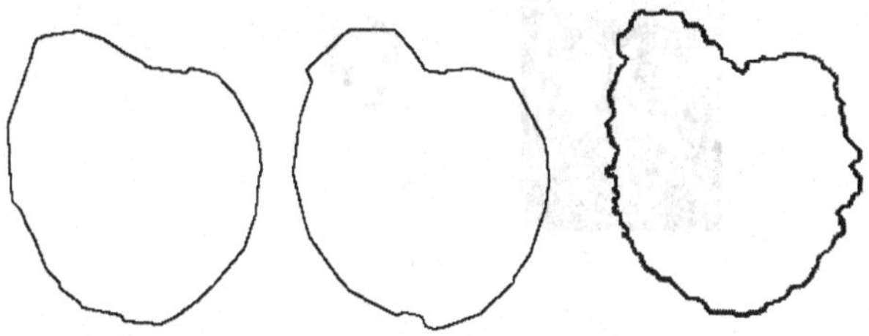

(a) (b) (c)

Figure 17.13 *Case study 3: Learned heart templates. (a) Using 20 angle samples drawn from a uniform distribution; (b) Using 20 angle samples drawn from a PUT distribution; (c) Using 100 angle samples drawn from a PUT distribution.*

Table 17.2 *Performance measure $\tilde{D}$ for three case studies.*

Case Study	Distribution for Angle Θ	$\tilde{D}$ ($\times 10^3$)
Star	Uniform (10 samples)	2.1
	PUT (10 samples)	1.6
	PUT (100 samples)	1.6
Brain	Uniform (20 samples)	0.5
	PUT (20 samples)	0.4
	PUT (100 samples)	0.4
Heart	Uniform (20 samples)	0.205
	PUT (20 samples)	0.211
	PUT (100 samples)	0.198

17.4.4 Case Study 4

The proposed scheme can also be applied to capture shapes of brain tumors. Figure 17.14(a) shows the MRI scan of a brain with a tumor. The learned template is shown in Figure 17.14(c).

17.4.5 Case Study 5

In this subsection, the results for modeling a car are presented; see Figure 17.15. Since there is no "marked" singularity in shape, we can select a uniform angle distribution. It is, however, sometimes reasonable to treat the vertices as critical points and hence, a *PUT* distribution may be used for weighting vertices more than other non-critical points.

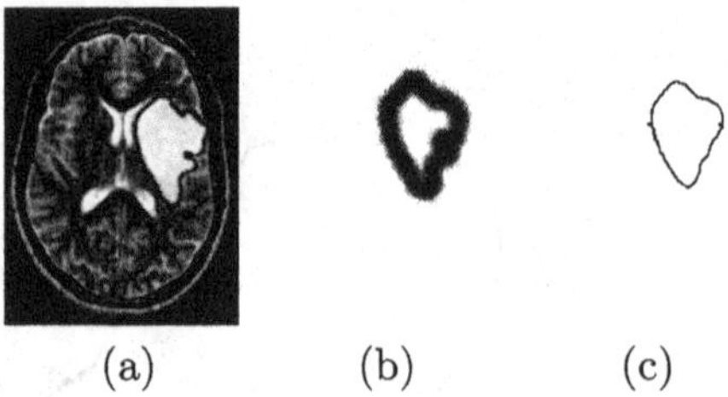

Figure 17.14 *Case study 4: (a) MRI scan of human brain; (b) Simulated realizations of tumor; (c) Learned template.*

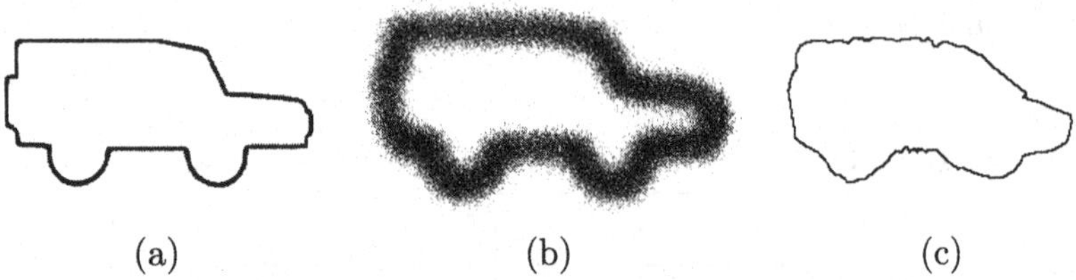

Figure 17.15 *Case study 5: (a) Ideal car shape; (b) Simulated realizations of car; (c) Learned template.*

17.4.6 Sampling from Models

To demonstrate that the distribution given by (17.11) really represents a particular class of shapes, we have simulated realizations from that distribution for a particular case of a circle with two singularities, see Figure 17.16(a).

This involves learning the conditional distribution, $p(r|\theta)$, from Figure 17.16(b), for which we used a $K^* = 3$ order polynomial. This was carried out for 100 Θ-samples drawn according to a piecewise uniform tapered distribution. Once the learning phase was complete, we proceeded to sample the distribution (17.11). A total of 10,000 $\tilde{\Theta}$-samples were drawn from the same piecewise tapered distribution, and then for each of them, the best approximation θ was selected and finally r was generated according to $p(r|\theta)$. The simulated realizations for our circular shape are displayed in Figure 17.16(c).

The comparison of Figure 17.16(b) and Figure 17.16(c) reveals that both are nearly identical. This clearly shows that any shape that lies in the permissible domain is well represented by the given model. Note that the use of the angle distribution given in Figure 17.6(b) is crucial if one is to preserve singularities.

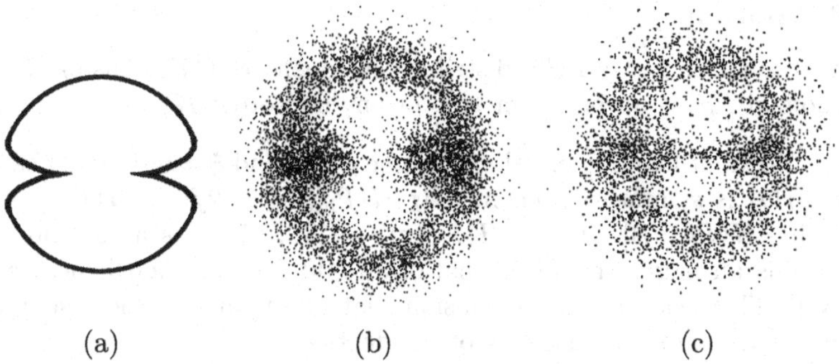

Figure 17.16 *Shape simulation according to the distribution (17.11): (a) Ideal shape; (b) Realizations; (c) Realizations simulated using parameters learned from (b).*

17.5 Conclusions

In this chapter, we discussed a novel approach to learning and modeling shapes using an extended class of flexible skew-symmetric distributions to capture the inherent variability with realizations. It involves the discretization of angle space, and the learning of the posterior density for the radii at given angles chosen according to some distribution $p(\theta)$. The shape may be recovered by capturing the local maxima (modes) of the resulting posterior for all angle samples. We presented several case studies related to different application areas using real images. We compared two possible distributions for angle sampling and discussed their role in capturing shape singularities.

Based on computer simulations, template learning by using flexible skew-symmetric distributions appears to be quite effective and robust in capturing the inherent variability of shapes. It can capture the shape singularities to some extent and may be applied to complex multi-loop templates using higher order polynomial P_{K^*} in (17.6). Using the learned model, we can simulate any shape realization, since the distribution given by (17.11) represents the entire class of shapes that lie in the permissible shape domain. The method can also be extended to 3-D shapes, which we are currently investigating.

Acknowledgments

This work was supported by an AFOSR grant F49620-98-1-0190 and by NSF grant CCR-9984067.

17.6 Appendix

Proposition 17.6.1 An FSS distribution, given by (17.5), using an odd polynomial P_K, is never symmetric for any odd order K.

Proof. Without loss of generality, assume $\sigma = 1$ and $\xi = 0$. Further assume that $p(r|\theta)$ is symmetric. Then, $2f(r)H(P_K(r)) = 2f(-r)H(P_K(-r)) = 2f(r)[1 - H(P_K(r))]$. Hence, $H(P_K(r)) = 1/2$, which is a contradiction, since H is an increasing functional of an odd function and hence cannot be even. This leads to the conclusion that two modes of the distribution $p(r|\theta)$ for the bimodal case cannot be identical. □

CHAPTER 18

An Astronomical Distance Determination Method Using Regression with Skew-Normal Errors

Laurent Eyer and Marc G. Genton

18.1 Introduction

One of the most fundamental long-standing problems of Astronomy has been the evaluation of distances to celestial objects. It took 16 centuries of investigations and human ingenuity to replace, *on observable grounds*, the paradigm of a celestial sphere of fixed stars with a moving and deep cosmos.

The first milestone was achieved by E. Halley in 1718. His line of thought was the following: the bright stars are probably the nearest; if they have a transverse motion, it is most likely to be observed in them. He therefore measured many stars and reported the detection of motion in four bright stars (Aldebaran, Arcturus, Betelgeuse, and Sirius), by comparing his observations with those of Ptolemy ($\sim$ 135 AC).

Since then, astronomers have struggled to find ways to determine the distances to the objects of interest. Huge errors have been made, scientific battles have been raging, sometimes over decades (distance to Sirius, distance to Nebulae, and in more recent years distance to the Magellanic Clouds or the Hubble constant problem).

The question of distance determination is still central for both galactic and extra-galactic astronomy and is giving rise to many studies. Even though the problems are often conceptually simple, difficulties arise from instrumentation, Earth atmosphere, astrophysics, and also from the statistical treatment of the data.

There are methods for the determination of distance that rely on geometry, and others that rely on the knowledge of absolute magnitude (logarithmic transformation of the luminosity) of stars, using them as so-called standard candles.

Here we show how we can use regression with errors having a skew-normal distribution to establish the absolute magnitude of a type of stars from a geometrical measurement of individual distances.

The structure of this chapter is as follows: we present a short introduction on the parallax in the next section, as well as a list of a few statistical problems that may be encountered. Then, in Section 18.3, we turn our attention to the Cepheids, a type of pulsating stars, and their period-luminosity relation. We try to calibrate the zero point of this relation in Section 18.4, using regression with skew-normal errors.

18.2 The Trigonometric Parallax

The parallactic effect is the apparent motion of a nearby (to the Sun) star with respect to a reference frame on the celestial sphere, caused by the revolution of the Earth around the Sun. This effect can be used to determine the distance of nearby stars. This is a purely geometrical measurement of the distance.

The parallactic motion is an ellipse in the absence of a proper motion of the observed star. The proper motion is the component of a star's velocity, relative to the Sun, perpendicular to the line of sight. If a star is located at a pole of the Earth's motion plane, then the resulting figure is a circle, whereas if it is located in the plane of the Earth's orbit, the apparent motion is a segment. The periodicity of the signal is one sidereal year.

The size of the ellipse is related to the inverse of the distance. The semi-major axis of the ellipse is called the parallax and noted ϖ. The relation between distance and parallax is $d = 1/\varpi$. If we express the angle ϖ in arcsec (second of arc) then its inverse gives the distance in parsec (1 parcsec = 3.26 light years).

It is difficult to measure the parallactic effect. This is the reason why the first parallax measured (one direct proof of heliocentric system) was only determined in the 19th century (1838) by F. Bessel. The measured effect was of 0.31 arcsec (the apparent diameter of Jupiter is about 50 arcsec at its best). It is smaller than the size of the stellar image (the stellar image size is caused by the atmosphere turbulences), which is typically larger than 1 arcsec.

Measurement of parallax is tricky because there are many factors such as the precession and nutation of the Earth's rotation axis, the aberration (combined effect of the finite light speed and the Earth motion), the differential chromatic refraction of the Earth atmosphere, the disentangling of proper motion or binary motion, and parallactic effect.

Furthermore, the absence of a really fixed reference system has been a problem. The far away extragalactic point-like sources, such as quasars, are intrinsically faint and have been difficult to use. So the measurement of parallax was generally relative to distant stars of our Galaxy that may show a systematic motion or residual parallaxes.

18.2.1 Astrometric Satellites

With the difficulties in measuring parallaxes from Earth and in achieving an equal precision over the whole celestial sphere, there has been a considerable effort from the European Space Agency (ESA) to determine positions of stars from space at the level of the milliarcsec (corresponding roughly to the angle of the height of a man at the distance of the Moon). ESA launched the Hipparcos satellite in 1989. The results of that very successful mission were delivered to scientists in 1997 (ESA, 1997). By determining the position of stars as a function of time, the Hipparcos mission determined precise coordinates, proper motion, and absolute parallax of about 120,000 preselected stars.

The Hipparcos mission raised some questions that are not solved to this day (2004). The distance of the Pleiades has been under debate: Hipparcos furnishing a distance of 118 ± 4 parsec while the traditional accepted distance is 132 ± 4 parsec (see Paczynski, 2004).

Because of the importance of distance determinations and their implications, ESA plans to launch in 2010 another satellite named GAIA. In addition to photometry and spectroscopy, very precise measurements of positions will be made (11 microarcsec at magnitude 15), leading to the determination of parallaxes. It is a survey mission that will measure about 1 billion stars (1 percent of the stars of the Milky Way). NASA is also planning a satellite, SIM, to measure very accurate positions (4 microarcsec) for a limited number of stars. SIM is scheduled for launch in 2009.

18.2.2 Some Statistical Aspects

In this section, we present several statistical aspects that can be intimately linked to each other. We can immediately foresee a difficulty: the quantity that is closest to the observations is the parallax. It is a reasonable assumption to say that a measured parallax is distributed according to a Gaussian, since it results from a least-square fit of the superposition of a proper motion and of parallactic motion. If ϖ is Gaussian then the distance is distributed according to a Cauchy distribution. The Cauchy distribution has no finite moments, e.g., no mean and no variance.

Parallax Is a Positive Quantity

Another aspect that is raising difficulties: the true parallax is always positive. Because we know in which direction the Earth orbits around the Sun, we know the direction of the projected motion of the star on the celestial sphere. There have been formulations that force the determined parallax to be positive. However, in the Hipparcos catalogue, for example, the choice of having negative measured parallaxes was done. If for individual stars it has no real significance, such negative parallaxes can nevertheless be combined in order to derive the mean distance of a group or a type of stars.

Parent Distribution

The question of determination of parallax suffers from another statistical difficulty. The distribution from which a star or a group of stars is drawn is not uniformly distributed in distance.

We can assume that the stars in the Sun's neighborhood form an exponential disk with scale height of 300 parsec. The number of stars is expected to grow as the cube of distance for a distance $\ll$ 300 parsec and as the square of the distance for a distance $\gg$ 300 parsec. This variation of the parent population (prior) can be large compared to the likelihood and sometimes cannot be neglected. The posterior probability distribution can be very asymmetrical.

As an example, let us divide a sample of objects into two classes, the faraway objects and the close objects. Because most objects are far away, they can strongly contaminate the sample of objects determined as near. Most objects that are truly close will be classified as close objects, most objects that are truly far will be classified as far, but it is possible that most objects that are determined as close can be truly faraway objects. The reservoir of far away "population" is so large that it is spilling, due to noise, over the population determined as close.

Formation of Samples

There is a well-known problem of bias due to selection of sample. When a sample is formed, the objects may be selected on their error level (noted, σ_ϖ) by removing supposed outliers or poor measurements. Furthermore the astronomical samples are often selected in a magnitude interval, faint objects being excluded: this induces a truncation in distance. All these procedures can obviously introduce a bias in the estimator. For example, to select a sample according to the smallness of σ_ϖ/ϖ can be a poor procedure.

18.3 Famous Standard Candles: the Cepheids

A variable star is a star whose luminosity is changing either for intrinsic reasons, like pulsation, or extrinsic reasons, like eclipsing binary systems. Here, we focus on periodic, pulsating stars called Cepheids.

One day in 1908, Henrietta Leavitt, an employee of Harvard Observatory, was ordering a type of variable stars, the Cepheids, from the Magellanic Clouds. The Magellanic Clouds are two small galaxies in the neighborhood of our own Galaxy; they are called the Small and Large Magellanic Clouds (abbreviated as SMC and LMC). She made the amazing discovery that these variable stars had a correlation between their period and their brightness. This was an intrinsic property of the stars since the stars composing one of these small galaxies can be considered at the same distance.

It is quite surprising that it was only in the 1960s that the reason for pulsation was really understood (see Gautschy, 1997). However, those stars have been used very rapidly to calibrate distances. Indeed, as mentioned earlier, another way to measure distance is to assume that the absolute magnitude of an object is known apriori and thus the measurement of its apparent magnitude permits us to determine the distance. As Leavitt found that there is a simple relation between the brightness of the variable stars and their period, the measurement of the pulsational period of a Cepheid and the knowledge of the period-luminosity relation permit us to determine the absolute magnitude of that particular object, and thus to determine its distance.

We show in Figure 18.1 the period-luminosity relation of Small Magellanic Cloud (SMC) Cepheids from a modern survey OGLE-II (Udalski, Soszynski, Szymanski, Kubiak, Pietrzynski, Wozniak, and Zebrun, 1999). The slope is well constrained.

There is one complication: if there are numerous Cepheids in the SMC

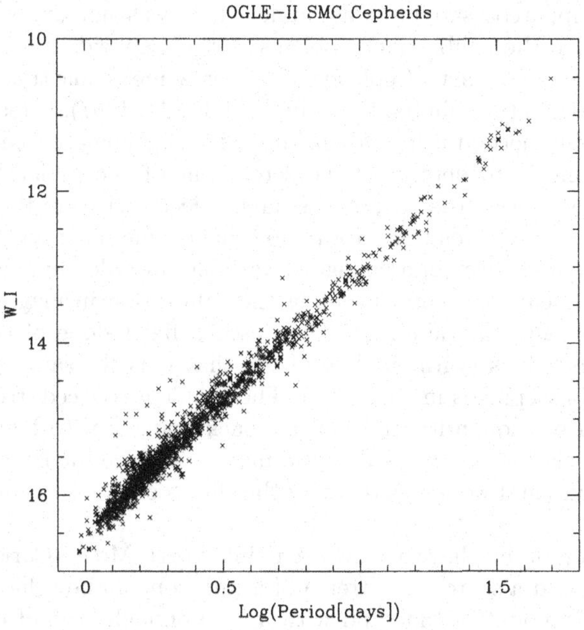

Figure 18.1 *Relation between period and luminosity of the Cepheid stars in the Small Magellanic Cloud. OGLE-II data (Udalski et al., 1999). W_I is the Wesenheit Index derived from the star magnitude and color.*

or LMC, we do not know the distance to those two galaxies. There are few Cepheids in the reach of instruments measuring parallax. So the zero point of the relation is not precisely calibrated. The question of the distance to the Magellanic Clouds has been a longstanding problem with a community schematically divided into two groups: one for the so-called short-distance scale and the other for the long-distance scale.

We have remarked that the unit for measuring distance is closely connected to the notion of parallax. We have to introduce another "unit" for distance measurement that is linked to the standard candles. Astronomers sometimes use the distance modulus, μ, instead of using distance in parsec. It is defined as the difference of absolute magnitude M to apparent magnitude m. It is related to the distance d (expressed in parsec) by $\mu = m - M = 5 \log d - 5$. As an example, the published distance moduli of LMC varies between 18.06 to 18.86 with typical estimated errors of 0.1–0.2 (Benedict, McArthur, Fredrick, Harrison, Lee, Slesnick, Rhee, Patterson, Nelan, Jefferys, van Altena, Shelus, Franz, Wasserman, Hemenway, Duncombe, Story, Whipple, and Bradley, 2002).

18.4 Calibration of the Period-Luminosity Relation

When the Hipparcos satellite was launched, it was not announced that it could resolve the calibration problems of the Cepheids, because of the "large" distance of Galactic Cepheids. However, some people tried to squeeze Hipparcos data to their limits. Feast and Catchpole (1997), hereafter noted FC, proposed a method using rather imprecise trigonometric parallaxes of Galactic Cepheids to determine the zero point of the period-luminosity relation. They showed that Hipparcos data indeed constrain the relations.

Instead of fitting a relation in a period-magnitude diagram, FC formulated the problem through a change of variable, in order to have the parallaxes as a linear parameter avoiding the calibration in magnitude with its associated asymmetric errors. They used a fixed slope of the Period-Luminosity relations (with the hypothesis that it is the same as the well studied one of Cepheids in the LMC). The problem reduced itself to compute a mean of a quantity and make a back transform to have the zero point of the relation. They could furthermore introduce the Hipparcos negative parallaxes and were able to study the effects of cuts by using different samples.

Modern computers have reached a point where Monte-Carlo methods can be used, and a search of extrema of a function can be achieved within a reasonable amount of time and with a high enough level of confidence. A discussion of the FC procedure was given by Pont (1999). He studied additional articles in the wake of the FC article by evaluating possible biases through Monte-Carlo simulations. His conclusion was that FC's result has a small, negligible bias; however, the estimation of the error is slightly higher than FC's estimation. FC give 0.10 and Pont evaluates the error at the level of 0.16.

In this chapter, we show that the calculation of the zero point of the period-luminosity relation of Cepheids can be computed using regression with errors having skew-normal distributions.

The method by FC has elegance and simplicity. Other authors (Madore and Freedman, 1998) derived the zero point by performing a weighted least squares regression in the period-magnitude diagram. This type of method is not to be recommended. However, we can stay in the physical parameters and derive the zero point and slope of the linear regression by modifying the error distributions. In case of negative observed parallaxes, it still may lead to bias due to selection effects, since, in such cases, magnitudes are not defined. If observed parallaxes are forced to be positive, the use of skewed distributions becomes natural.

18.4.1 Regression with Skew-Normal Errors

Our idea is to keep the formulation in the period-magnitude diagram; that is, we want to perform a linear regression allowing for skewness in the errors, see Sahu, Dey, and Branco (2003). Let $(M_1, x_1), \ldots, (M_n, x_n)$ be a sample where M_i denotes the absolute V magnitude and x_i the logarithm of the period. We consider the linear regression model:

$$M_i = ax_i + b + \xi_i + \delta_i|Z_i| + \sigma\epsilon_i, \qquad i = 1, \ldots, n, \tag{18.1}$$

where a is the slope, b is the intercept, σ is a scale parameter, and ξ_i and δ_i are location and skewness parameters for each observation. The distributions of the errors Z_i and ϵ_i are independent standard normal $N(0, 1)$. Therefore, the distribution of $\xi_i + \delta_i|Z_i| + \sigma\epsilon_i$ is skew-normal. One problem is that the skewness of the error distributions depends on the parallax ϖ and its accuracy σ_ϖ. Stronger asymmetries result from large ratios σ_ϖ/ϖ. Therefore, we model the skewness parameter δ_i for each observation as

$$\delta_i = \delta\frac{\sigma_{\varpi i}}{\varpi_i}, \qquad i = 1, \ldots, n, \tag{18.2}$$

where δ is a common skewness parameter for all observations. Note that the variance of the errors will be different for each observation because it depends on the skewness parameter δ_i. In the case of regression with symmetric distributions of the errors, it is common to request that the mean of the errors is zero, and this is equivalent to saying that the median or the mode (in case of unimodal densities) of the errors is zero. The case of symmetric errors corresponds to $\delta = 0$. When dealing with skewed distributions of the errors, the above three requirements are not equivalent and lead to different models. We investigate each of them.

In order to impose a zero mean for the distribution of the errors, we must choose

$$\xi_i = -\sqrt{\frac{2}{\pi}}\delta_i, \qquad i = 1, \ldots, n. \tag{18.3}$$

We estimate the parameters a, b, σ, and δ by maximizing the likelihood function $L(a, b, \sigma, \delta)$ which is equal to

$$\prod_{i=1}^{n}\left[\frac{2}{\sqrt{\sigma^2+\delta_i^2}}\phi\left(\frac{M_i - ax_i - b - \xi_i}{\sqrt{\sigma^2+\delta_i^2}}\right)\Phi\left(\frac{\delta_i(M_i - ax_i - b - \xi_i)}{\sigma\sqrt{\sigma^2+\delta_i^2}}\right)\right], \tag{18.4}$$

where δ_i is given by (18.2).

Because there is no explicit formula for the mode and median of a skew-normal distribution, we need to impose implicit conditions on the location parameters ξ_i. For example, we impose that the derivative of the pdf in (18.4) is zero at the mode. This yields the implicit condition:

$$\frac{\delta_i^2}{\sigma^2}\phi(u) = u\Phi(u), \tag{18.5}$$

where $u = -\delta_i\xi_i/(\sigma\sqrt{\sigma^2+\delta_i^2})$. For the median, we impose that the cumulative distribution function is equal to 1/2 at the median.

The estimation of standard errors of the estimates obtained from the likelihood L are computed with the usual sandwich matrix formulae; see, e.g., Stefanski and Boos (2002).

18.4.2 Discussion

For comparison reasons, we have computed the intercept with several methods.

Fixed Slope

We reproduce the result of Feast and Catchpole (1997, noted FC). We use ordinary and weighted least squares fit (noted OLS and WLS, respectively) and finally with the maximum likelihood method built on regression with skew-normal distributions of the errors and different centering rules (on the mean, mode, and median). We also compute the solution for an identical skewness parameter $\delta_i = \delta$.

From Table 18.1, we remark that our estimation does not agree with that of FC. It is surprising that all centering rules give a value larger than the one given by FC. We plan to study these differences by Monte-Carlo simulations. The error on the intercept is larger than FC and the estimation of Pont (1999) established with Monte-Carlo simulations.

The situation can be visualized in Figure 18.2. As we remark from that figure, the data points are very scattered with large error bars (derived from the transformed one-sigma limits on the parallaxes).

Table 18.1 *Fixed slope: a=slope; b=intercept (symmetric case); regression with skew-normal errors (S): (1) mean centered; (2) mode centered; (3) median centered; "s" denotes the constraint $\delta_i = \delta$.*

Meth.	a	$\hat{b}$	$se(\hat{b})$	$\hat{\sigma}$	$se(\hat{\sigma})$	$\hat{\delta}$	$se(\hat{\delta})$
FC:	−2.81	−1.430	0.100	–	–	–	–
OLS:	−2.81	−1.086	0.219	1.115	–*	–	–
WLS:	−2.81	−1.149	0.122	0.531	–*	–	–
S(1):	−2.81	−0.821	0.180	0.399	0.170	2.550	0.264
S(2):	−2.81	−0.766	0.172	0.667	0.111	−1.590	0.287
S(3):	−2.81	−0.668	0.198	0.618	0.072	−2.445	0.709
S(1s):	−2.81	−1.096	0.247	0.682	0.157	−1.399	0.457
S(2s):	−2.81	−0.801	0.303	0.682	0.172	−1.399	0.453
S(3s):	−2.81	−0.392	0.465	0.682	0.158	−1.399	0.459

(* not given for OLS/WLS)

Table 18.2 *Free slope: a=slope; b=intercept (symmetric case); regression with skew-normal errors (S): (1) mean centered; (2) mode centered; (3) median centered; "s" denotes the constraint $\delta_i = \delta$.*

Meth.	$\hat{a}$	$se(\hat{a})$	$\hat{b}$	$se(\hat{b})$	$\hat{\sigma}$	$se(\hat{\sigma})$	$\hat{\delta}$	$se(\hat{\delta})$
OLS:	−1.665	0.987	−2.025	0.837	1.108	–*	–	–
WLS:	−1.860	0.605	−1.908	0.498	0.516	–*	–	–
S(1):	−2.139	0.295	−1.376	0.371	0.370	0.161	2.502	0.301
S(2):	−2.114	0.505	−1.341	0.541	0.654	0.170	−1.547	0.451
S(3):	−2.093	0.553	−1.276	0.550	0.609	0.096	−2.369	0.707
S(1s):	−1.920	0.365	−1.825	0.415	0.717	0.142	−1.289	0.403
S(2s):	−1.920	0.344	−1.591	0.376	0.717	0.147	−1.289	0.398
S(3s):	−1.920	0.355	−1.244	0.363	0.717	0.148	−1.289	0.390

(* not given for OLS/WLS)

Free Slope

From Table 18.2, we deduce the slope and the zero-point when the constraint of fixed slope is relaxed.

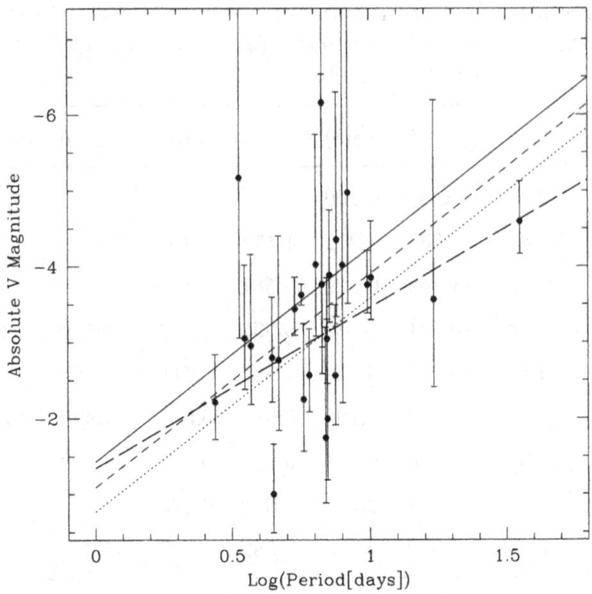

Figure 18.2 *Period luminosity relations of 26 Galactic Cepheids with absolute magnitude derived from Hipparcos parallaxes. The error bars are the transformation of the one-sigma limits in parallax into absolute magnitude. Different regression lines are presented. Solid line: Solution by Feast and Catchpole (1997). Short dashed line: Ordinary least-squares with fixed slope. Dotted line: Maximum likelihood with skew-normal distributions and fixed slope (mode centered). Long dashed line: Maximum likelihood with skew-normal distributions (mode centered) to find the regression line.*

The method of FC is more difficult to adapt to this case. Arenou and Luri (1999) proposed a method based on the so-called Absolute Based Luminosity, in the spirit of the FC method. Knapp, Pourbaix, Platais, and Jorissen (2003) used that method for a sample of Long Period Variable stars.

We computed OLS and WLS solutions, as well as with the maximum likelihood method built on regression with skew-normal distributions of the errors and different centering rules (on the mean, mode, and median). The maximum likelihood is easily generalized to this case. The slope is systematically larger than the one derived from the LMC.

We suffer from the small number of measurements. The data does not provide sufficient constraints. Even though the stars of the SMC are metal deficient compared to the stars of our Galaxic disc, which could affect the slope of the period-luminosity relation, it should not be markedly different.

FC are able to study their results with samples of different sizes (including those stars with negative parallaxes). In our case, the bias in our procedure should be evaluated with Monte-Carlo simulations.

We also await future better measurement of parallaxes to calibrate this zero point with our method.

Acknowledgments

We would like to thank Dr. D. Pourbaix for informative and stimulating discussions.

CHAPTER 19

On a Bayesian Multivariate Survival Model with a Skewed Frailty

Sujit K. Sahu and Dipak K. Dey

19.1 Introduction

Multivariate survival data arise when each study subject may experience multiple events or when study subjects are clustered into groups. Examples of such data include the recurrence times of a certain disease and the survival times of members of a family or litter.

Suppose that the survival time of the j-th subject ($j = 1, \dots, m$) in the i-th group ($i = 1, \dots, n$) is denoted by T_{ij}. The conditional hazard function of T_{ij} is often modeled as the product of: (i) an individual level random effect called the frailty, (ii) a baseline hazard function, and (iii) a proportional hazard model, which takes into account the effect of the covariates. That is, given the covariates $\mathbf{z}_{ij}$ and the unobserved frailty parameters w_i, the hazard function is modeled as

$$h(t_{ij}|\mathbf{z}_{ij}, w_i) = w_i h_o(t_{ij}) \exp(\boldsymbol{\beta}^T \mathbf{z}_{ij}), \tag{19.1}$$

where $h_o(\cdot)$ is the baseline hazard function and $\boldsymbol{\beta}$ is the regression parameter. Observe that the frailty parameter w_i can be specified equivalently by the one-to-one transformation $b_i = \log w_i$ as well.

The main focus of the current chapter is to develop flexible models for the frailty and the baseline hazard function. In particular, we adopt the following two new approaches to modeling.

- We develop a class of log-skew-t frailty distribution for modeling the dependence. The class includes the log-skew-normal distribution along with other heavy tailed distributions such as the log-skew-t distribution as special cases. These heavy-tailed distributions allow modeling of extreme observations and as a result provide very flexible models for frailty.
- We propose a correlated prior process for the baseline hazard function, which jumps according to a time-homogeneous Poisson process. Thus the number and positions of the jump times are not fixed in advance, but are estimated using data and prior assumptions.

Frailty is a random effect common to the individuals of the same group or cluster; see, for example, the recent book by Ibrahim, Chen, and Sinha (2001b) for a general introduction and early references. A convenient choice of the frailty distribution is the gamma distribution since it provides conjugate sampling distributions for Gibbs sampling (Gelfand and Smith, 1990). However, it has many drawbacks; see, e.g., Hougaard (1986). For example, the gamma distribution attenuates the covariate effect. To overcome such problems, Hougaard (1986) used the positive stable frailty distributions, which have been followed up by Qiou, Ravishanker, and Dey (1999) in the Bayesian context.

In this chapter, we consider an alternative class of frailty distributions using the log-skew-t distributions, following Sahu, Dey, and Branco (2003). The class *without the log transformation* reduces to the family of normal (t or Cauchy) distributions for particular values of the parameters. On the other extreme, it behaves as a half-normal (half-t or half-Cauchy) distribution by placing all its mass on the positive side of the real line. Thus the family of distributions is quite flexible and general, and the existence of the variance of the frailty distribution is not assumed since the degrees of freedom of the t distribution can be less than two.

In the Bayesian setup of the current chapter, a suitable stochastic process needs to be considered for the baseline hazard function. Several parametric and nonparametric models are available; see, e.g., Sinha and Dey (1997) for a review. Here we adopt a piecewise constant baseline hazard function. This choice is popular when modeling univariate survival data; see, for example, Gamerman (1991), Arjas and Gasbarra (1994), and McKeague and Tighiouart (2000).

The correlated prior process imposes smoothness on the baseline hazard function in adjacent intervals. In particular, we generalize a first order autoregressive process considered in Sahu, Dey, Aslanidou, and Sinha (1997). In addition, we assume that the endpoints of the interval themselves form a time-homogeneous Poisson process. This introduces further flexibility since the number of endpoints where jumps are allowed to occur is left unknown.

The full Bayesian model is rather complex and does not allow fitting and comparison using analytic methods. The straightforward Gibbs sampler is also not able to handle the computations since the parameter space is of varying dimension. Thus we develop Bayesian computation methods using the reversible jump MCMC method; see, for example, Green (1995).

A natural next step after Bayesian model fitting is to investigate the issues relating to model adequacy and model choice. Here we adopt familiar predictive Bayesian model choice criteria and adequacy checks for comparing different models. We do not consider the Bayes factor because it is not meaningful for models accommodating improper prior distributions and it is more computationally expensive. Instead, we use the easily computed pseudo-Bayes factors – see, e.g., Geisser and Eddy (1979) – which are interpretable even when some prior distributions are vague. In this chapter we do adopt vague prior distributions for some parameters; see Section 19.4.

The remainder of the chapter is organized as follows. Section 19.2 introduces the frailty distributions. The baseline hazard function is discussed in Section 19.3. The likelihood and prior specifications are discussed in Section 19.4. Section 19.5 develops computing methods and may be omitted without loss on a first reading. In Section 19.6 we provide two numerical examples. We conclude with a few summary remarks in Section 19.7.

19.2 Frailty Models

19.2.1 Frailty Distributions

The popular gamma frailty distribution assumes that W_i, $i = 1, \ldots, n$ are independent and identically distributed (iid), each having the gamma probability density function (pdf)

$$f(w|\eta) = \frac{\eta^\eta}{\Gamma(\eta)} w^{\eta-1} \exp(-w\eta), \; w > 0. \tag{19.2}$$

Here η^{-1} is the unknown variance of the W_i, and the strength of the association between the survival times is a non-increasing function of η.

The positive stable distribution has been suggested as a suitable frailty distribution on the grounds that it preserves the proportional hazard assumption in the marginal model. A good introduction to this distribution is given by Hougaard (2000). The distribution is specified by its Laplace transform $\mathrm{E}[\exp(-sW)] = \exp(-s^\alpha)$, where $0 < \alpha \leq 1$. The components of the multivariate survival distribution induced by the stable frailty are locally independent when $\alpha = 1$. Other values of α introduce dependence among the components.

The pdf of the positive stable distribution is non-standard. We adopt the version obtained by Buckle (1995), see also Ravishanker and Dey (2000). The pdf is given by

$$f(w|\alpha) = \frac{\alpha\, w^{1/(\alpha-1)}}{|\alpha - 1|} \int_{-1/2}^{1/2} \exp\left\{ - \left| \frac{w}{d_\alpha(y)} \right|^{\alpha/(\alpha-1)} \right\} \left| \frac{1}{d_\alpha(y)} \right|^{\alpha/(\alpha-1)} dy, \tag{19.3}$$

where

$$d_\alpha(y) = \frac{\sin(\pi\alpha y + s_\alpha)}{\cos(\pi y)} \left[\frac{\cos(\pi y)}{\cos\{\pi(\alpha-1)y + s_\alpha\}} \right]^{(\alpha-1)/\alpha},$$

and $s_\alpha = \min(\alpha, 2-\alpha)\pi/2$ and $0 < \alpha \leq 1$. It is well-known that the mean and variance of the positive stable distribution do not always exist. However, the mean and variance of $B = \log W$ are available in closed form and are given by

$$\mathrm{E}(B) = -\left(\frac{1}{\alpha} - 1\right)\psi(1), \quad \mathrm{Var}(B) = \left(\frac{1}{\alpha^2} - 1\right)\frac{\pi^2}{6},$$

where $\psi(x)$ is the digamma function. These expressions will be useful when comparing the different frailty distributions.

We contrast the above two frailty distributions with a new class of frailty distributions called the log-skew-t distributions obtained by Sahu *et al.* (2003). We suppose that the frailty parameters $B_i, (= \log W_i)\ i = 1, \ldots, n$ are iid for every group with the following pdf,

$$\begin{aligned} f_B(b|\delta,\nu) &= 2\,(1+\delta^2)^{-1/2}\frac{\Gamma\left(\frac{\nu+1}{2}\right)}{\Gamma(\nu/2)(\nu\pi)^{1/2}}\left[1+\frac{b^2}{\nu(1+\delta^2)}\right]^{-(\nu+1)/2} \\ &\quad \times T_{\nu+1}\left[\sqrt{q(b)}\,\delta\,\frac{b}{\sqrt{1+\delta^2}}\right], \end{aligned} \tag{19.4}$$

where

$$q(b) = \frac{\nu+1}{\nu + b^2/(1+\delta^2)},$$

and $T_m(\cdot)$ is the cumulative distribution function (cdf) of the standard t distribution with m degrees of freedom (df). The parameters δ and ν influence the shape of the distribution as discussed below.

First, observe that with $\delta = 0$ the above pdf reduces to the standard t density with ν df. In addition, if $\nu \to \infty$ then it approaches the normal distribution. Second, with $\delta = 0$ and $\nu = 1$, (19.4) is the pdf of the Cauchy distribution. Clearly, the mean and variance of B do not exist in this case. Third, skewed distributions emerge for non-zero values of δ. For positive values of δ the distribution is positively skewed and for negative values it is negatively skewed. The mean of the distribution exists if $\nu > 1$ and the variance exists if $\nu > 2$. We note the mean and variance of B for future reference. These are given by

$$\mathrm{E}(B) = \left(\frac{\nu}{\pi}\right)^{1/2}\frac{\Gamma[(\nu-1)/2]}{\Gamma(\nu/2)}\delta, \text{ when } \nu > 1,$$

and

$$\mathrm{Var}(B) = (1+\delta^2)\frac{\nu}{\nu-2} - \frac{\nu}{\pi}\left(\frac{\Gamma[(\nu-1)/2]}{\Gamma(\nu/2)}\right)^2\delta^2, \text{ when } \nu > 2. \tag{19.5}$$

The pdf of $W = \exp(B)$ implied by (19.4) is given by

$$f_W(w|\delta,\nu) = \frac{1}{w} f_B\left(\log w|\delta,\nu\right), \tag{19.6}$$

where $f_B(\cdot|\delta,\nu)$ is given in (19.4).

19.2.2 Comparison of Frailty Distributions

Frailty models are usually compared by studying the dependence structures induced by them. However, there are two types of dependence structures to consider: global and local. Measures for local dependence structures include the cross-ratio function, which is often defined by taking an appropriate ratio of the derivatives of the joint and marginal survival function. The quantities required to form the cross-ratio function are not available in analytic closed form under the proposed log-skew-t frailty distributions, thus limiting the scope of theoretical comparisons to be made.

To study the frailty models using a global measure of dependence we use the correlation of log-survival times, though there are other measures available. It is not meaningful to study the correlation between the survival times themselves since the moments of the frailty distribution of W do not always exist. To simplify the exposition, suppose that the survival times are bivariate with two components denoted by T_1 and T_2. Assuming that the baseline hazard function is Weibull, Hougaard (2000, p. 227) showed that

$$\text{Corr}(\log T_1, \log T_2) = \frac{\text{Var}(B)}{\text{Var}(B) + \pi^2/6}. \tag{19.7}$$

For the gamma frailty model $\text{Var}(\log W) = \psi'(\eta)$. For the stable frailty model $\text{Corr}(\log T_1, \log T_2)$ is given by $1 - \alpha^2$. The correlation of the log-survival times under the log-skew-t model is obtained using (19.5).

The above analytical results facilitate comparisons between the frailty distributions. To illustrate, we set $\nu = 8$ to have a moderate tail in the log-skew-t frailty distribution. Now we equate $\text{Var}(\log W)$ under the three different frailty models. In addition, we suppose that the $\text{E}(\log W)$ under the stable and the log-skew-t models are equal. Solving these equations, we obtain $\eta = 1.14$, $\alpha = 0.74$, and $\delta = 0.23$. The resulting $\text{Corr}(\log T_1, \log T_2)$ is $1 - 0.74^2 \approx 0.45$.

Since the tail of the frailty distribution plays an important role in dictating the dependence structure, we investigate the shape and tail of the frailty densities for the above parameter values. The densities are plotted in Figure 19.1. The tail of the log-skew-t density is heavier than the gamma but lighter than the stable density (19.3). Thus it is seen that for the same amount of correlations between the log survival times, the log-skew-t distribution provides a flexible alternative to the heavy tailed stable distribution and light tailed gamma distributions.

Note that if W has the gamma distribution (19.2) then $\text{E}(\log W) = \psi(\eta) - \log \eta$, which is always non-positive. However, under the stable frailty model $\text{E}(\log W)$ is always non-negative. This is the reason for not equating $\text{E}(\log W)$ for the gamma case with that of the other frailty distributions.

The proposed log-skew-t frailty distribution has one more parameter than the gamma or stable frailty distribution and an anonymous referee has

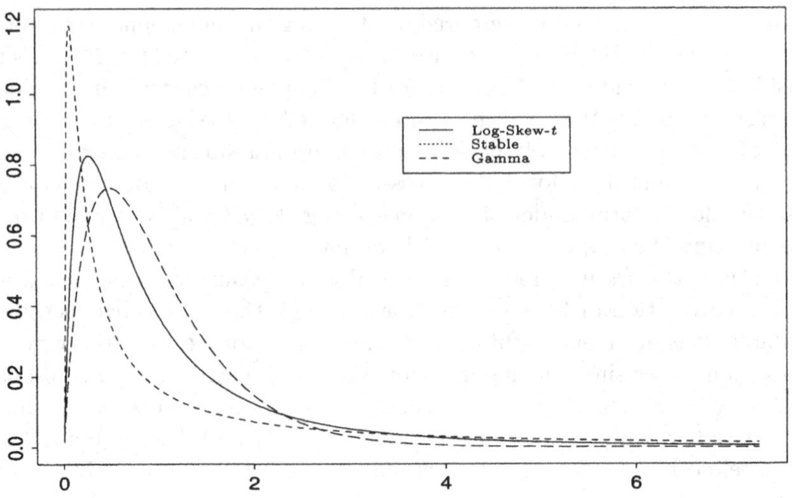

Figure 19.1 *Densities of frailty distributions. The graphs are for (i) log-skew-t distribution with* $\delta = 0.23$ *and* $\nu = 8$, *(ii) positive stable with* $\alpha = 0.74$, *and (iii) gamma distribution with* $\eta = 1.14$.

noted that the flexibility in modeling arises because of this extra parameter. This is indeed true and as a result some model comparison measures that penalize complexity of a model are to be adopted in practical situations. Such comparisons have been made in the real data examples in Section 19.6.

19.3 Baseline Hazard Function

The assumption of correlated prior processes for the baseline hazard function is very common in survival analysis. See, for example, Gamerman (1991), Arjas and Gasberra (1994), and Sinha and Dey (1997) for a review. The setup is as follows.

Suppose that time is divided into g pre-specified intervals $I_k = (\tau_{k-1}, \tau_k]$ for $k = 1, \ldots, g$ where $0 = \tau_0 < \tau_1 < \ldots < \tau_g < \infty$, τ_g being the last survival or censored time. Assume that the baseline hazard is constant within each interval. That is,

$$h_o(t_{ij}) = h_k, \quad \text{for} \quad t_{ij} \in I_k. \tag{19.8}$$

Following Sahu *et al.* (1997) we assume a martingale process prior for $\lambda_k = \log h_k$. We assume that

$$\lambda_k | \lambda_1, \ldots, \lambda_{k-1} \sim N(\lambda_{k-1}, \sigma^2), \tag{19.9}$$

with $\lambda_0 = 0$. Let $\boldsymbol{\lambda} = (\lambda_1, \ldots, \lambda_g)^T$. See, e.g., Gamerman (1991), and Arjas and Gasbarra (1994) for more general models.

Several authors have suggested different choices for the number of grid points g. Some early references include Breslow (1974), and Kalbfleisch and Prentice (1973). More recent solutions to this problem suggest leaving g unspecified; see, for example, Arjas and Heikkinen (1997), and McKeague and Tighiouart (2000). Following this last article we assume that the jump times $\tau_1, \tau_2, \ldots$ form a time-homogeneous Poisson process. Equivalently, this leads to the assumption that g follows the Poisson distribution with a suitable parameter that will be discussed later. This has several advantages over a fixed value of g. One such advantage is that the number and positions of the grid points need not be fixed in advance.

19.4 Model Specification

19.4.1 Likelihood Specification

We only consider right-censored survival data and assume that the censoring is non-informative. Let δ_{ij} denote the indicator variable taking value 1 if the j-th subject ($j = 1, \ldots, m$) of the i-th group ($i = 1, \ldots, n$) fails and value 0 otherwise. Hence t_{ij} is a failure time if $\delta_{ij} = 1$ and a censoring time otherwise. Recall that $\mathbf{z}_{ij}$ is the co-variate vector for subject j in the i-th group. Thus the triplet $(t_{ij}, \delta_{ij}, \mathbf{z}_{ij}^T)$ is observed for all i and j. Let $(\mathbf{x}, \mathbf{z})$ denote the collection of all such triplets $(t_{ij}, \delta_{ij}, \mathbf{z}_{ij}^T)$. Let $\mathbf{w}$ denote the vector of unobserved w_i's. Finally, let $\theta_{ij} = \exp(\boldsymbol{\beta}^T \mathbf{z}_{ij})$ for all i and j.

The likelihood is derived as follows. The j-th subject of the i-th group has a constant hazard of $h_{ij} = w_i h_k \theta_{ij}$ in the k-th interval ($k = 1, \ldots, g$) given the unobserved frailty w_i. If the subject has survived beyond the k-th interval, i.e., $t_{ij} > \tau_k$, the likelihood contribution is $\exp\{-h_k \Delta_k \theta_{ij} w_i\}$ where $\Delta_k = \tau_k - \tau_{k-1}$. If the subject has failed or was censored in the k-th interval, i.e., $\tau_{k-1} < t_{ij} \leq \tau_k$ then the likelihood contribution is

$$(h_k \theta_{ij} w_i)^{\delta_{ij}} \exp\{-h_k(t_{ij} - \tau_{k-1})\theta_{ij} w_i\}.$$

Recall that $\lambda_k = \log h_k$ for all values of k. Hence, given the number of intervals g, we arrive at the following likelihood function $L_g(\boldsymbol{\beta}, \boldsymbol{\lambda}, \mathbf{w}, \mathbf{x}, \mathbf{z})$ say,

$$\prod_{i=1}^{n} \prod_{j=1}^{m} \left\{ \prod_{k=1}^{g_{ij}} \exp(-h_k \Delta_k \theta_{ij} w_i) \right\} \left(h_{g_{ij}+1} \theta_{ij} w_i\right)^{\delta_{ij}} \times \exp\left\{-h_{g_{ij}+1}(t_{ij} - \tau_{g_{ij}})\theta_{ij} w_i\right\}, \qquad (19.10)$$

where g_{ij} is such that $t_{ij} \in (\tau_{g_{ij}}, \tau_{g_{ij}+1}] = I_{g_{ij}+1}$.

19.4.2 Prior Specification

The joint prior distribution of all the parameters is given by

$$\prod_{i=1}^{n} [\pi(w_i|\cdots)] \, \pi(\cdots) \; \pi(\boldsymbol{\beta}) \; \pi(g) \; \pi(\boldsymbol{\lambda}|g), \tag{19.11}$$

where $\cdots$ denote the hyper-parameters of the frailty distribution. We assume vague prior distributions for the components of $\boldsymbol{\beta}$. Thus each component of $\boldsymbol{\beta}$ is assumed to follow the normal distribution with mean zero and a large variance (10^4) independently. The frailty distribution can be any one of the gamma, stable, and log-skew-t distributions given in (19.2), (19.3), and (19.6), respectively. In the gamma frailty case the hyper-parameter is η. We assume that η follows a gamma distribution with mean 1 (for identifiability) and large variance to incorporate flexibility, $Gamma(\phi, \phi)$ say with a small choice of ϕ. For the parameter α in the stable frailty model we specify the beta prior distribution since the beta family of distributions is quite flexible.

Under the log-skew-t frailty distribution given by (19.6), we have two hyper-parameters, δ and ν. We assume that δ follows the uniform distribution in an interval $[0, q]$ with pre-specified q. We restrict δ to a bounded interval in order to avoid the conflict between fat tail and skewness in (19.6). Negative values of δ are not considered since we intend to have a positively skewed t distribution.

We now specify a suitable prior distribution for ν, the degrees of freedom parameter. We assume that $\nu > 1$ so that the underlying skew-t distribution has finite mean. Observe that the log-skew-t frailty distribution (19.6) is well-defined for positive values of ν. Thus we treat ν as a continuous random variable taking values greater than 1. We propose the exponential distribution with mean κ but truncated to be greater than 1 as the prior distribution for ν. Finally, we recommend a moderate value of κ (between 5 and 15) so that the heavy tailed log-skew-t distributions are put forward as prior distributions.

Earlier, we have assumed that g follows the Poisson distribution. The parameter of this distribution will be specified in the next subsection. Given g, the martingale specification (19.9) implies that

$$\boldsymbol{\lambda} \sim N_g\left(\mathbf{0}, \sigma^2 C^{-1}\right), \tag{19.12}$$

where the $g \times g$ matrix C has all elements zero except for $c_{kk} = 2$, $k = 1, \ldots, g$ and $c_{k,k+1} = -1$, $k = 1, \ldots, g-1$ and $c_{k-1,k} = -1$ for $k = 2, \ldots, g$.

Let $\boldsymbol{\zeta}$ denote the collection of parameters, for which the prior distribution is given by (19.11). The joint posterior density of $\boldsymbol{\zeta}$ is simply proportional to the likelihood (19.10) times the prior (19.11), i.e.,

$$\pi(\boldsymbol{\zeta}|\mathbf{x},\mathbf{z}) \propto L_g(\boldsymbol{\beta},\boldsymbol{\lambda},\mathbf{w},\mathbf{x},\mathbf{z}) \times \prod_{i=1}^{n} [\pi(w_i|\cdots)]\, \pi(\cdots)\ \pi(\boldsymbol{\beta})\ \pi(g)\ \pi(\boldsymbol{\lambda}|g). \tag{19.13}$$

19.4.3 Hyper-Parameter Values and Prior Sensitivity

Several hyper-parameters and simulation constants need to be specified in order to successfully adopt the Bayesian approaches. The aim is to keep the prior distributions as vague as possible so that the data, rather than the prior, drive the inference. Also, sensitivity of the assumed values to statistical inference needs to be checked. The adopted hyper-parameter values for our two examples in Section 19.6 are discussed below.

For the gamma distribution prior on η of the gamma frailty model, ϕ was taken to be 0.001. This specifies a diffuse but proper prior distribution for the inverse-variance parameter. The skewness parameter δ was given a uniform prior in $[0, 5]$. The resulting interval was large enough to capture the skewness in the frailty distribution. The degrees-of-freedom parameter ν was given an exponential prior distribution (truncated to be greater than 1) with mean 10. This is to specify a moderate to heavy tail of the assumed t distribution.

Recall that for the stable frailty model we have assumed the beta distribution for the parameter α. We have experimented with many combinations of values for the parameters of the beta distribution including the case for uniform distribution. In the latter case it was difficult to sample from the resulting full conditional distribution due to computer underflow problems similar to the ones reported in Buckle (1995). As a result we work with informative beta prior distribution which is further justified by the fact that in both of our examples there is only weak association present between the multivariate survival times. In particular, we assume the parameters of the beta distribution to be proportional to a and b, say where $a/(a+b) = 3/4$. This gives the *apriori* mean of α to be 3/4 which leads to moderate correlations between the log survival times as discussed earlier in Section 19.2.2.

The number of jump times g was assumed to have a Poisson distribution with mean 10, and truncated in the interval 1 to 20. The prior variance σ^2 of the log baseline hazard levels was assumed to be 1. Bayesian inference was largely insensitive to changes in the values of these hyper-parameters. In particular we have always obtained a robust posterior distribution of g with about 6–8% acceptance in the reversible jump steps. The values $g = 7, \ldots, 15$ always accounted for more than 90% of the probability mass.

We have adopted the pseudo-Bayes factor (see, e.g., Geisser and Eddy, 1979; Gelfand and Dey, 1994) for model comparison. The cross-validation predictive densities known as conditional predictive ordinates (CPO) have also been used; see, e.g., Sahu *et al.* (1997) for a definition. The CPOs measure the influence of individual observations and are often used as predictive model checking tools.

19.5 Reversible Jump Steps

Observe that the dimension of $\boldsymbol{\zeta}$ changes as the number of jump times g associated with the baseline hazard function changes. Hence we fit the full Bayesian model using the reversible jump Markov chain Monte Carlo method; see Green (1995), and McKeague and Tighiouart (2000). The algorithm is described as follows.

Suppose that the Markov chain is at a current state x and it is intended to move to a new state y where the dimension of x and y can be different. The new point y and its acceptance probability are obtained as follows. The move will be implemented by drawing u and v from continuous random variables U and V of appropriate dimensions, such that the dimension of (x, u) is the same as the dimension of (y, v). Further, the proposal point y is obtained using a one-to-one transformation $(y, v) = T(x, u)$. The proposed move is then accepted with probability

$$\alpha\{(x, u), (y, v)\} = \min\left\{1, \frac{\pi(y)q_2(v)}{\pi(x)q_1(u)}\left|\frac{\partial T(x, u)}{\partial(x, u)}\right|\right\}, \tag{19.14}$$

where $\pi(\cdot)$ is the posterior distribution and $q_1(u)$ and $q_2(v)$ are the densities of U and V, respectively.

It is straightforward to check that the acceptance probability in (19.14) is the usual Metropolis-Hastings acceptance probability for accepting the proposal (y, v) when the current point is (x, u).

We consider the following three move types for our problem:

(a) updating all the parameters in $\boldsymbol{\zeta}$ except for g,

(b) birth of a new jump time,

(c) death of an existing jump time.

The move type (a) does not involve dimension change and is accomplished by the usual Gibbs sampling steps. Sahu *et al.* (1997) provided details for the steps involving $\boldsymbol{\beta}, \mathbf{w}$ and η for the gamma frailty model. The integral in the density of the stable frailty model (19.3) is calculated using the Newton-Cotes numerical quadrature formula; see, e.g., Forsythe, Malcolm, and Moler (1977). Under the log-skew-t frailty model the calculations make use of appropriate Metropolis-Hastings steps. The remaining two move types change the dimension of $\boldsymbol{\lambda}$ by 1. These are detailed as follows.

Consider the birth move (b) first. A new jump time τ^* is drawn uniformly in $(\tau_0, \tau_{\max})$, where $\tau_{\max}$ is the maximum observed survival time. Suppose that $\tau^* \in (\tau_{k-1}, \tau_k)$. Now we need to generate two new log baseline hazard rates λ'_k and λ'_{k+1} in the proposal when the current point is λ_k. Since there is one degree of freedom for proposing the move we simulate u uniformly in $(-\epsilon, \epsilon)$ for some $\epsilon > 0$. Now λ'_k is taken to be a convex combination of λ_{k-1} and $\lambda_k + u$, and λ'_{k+1} is taken to be a convex combination of $\lambda_k - u$ and λ_{k+1}. In particular we take

$$\lambda'_k = \frac{\tau^* - \tau_{k-1}}{\tau_k - \tau_{k-1}}\lambda_{k-1} + \frac{\tau_k - \tau^*}{\tau_k - \tau_{k-1}}(\lambda_k + u), \tag{19.15}$$

$$\lambda'_{k+1} = \frac{\tau^* - \tau_{k-1}}{\tau_k - \tau_{k-1}}(\lambda_k - u) + \frac{\tau_k - \tau^*}{\tau_k - \tau_{k-1}}\lambda_{k+1}. \tag{19.16}$$

Now we set $\lambda'_i = \lambda_i$ for $1 \leq i \leq k-1$ and $\lambda'_{i+2} = \lambda_{i+1}$ for $k \leq i \leq g-1$. Further, we let $\tau'_k = \tau^*$ and set $\tau'_i = \tau_i$ for $0 \leq i \leq k-1$ and $\tau'_{i+1} = \tau_i$ for $k \leq i \leq g$.

Let $\boldsymbol{\zeta}'$ be the proposed parameter vector where $\boldsymbol{\lambda}$ and g is replaced by $\boldsymbol{\lambda}'$ and $g' = g + 1$, respectively. The remaining parameters in $\boldsymbol{\zeta}$ and $\boldsymbol{\zeta}'$ are kept the same. Observe that the Jacobian for this type of move is

$$\frac{2\,\Delta'_k\,\Delta'_{k+1}}{\Delta_k^2}.$$

The pdf $q_1(u)$ in (19.14) is $\frac{1}{2\epsilon}I(-\epsilon < u < \epsilon)$ and the generation of v is not required.

Now the ratio of the full posterior density (19.13) evaluated at $\boldsymbol{\zeta}'$ and $\boldsymbol{\zeta}$ is calculated. Observe that there are some obvious cancellations in the ratio since the parameters $\boldsymbol{\beta}, \mathbf{w}$ and the hyper-parameters of the frailty distribution $\cdots$ are unchanged in $\boldsymbol{\zeta}'$. The density ratio is multiplied by the Jacobian and divided by $q_1(u)$. Finally the acceptance probability (19.14) is calculated by taking the minimum.

The move type (c) is now straightforward. The proposal is generated as follows. An index k is randomly selected from $\{1, \ldots, g-1\}$ and the corresponding jump time τ_k is removed. The remaining jump times are re-labeled as: $\tau'_i = \tau_i$ for $0 \leq i \leq k-1$ and $\tau'_i = \tau_{i+1}$ for $k \leq i \leq g-1$. Now λ_k and λ_{k+1} are to be combined to obtain a new log baseline hazard level λ'_k. Toward this end, we obtain solutions of (19.15) and (19.16) for λ'_k and u (which appear on the right-hand sides of the equations). If u falls outside the interval $(-\epsilon, \epsilon)$ then the move is rejected forthwith since the corresponding birth move would not be reversible. Otherwise, we set $\lambda'_i = \lambda_i$ for $1 \leq i \leq k-1$ and $\lambda'_i = \lambda_{i+1}$ for $k+1 \leq i \leq g$. The acceptance probability is calculated using the inverse ratio in (19.14) and the fate of the move is decided accordingly.

19.6 Examples

19.6.1 Kidney Infection Data Example

McGilchrist and Aisbett (1991) analyzed time to first and second recurrence of infection in 38 kidney patients on dialysis using a Cox proportional hazard model with a multiplicative frailty parameter for each patient. The primary co-variate is the sex of the patients. There were 28 female patients, each with two recurrence times, some of which were censored.

Table 19.1 shows the posterior mean, standard deviation, and 95% credible intervals for all the parameters. Although the parameter β is significant under all three models, it is farthest from zero and has the smallest variance under the log-skew-t model. That is, the covariate effect is most strongly pronounced under the log-skew-t model. The skewness parameter δ under this model is also significant and the estimate of the degrees of freedom parameter ν suggests that a log-t model is better than a log-normal model.

To further quantify the difference between the frailty models we also estimate the predictive survival curve for a typical female patient under the three models. Figure 19.2 plots the curves. The Kaplan-Meier estimate (Kaplan and Meier, 1958) of the survival function is also plotted in the same graph. The Kaplan-Meier estimates ignore the dependence present in the data and are to be *used as rough guide* only. The observed survival times are shown as points on the time axis.

The majority of the observed survival times are below 200. Around time= 200, the Kaplan-Meier estimate shows a sharp decrease in the survival function as the data suggest. The predictive survival function under the log-skew-t model adapts to this most rapidly. Other models follow suit, but at a slower pace than the log-skew-t model.

Table 19.1 *Parameter estimates from the gamma, stable, and log-skew-t model. Posterior means are followed by posterior standard deviations in the first row. 95% credible intervals are shown in the second row. φ is η^{-1} for the gamma model, α for the stable model, and δ for the log-skew-t model.*

	β	φ	ν
Gamma	−1.95 (0.51)	0.58 (0.31)	–
	(−3.02, −1.02)	(0.13, 1.31)	–
Stable	−2.14 (0.55)	0.78 (0.0091)	–
	(−3.20, −0.48)	(0.76, 0.81)	–
Log-Skew-t	−2.34 (0.58)	0.34 (0.26)	13.4 (9.0)
	(−3.47, −1.25)	(0.01, 0.96)	(2.98, 37.10)

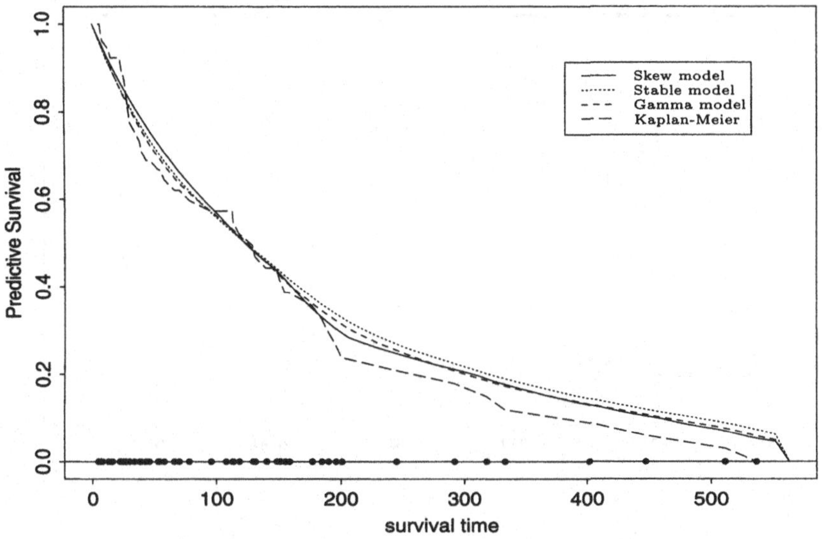

Figure 19.2 *Predictive survival curves for a typical female patient for the kidney infection example.*

The last issue we investigate is whether anything is gained by having an unknown number of jump times. We compare two versions of the log-skew-t model. The first is as described above and the second version keeps g fixed at 10, which was the choice adopted by other authors; see, e.g., Sahu *et al.* (1997). Here the endpoints are placed at equal time intervals.

The posterior mean estimates of the log baseline hazard function under the fixed g and the random g model are plotted in Figure 19.3. Survival times are plotted as points on top of the graph. Some interesting conclusions can be made from the plot. The peaks in the hazard functions are caused by observed survival times. Similarly the troughs are seen in the intervals where there are no observed survival times. The baseline hazard function for the model with random g quickly adapts to an occurrence of a failure while the function corresponding to the fixed case does not. The baseline hazard function in the fixed g case remains unchanged even after a failure has occurred in some intervals. As a result the function for the random g case is seen to be much smoother. That is, the random g model is more flexible.

To compare the fixed and random models using the pseudo-Bayes factor we have also calculated the pseudo-marginal log-likelihood under the two models. The log pseudo-Bayes factor for the random g model is 1.3 which

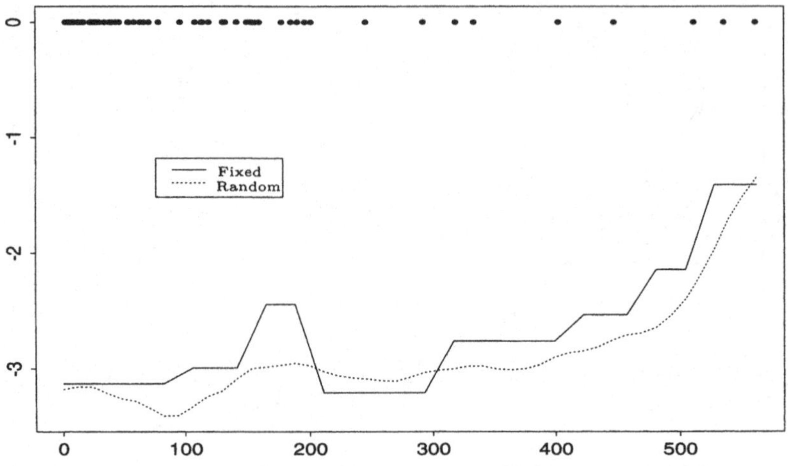

Figure 19.3 *Posterior mean estimate of the log base line hazard function for fixed g and random g model for the kidney infection example.*

suggests somewhat strong enough evidence in favor of the random g model. In conclusion, the random g model is much better than the fixed g model.

19.6.2 Litters Example

We consider the rat tumor data first studied by Mantel, Bohidar, and Ciminera (1977). There were 50 litters of rat with each litter consisting of three rats. A randomly selected rat from each litter was given a drug and the other two were selected as controls and were given a placebo. The survival time to develop tumor for each rat was recorded.

The parameter estimates are given in Table 19.2. Note that, as expected, for the gamma frailty model the estimate of β is closest to zero. That is, the covariate effect is attenuated under the gamma frailty model. To see this more clearly, the kernel density estimates of the posterior for β under the three models are plotted in Figure 19.4. The right tail of this density is heavier under the stable frailty model. The figure also shows that the estimate from the proposed log-skew-t model lies in between the estimates from the gamma and the stable frailty model.

The dependence parameter α under the stable model is estimated to be 0.70, which shows moderate local dependence between the component survival times. The estimate of variance of the frailty distribution under the gamma model is also significant. Under the log-skew-t model the estimate of

Table 19.2 *Parameter estimates from the gamma, stable, and log-skew-t model for the litters example. Posterior means are followed by (standard deviations) in the first row. 95% credible intervals are shown in the second row. φ is η^{-1} for the gamma model, α for the stable model, and δ for the log-skew-t model.*

	β	φ	ν
Gamma	0.69 (0.31)	0.51 (0.45)	-
	(0.09, 1.30)	(0.06, 1.69)	-
Stable	0.78 (0.32)	0.70 (0.0061)	-
	(0.13, 1.41)	(0.69, 0.71)	-
Log-Skew-t	0.71 (0.31)	0.33 (0.25)	15.6 (9.6)
	(0.09, 1.31)	(0.01, 0.95)	(3.32, 40.02)

the skewness parameter δ is also significant and the estimate of the degrees of freedom shows that the frailty distribution has a moderate tail, which agrees with the stable frailty model.

It is of interest to check whether the form of the assumed frailty distri-

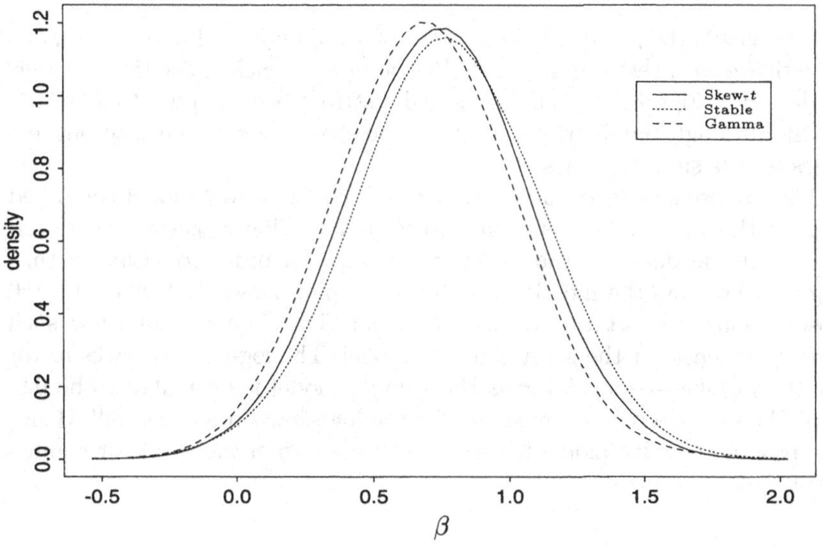

Figure 19.4 *The kernel density estimates of the posterior for β under the three different frailty distributions for the litters example.*

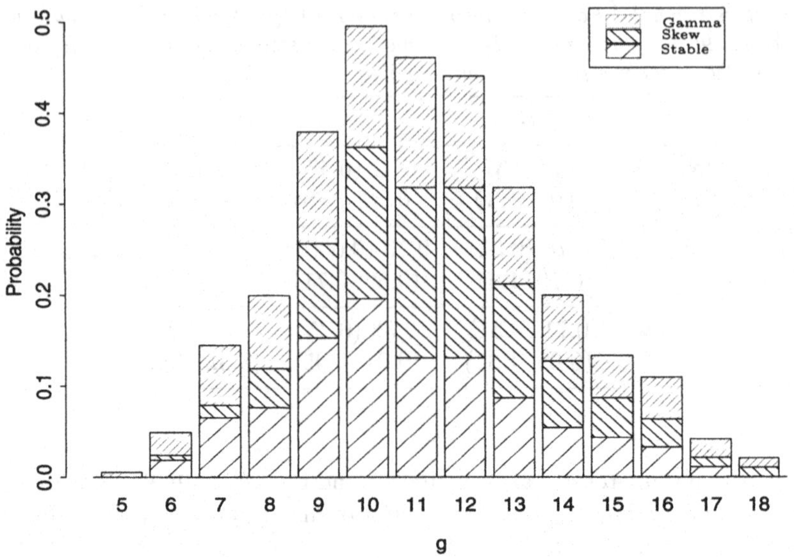

Figure 19.5 *The posterior distribution of g under the three models for the litters example.*

bution affects the posterior distribution of the number of jump times, g. To investigate this, the posterior distribution of g for each of the three models is plotted in Figure 19.5. The estimated distributions seem not to differ too much, although the distribution under the skewed model tends to put less mass on the smaller values of g.

The log-pseudo-Bayes factor in favor of the log-skew-t model compared against the stable model is estimated to be 2.3. This suggests strong preference of the data for the log-skew-t model. In order to visualize this, Figure 19.6 plots the individual CPOs. The plot shows that 93 of the 150 observations support the log-skew-t model. The data do not show such strong evidence for the gamma frailty model. The log-pseudo-Bayes factor for the log-skew-t model versus the gamma model is estimated to be 0.9. This shows some positive evidence for the log-skew-t model as well. Hence the proposed frailty model is seen to be better than the stable or gamma frailty models.

19.7 Conclusion

In this chapter we have extended the multivariate survival models in two directions. The log-skew-t frailty distribution adopted here is shown to be

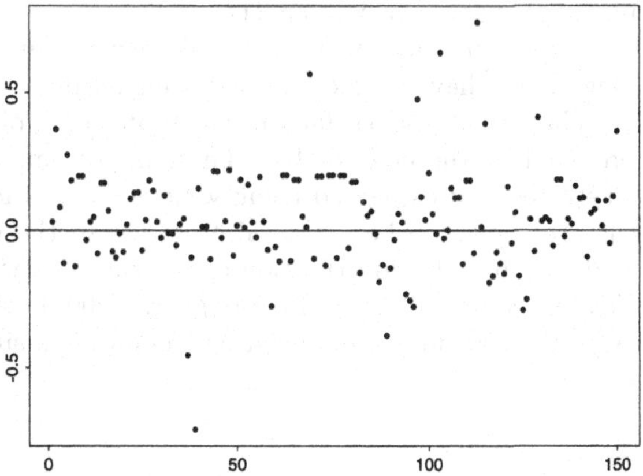

Figure 19.6 *The CPO plot for comparing the log-skew-t model versus the stable model. 93 out of 150 observations support the skewed model. The log-pseudo-Bayes factor is 2.3.*

more flexible than the popular gamma and stable frailty distributions. The new development is also shown to provide better model fit according to some well-known predictive Bayesian model checking and selection criteria in our examples. Also we have shown in our examples that the gamma frailty distribution attenuates the covariate effect much more than the proposed log-skew-t frailty distribution. These reasons justify the consideration of the log-skew-t model.

The new frailty distributions cannot take multimodal shape, however. If obtaining multimodal frailty distributions is desired, then mixture distributions (Ravishanker and Dey, 2000) or a nonparametric specification in infinite dimensional parameter space (Walker and Mallick, 1997) should be considered. Recently, Ma and Genton (2004) showed that another possibility is to use higher order (odd) polynomials in the skewing function to achieve multimodality.

The baseline hazard function conditional on the frailty distribution is modeled using a flexible martingale process. It imposes smoothing using its neighbors. The choice of jump times of the baseline hazard function is also made quite flexible using a time-homogeneous Poisson process. This removes the ad-hoc assumptions often made when specifying the number and position of the jump times.

We have developed the powerful reversible jump Markov chain Monte Carlo method for multivariate survival analysis. Our methods can be extended to perform data analysis in many other scenarios including the models with time-dependent covariate effects.

We have illustrated our methods using two well-known examples. Various aspects of the new models have been quantified using output of our MCMC implementation. The pseudo-Bayes factors show an order of magnitude improvement provided by the new models. These improvements in model fit have been illustrated and explained using various diagnostic plots. The proposed models are shown to be viable alternatives to the gamma and stable frailty models. Thus the contribution of the current chapter can be seen in the following comment made by Hougaard (2000): "Finding the right tools for a given problem is more exciting than using a single tool for all problems."

CHAPTER 20

Linear Mixed Effects Models with Flexible Generalized Skew-Elliptical Random Effects

Yanyuan Ma, Marc G. Genton, and Marie Davidian

20.1 Introduction

The linear mixed effects model is a popular framework for analysis of continuous longitudinal responses from a sample of individuals in biomedical, agricultural, environmental, and social science applications. Inter-individual heterogeneity in features of the longitudinal profiles, such as (error-free) baseline response or rate of change, is represented explicitly by individual-specific random effects, so that inference on questions involving population or individual characteristics is possible. A standard assumption is that the random effects and within-individual errors are normally distributed, and software such as SAS `proc mixed` (Littell, Milliken, Stroup, and Wolfinger, 1996) or Splus `lme()` (Pinheiro and Bates, 2000) incorporates this assumption. Although normality of within-individual deviations may be a reasonable model for many continuous measurements, perhaps on a transformed scale, normality of the random effects implies a belief about inter-individual heterogeneity that may be unrealistic. For example, a common feature in biological contexts is skewness or heavy-tailedness of features underlying individual profiles.

Accordingly, relaxation of this assumption on the random effects distribution may yield valuable insight on such features in the population. Several approaches have been advocated, based on replacing the assumption of normality by a weaker assumption that the random effects have only a "smooth" density that may be skewed or multimodal. For practical implementation, this involves representing the density by a flexible form and estimating the density along with other model parameters. For example, Zhang and Davidian (2001) use the seminonparametric (SNP) density representation proposed by Gallant and Nychka (1987) to characterize the random effects density; an appealing feature of this approach is that the likelihood for all model parameters may be expressed in a closed form. Alternatively, Verbeke and Lesaffre (1996) adopt a mixture of normals representation and carry out inference via an EM algorithm (see Verbeke and Molenberghs, 2000, Chap. 12). Magder and Zeger (1996) use a form of non-parametric maximum likelihood based normal densities, and Tao, Palta, Yandell, and Newton (1999) estimate the density of a scalar random effect via a predictive recursive algorithm.

In this chapter, we take an alternative approach that may offer advantages for practical interpretation and performance over these techniques and represent the random effects density by that of a flexible generalized skew-elliptical distribution ($FGSE$). The $FGSE$ family is able to take into account heavy-tailedness, skewness, and multimodality, has a form that permits insight into properties of the true density, and yields density estimates that are not subject to artifactual anomalies sometimes seen with competing approaches such as the SNP or mixtures of normals.

In Section 20.2, we introduce the linear mixed effects model and describe the $FGSE$ class. We describe two approaches to implementation in Section 20.3, one based on a Monte Carlo EM algorithm and the other on Markov chain Monte Carlo (MCMC) techniques. Section 20.4 presents results of several simulation studies demonstrating performance, and the approach is applied to data from a study of cholesterol in Section 20.5.

20.2 Flexible Generalized Skew-Elliptical Distributions and the Linear Mixed Effects Model

Consider a linear mixed effects model of the form

$$Y_{ij} = \mathbf{x}_{ij}^T \boldsymbol{\beta} + \mathbf{s}_{ij}^T \boldsymbol{b}_i + e_{ij}, \tag{20.1}$$

where Y_{ij}, $i = 1, \ldots, m$, $j = 1, \ldots, n_i$, is the response of individual i at time t_{ij}; $\boldsymbol{\beta}$ is a p-dimensional vector of fixed effects associated with the vector $\mathbf{x}_{ij}$ of covariates; $\boldsymbol{b}_i$ are q-dimensional subject-specific random effects associated with the vector $\mathbf{s}_{ij}$ of covariates; $\boldsymbol{b}_i$ are mutually independent; and $e_{ij} \sim N(0, \sigma_e^2)$ are mutually independent errors, independent of the $\boldsymbol{b}_i$. Thus, there are $N = \sum_{i=1}^m n_i$ total observations. A standard but possibly restrictive assumption is that the random effects $\boldsymbol{b}_i$ have a multivariate normal distribution. In order to allow flexibility in characterizing features of population longitudinal behavior, we assume instead that the $\boldsymbol{b}_i$ have a distribution with density that may be represented by a density in the class of flexible generalized skew-elliptical ($FGSE$) distributions.

Densities in the $FGSE$ class are defined by

$$2f(\mathbf{x} - \boldsymbol{\xi})H(P_K(\mathbf{x} - \boldsymbol{\xi})), \tag{20.2}$$

where $\boldsymbol{\xi}$ is a $(q \times 1)$ location parameter; f is a multivariate elliptically contoured density centered at zero; H is the cumulative distribution function of any continuous random variable, symmetric about zero; and P_K is an odd polynomial of order K. The function $\pi_K(\mathbf{x}) = H(P_K(\mathbf{x}))$ is called a skewing function and satisfies $\pi_K(-\mathbf{x}) = 1 - \pi_K(\mathbf{x})$, thus ensuring that the integral of (20.2) equals one. When $P_K(\mathbf{x}) \equiv 0$, then $H(0) = 1/2$, and (20.2) reduces to the elliptically contoured density f. Ma and Genton (2004) study the class (20.2) and show that it is able to represent skewness, tail behavior, and multimodality well in practice, where K acts as a tuning parameter to control the degree of flexibility and imparts certain structural properties discussed below.

The $FGSE$ representation has a number of properties that make it an attractive practical tool. In the particular case where $f = \phi_q$ and $H = \Phi$, where ϕ_q is the q-variate normal density and Φ is the univariate standard normal cumulative distribution function (cdf), (20.2) defines the flexible generalized skew-normal ($FGSN$) distributions. Similarly, define the flexible generalized skew-t distributions ($FGST$) by taking $f = t_q$ and $H = T$, the q-dimensional multivariate t density and the univariate t cdf, respectively; alternatively, Φ or any other symmetric cdf could be used. Thus, note that when $P_K(\mathbf{x}) \equiv 0$, the $FGSN$ reduces to the usual assumption of normality for the random effects in (20.1), while under these conditions, the $FGST$ yields multivariate-t random effects that have been used in the linear mixed model context, for example, by Wakefield, Smith, Racine-Poon, and Gelfand (1994) to allow detection and accommodation of "outlying" individuals. Thus, (20.2) includes popular mixed model assumptions as special cases, but enhances these by embedding them in a more general framework.

It can be shown that univariate $FGSN$ distributions with the order $K = 3$ may have at most two modes, whereas they are unimodal for $K = 1$, so that the value of K required to achieve an adequate representation is informative in this case. Moreover, a straightforward means of identifying properties of the true distribution emerges. As an example, the standard $FGSN$ density with $K = 3$, $q = 2$, and $\boldsymbol{\xi} = \mathbf{0}$ has the form $2\phi_2(x_1, x_2)\Phi(\alpha_1 x_1 + \alpha_2 x_2 + \alpha_3 x_1^3 + \alpha_4 x_2^3 + \alpha_5 x_1^2 x_2 + \alpha_6 x_1 x_2^2)$. Here, the first term governs tail behavior, while the second controls skewness; additional flexibility can be imposed on the tail behavior by replacing ϕ_2 with the bivariate t, for instance. In particular, if $\alpha_3 = \cdots = \alpha_6 = 0$, then the density reduces to the bivariate skew-normal density of Azzalini and Dalla Valle (1996), which is known to be unimodal; if in addition $\alpha_1 = \alpha_2 = 0$, then the density is symmetric (normal). Hence, an appealing feature of this formulation is that questions regarding skewness and multimodality may be straightforwardly expressed in terms of parameters in P_K.

In general, inference for the $FGSE$ distribution can be carried out by maximizing the corresponding likelihood function for a given order K; there are no constraints on the parameters of the skewing function π_K, so that standard optimization techniques can be used. Because for a given elliptically contoured density f and skewing function π_K, $FGSE$ models are nested when K decreases, likelihood ratio tests may be used to identify an appropriate value of K and the necessity of the presence of individual terms in the skewing function; model selection criteria such as Akaike's Information Criterion (AIC), Schwarz's Bayesian Information Criterion (BIC), and the Hannan-Quinn criterion (HQ), (see, for example, Zhang and Davidian, 2001), may be applied, as we demonstrate shortly. Preliminary exploration by Ma and Genton (2004) fitting $FGSE$ densities directly to data shows that $K = 3$ seems to provide enough flexibility to approximate a wide variety of densities and is free from artifacts that affect other density representations like SNP.

The $FGSE$ distribution possesses a further attractive property that may be exploited in the linear mixed effects model context: it is straightforward to simulate pseudo-realizations from the $FGSE$ as long as it is possible to simulate from the elliptically contoured density f. In particular, simulate a random vector $\mathbf{y}$ with density f and a random variable $U \sim Uniform[0,1]$. If $U \leq \pi_K(\mathbf{y} - \boldsymbol{\xi})$, then we set the random vector $\mathbf{x} = \mathbf{y}$. Otherwise, the relation $U > \pi_K(\mathbf{y} - \boldsymbol{\xi})$ holds, and we set $\mathbf{x} = 2\boldsymbol{\xi} - \mathbf{y}$. The random vector $\mathbf{x}$ has the $FGSE$ distribution of order K with density given in Equation (20.2). Consequently, implementation via a Monte Carlo EM algorithm is feasible and is elucidated in Section 20.3.1, with intractable integrations handled via Monte Carlo. Alternatively, placing the $FGSE$ distributions in linear mixed models in a Bayesian framework, this feature may be exploited to develop an MCMC implementation, see Section 20.3.2.

We use $FGSN_q(\boldsymbol{\xi}, \Omega, P_K)$ to represent a q-variate flexible generalized skew-normal distribution, where the normal distribution component has location parameter $\boldsymbol{\xi}$ and scale matrix Ω, and the skewing function has polynomial P_K. Similarly, $FGST_q(\boldsymbol{\xi}, \Omega, \nu, P_K)$ denotes a q-variate flexible generalized skew-t distribution, where the associated t distribution has location parameters $\boldsymbol{\xi}$, scale matrix Ω, and degrees of freedom ν; and the skewing function component has a polynomial P_K, respectively. In this case, small degrees of freedom ν indicate heavy-tailedness. In this chapter, we consider primarily linear mixed effect models whose random effects are represented by $FGSN_q$ and $FGST_q$ distributions, although other $FGSE$ distributions can also be implemented.

Summarizing, we propose to fit the linear mixed effects model given in (20.1) where the distribution of the $\boldsymbol{b}_i$ is assumed to have a density belonging to a class whose elements can be approximated by the $FGSN_q(\boldsymbol{\xi}, \Omega, P_K)$ or $FGST_q(\boldsymbol{\xi}, \Omega, \nu, P_K)$ densities for suitably chosen K. Because these representations are not restricted to have mean zero, to ensure identifiability, even when $K = 0$, we require that $\mathbf{x}_{ij}$ does not contain $\mathbf{s}_{ij}$; however, this does not involve any restriction over the usual linear mixed model, as we illustrate in Sections 20.4 and 20.5. Interest focuses on estimation of the fixed effects $\boldsymbol{\beta}$, the intra-individual error variance σ_e^2, and the distribution of the $\boldsymbol{b}_i$, including features such as its mean and variance, which may be derived as functions of $\boldsymbol{\xi}$, Ω, and ν in each case. The general strategy is to fit the model for several fixed values $K = 0, 1, 3, \dots, K_{max}$ and select an appropriate value of K via inspection of the AIC, BIC, or HQ criteria; $K = 0$ yields the normal or t assumptions on the $\boldsymbol{b}_i$, respectively, and odd $K > 0$ involves more complexity but provides greater flexibility. In accordance with Ma and Genton (2004), we have found $K_{max} = 3$ is sufficiently large to provide an adequate approximation to the true, underlying density. Specifically, then, to fit the model for fixed K, letting $\boldsymbol{\alpha}$ denote the coefficients in the polynomial P_K and writing $\boldsymbol{\theta} = (\boldsymbol{\xi}^T, \text{vec}(\Omega), \boldsymbol{\alpha}^T)^T$ or $(\boldsymbol{\xi}^T, \text{vec}(\Omega), \boldsymbol{\alpha}^T, \nu)^T$ depending on whether the $FGSN$ or $FGST$ representation is adopted, we propose estimating jointly $\boldsymbol{\delta} = (\boldsymbol{\beta}^T, \sigma_e, \boldsymbol{\theta}^T)^T$. In the next section, we describe implementation using the EM algorithm and MCMC techniques.

For brevity in the sequel, for each individual i, $i = 1, \dots, m$, collect Y_{ij}, $j = 1, \dots, n_i$ into a vector $\mathbf{y}_i$. Also, let $\boldsymbol{b}$ denote the $(mq \times 1)$ vector of concatenated $\boldsymbol{b}_i$, and let $\mathbf{y} = (\mathbf{y}_1^T, \dots, \mathbf{y}_m^T)^T$ $(N \times 1)$, and denote the data by y_{ij}.

20.3 Implementation and Inference

20.3.1 Maximum Likelihood via the EM Algorithm

We illustrate implementation of likelihood inference via the EM algorithm in the case where the random effects density is represented by that of the $FGSN_q$ distribution. The likelihood for $\boldsymbol{\delta} = (\boldsymbol{\beta}^T, \sigma_e, \boldsymbol{\theta}^T)^T$ for a fixed choice of K dictating the elements of $\boldsymbol{\theta}$ depending on the observed data $\mathbf{y}$ is

$$\ell(\boldsymbol{\delta}|\mathbf{y}) = \int \prod_{i=1}^{m} f(\mathbf{y}_i|\boldsymbol{b}_i, \boldsymbol{\delta}) f(\boldsymbol{b}_i|\boldsymbol{\delta}) d\boldsymbol{b}_1 \dots d\boldsymbol{b}_m = \int f(\mathbf{y}|\boldsymbol{b}, \boldsymbol{\delta}) f(\boldsymbol{b}|\boldsymbol{\delta}) d\boldsymbol{b}, \tag{20.3}$$

where

$$f(\mathbf{y}_i|\boldsymbol{b}_i, \boldsymbol{\delta}) = \prod_{j=1}^{n_i} \frac{1}{\sigma_e} \phi\left(\frac{y_{ij} - \mathbf{x}_{ij}^T \boldsymbol{\beta} - \mathbf{s}_{ij}^T \boldsymbol{b}_i}{\sigma_e}\right),$$

$$f(\boldsymbol{b}_i|\boldsymbol{\delta}) = \frac{2}{\sqrt{(2\pi)^q \det(\Omega)}} \exp\left\{-\frac{(\boldsymbol{b}_i - \boldsymbol{\xi})^T \Omega^{-1} (\boldsymbol{b}_i - \boldsymbol{\xi})}{2}\right\} \Phi(P_K(\boldsymbol{b}_i - \boldsymbol{\xi})),$$

$f(\mathbf{y}|\boldsymbol{b}, \boldsymbol{\delta}) = \prod_{i=1}^{m} f(\mathbf{y}_i|\boldsymbol{b}_i, \boldsymbol{\delta})$, and $f(\boldsymbol{b}|\boldsymbol{\delta}) = \prod_{i=1}^{m} f(\boldsymbol{b}_i|\boldsymbol{\theta})$. Maximizing the logarithm of (20.3) in $\boldsymbol{\delta}$ via the EM algorithm requires iteration between calculating the expectation of $\log(f(\mathbf{y}, \boldsymbol{b}|\boldsymbol{\delta}))$ conditional on $\mathbf{y}$ and the previous iterate (the E-step) and maximizing this expectation in $\boldsymbol{\delta}$ (the M-step).

If we denote the iterate from the rth step as $\hat{\boldsymbol{\delta}}^{(r)}$, then the conditional expectation in the E-step has the form

$$Q(\boldsymbol{\delta}|\hat{\boldsymbol{\delta}}^{(r)}) = \mathrm{E}[\log\{f(\mathbf{y}, \boldsymbol{b}|\boldsymbol{\delta})\}|\mathbf{y}, \hat{\boldsymbol{\delta}}^{(r)}] = \int \log\{f(\mathbf{y}, \boldsymbol{b}, |\boldsymbol{\delta})\} f(\boldsymbol{b}|\mathbf{y}, \hat{\boldsymbol{\delta}}^{(r)}) d\boldsymbol{b}$$

$$= \sum_{i=1}^{m} \int \left(\log\left[\frac{2\exp\{-(\boldsymbol{b}_i - \boldsymbol{\xi})^T \Omega^{-1}(\boldsymbol{b}_i - \boldsymbol{\xi})/2\}\,\Phi(P_K(\boldsymbol{b}_i - \boldsymbol{\xi}))}{\sqrt{(2\pi)^q \det(\Omega)}}\right] \right.$$

$$\left. + \sum_{j=1}^{n_i} \log\left\{\frac{1}{\sigma_e}\phi\left(\frac{y_{ij} - \mathbf{x}_{ij}^T\boldsymbol{\beta} - \mathbf{s}_{ij}^T\boldsymbol{b}_i}{\sigma_e}\right)\right\}\right) f(\boldsymbol{b}_i|\mathbf{y}_i, \hat{\boldsymbol{\delta}}^{(r)})\, d\boldsymbol{b}_i, \qquad (20.4)$$

where $f(\boldsymbol{b}_i|\mathbf{y}_i, \boldsymbol{\delta})$ is the conditional density of $\boldsymbol{b}_i$ given $\mathbf{y}_i$. The M-step thus maximizes (20.4) evaluated at the data to obtain the update $\hat{\boldsymbol{\delta}}^{(r+1)}$. Iteration between the E-step and M-step proceeds until convergence. The starting values of the iteration are often chosen to be the corresponding estimates under a normal assumption of the random effects, where the starting values for the coefficients of P_K are set to be 0.

Note that the first term in the integrand involves $\boldsymbol{\theta}$ only, and the second involves $\boldsymbol{\beta}$ and σ_e^2 only, which leads to the following. It may be shown that

$$f(\boldsymbol{b}_i|\mathbf{y}_i, \hat{\boldsymbol{\delta}}^{(r)}) \propto \exp\left\{-\frac{(\boldsymbol{b}_i - \hat{\boldsymbol{\eta}}_i^{(r)})^T \hat{\Omega}_i^{(r)-1}(\boldsymbol{b}_i - \hat{\boldsymbol{\eta}}_i^{(r)})}{2}\right\} \Phi(P_K(\boldsymbol{b}_i - \hat{\boldsymbol{\xi}}^{(r)})), \qquad (20.5)$$

where $\hat{\Omega}_i^{(r)-1} = \hat{\Omega}^{(r)-1} + \hat{\sigma}_e^{(r)-2}\sum_{j=1}^{n_i}\mathbf{s}_{ij}\mathbf{s}_{ij}^T$, and $\hat{\boldsymbol{\eta}}_i^{(r)} = \hat{\Omega}_i^{(r)}(\hat{\Omega}^{(r)-1}\hat{\boldsymbol{\xi}}^{(r)} + \hat{\sigma}_e^{(r)-2}\sum_{j=1}^{n_i}(y_{ij} - \mathbf{x}_i^T\boldsymbol{\beta})\mathbf{s}_{ij})$. Assume a Cholesky decomposition $\hat{\Omega}_i^{(r)-1} = (\hat{V}_i^{(r)})^T\hat{V}_i^{(r)}$, and $\hat{\mathbf{h}}_i^{(r)} = (\hat{V}_i^{(r)}\boldsymbol{b}_i - \hat{V}_i^{(r)}\hat{\boldsymbol{\eta}}_i^{(r)})/\sqrt{2}$. It may be verified that $\boldsymbol{\beta}$ and σ_e^2 that maximize (20.4) must satisfy

$$\hat{\boldsymbol{\beta}}^{(r+1)} = \left(\sum_{i=1}^{m}\sum_{j=1}^{n_i}\mathbf{x}_{ij}\mathbf{x}_{ij}^T\right)^{-1}\sum_{i=1}^{m}\sum_{j=1}^{n_i}\mathbf{x}_{ij}(y_{ij} - \mathbf{s}_{ij}^T\hat{\boldsymbol{B}}_i^{(r)}/\hat{A}_i^{(r)}), \qquad (20.6)$$

$$\hat{\sigma}_e^{(r+1)2} = \frac{1}{N}\left\{\sum_{i=1}^{m}\frac{1}{\hat{A}_i^{(r)}}\sum_{j=1}^{n_i}\mathbf{s}_{ij}^T\hat{C}_i^{(r)}\mathbf{s}_{ij} - \sum_{i=1}^{m}\frac{2\hat{\boldsymbol{B}}_i^{(r)T}}{\hat{A}_i^{(r)}}\sum_{j=1}^{n_i}\mathbf{s}_{ij}(y_{ij} - \hat{\boldsymbol{\beta}}^{(r+1)T}\mathbf{x}_{ij})\right.$$

$$\left. + \sum_{i=1}^{m}\sum_{j=1}^{n_i}(y_{ij} - \hat{\boldsymbol{\beta}}^{(r+1)T}\mathbf{x}_{ij})^2\right\}, \qquad (20.7)$$

and $\boldsymbol{\theta}$ maximizes

$$\sum_{i=1}^{m}\frac{1}{\hat{A}_i^{(r)}}\left\{\boldsymbol{\xi}^T\Omega^{-1}\hat{\boldsymbol{B}}_i^{(r)} + \hat{E}_i^{(r)} + \Omega^{-1}\odot\hat{C}_i^{(r)}\right\} - \frac{m}{2}(\log\det(\Omega) + \boldsymbol{\xi}^T\Omega^{-1}\boldsymbol{\xi}), \qquad (20.8)$$

where

$$\begin{aligned}
\hat{A}_i^{(r)} &= \int \exp(-\hat{\mathbf{h}}_i^{(r)T}\hat{\mathbf{h}}_i^{(r)})\Phi(P_K(\boldsymbol{b}_i - \hat{\boldsymbol{\xi}}^{(r)}))\, d\boldsymbol{b}_i,\\
\hat{\boldsymbol{B}}_i^{(r)} &= \int \boldsymbol{b}_i \exp(-\hat{\mathbf{h}}_i^{(r)^T}\hat{\mathbf{h}}_i^{(r)})\Phi(P_K(\boldsymbol{b}_i - \hat{\boldsymbol{\xi}}^{(r)}))\, d\boldsymbol{b}_i,\\
\hat{C}_i^{(r)} &= \int \boldsymbol{b}_i\boldsymbol{b}_i^T \exp(-\hat{\boldsymbol{b}}_i^{(r)T}\hat{\mathbf{h}}_i^{(r)})\Phi(P_K(\boldsymbol{b}_i - \hat{\boldsymbol{\xi}}^{(r)}))\, d\boldsymbol{b}_i,\\
\hat{E}_i^{(r)} &= \int \log[\Phi(P_K(\boldsymbol{b}_i - \hat{\boldsymbol{\xi}}^{(r)}))] \exp(-\hat{\mathbf{h}}_i^{(r)T}\hat{\mathbf{h}}_i^{(r)})\Phi(P_K(\boldsymbol{b}_i - \hat{\boldsymbol{\xi}}^{(r)}))\, d\boldsymbol{b}_i,
\end{aligned}$$

and the operator $\odot$ means taking the Hadamard product of two matrices and summing up all the entries in the resulting matrix. The M-step may thus be carried out by calculating (20.6), (20.7) and solving (20.8). To accomplish this, an evaluation of the integrals in the above expressions is required, which may be achieved via Gauss-Hermite quadrature (e.g., Monahan, 2001, Chap. 10).

Alternatively, Monte Carlo integration may be employed, which yields a so-called Monte Carlo EM algorithm (e.g., Booth and Hobert, 1999). Examination of (20.5) shows that $f(\boldsymbol{b}_i|\mathbf{y}_i, \boldsymbol{\delta})$ is not a member of the *FGSN* family, because the polynomial in the skewing function is not an odd polynomial when re-centered at $\boldsymbol{\eta}_i$. However, it has a similar stochastic representation (Arnold and Beaver, 2002) that allows straightforward generation of independent samples from $f(\boldsymbol{b}_i|\mathbf{y}_i, \hat{\boldsymbol{\delta}}^{(r)})$ that may then be used to approximate the integrals in (20.4). In particular, to generate a sample from $f(\boldsymbol{b}_i|\mathbf{y}_i, \hat{\boldsymbol{\delta}}^{(r)})$, first generate $\mathbf{w}$ from $N_q(\hat{\boldsymbol{\eta}}_i^{(r)}, \hat{\Omega}_i^{(r)})$, and U from the standard normal distribution, respectively. If $P_K(\mathbf{w}) > U$, where P_K is evaluated at $\hat{\boldsymbol{\alpha}}^{(r)}$, then $\mathbf{w}$ is a sample from $f(\boldsymbol{b}_i|\mathbf{y}_i, \hat{\boldsymbol{\delta}}^{(r)})$. Although this scheme has the flavor of rejection sampling, we have found in practice that it is considerably more efficient than the rejection sampling procedure suggested in the context of the Monte Carlo EM by Booth and Hobert (1999) or rejection sampling using the $N_q(\hat{\boldsymbol{\eta}}_i^{(r)}, \hat{\Omega}_i^{(r)})$ density as the envelope function.

Exploiting these developments leads to an algorithm that is simpler to implement than that given above, as it avoids numerical integration or joint solution of (20.6)–(20.8). At the rth iteration, generate L samples $\boldsymbol{b}_i^{(1)}, \boldsymbol{b}_i^{(2)}, \ldots, \boldsymbol{b}_i^{(L)}$ from $f(\boldsymbol{b}_i|\mathbf{y}_i, \hat{\boldsymbol{\delta}}^{(r)})$ for each i as described above. Then we may approximate $Q(\boldsymbol{\delta}|\hat{\boldsymbol{\delta}}^{(r)})$ in (20.4) directly by

$$Q_L(\boldsymbol{\delta}|\hat{\boldsymbol{\delta}}^{(r)}) = L^{-1}\sum_{l=1}^{L}\sum_{i=1}^{m}\sum_{j=1}^{n_i}\log\left\{\frac{1}{\sigma_e}\phi\left(\frac{y_{ij} - \mathbf{x}_{ij}^T\boldsymbol{\beta} - \mathbf{s}_{ij}^T\boldsymbol{b}_i^{(l)}}{\sigma_e}\right)\right\} \qquad (20.9)$$

$$+L^{-1}\sum_{l=1}^{L}\sum_{i=1}^{m}\log\left[\frac{2\exp\{-(\boldsymbol{b}_i^{(l)} - \boldsymbol{\xi})^T\Omega^{-1}(\boldsymbol{b}_i^{(l)} - \boldsymbol{\xi})/2\}\Phi(P_K(\boldsymbol{b}_i^{(l)} - \boldsymbol{\xi}))}{\sqrt{(2\pi)^q\det(\Omega)}}\right].$$

To implement the M-step, we may maximize each term in (20.9). Maximizing the first term, we obtain the updates $\hat{\beta}^{(r+1)} = A^{-1}\boldsymbol{B}$ and

$$(\hat{\sigma}_e^{(r+1)})^2 = L^{-1}N^{-1}\sum_{l=1}^{L}\sum_{i=1}^{m}\sum_{j=1}^{n_i}(y_{ij} - \mathbf{x}_{ij}^T\beta - \mathbf{s}_{ij}^T\boldsymbol{b}_i^{(l)})^2,$$

where $A = \sum_{l=1}^{L}\sum_{i=1}^{m}\sum_{j=1}^{n_i}\mathbf{x}_{ij}\mathbf{x}_{ij}^T$ and $\boldsymbol{B} = \sum_{l=1}^{L}\sum_{i=1}^{m}\sum_{j=1}^{n_i}\mathbf{x}_{ij}(y_{ij} - \mathbf{s}_{ij}^T\boldsymbol{b}_i)$. The second term in (20.9) may be maximized directly to obtain the updates $\hat{\boldsymbol{\xi}}^{(r+1)}, \hat{\Omega}^{(r+1)}$, and $\hat{\boldsymbol{\alpha}}^{(r+1)}$. Booth and Hobert (1999) discuss selection of a suitable value for L at each iteration and stopping criteria for this algorithm; see also Chen, Zhang, and Davidian (2002).

In either case, final estimates $\hat{\boldsymbol{\delta}}_K$ for each $K = 0, 1, 3, \ldots, K_{max}$ may be obtained by iterating the algorithm for each K until convergence. The value of K that achieves a balance between flexibility and parsimony may be selected by evaluating one of the information criteria AIC $= -\log\{\ell(\hat{\boldsymbol{\delta}}_K|\mathbf{y})\} + p_K$, BIC $= -\log\{\ell(\hat{\boldsymbol{\delta}}_K|\mathbf{y})\} + \log(N)p_K/2$, or HQ $= -\log\{\ell(\hat{\boldsymbol{\delta}}_K|\mathbf{y})\} + \log\{\log(N)\}p_K$ for each K and choosing the value of K that minimizes the criterion, where p_K is the number of parameters in the model under K. Here, $\ell(\hat{\boldsymbol{\delta}}_K|\mathbf{y})$ may be evaluated directly by Markov chain Monte Carlo simulation of the integrals in (20.3) evaluated at the converged value $\hat{\boldsymbol{\delta}}_K$.

Following convergence of either EM algorithm, standard errors may be calculated using the empirical sandwich variance estimator (e.g., Stefanski and Boos, 2002)

$$V(\mathbf{y}, \hat{\boldsymbol{\delta}}_K) = A(\mathbf{y}, \hat{\boldsymbol{\delta}}_K)^{-1}B(\mathbf{y}, \hat{\boldsymbol{\delta}}_K)\{A(\mathbf{y}, \hat{\boldsymbol{\delta}}_K)^{-1}\}^T,$$

where

$$A(\mathbf{y}, \hat{\boldsymbol{\delta}}_K) = -\left.\left[\frac{\partial^2}{\partial\boldsymbol{\delta}\,\partial\boldsymbol{\delta}^T}\log\{\ell(\boldsymbol{\delta}|\mathbf{y})\}\right]\right|_{\boldsymbol{\delta}=\hat{\boldsymbol{\delta}}_K}$$

and

$$B(\mathbf{y}, \hat{\boldsymbol{\delta}}_K) = \left.\left[\frac{\partial}{\partial\boldsymbol{\delta}}\log\{\ell(\boldsymbol{\delta}|\mathbf{y})\}\right]\right|_{\boldsymbol{\delta}=\hat{\boldsymbol{\delta}}_K}\left.\left[\frac{\partial}{\partial\boldsymbol{\delta}}\log\{\ell(\boldsymbol{\delta}|\mathbf{y})\}\right]^T\right|_{\boldsymbol{\delta}=\hat{\boldsymbol{\delta}}_K}$$

To calculate the derivatives of $\log\{\ell(\boldsymbol{\delta}|\mathbf{y})\}$, we take the corresponding derivatives of (20.3), with the integrals calculated through Monte Carlo simulation. Had the random effects truly come from an $FGSN$ distribution, $A(\mathbf{y}, \hat{\boldsymbol{\delta}}_K)^{-1}$ would have been a good approximation of $V(\mathbf{y}, \hat{\boldsymbol{\delta}}_K)$. Because $FGSN$ cannot approximate an arbitrary distribution, using the sandwich estimator will provide more precise estimation of the true variance.

20.3.2 Bayesian Inference via Markov Chain Monte Carlo Simulation

Implementation of the EM algorithm becomes more difficult if the $FGSN_q$ representation is replaced by $FGST_q$. In particular, in the $FGSN$ case just described, the conditional distribution of $\boldsymbol{b}_i$ given $\mathbf{y}_i$ takes the form of the product of a normal density and a "skewing function" as in (20.5), which admits a stochastic representation; however, in the case of the $FGST$, this pleasing property no longer holds. Thus, to carry out inference when the random effects density is represented by the $FGST_q$ density, an alternative approach is desirable. As we now demonstrate, taking a Bayesian perspective and implementing this via an MCMC procedure is one convenient way to achieve this.

It is instructive to first consider this approach in the $FGSN$ setting. To obtain a fully Bayesian specification, let $p(\boldsymbol{\delta})$ denote a prior distribution for $\boldsymbol{\delta}$. Following similar implementations in the literature (e.g., Wakefield *et al.*, 1994), it is natural to take, in obvious notation and parameterizing in terms of $\tau = (\sigma_e^2)^{-1}$ and $\Lambda = \Omega^{-1}$, $p(\boldsymbol{\delta}) = p(\boldsymbol{\beta})p(\tau)p(\boldsymbol{\xi})p(\Lambda)p(\boldsymbol{\alpha})$, where $p(\boldsymbol{\beta})$ is a normal density with known hyperparameters $\boldsymbol{\beta}_0, \Sigma_0$; $p(\tau) = 1$; $p(\boldsymbol{\xi})$ is normal with known hyperparameters $\boldsymbol{\xi}_0, \Omega_0$; $p(\Lambda)$ is Wishart with known hyperparameters ρ_0, R_0, and $p(\boldsymbol{\alpha})$ is normal with known hyperparameters $\mathbf{0}, \sigma^2 I_c$, where $c = \dim(\boldsymbol{\alpha})$. Then the posterior density of $\boldsymbol{\delta}$ and $\boldsymbol{b}$ is proportional to

$$p(\boldsymbol{\delta}) \prod_{i=1}^{m} \prod_{j=1}^{n_i} \frac{1}{\sigma_e} \phi\left(\frac{y_{ij} - \mathbf{x}_{ij}^T \boldsymbol{\beta} - \mathbf{s}_{ij}^T \boldsymbol{b}_i}{\sigma_e}\right)$$
$$\times \prod_{i=1}^{m} \frac{2}{\sqrt{\det(\Omega)}} \exp\left\{-\frac{(\boldsymbol{b}_i - \boldsymbol{\xi})^T \Lambda (\boldsymbol{b}_i - \boldsymbol{\xi})}{2}\right\} \Phi(P_K(\boldsymbol{b}_i - \boldsymbol{\xi})).$$

It is then straightforward to derive an algorithm to sample from the posterior using Markov chain Monte Carlo techniques. This is facilitated by identifying the full conditional distributions of each component where possible, as we now describe.

Under these specifications, the full conditional distributions for τ, $\boldsymbol{\beta}$, and Λ are straightforward to derive and are given by gamma, normal, and Wishart distributions proportional to

$$\tau^{N/2} \exp\left\{-\frac{\tau}{2} \sum_{i=1}^{m} \sum_{j=1}^{n_i} (y_{ij} - \mathbf{x}_{ij}^T \boldsymbol{\beta} - \mathbf{s}_{ij}^T \boldsymbol{b}_i)^2\right\},$$

$$\exp\left\{-\frac{1}{2}(\boldsymbol{\beta} - \boldsymbol{\eta})^T M (\boldsymbol{\beta} - \boldsymbol{\eta})\right\}, \text{ and}$$

$$\{\det(\Lambda)\}^{\frac{m+\rho_0-q-1}{2}} \exp\left(-\frac{1}{2}\mathrm{tr}\left[\left\{R_0^{-1} + \sum_{i=1}^{m} (\boldsymbol{b}_i - \boldsymbol{\xi})(\boldsymbol{b}_i - \boldsymbol{\xi})^T\right\} \Lambda\right]\right),$$

respectively, where $M = \Sigma_0^{-1} + \tau \sum_{i=1}^{m} \sum_{j=1}^{n_i} \mathbf{x}_{ij}\mathbf{x}_{ij}^T$, and $\boldsymbol{\eta} = M^{-1}\{\Sigma_0^{-1}\boldsymbol{\beta}_0 + \tau \sum_{i=1}^{m} \sum_{j=1}^{n_i} (y_{ij} - \mathbf{s}_{ij}^T \boldsymbol{b}_i)\mathbf{x}_{ij}\}$. Each $\boldsymbol{b}_i$ has full conditional distribution proportional to

$$\prod_{j=1}^{n_i} \frac{1}{\sigma_e} \phi\left(\frac{y_{ij} - \mathbf{x}_{ij}^T \boldsymbol{\beta} - \mathbf{s}_{ij}^T \boldsymbol{b}_i}{\sigma_e}\right) \exp\left\{-\frac{(\boldsymbol{b}_i - \boldsymbol{\xi})^T \Lambda (\boldsymbol{b}_i - \boldsymbol{\xi})}{2}\right\} \Phi(P_K(\boldsymbol{b}_i - \boldsymbol{\xi})),$$

which has exactly the same form as $p(\boldsymbol{b}_i|\mathbf{y}_i, \boldsymbol{\delta})$ given in (20.5). As pointed out in the paragraph before (20.9), we have an efficient way to generate samples from such a distribution. Alternatively, we can use Metropolis-Hastings with a random walk chain to generate a sample. That is, at the rth iteration in the chain, we generate a sample $\boldsymbol{b}_i^*$ from $N_q(\boldsymbol{b}_i^{(r)}, V)$ and a random number U from a uniform distribution on the interval $[0, 1]$, then set the new sample $\boldsymbol{b}_i^{(r+1)}$ to be either $\boldsymbol{b}_i^*$ or $\boldsymbol{b}_i^{(r)}$ depending on whether

$$U < \frac{\exp(-(\boldsymbol{b}_i^* - \boldsymbol{\eta}_i^{(r)})^T \Omega_i^{(r)-1} (\boldsymbol{b}_i^* - \boldsymbol{\eta}_i^{(r)})/2) \Phi(P_K(\boldsymbol{b}_i^* - \boldsymbol{\xi}^{(r)}))}{\exp(-(\boldsymbol{b}_i^{(r)} - \boldsymbol{\eta}_i^{(r)})^T \Omega_i^{(r)-1} (\boldsymbol{b}_i^{(r)} - \boldsymbol{\eta}_i^{(r)})/2) \Phi(P_K(\boldsymbol{b}_i^{(r)} - \boldsymbol{\xi}^{(r)}))},$$

or not. Here, $\boldsymbol{\eta}_i^{(r)}$ and $\Omega_i^{(r)}$ are of the form given right after (20.5), and V is a proposal covariance matrix that is selected after generating an initial MCMC chain to achieve a good convergence. The conditional density of $\boldsymbol{\xi}$ is proportional to

$$\exp\left\{-\frac{(\boldsymbol{\xi} - \tilde{\boldsymbol{\eta}})^T \tilde{\Omega}^{-1} (\boldsymbol{\xi} - \tilde{\boldsymbol{\eta}})}{2}\right\} \prod_{i=1}^{m} \Phi(P_K(\boldsymbol{b}_i - \boldsymbol{\xi})),$$

where $\tilde{\Omega}^{-1} = \Omega_0^{-1} + m\Omega^{-1}$ and $\tilde{\boldsymbol{\eta}} = \tilde{\Omega}(\Omega_0^{-1}\boldsymbol{\xi}_0 + \Omega^{-1}\sum_{i=1}^m \boldsymbol{b}_i)$. We use a Metropolis-Hastings method, i.e., at the rth iteration generate a sample $\boldsymbol{\xi}^*$ from $N_q(\tilde{\boldsymbol{\eta}}^{(r)}, \tilde{\Omega}^{(r)})$ and a random number U from a uniform distribution on the interval $[0, 1]$, then set the new sample $\boldsymbol{\xi}^{(r+1)}$ to be either $\boldsymbol{\xi}^*$ or $\boldsymbol{\xi}^{(r)}$ depending on whether $U < \prod_{i=1}^m [\Phi(P_K(\boldsymbol{b}_i^{(r)} - \boldsymbol{\xi}^*))/\Phi(P_K(\boldsymbol{b}_i^{(r)} - \boldsymbol{\xi}^{(r)}))]$ or not. The conditional distribution of the coefficients $\boldsymbol{\alpha}$ of the polynomial P_K is proportional to $p(\boldsymbol{\alpha})\prod_{i=1}^m \Phi(P_K(\boldsymbol{b}_i - \boldsymbol{\xi}))$. We adopt a Metropolis-Hastings method with a random walk chain to generate samples for $\boldsymbol{\alpha}$'s. Specifically, we generate a random sample $\boldsymbol{\alpha}^*$ from the proposal density $N_c(\boldsymbol{\alpha}^{(r)}, V)$ and a random number U from a uniform distribution on $[0, 1]$, then set the new sample $\boldsymbol{\alpha}^{(r+1)}$ to be either $\boldsymbol{\alpha}^*$ or $\boldsymbol{\alpha}^{(r)}$ depending on whether $U < [p(\boldsymbol{\alpha}^*)\prod_{i=1}^m \Phi(P_K(\boldsymbol{b}_i - \boldsymbol{\xi}))]/[p(\boldsymbol{\alpha}^{(r)})\prod_{i=1}^m \Phi(P_K(\boldsymbol{b}_i - \boldsymbol{\xi}))]$ or not. Notice that the coefficients in the polynomial P_K are set to be $\boldsymbol{\alpha}^*$ and $\boldsymbol{\alpha}^{(r)}$, respectively, in the numerator and denominator, the matrix V is selected after some initial running of the MCMC chain to achieve a good convergence.

Armed with these developments, it is straightforward to generalize the foregoing algorithm to the case where the $\boldsymbol{b}_i$ are represented by the *FGST* model. As described by Racine-Poon (1992), Wakefield *et al.* (1994), and Wakefield (1996), a q-variate t distribution with ν degrees of freedom may be represented as a scale mixture of normal distributions, given by

$$t_q(\mathbf{x};\boldsymbol{\xi},\Omega,\nu) = \frac{\Gamma(\frac{q+\nu}{2})}{\pi^{q/2}\{\det(\Omega)\}^{1/2}\Gamma(\nu/2)}\{1+(\mathbf{x}-\boldsymbol{\xi})^T\Omega^{-1}(\mathbf{x}-\boldsymbol{\xi})\}^{-(q+\nu)/2}$$
$$= \int \frac{\lambda^{q/2}}{\sqrt{(2\pi)^q\det(\Omega)}}\exp\{-\frac{\lambda}{2}(\mathbf{x}-\boldsymbol{\xi})^T\Omega^{-1}(\mathbf{x}-\boldsymbol{\xi})\}\chi^2_\nu(\lambda)d\lambda,$$

where χ^2_ν is the chi-square density with ν degrees of freedom. We adopt the same prior as in the $FGSN$ case for $\boldsymbol{\delta}$, and set the prior $p(\nu) = 1$ if $\nu > 0$ and $p(\nu) = 0$ if $\nu \leq 0$. The posterior density of $\boldsymbol{\delta}$, ν, $\boldsymbol{b}$ and the λ_i is proportional to

$$p(\nu)p(\boldsymbol{\delta})\prod_{i=1}^{m}\prod_{j=1}^{n_i}\frac{1}{\sigma_e}\phi\left(\frac{y_{ij}-\mathbf{x}_{ij}^T\boldsymbol{\beta}-\mathbf{s}_{ij}^T\boldsymbol{b}_i}{\sigma_e}\right)$$
$$\times\prod_{i=1}^{m}\frac{\lambda_i^{q/2}}{\sqrt{\det(\Omega)}}\exp\left\{-\frac{\lambda_i}{2}(\boldsymbol{b}_i-\boldsymbol{\xi})^T\Lambda(\boldsymbol{b}_i-\boldsymbol{\xi})\right\}\chi^2_\nu(\lambda_i)\Phi(P_K(\boldsymbol{b}_i-\boldsymbol{\xi})).$$

Under these specifications, the conditional distributions for τ, $\boldsymbol{\beta}$ and $\boldsymbol{\alpha}$ are identical to those in the $FGSN$ setting. That for Λ changes slightly and is now a Wishart distribution proportional to

$$\det(\Lambda)^{\frac{m+\rho_0-q-1}{2}}\exp\left(-\frac{1}{2}\mathrm{tr}\left[\left\{R_0^{-1}+\sum_{i=1}^{m}\lambda_i(\boldsymbol{b}_i-\boldsymbol{\xi})(\boldsymbol{b}_i-\boldsymbol{\xi})^T\right\}\Lambda\right]\right).$$

Each $\boldsymbol{b}_i$ has a full conditional distribution slightly different from that for $FGSN$ and is proportional to

$$\prod_{j=1}^{n_i}\frac{1}{\sigma_e}\phi\left(\frac{y_{ij}-\mathbf{x}_{ij}^T\boldsymbol{\beta}-\mathbf{s}_{ij}^T\boldsymbol{b}_i}{\sigma_e}\right)\exp\left\{\frac{(\boldsymbol{b}_i-\boldsymbol{\xi})^T\lambda_i\Lambda(\boldsymbol{b}_i-\boldsymbol{\xi})}{-2}\right\}\Phi(P_K(\boldsymbol{b}_i-\boldsymbol{\xi})).$$

Random samples from this distribution can be generated in the same way as in the $FGSN$ case. The conditional distribution of $\boldsymbol{\xi}$ is proportional to

$$\exp\left\{-\frac{(\boldsymbol{\xi}-\tilde{\boldsymbol{\eta}})^T\tilde{\Omega}^{-1}(\boldsymbol{\xi}-\tilde{\boldsymbol{\eta}})}{2}\right\}\prod_{i=1}^{m}\Phi(P_K(\boldsymbol{b}_i-\boldsymbol{\xi})),$$

which has the same appearance as for $FGSN$ except that here $\tilde{\Omega}^{-1} = \Omega_0^{-1}+\Omega^{-1}\sum_{i=1}^m\lambda_i$ and $\tilde{\boldsymbol{\eta}} = \tilde{\Omega}(\Omega_0^{-1}\boldsymbol{\xi}_0+\Omega^{-1}\sum_{i=1}^m\lambda_i\boldsymbol{b}_i)$. Random samples can again be generated in the same way as in the $FGSN$ setting as well. The conditional distribution of each λ_i is a gamma distribution proportional to

$$\lambda_i^{(q+\nu)/2-1}\exp[-\lambda_i\{(\boldsymbol{b}_i-\boldsymbol{\xi})^T\Omega^{-1}(\boldsymbol{b}_i-\boldsymbol{\xi})+1\}/2].$$

The conditional distribution of ν is of a relatively complex form that is proportional to

$$f(\nu) = p(\nu)2^{-m\nu/2}\Gamma(\frac{\nu}{2})^{-m}(\prod_{i=1}^{m}\lambda_i)^{\nu/2-1}.$$

To generate a random sample from this distribution, we use Metropolis-Hastings with random walk chain again. We generate a random sample ν^* of the proposal density $N(\nu^{(r)}, \tilde{\sigma}^2)$ and a random number U from a uniform distribution on $[0, 1]$, then set $\nu^{(r+1)}$ to be equal to either ν^* or $\nu^{(r)}$ depending on whether $U < f(\nu^*)/f(\nu^{(r)})$ or not; here $\tilde{\sigma}^2$ is a proposal variance that is selected after generating the initial MCMC chain to achieve a good convergence.

For both of the *FGSN* or *FGST* cases, then, the foregoing developments may be used to generate draws from the posterior distribution that may be used to make inference according to the usual Bayesian paradigm; e.g., posterior standard deviations may be calculated to characterize variation in the estimates. As in the case of likelihood inference, the preceding algorithms take K to be fixed. Thus, to choose a suitable K, a separate analysis may be run for each $K = 0, 1, 3, \ldots, K_{max}$, and selection may be made according to the information criteria as described earlier.

In practice, we have found that these developments are also useful in the context of the following strategy for deducing whether the *FGST* or *FGSN* model provides a better representation of the apparent features of the random effects distribution. We suggest running a preliminary Markov chain assuming the *FGST* model, as described above. If the posterior estimate for ν is very large, then there is little evidence to suggest heavy-tailedness of the true underlying density, indicating that the *FGSN* model likely provides adequate flexibility and that either likelihood-based inference via the EM algorithm in Section 20.3.1 or fitting via the above MCMC approach may be implemented. Alternatively, if ν is sufficiently small to suggest heavy-tail behavior, suggesting adoption of the *FGST* model to provide adequate latitude to accommodate this feature, the procedure outlined above may be used to implement the fitting.

20.4 Simulation Results

To assess the performance of the proposed model and methods, we conducted three simulation studies. In all cases, we took the linear mixed model to be

$$Y_{ij} = t_{ij}\beta_1 + w_i\beta_2 + b_i + e_{ij}, \tag{20.10}$$

where for $j = 1 \ldots, 5$, $t_{ij} = j - 3$; $\beta_1 = 2$ and $\beta_2 = 1$; and $e_{ij} \sim N(0, 0.5^2)$. Note that there is no need for an intercept term, as the mean of b_i is non-zero; here $x_{ij} = t_{ij}$ and $s_{ij} = 1$, satisfying the requirement that the x_{ij} do not contain s_{ij}. In each simulation, 100 Monte Carlo data sets were

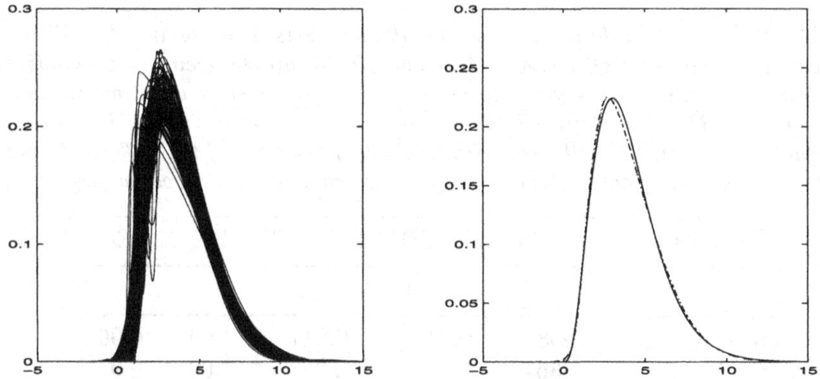

Figure 20.1 *Results, simulation with gamma-distributed random effects. The left-hand panel shows 100 estimated densities favored by BIC. The right-hand panel shows the true distribution (solid line) and the average of the 100 densities favored by BIC (dashed-dotted) and HQ (dotted).*

simulated from (20.10) according to the additional specifications described below, and (20.10) was fit to each data set via the strategy outlined at the end of Section 20.3.2. In particular, for each data set in each of the three simulation scenarios discussed below, we assumed the $FGST$ model with $K = 1$ and obtained samples from the posterior distribution for ν via the MCMC algorithm with 3,000 samples. In each scenario, the mean of this distribution was ≥ 400 with a 0.025 quantile above 300 for each data set, suggesting negligible heavy tail behavior. We thus adopted the $FGSN$ model in all three simulations, implemented by the Monte Carlo EM algorithm described in Section 20.3.1, with standard errors calculated by the empirical sandwich method.

As an advantage of the $FGSE$ representation for the random effects is its propensity for accommodating skewness, in the first simulation we generated the b_i according to a $Gamma(4, 1)$ distribution with probability density $1/6x^3 \exp(-x)$, which yields a highly skewed distribution as suggested by Figure 20.1. Here, we took $m = 200$ with $w_i = 1$ if $i \leq 100$ and $w_i = 0$ if $i > 100$, so that w_i represents an individual-level covariate, e.g., a treatment indicator.

For the EM algorithm, we took $K_{max} = 3$. Of the 100 simulated data sets, BIC chooses $K = 1$ in 96 data sets and $K = 3$ in the remainder, while HQ selects $K = 1$ in 82 data sets and $K = 3$ in the remainder. Neither criterion selects $K = 0$, suggesting strong evidence of skewness. For a small proportion of the data sets, a bimodal estimate of the density is obtained, as seen in the left-hand panel of Figure 20.1; however, on average, the estimated densities with the selected K track the true density remarkably well,

Table 20.1 *Monte Carlo results based on 100 data sets, true Gamma*$(4,1)$ *distribution for the random effects. MC Mean and MC SD are the average and standard deviation of the estimates, respectively; AVE SE is the average of estimated standard errors; RE is the Monte Carlo mean squared error for the indicated fit divided by that for* $K = 0$*; EC is the empirical coverage probability of the 95% confidence intervals of the estimates. True values of the parameters are in parentheses.*

Parameter	MC Mean	MC SD	AVE SE	RE	EC
		$K = 0$			
$\beta_1(2)$	1.998	0.010	0.011	1.00	0.96
$\beta_2(1)$	1.036	0.285	0.288	1.00	0.97
$E(b)(4)$	3.969	0.180	0.200	1.00	0.96
$Var(b)(4)$	3.928	0.521	0.487	1.00	0.90
$\sigma_e(0.5)$	0.501	0.013	0.012	1.00	0.94
		BIC			
$\beta_1(2)$	1.998	0.010	0.012	1.00	0.97
$\beta_2(1)$	1.023	0.252	0.220	0.77	0.94
$E(b)(4)$	3.980	0.171	0.187	1.00	0.96
$Var(b)(4)$	3.918	0.528	0.448	0.89	0.87
$\sigma_e(0.5)$	0.501	0.013	0.013	1.03	0.92
		HQ			
$\beta_1(2)$	1.998	0.010	0.012	1.00	0.97
$\beta_2(1)$	1.023	0.252	0.217	0.77	0.94
$E(b)(4)$	3.983	0.173	0.226	1.00	0.96
$Var(b)(4)$	3.917	0.528	0.514	0.91	0.87
$\sigma_e(0.5)$	0.501	0.013	0.013	1.04	0.92

as seen in the right-hand panel. Table 20.1 shows the numerical results for $K = 0$ (normality) and for the estimates where K was chosen by BIC and HQ, respectively. All estimators are approximately unbiased. In most cases, the average of estimated standard errors agrees well with the Monte Carlo standard deviation. As found by other authors (e.g., Tao *et al.*, 1999; Zhang and Davidian, 2001), efficiency of estimation on β_2 associated with the individual-level covariate w_i is degraded when normality is assumed $(K = 0)$ relative to allowing a more flexible representation via the $FGSN$. Thus, although unbiased estimation is still possible under normality, failure to take appropriate account of the true features of the random effects leads to less precise inference on what are usually quantities of key interest

Table 20.2 *Monte Carlo results based on 100 data sets, true Normal(4, 4) distribution for the random effects. Entries are as in Table 20.1.*

Parameter	MC Mean	MC SD	AVE SE	RE	EC
		$K = 0$ and BIC			
$\beta_1(2)$	1.999	0.012	0.011	1.00	0.96
$\beta_2(1)$	1.034	0.295	0.281	1.00	0.92
$E(b)(4)$	4.006	0.201	0.199	1.00	0.93
$Var(b)(4)$	3.912	0.339	0.391	1.00	0.95
$\sigma_e(0.5)$	0.499	0.012	0.012	1.00	0.95
		HQ			
$\beta_1(2)$	1.999	0.012	0.011	1.00	0.96
$\beta_2(1)$	1.035	0.295	0.281	1.00	0.92
$E(b)(4)$	4.006	0.201	0.199	1.00	0.93
$Var(b)(4)$	3.912	0.339	0.390	1.06	0.95
$\sigma_e(0.5)$	0.499	0.012	0.012	1.00	0.95

such as treatment effects. Some mild efficiency loss is also associated with estimation of the inter-individual variance $Var(b_i)$.

The second simulation was identical to the first except that the true distribution of the random effects was instead $N(4, 4)$. Here, then, there is no need for greater flexibility, and the hope would be that the proposed methods would identify this. We again took $K_{max} = 3$. The BIC criterion selected $K = 0$ for all 100 simulated data sets, and HQ chose $K = 0$ for 98 data sets and $K = 1$ for the remaining two, demonstrating that the methods will not in general impose skewness or multimodality when these features are not present; all estimates are unimodal and essentially normal, so we do not plot them for brevity. The results in Table 20.2 show that there is no efficiency loss associated with using the more flexible methods.

In the third simulation, we investigated the ability of the methods to distinguish bimodality of the underlying random effects density. The model was the same as before except that we took $m = 100$ and let $w_i = 1$ if $i \leq 50$ and $w_i = 0$ if $i > 50$. The true distribution of the random effects was a mixture of normals of the form $b_i \sim 0.7N(-3, 1) + 0.3N(2, 1)$, which exhibits perceptible bimodality, as reflected in the estimates shown in Figure 20.2, suggesting that $K \geq 3$ would be required to represent this feature. Indeed, the BIC criterion selects $K = 3$ for 98 of the 100 simulated data sets and $K = 5$ for the remaining two and yields no unimodal estimates, as shown in the left-hand panel of the figure. Similarly, HQ chooses $K = 3$ and

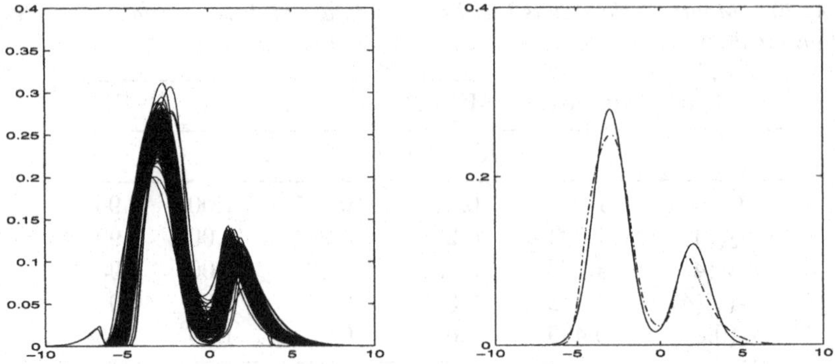

Figure 20.2 *Results, simulation with random effects distributed according to a bimodal mixture of normals. The left-hand panel shows 100 estimated densities favored by BIC. The right-hand panel shows the true distribution (solid line) and the average of the 100 densities favored by BIC (dashed-dotted) and HQ (dotted).*

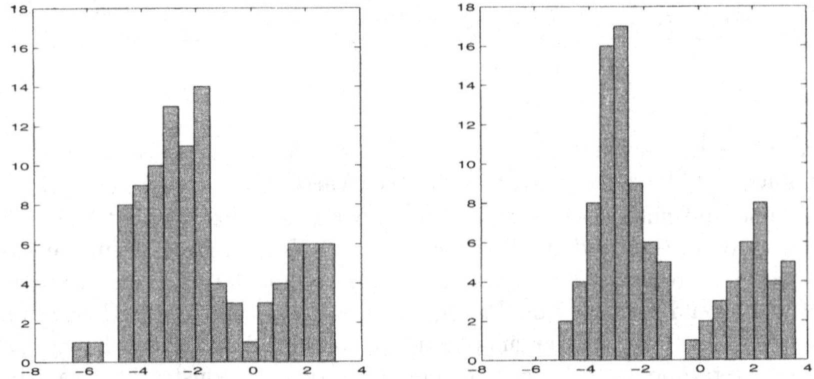

Figure 20.3 *Left panel: the histogram of the random effects b_i's. The selected model yields $K = 5$. Right panel: the histogram of the random effects b_i's. The selected model yields $K = 3$.*

$K = 5$ in 92 and 8 of the data sets, respectively, and admits no unimodal estimates. Thus, we see that the proposed approach accurately identifies the departure from unimodality in every case through choices of $K \geq 3$. The right-hand panel of Figure 20.2 shows that, on average, the $FGSN$ representation tracks the true bimodal features well. From Table 20.3, all estimators for β_1 and β_2 are essentially unbiased; however, dramatic efficiency losses result for estimation of β_2 and the mean of the random effects (intercept) when normality is assumed ($K = 0$) relative to taking the bimodality into account. Although a byproduct of the proposed methods is

Table 20.3 *Monte Carlo results based on 100 data sets, true bimodal mixture of normal* $0.7N(-3,1)+0.3N(2,1)$ *distribution for the random effects. Entries are as in Table 20.1.*

Parameter	MC Mean	MC SD	AVE SE	RE	EC
		$K=0$			
$\beta_1(2)$	2.001	0.015	0.016	1.00	0.96
$\beta_2(1)$	1.121	0.525	0.516	1.00	0.96
$E(b)(-1.5)$	−1.577	0.370	0.352	1.00	0.92
$Var(b)(6.25)$	6.087	0.685	0.663	1.00	0.91
$\sigma_e(0.5)$	0.498	0.018	0.174	1.00	0.92
		BIC			
$\beta_1(2)$	2.001	0.015	0.017	1.00	0.97
$\beta_2(1)$	0.989	0.269	0.237	0.25	0.93
$E(b)(-1.5)$	−1.656	0.253	0.322	0.62	0.95
$Var(b)(6.25)$	6.239	0.763	1.230	1.18	1.00
$\sigma_e(0.5)$	0.497	0.018	0.019	1.00	0.93
		HQ			
$\beta_1(2)$	2.001	0.015	0.016	1.00	0.97
$\beta_2(1)$	0.991	0.308	0.261	0.33	0.93
$E(b)(-1.5)$	−1.637	0.271	0.327	0.65	0.95
$Var(b)(6.25)$	6.225	0.757	1.183	1.16	1.00
$\sigma_e(0.5)$	0.499	0.018	0.019	1.01	0.93

some efficiency loss in estimation of the variance of b_i, note that under normality the estimator of this quantity is noticeably biased. As observed in the previous two simulations and by other authors (e.g., Zhang and Davidian, 2001), estimation of β_1 and σ_e^2 is insensitive to the representation of the random effects, which has been speculated to result from the fact that these parameters pertain to within-individual characteristics and would thus be less influenced by assumptions on the population. From the left-hand panel of Figure 20.2, a few of the data sets admit tri-modal estimated densities, all of which are associated with the BIC criterion selecting $K=5$. For each of these, we inspected the 100 generated b_i and found that their empirical distributions show a small group on the left. A typical histogram for such a data set is displayed in the left-hand panel of Figure 20.3; for comparison, a histogram for a data set for which $K=3$ was selected is presented on the right and shows no such "third group."

20.5 Application to Cholesterol Data

To illustrate how the methods would be used in practice, we apply them to longitudinal data on cholesterol levels collected as part of the famed Framingham heart study, reported in Zhang and Davidian (2001). We adopt the same linear mixed effects model used by these authors, given by

$$Y_{ij} = \beta_1 \text{age}_i + \beta_2 \text{sex}_i + b_{i1} + b_{i2} t_{ij} + e_{ij}, \tag{20.11}$$

where Y_{ij} is the measured cholesterol level at jth time for individual i divided by 100, t_{ij} is (time − 5)/10, $e_{ij} \sim N(0, \sigma_e^2)$ as in (20.1), and age_i and sex_i are the age and gender of the ith participant at baseline.

To establish the appropriate representation for the density of the random effects $\boldsymbol{b}_i = (b_{i1}, b_{i2})^T$, we carried out a preliminary fit via the Bayesian approach using MCMC techniques discussed in Section 20.3.2 using the $FGST$ model, $K = 1$, and three chains of length 1,000,000 with starting values 1, 10, and 30, respectively, for ν. The posterior distribution for ν has several modes, with the dominant mode centered above 5,000, suggesting little evidence of heavy-tailedness; accordingly, we adopted the $FGSN$ model as adequate to capture the underlying features. We fit (20.11) using the Monte Carlo EM algorithm described in Section 20.3.1 and took $K_{max} = 3$.

The results are shown in Table 20.4. From the bottom part of the table, all three model selection criteria favor $K = 1$, supporting the contention of a departure from normality. From the discussion in Section 20.2, these results

Table 20.4 *Estimated parameters, and model selection result for Framingham Cholesterol Data.*

	$K = 0$	$K = 1$	$K = 3$
β_1	0.0184 (0.0034)	0.0149 (0.0033)	0.0156 (0.0033)
β_2	−0.0630 (0.0540)	−0.0491 (0.0506)	−0.0531 (0.0508)
$\text{E}(b_{i1})$	1.5968 (0.1580)	1.7441 (0.1500)	1.6901 (0.1478)
$\text{E}(b_{i2})$	0.2817 (0.0255)	0.2846 (0.0268)	0.2770 (0.0394)
$\text{Var}(b_{i1})$	0.1412 (0.0153)	0.1351 (0.0156)	0.1973 (0.0588)
$\text{Var}(b_{i2})$	0.0380 (0.0116)	0.0434 (0.0115)	0.0376 (0.0132)
$\text{Cov}(b_{i1}, b_{i2})$	0.0314 (0.0101)	0.0282 (0.0111)	0.0292 (0.0141)
σ_e	0.2084 (0.0042)	0.2071 (0.0041)	0.2074 (0.0041)
−loglikelihood	160.994	152.033	151.610
AIC	0.162	0.155	0.159
BIC	0.181	0.179	0.192
HQ	0.169	0.164	0.171

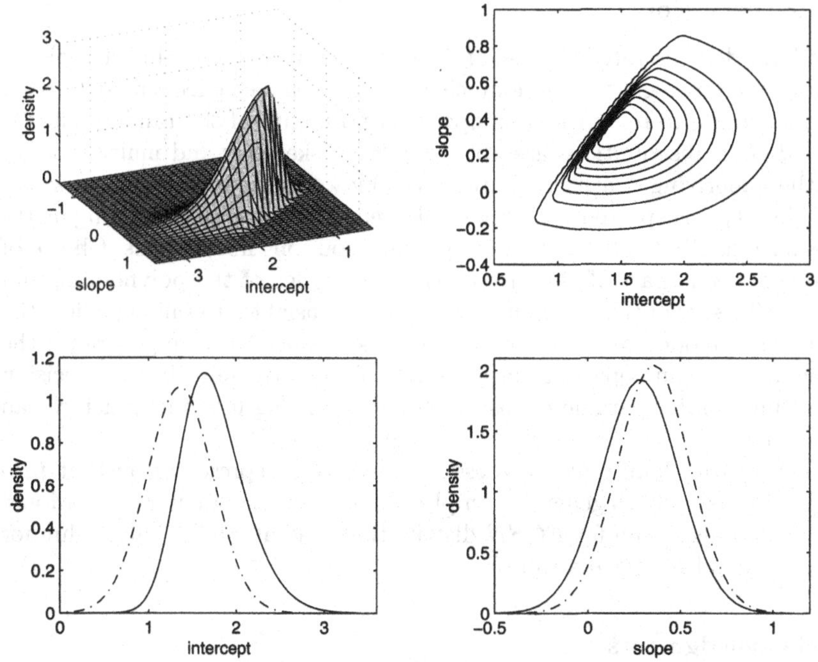

Figure 20.4 *Estimated random effect density for the cholesterol data,* $K = 1$. *Upper left panel: estimated density. Upper right panel: contour plot of estimated density. Lower left panel: estimated marginal density of intercept, and lower right panel: estimated marginal density of slope (solid lines) with estimated normal density using* $K = 0$ *(dash-dotted line) superimposed.*

suggest that, although skewness may be present, there is no indication of multimodality. Estimates of the individual-level covariate coefficients β_1 and β_2 differ as well, reflecting the qualitative behavior seen in the simulations.

Figure 20.4 shows the estimated random effect distribution when $K = 1$. The estimate is unimodal, with obvious skewness that is particularly evident in the associated marginal density for intercept (baseline cholesterol). Zhang and Davidian (2001) fitted model (20.11) to these data but representing the random effects density by the SNP density model of Gallant and Nychka (1987), which is known to produce artifactual modes when fitted to skewed distributions. Comparing to their Figure 2, we see that while the SNP estimate suggests the presence of a second mode, which the authors speculate may be associated with a subpopulation of individuals with higher baseline cholesterol, the $FGSN$ representation attributes this feature to skewness of the underlying density.

20.6 Discussion

We have demonstrated the use of flexible generalized skew-elliptical distributions to represent the random effects in linear mixed effects models when one is willing to relax the assumption of normality of the random effects. The $FGSE$ distributions are able to capture skewness and multimodality of the underlying density and are especially well suited to situations where the density is suspected to be highly skewed, in which case competing methods such as SNP and mixtures may admit spurious fluctuations. Choice of the symmetric part of the distribution, the order of the polynomial, and the coefficients of the polynomial can provide insight into tail behavior, the number of modes, and the level of skewness, respectively. In general, if the true distribution departs from normality, impressive gains in the precision of estimation for parameters associated with among-individual factors can be realized.

The $FGSE$ family also possesses a stochastic representation that provides an easy way to generate samples. This facilitates an attractive Monte Carlo EM algorithm for $FGSN$ distributions and an MCMC algorithm for more general $FGST$ distributions.

Acknowledgments

This work was supported by NIH grants AI31789, CA085848, and GM67299.

References

Adcock, C. J. and Clark, E. (1999). Beta lives - some statistical perspectives on the capital asset pricing model. *Europ. J. Finance* **5**, 213–224.

Adcock, C. J. and Shutes, K. (2000). Fat tails and the capital asset pricing model. In *Advances in Quantitative Asset Management*, C. Dunis (ed.), Boston: Kluwer Academic Press.

Adcock, C. J. and Shutes, K. (2001). Portfolio selection based on the multivariate-skew normal distribution. In *Financial Modelling*, A. Skulimowski (ed.), Krakow: Progress & Business Publishers.

Agresti, A. (1990). *Categorical Data Analysis.* New York: Wiley.

Ahn, S. (2001). Firm dynamics and productivity growth: a review of micro evidence from OECD countries, *Technical Report ECO/WKP(2001)23*, OECD, Paris.

Aigner, D. J., Lovell, C. A. K. and Schmidt, P. (1977). Formulation and estimation of stochastic frontier production function model. *J. Econometrics* **12**, 21–37.

Akaike, H. (1973). Information theory and an extension of the maximum likelihood principle. In *International Symposium on Information Theory*, B. N. Petrov, F. Csaki (eds.), Budapest: Akademia Kiado, 267–281.

Albert, J. H. and Chib, S. (1995). Bayesian residual analysis for binary response regression models. *Biometrika* **82**, 747–759.

Albert, J. H. and Chib, S. (1993). Bayesian analysis of binary and polychotomous response data. *J. Amer. Statist. Assoc.* **88**, 669–679.

Amemiya, T. (1981). Qualitative response models: a survey. *J. Econ. Lit.* **19**, 1483–1536.

Ammann, C. M., Kiehl, J. T., Zender, C. S., Otto-Bliesner, B. L. and Bradley, R. S. (2002). Coupled simulations of the 20th-century including external forcing. *J. Climate*, revised.

Andel, J., Netuka, I. and Zvara, K. (1984). On threshold autoregressive processes. *Kybernetika* **20**, 89–106.

Andrews, D. F. and Mallows, C. L. (1974). Scale mixture of normal distributions. *J. R. Stat. Soc. Ser. B* **36**, 99–102.

Aparicio Acosta, F. and Estrada, J. (2000). Empirical distributions of stock returns: European Securities markets. *Europ. J. Finance* **7**, 1–21.

Aranda-Ordaz, F. J. (1981). On two families of transformations to additivity for binary response data. *Biometrika* **68**, 357–363.

Arellano-Valle, R. B., del Pino, G. and San Martín, E. (2002). Definition and probabilistic properties of skew-distributions. *Statist. Probab. Lett.* **58**, 111–121.

Arellano-Valle, R. B. and Genton, M. G. (2003). On fundamental skew distributions. *Institute of Statistics Mimeo Series* **2551**, *J. Multivariate Anal.*, to appear.

Arellano-Valle, R. B., Gómez, H. W. and Quintana, F. A. (2004a). Statistical inference for a general class of asymmetric distributions. *J. Statist. Plann. Inference*, to appear.

Arellano-Valle, R. B., Gómez, H. W. and Quintana, F. A. (2004b). A new class of skew-normal distributions. *Comm. Statist. Theory Methods*, to appear.

Arenou, F. and Luri, X. (1999). The Cepheid distance scale after Hipparcos. *ASP Conference Series* **167**, 13–32.

Arjas, E. and Gasbarra, D. (1994). Nonparametric Bayesian inference for right-censored survival data, using the Gibbs sampler. *Statist. Sinica* **4**, 505–524.

Arjas, E. and Heikkinen, J. (1997). An algorithm for nonparametric Bayesian estimation of a Poisson intensity. *Comput. Statist.* **12**, 385–402.

Arnold, B. C. and Beaver, R. J. (2000a). Hidden truncation models. *Sankhyă Ser. A* **62**, 23–35.

Arnold, B. C. and Beaver, R. J. (2000b). The skew-Cauchy distribution. *Statist. Probab. Lett.* **49**, 285–290.

Arnold, B. C. and Beaver, R. J. (2000c). Some skewed multivariate distributions. *Amer. J. Math. Management Sci.* **20**, 27–38.

Arnold, B. C. and Beaver, R. J. (2002). Skew multivariate models related to hidden truncation and/or selective reporting. *Test* **11**, 7–54.

Arnold, B. C. and Beaver, R. J. (2003). An alternative construction of skewed multivariate distributions. Technical Report 270, Department of Statistics, University of California, Riverside.

Arnold, B. C., Beaver R. J., Groeneveld R. A. and Meeker W. Q. (1993). The nontruncated marginal of a truncated bivariate normal distribution. *Psychometrika* **58**, 471–478.

Arnold, B. C., Castillo, E. and Sarabia, J. M. (2002). Conditionally specified multivariate skewed distributions. *Sankhyă Ser. A* **64**, 206–226.

Australian Bureau of Statistics (1995). *How Australians Measure Up*. Canberra: ABS Publications.

Azzalini, A. (1985). A class of distributions which includes the normal ones. *Scand. J. Statist.* **12**, 171–178.

Azzalini, A. (1986). Further results on a class of distributions which includes the normal ones. *Statistica* **46**, 199–208.

Azzalini, A. (2001). A note on regions of given probability of the skew-normal distribution. *Metron* **59**, 27–34.

Azzalini, A. and Capitanio, A. (1999). Statistical applications of the multivariate skew-normal distribution. *J. R. Stat. Soc. Ser. B* **61**, 579–602.

Azzalini, A. and Capitanio, A. (2003). Distributions generated by perturbation of symmetry with emphasis on a multivariate skew t distribution. *J. R. Stat. Soc. Ser. B* **65**, 367–389.

Azzalini, A. and Dalla Valle, A. (1996). The multivariate skew-normal distribution. *Biometrika* **83**, 715–726.

Ball, L. and Mankiw, N. G. (1995). Relative-price changes as aggregate supply shocks. *Quart. J. Econ.* **110**, 161–193.

Bartolucci, F., De Luca, G. and Loperfido, N. (2000). A generalization for skewness in the basic stochastic volatility model. *Proceedings of the 15th International Workshop on Statistical Modelling*, 140–145.

Basu, S. and Mukhopadhyay, S. (2000). Binary response regression with normal scale mixture links. In *Generalized Linear Models: A Bayesian Perspective*, D. K. Dey, S. K. Ghosh, B. K. Mallick (eds.), New York: Marcel Dekker, Inc., 231–241.

Basu, D. and Pereira, C. A. B. (1983). Conditional independence in statistics. *Sankhyă Ser. A* **45**, 324–337.

Bayarri, M. J. and DeGroot, M. (1992). A BAD view of weighted distributions and selection models. In *Bayesian Statistics* **4**, J. M. Bernardo, J. O. Berger, A. P. Dawid, and A. F. M. Smith (eds.), London: Oxford University Press, 17–29.

Benedict, G. F., McArthur, B. E., Fredrick, L. W., Harrison, T. E., Lee, J., Slesnick, C. L., Rhee, J., Patterson, R. J., Nelan, E., Jefferys, W. H., van Altena, W., Shelus, P. J., Franz, O. G., Wasserman, L. H., Hemenway, P. D., Duncombe, R. L., Story, D., Whipple, A. L. and Bradley, A. J. (2002). Astrometry with the Hubble space telescope: A parallax of the fundamental distance calibrator RR Lyrae. *Astron. J.* **123**, 473–484.

Berger, J. O. and Bernardo, J. M. (1992). On the development of reference priors. In *Bayesian Statistics* **4**, J. M. Bernardo, J. O. Berger, A. P. Dawid, and A. F. M. Smith (eds.), London: Oxford University Press, 35–60.

Berger, J. O., Rios Insua, D. and Ruggeri, F. (2000). Bayesian robustness. In *Robust Bayesian Analysis*, Lecture Notes in Statistics **152**, New York: Springer, 1–32.

Berger, J. O., De Oliveira, V. and Sansó, B. (2001). Objective Bayesian analysis of spatially correlated data. *J. Amer. Statist. Assoc.* **96**, 1361–1374.

Bernardo, J. M. and Smith, A. F. M. (1994). *Bayesian Theory*. New York: Wiley.

Bhaumik, A., Chen, M.-H. and Dey, D. K. (2004). Dynamic generalized linear models for correlated binary responses with skewed links. Technical Report, Department of Statistics, University of Connecticut.

Bickel, P., Klaassen, C. A. J., Ritov, Y. and Wellner, J. A. (1993). *Efficient and Adaptive Inference in Semiparametric Models*. Baltimore: Johns Hopkins University Press.

Birnbaum, Z. W. (1950). Effect of linear truncation on a multinormal population. *Ann. Math. Statistics* **21**, 272–279.

Black, F. and Scholes, M. (1973). The pricing of options and corporate liabilities. *J. Polit. Economy* **81**, 637–659.

Blake, A. and Isard, M. (1998). *Active Contours*. New York: Springer.

Blattberg, R. and Gonedes, N. (1974). A comparison of the stable and student distributions as statistical models for stock prices. *J. Bus.* **47**, 244–280.

Bollerslev, T. (1986). Generalized autoregressive conditional heteroskedasticity. *J. Econometrics* **31**, 307–327.

Bollerslev, T. (1987). A conditionally heteroskedastic time series model for speculative prices and rates of returns. *Rev. Econ. Statist.* **6**, 542–547.

Bollerslev, T. (1990). Modelling the coherence in short-run nominal exchange rates: a multivariate generalized ARCH approach. *Rev. Econ. Statist.* **72**,

498–505.

Bollerslev, T., Chou, R. and Kroner, K. (1992). ARCH modelling in finance. *J. Econometrics* **52**, 5–59.

Bollerslev, T. and Engle, R. F. (1993). Common persistence in conditional variances. *Econometrica* **61**, 167–186.

Booth, J. G. and Hobert, J. P. (1999). Maximizing generalized linear mixed model likelihoods with an automated Monte Carlo EM Algorithm. *J. R. Stat. Soc. Ser. B* **61**, 265–285.

Boville, B. A. and Gent, P. R. (1998). The NCAR climate system model, version one. *J. Climate* **11**, 1115–1130.

Branco, M. D. and Dey, D. K. (2001). A general class of multivariate skew-elliptical distributions. *J. Multivariate Anal.* **79**, 99–113.

Breslow, N. E. (1974). Covariance analysis of censored survival data. *Biometrics* **30**, 89–99.

Brockwell, P. J. and Davis, R. A. (1991). *Time Series: Theory and Methods.* New York: Springer.

Brown, K. C., Harlow, W. V. and Tinic, S. M. (1988). Risk aversion, uncertain information, and market efficiency. *J. Finan. Econ.* **22**, 355–385.

Brown, L. (1971). Admissible estimators, recurrent diffusions and insoluble boundary-value problems. *Ann. Math. Statist.* **42**, 855–903.

Buckle, D. J. (1995). Bayesian inference for stable distributions. *J. Amer. Statist. Assoc.* **90**, 605–613.

Cain, M. (1994). The moment-generating function of the minimum of bivariate normal random variables. *Amer. Statist.* **48**, 124–125.

Campbell, J. Y. and Hentschel, L. (1992). No news is a good news: an asymmetric model of changing volatility in stock returns. *J. Finan. Econ.* **31**, 281–331.

Capitanio, A., Azzalini, A. and Stanghellini, E. (2003). Graphical models for skew-normal variates. *Scand. J. Statist.* **30**, 129–144.

Carlin, B., Polson, N. G. and Stoffer, D. S. (1992). A Monte Carlo approach to nonnormal and nonlinear state-space modeling, *J. Amer. Statist. Assoc.* **87**, 493–500.

Catchpole, E. A. and Morgan, B. T. J. (1997). Detecting parameter redundancy. *Biometrika* **84**, 187–196.

Caudill, S. M. (1993). Estimating the cost of partial coverage rent controls: a stochastic frontier approach. *Rev. Econ. Statist.* **75**, 727–731.

Chamberlain, G. (1983). A characterization of the distributions that imply mean-variance utility functions. *J. Econ. Theory* **29**, 185–201.

Chen, M.-H. (1998). Bayesian analysis of correlated mixed categorical data by incorporating historical prior information. *Comm. Statist. Theory Methods* **27**, 1341–1361.

Chen, M.-H. and Deely, J. J. (1996). Bayesian analysis for a constrained linear multiple regression problem for predicting the new crop of apples. *J. Agric. Biol. Environ. Stat.* **1**, 467–489.

Chen, M.-H. and Dey, D. K. (1998). Bayesian modeling of correlated binary responses via scale mixture of multivariate normal link functions. *Sankhyă Ser. A* **60**, 322–343.

Chen, M.-H. and Dey, D. K. (2000a). A unified Bayesian analysis for correlated ordinal data models. *Braz. J. Probab. Stat.* **14**, 87–111.

Chen, M.-H. and Dey, D. K. (2000b). Bayesian data analysis of longitudinal binary response models. In *Proceedings of the Third International Triennial Calcutta Symposium on Probability and Statistics*, J. K. Ghosh, P. K. Sen, B. K. Sinha (eds.), New York: Oxford University Press, 100–123.

Chen, M.-H. and Dey, D. K. (2003). Variable selection for multivariate logistic regression models. *J. Statist. Plann. Inference* **111**, 37–55.

Chen, M.-H., Dey, D. K. and Ibrahim, J. G. (2004). Bayesian criterion based model assessment for categorical data. *Biometrika* **91**, 41–63.

Chen, M.-H., Dey, D. K. and Shao, Q.-M. (1999). A new skewed link model for dichotomous quantal response data. *J. Amer. Statist. Assoc.* **94**, 1172–1186.

Chen, M.-H., Ibrahim, J. G. and Yiannoutsos, C. (1999). Prior elicitation, variable selection, and Bayesian computation for logistic regression models. *J. R. Stat. Soc. Ser. B* **61**, 223–242.

Chen, M.-H. and Shao, Q.-M. (1998). Monte Carlo methods on Bayesian analysis of constrained parameter problems. *Biometrika* **85**, 73–87.

Chen, M.-H. and Shao, Q.-M. (1999a). Properties of prior and posterior distributions for multivariate categorical response data models. *J. Multivariate Anal.* **71**, 277–296.

Chen, M.-H. and Shao, Q.-M. (1999b). Existence of Bayes estimators for the polychotomous quantal response models. *Ann. Inst. Statist. Math.* **51**, 637–656.

Chen, M.-H. and Shao, Q.-M. (2001). Propriety of posterior distribution for dichotomous quantal response models with general link functions. *Proc. Amer. Math. Soc.* **129**, 293–302.

Chen, M.-H., Shao, Q.-M. and Ibrahim, J. G. (2000). *Monte Carlo Methods in Bayesian Computation.* New York: Springer.

Chen, J., Zhang, D. and Davidian, M. (2002). Generalized linear mixed models with flexible distributions of random effects for longitudinal data. *Biostatistics* **3**, 347–360.

Chen, Z. (2001). On modeling discrete choice data. Unpublished Ph.D. Dissertation, Department of Statistics, University of Connecticut.

Chib, S. and Greenberg, E. (1995). Understanding the Metropolis-Hastings algorithm. *Amer. Statist.* **49**, 327–335.

Chib, S. and Greenberg, E. (1998). Bayesian analysis of multivariate probit models. *Biometrika* **85**, 347–361.

Chib, S. and Jeliazkov, I. (2001). Marginal likelihood from the Metropolis-Hastings output. *J. Amer. Statist. Assoc.* **96**, 270–281.

Chiogna, M. (1997). Notes on estimation problems with scalar skew-normal distributions. *Technical Report 1997.15*, Department of Statistical Sciences, University of Padua, Padua.

Chiogna, M. (1998). Some results on the scalar skew-normal distribution. *J. Ital. Statist. Soc.* **7**, 1–13.

Choy, S. T. B. (1995). *Robust Bayesian Analysis Using Scale Mixture of Normals Distributions.* Ph.D. dissertation, Department of Mathematics, Imperial College, London.

Christie, A. (1982). The stochastic behavior of common stock variances: value, leverage and interest rate effects. *J. Finan. Econ.* **10**, 407–432.

Cliff, A. D. and Ord, J. K. (1981). *Spatial Processes: Models and Applications.* London: Pion.

Clyde, M. A. (1999). Bayesian model averaging and model search strategies. In *Bayesian Statistics* **6**, J. M. Bernardo, J. O. Berger, A. P. Dawid, A. F. M. Smith (eds.), Oxford: Oxford University Press, 157–185.

Coles, S. (2001). *An Introduction to Statistical Modelling of Extreme Values.* Springer Series in Statistics, New York: Springer.

Cook, R. D. and Weisberg, S. (1994). *An Introduction to Regression Graphics.* New York: Wiley.

Cootes, T. F., Edwards, G. J. and Taylor, C. J. (1998). Active appearance models. *Proc. European Conference on Computer Vision* **2**, 484–498.

Cootes, T. F., Taylor, C. J., Cooper, D. H. and Graham, J. (1995). Active shape models – their training and application. *Computer Vision and Image Understanding* **61**, 38–59.

Copas, J. B. and Li, H. G. (1997). Inference for non-random samples (with discussion). *J. R. Statist. Soc. B* **59**, 55–95.

Corcoran, J. N. and Schneider, U. (2003). Coupling methods for perfect simulation. *Probability in the Enginneering and Informational Sciences* **17**, 277–303.

Cressie, N. (1993). *Statistics for Spatial Data.* New York: Wiley.

Czado, C. (1992). On link selection in generalized linear models. In *Advances in GLIM and Statistical Modelling, Lecture Notes in Statistics* **78**, New York: Springer.

Czado, C. (1994a). Parametric link modification of both tails in binary regression. *Statistical Papers* **35**, 189–201.

Czado, C. (1994b). Bayesian inference of binary regression models with parametric link. *J. Statist. Plann. Inference* **41**, 121–140.

Czado, C. and Santner, T. J. (1992). The effect of link misspecification on binary regression inference. *J. Statist. Plann. Inference* **33**, 213–231.

David, H.A. (1981). *Order Statistics.* 2nd edition, New York: Wiley.

Dalla Valle, A. (2001). A test for the hypothesis of skew-normality in a population. *Working Paper 2001.2* Department of Statistics, University of Padua, Padua.

De Forest, E. (1882). On an unsymmetrical probability curve. *Analyst* **9**, 135–142, 161–168.

De Forest, E. (1883). On an unsymmetrical probability curve. *Analyst* **10**, 1–7, 67–74.

De Oliveira, V., Kedem, B. and Short, D. S. (1997). Bayesian prediction of transformed Gaussian random fields. *J. Amer. Statist. Assoc.* **92**, 1422–1433.

Dey, D. K. and Chen, M.-H. (2000). Bayesian model diagnostics for correlated binary data. In *Generalized Linear Models: A Bayesian Perspective*, D. K. Dey, S. K. Ghosh, B. K. Mallick (eds.), New York: Marcel Dekker, Inc., 313–327.

Dey, D. K. and Liu, J.-F. (2003). A new construction for skew multivariate distributions. *J. Multivariate Anal.*, to appear.

Dey, D. K., Chen, M.-H. and Chang, H. (1997). Bayesian approach for nonlinear random effects models. *Biometrics* **53**, 1239–1252.

Dickey, J. (1968). Three multidimensional integral identities with Bayesian applications. *Ann. Math. Statist.* **39**, 1615–1628.

Domínguez-Molina, J. A., González-Farías, G. and Gupta, A. K. (2003). The multivariate closed skew normal distribution. Department of Mathematics and Statistics, Bowling Green State University, Technical Report No. 03–12.

Domínguez-Molina, J. A., González-Farías, G. and Ramos-Quiroga, R. (2003). Skew-normality in stochastic frontier analysis. Comunicación Técnica No. I–03–18/06–10–2003, (PE/CIMAT).

Eaton, M. (1983). *Multivariate Statistics.* New York: Wiley.

Edgeworth, F. Y. (1886). The law of error and the elimination of chance. *Philosophical Magazine* **21**, 308–324.

Engle, R. F. (1982). Autoregressive conditional heteroskedasticity with estimates of the variance of UK inflation. *Econometrica* **50**, 987–1008.

Engle, R. F. and Gonzalez-Rivera, G. (1991). Semiparametric ARCH models. *J. Bus. Econ. Statist.* **9**, 345–359.

Engle, R. F. and Ng, V. K. (1993). Measuring and testing the impact of news on volatility. *J. Finance* **48**, 213–238.

Erkanli, A., Stangl, D. and Müller, P. (1993). Analysis of ordinal data by the mixture of probit links. *ISDS Discussion Paper 93–01*, Institute of Statistics and Decision Sciences, Duke University.

ESA (1997). The Hipparcos and Tycho catalogues. *ESA*, SP–1200.

Eun, C. S. and Shim, S. (1989). International transmission of stock markets movements. *J. Finan. Quant. Anal.* **24**, 241–256.

Fahrmeir, L. and Tutz, G. (1997). *Multivariate Statistical Modelling Based on Generalized Linear Models.* New York: Wiley.

Fang, K.-T., Kotz, S. and Ng, K. W. (1990). *Symmetric Multivariate and Related Distributions.* New York: Chapman and Hall.

Fang, K.-T. and Zhang, Y.-T. (1990). *Generalized Multivariate Analysis.* New York: Springer-Verlag.

Feast, M. W. and Catchpole, R. M. (1997). The Cepheid period-luminosity zero-point from Hipparcos trigonometrical parallaxes. *Monthly Notices of the Royal Astronomical Society* **286**, L1–L5.

Feller, W. (1971). *An Introduction to Probability Theory, Vol. II.* New York: Wiley.

Ferguson, T. S. (1973). A Bayesian analysis of some non-parametric problems. *Ann. Statist.* **1**, 209–230.

Fernández, C., Osiewalsky, J. and Steel, M. F. J. (1995). Modeling and inference with ν-spherical distributions. *J. Amer. Statist. Assoc.* **90**, 1331–1340.

Fernández, C. and Steel, M. F. J. (1998). On Bayesian modeling of fat tails and skewness. *J. Amer. Statist. Assoc.* **93**, 359–371.

Ferreira, J. T. A. S. and Steel, M. F. J. (2003). Bayesian multivariate regression analysis with a new class of skewed distributions, *Research Report 419*, Dept. of Statistics, University of Warwick.

Firth, D. (1993). Bias reduction of maximum likelihood estimates. *Biometrika* **80**, 27–38.

Forsythe, G. E., Malcolm, M. A. and Moler, C. B. (1977). *Computer Methods for Mathematical Computation.* Englewood Cliffs, NJ: Prentice-Hall.

French, K. R., Schwert, W. G. and Stambaugh, R. F. (1987). Expected stock returns and volatility. *J. Finan. Econ.* **19**, 3–29.

Gallant, A. R., Hsieh, D. and Tauchen, G. (1989). On fitting a recalcitrant series:

The Pound/Dollar exchange rate, 1947–83. *Unpublished manuscript.*

Gallant, A. R. and Nychka, D. W. (1987). Seminonparametric maximum likelihood estimation. *Econometrica* **55**, 363–390.

Gamerman, D. (1991). Dynamic Bayesian models for survival data. *Appl. Statist.* **40**, 63–79.

Gautschy, A. (1997). The development of the theory of stellar pulsations. *Vistas in Astronomy* **41**, 95–115.

Geisser, S. (1993). *Predictive Inference: An Introduction.* London: Chapman and Hall.

Geisser, S. and Eddy, W. (1979). A predictive approach to model selection. *J. Amer. Statist. Assoc.* **74**, 153–160.

Gelfand, A. E. (1996). Model determination using sampling-based methods. In *Markov Chain Monte Carlo in Practice*, W. R. Gilks, S. Richardson, D. J. Spiegelhalter (eds.), London: Chapman & Hall, 145–161.

Gelfand, A. E. and Dey, D. K. (1994). Bayesian model choice: asymptotics and exact calculations. *J. Roy. Statist. Soc. Ser. B* **56**, 501–514.

Gelfand, A. E., Dey, D. K. and Chang, H. (1992). Model determinating using predictive distributions with implementation via sampling-based methods (with discussion). In *Bayesian Statistics* **4**, J. M. Bernado, J. O. Berger, A. P. Dawid, A. F. M. Smith (eds.). Oxford: Oxford University Press, 147–167.

Gelfand, A. E. and Ghosh, S. K. (1998). Model choice: A minimum posterior predictive loss approach. *Biometrika* **85**, 1–13.

Gelfand, A. E. and Smith, A. F. M. (1990). Sampling-based approaches to calculating marginal densities. *J. Amer. Statist. Assoc.* **85**, 289–309.

Gelman, A., Meng, X. L. and Stern, H. S. (1996). Posterior predictive assessment of model fitness via realized discrepancies (with discussion). *Statist. Sinica* **6**, 733–807.

Genton, M. G. (1998). Variogram fitting by generalized least squares using an explicit formula for the covariance structure. *Math. Geol.* **30**, 323–345.

Genton, M. G., He, L. and Liu, X. (2001). Moments of skew-normal random vectors and their quadratic forms. *Statist. Probab. Lett.* **51**, 319–325.

Genton, M. G. and Loperfido, N. (2002). Generalized skew-elliptical distributions and their quadratic forms. *Institute of Statistics Mimeo Series* **2539**, to appear in *Annals of the Institute of Statistical Mathematics.*

Genton, M. G. and Thompson, K. R. (2003). Skew-elliptical time series with application to flooding risk. In *Time Series Analysis and Applications to Geophysical Systems*, D. R. Brillinger, E. A. Anderson, F. P. Schoenberg (eds.), IMA Volume in Mathematics and Its Applications **139**, New York: Springer, 169–186.

Genz, A. (1992). Numerical computation of multivariate Normal probabilities. *J. Comp. Graph. Stat.* **1**, 141–149.

George, I. E. and McCulloch, R. E. (1993). Variable selection via Gibbs sampling. *J. Amer. Statist. Assoc.* **88**, 881–889.

George, I. E., McCulloch, R. E. and Tsay, R. S. (1996). Two approaches to Bayesian model selections with applications. In *Bayesian Analysis in Econometrics and Statistics – Essays in Honor of Arnold Zellner*, D. A. Berry, K. A. Chaloner, J. K. Geweke (eds.), New York: Wiley, 339–348.

Geroski, P. A., Lazarova, S., Urga, G. and Walters, C. F. (2003). Are differences in firm size transitory or permanent? *J. Appl. Econ.* **18**, 47–59.

Ghysels, E., Harvey, A. C. and Renault, E. (1996). *Stochastic volatility.* In *Statistical Methods in Finance*, C. R. Rao, G. S. Maddala (eds.), Amsterdam: North-Holland.

Gibrat, R. (1931). *Les inégalités économiques; applications: aux inégalités des richesses, à la concentration des entreprises, aux populations des villes, aux statistiques des familles, etc., d'une loi nouvelle, la loi de l'effet proportionnel*, Paris: Librairie du Recueil Sirey.

Gilks, W. R., Richardson, S. and Spiegelhalter, D. (1996). *Markov Chain Monte Carlo in Practice.* London: Chapman & Hall.

González-Farías, G., Domínguez-Molina, J. A. and Gupta, A. K. (2003). Additive properties of skew normal random vectors. *J. Statist. Plann. Inference*, to appear.

Gorsich, D. J., Genton, M. G. and Strang, G. (2002). Eigenstructures of spatial design matrices. *J. Multivariate Anal.* **80**, 138–165.

Green, P. J. (1995). Reversible jump Markov chain Monte Carlo computation and Bayesian model determination. *Biometrika* **82**, 711–732.

Grenander, U. (1976). *Pattern Synthesis.* Lectures in Pattern Theory, New York: Springer.

Grenander, U. (1978). *Pattern Analysis.* Lectures in Pattern Theory, New York: Springer.

Grenander, U. (1996). *Elements of Pattern Theory.* Baltimore, MD: John Hopkins University Press.

Guerrero, V. M. and Johnson, R. A. (1982). Use of the Box-Cox transformation with binary response models. *Biometrika* **69**, 309–314.

Guo, W., Wang, Y. and Brown, M. (1999). A signal extraction approach to modeling hormones time series with pulses and a changing baseline. *J. Amer. Stat. Assoc.* **94**, 746–756.

Gupta, A. K. and Chen, T. (2001). Goodness-of-fit tests for the skew-normal distribution. *Comm. Statist. Simulation Computation* **30**, 907–930.

Gupta, A. K., González-Farías, G. and Domínguez-Molina, J. A. (2004). A multivariate skew normal distribution. *J. Multivariate Anal.* **89**, 181–190.

Gupta, R. C. and Brown, N. (2001). Reliability studies of the skew-normal distribution and its application to a strength-stress model. *Comm. Statist. Theory Methods* **30**, 2427–2445.

Hall, B. H. (1993a). Industrial research during the 1980's: Did the rate of return fall? *Brookings Pap. Econ. Act.* **2**, 289–344.

Hall, B. H. (1993b). The stock market's valuation of R&D investment during the 1980's. *Amer. Econ. Rev.* **83**, 259–263.

Handcock, M. S. and Stein, M. L. (1993). A Bayesian analysis of kriging. *Technometrics* **35**, 403–410.

Harrison, P. J. and Stevens, C. F. (1971). A Bayesian approach to short-term forecasting. *Operational Res. Quart.* **22**, 341–362.

Harrison, P. J. and Stevens, C. F. (1976). Bayesian forecasting. *J. Roy. Statist. Soc. Ser. B* **38**, 205–247.

Harvey, C. R. and Siddique, A. (1999). Autoregressive conditional skewness. *J.*

Finan. Quant. Anal. **34**, 465–487.

Healy, M. J. R. (1968). Multivariate normal plotting. *Appl. Statist.* **17**, 157–161.

Heckman, J. (1979). Sample selection bias as a specification error. *Econometrica* **47**, 153–161.

Henze, N. (1986). A probabilistic representation of the "skew-normal" distribution. *Scand. J. Statist.* **13**, 271–275.

Hill, M. A. and Dixon, W. J. (1982). Robustness in real life: A study of clinical laboratory data. *Biometrics* **38**, 377–396.

Hong, C. (1988). Options, volatilities and the edge strategy. *Unpublished Ph.D. dissertation*, University of California, San Diego, Dept of Economics.

Horn, R. A. and Johnson, C. R. (1991). *Topics in Matrix Analysis.* New York: Cambridge University Press.

Hougaard, P. (1986). A class of multivariate failure time distributions. *Biometrika* **73**, 671–678.

Hougaard, P. (2000). *Analysis of Multivariate Survival Data.* New York: Springer.

Hsieh, D. (1989). Modelling heteroskedasticity in daily foreign-exchange rates. *J. Bus. Econ. Statist.* **7**, 307–317.

Ibrahim, J. G. and Chen, M.-H. (2000). Power prior distributions for regression models. *Statist. Sci.* **15**, 46–60.

Ibrahim, J. G., Chen, M.-H. and Sinha, D. (2001a). Criterion based methods for Bayesian model assessment. *Statist. Sinica* **11**, 419–443.

Ibrahim, J. G., Chen, M.-H. and Sinha, D. (2001b). *Bayesian Survival Analysis.* New York: Springer.

Ibrahim, J. G. and Laud, P. W. (1994). A predictive approach to the analysis of designed experiments. *J. Amer. Statist. Assoc.* **89**, 309–319.

Intergovernmental Panel on Climate Change (2001). Climate Change 2001: The Scientific Basis. Contribution of Working Group I to the Third Assessment Report of the Intergovernmental Panel on Climate Change. J. T. Houghton, Y. Ding, D. J. Griggs, M. Noguer, P. J. van der Linden, D. Xiaosu (eds.), Cambridge, UK: Cambridge University Press.

Ishwaran, H. and James, L. F. (2001). Gibbs sampling methods for stick-breaking priors. *J. Amer. Statist. Assoc.* **96**, 161–173.

Jones, M. C. (2002). Marginal replacement in multivariate densities, with application to skewing spherically symmetric distributions. *J. Multivariate Anal.* **81**, 85–99.

Kalbfleisch, J. D. and Prentice, R. L. (1973). Marginal likelihoods based on Cox's regression and life model. *Biometrika* **60**, 267–278.

Kalbfleish, J. G. (1985). *Probability and Statistical Inference.* Vol. 2. New York: Springer.

Kaplan, E. L. and Meier, P. (1958). Nonparametric estimation from incomplete observations. *J. Amer. Statist. Assoc.* **53**, 457–481.

Karolyi, G. A. (1995). A multivariate *GARCH* model of international transmissions of stock returns and volatility: the case of the United States. *J. Bus. Econ. Statist.* **13**, 11–24.

Kass, M., Witkin, A. and Terzopoulos, D. (1987). Snakes: active contour models. *Proc. First IEEE International Conference on Computer Vision*, 259–268.

Kass, R. E. and Raftery, A. E. (1995). Bayes factors. *J. Amer. Statist. Assoc.* **90**,

773–795.

Kendall, D. G. (1989). A survey of the statistical theory of shape. *Statist. Sci.* **4**, 87–99.

Kim, H. and Mallick, B. K. (2001). Analyzing spatial data using skew-Gaussian processes. In *Spatial Cluster Modelling*, A. B. Lawson, D. G. T. Denison (eds.), 163–173.

Kim, H. and Mallick, B. K. (2003). A note on Bayesian spatial prediction using the elliptical distribution. *Statist. Probab. Lett.* **64**, 271–276.

Kim, H. and Mallick, B. K. (2004). A Bayesian prediction using the skew-Gaussian processes. *J. Statist. Plann. Inference* **120**, 85–101.

Kitagawa, G. and Gersch, W. (1984). A smoothness priors state-space modeling of times series with trend and seasonality. *J. Amer. Statist. Assoc.* **79**, 378–389.

Knapp, G. R., Pourbaix, D., Platais, I. and Jorissen, A. (2003). Reprocessing the Hipparcos data of evolved stars. III. Revised Hipparcos period-luminosity relationship for galactic long-period variable stars. *Astronomy and Astrophysics* **403**, 993–1002.

Koop, G., Osiewalski, J., Steel, M. and van den Broeck, J. (1994). Stochastic frontier models: A Bayesian perspective. *J. Econometrics* **61**, 273–303.

Kotz, S., Balakrishnan, N. and Johnson, N. L. (2000). *Continuous Multivariate Distributions.* Volume 1: Models and Applications, 2nd edition. Wiley Series in Probability and Statistics, New York: John Wiley & Sons.

Kumbhakar, S. C. and Lovell, C. A. (2000). *Stochastic Frontier Analysis.* Cambridge, UK: Cambridge University Press.

Laud, P. W. and Ibrahim, J. G. (1995). Predictive model selection. *J. R. Stat. Soc. Ser. B* **57**, 247–262.

Leadbetter, M. R., Lindgren, G. and Rootzén, H. (1983). *Extremes and Related Properties of Random Sequences and Processes.* Springer Series in Statistics, New York: Springer.

Lee, J. and Berger, J. O. (2001). Semiparametric Bayesian analysis of selection models. *J. Amer. Statist. Assoc.* **96**, 1397–1409.

Lehmann, E. L and Casella, G. (1998). *Theory of Point Estimation.* New York: Springer.

Lietchy, M. W. (2003). Covariance matrices and skewness: modeling and applications in finance. Doctoral Dissertation, Duke University, Durham, NC.

Ling, W. L., Engle, R. F. and Ito, T. (1994). Do Bulls and Bears move across borders? International transmission of stock returns and volatility. *Rev. Finan. Stud.* **7**, 507–538.

Lintner, J. (1965). Security prices, risk and maximal gains from diversification. *J. Finance* **20,** 347–400.

Liseo, B. (1990). The skew-normal class of densities: inferential aspects from a Bayesian viewpoint (in Italian). *Statistica* **50**, 59–70.

Liseo, B. and Loperfido, N. (2003a). A Bayesian interpretation of the multivariate skew-normal distribution. *Statist. Probab. Lett.* **61**, 395–401.

Liseo, B. and Loperfido, N. (2003b). A note on reference priors for the scalar skew-normal distribution. *J. Statist. Plann. Inference*, to appear.

Liseo, B. and Loperfido, N. (2003c). Integrated likelihood inference for the shape parameter of the multivariate skew-normal distribution. *Proceedings of the*

2003 Meeting of the Italian Society of Statistics.

Liseo, B. and Macaro, C. (2003). Bayesian analysis of skew-in-mean GARCH models. *Tech. Report, Dip. Studi Geoeconomici, Univ. di Roma "La Sapienza,"* under review.

Littell, R. C., Milliken, G. A., Stroup, W. W. and Wolfinger, R. D. (1996). *SAS System for Mixed Models.* Cary, NC: SAS Institute.

Liu, J.-F. and Dey, D. (2003). Modeling random effects for multilevel binomial regression models. Technical Report, Department of Statistics, University of Connecticut.

Lo, A. Y. (1984). On a class of Bayesian nonparametric estimates: I. Density estimates. *Ann. Statist.* **12**, 351–357.

Loperfido, N. (2001). Quadratic forms of skew-normal random vectors. *Statist. Probab. Lett.* **54**, 381–387.

Loperfido, N. (2002). Statistical implications of selectively reported inferential results. *Statist. Probab. Lett.* **56**, 13–22.

Loperfido, N. (2003). Robust Bayesian analysis for location-scale models with skew-symmetric priors. Technical Report, Università di Urbino.

Louis, H., Blenman, L. P. and Thatcher, J. S. (1999). Interest rate parity and the behavior of the Bid-Ask Spread. *J. Finan. Res.* **22**, 189–206.

Ludbrook, J. and Dudley, H. (1998). Why permutation tests are superior to t and F tests in biomedical research. *Amer. Statist.* **52**, 127–132.

Ma, Y. and Genton, M. G. (2004). A flexible class of skew-symmetric distributions. *Scand. J. Statist.*, to appear.

Ma, Y., Genton, M. G. and Tsiatis, A. A. (2003). Locally efficient semiparametric estimators for generalized skew-elliptical distributions. *Institute of Statistics Mimeo Series* **2550**, under review.

Madger, L. S. and Zeger, S. L. (1996). A smooth nonparametric estimate of a mixing distribution using mixtures of Gaussians. *J. Amer. Statist. Assoc.* **91**, 1141–1151.

Madigan, D. and Raftery, A. E. (1994). Model selection and accounting for model uncertainty in graphical models using Occam's window. *J. Amer. Statist. Assoc.* **89**, 1535–1546.

Madore, B. F. and Freedman, W. L. (1998). Hipparcos parallaxes and the Cepheid distance scale. *Astrophys. J.* **492**, 110–115.

Mantel, N., Bohidar, N. R. and Ciminera, J. L. (1977). Mantel-Haenszel analysis of litter matched time-to-response data, with modifications for recovery of interlitter information. *Cancer Res.* **37**, 3863–3868.

Mardia, K. V. (1970). Measures of multivariate skewness and kurtosis with applications. *Biometrika* **57**, 519–530.

Mardia, K. V., Kent, J. T. and Bibby, J. M. (1979). *Multivariate Analysis.* London: Academic Press.

Markowitz, H. M. (1952). Portfolio selection. *J. Finance* **7**, 77–91.

McCullagh, P. (1980). Regression models for ordinal data. *J. R. Statist. Soc. B* **42**, 109–142.

McCullagh, P. and Nelder, J. A. (1989). *Generalized Linear Models.* 2nd edition, London: Chapman & Hall.

McDonald, J. B. (1996). Probability distributions for financial models. In *Hand-*

book of Statistics: Statistical Methods in Finance **14**, G. S. Maddala, C. R. Rao (eds.), Amsterdam: Elsevier Science.

McDonald, J. B. and Nelson, R. D. (1989). Alternative beta estimation for the market model using partially adaptive techniques. *Comm. Statist. Theory Methods* **18**, 4039–4058.

McDonald, J. B. and Newey, W. K. (1988). Partially adaptive estimation of regression models via the generalized T distribution. *Econ. Theory* **4**, 428–457.

McDonald, J. B. and Xu, Y. J. (1995). A generalization of the beta distribution with applications. *J. Econometrics* **66**, 133–152.

McGilchrist, C. A. and Aisbett, C. W. (1991). Regression with frailty in survival analysis. *Biometrics* **47**, 461–466.

McKeague, I. W. and Tighiouart, M. (2000). Bayesian estimators for conditional hazard functions. *Biometrics* **56**, 1007–1015.

Meeusen, W. and Van den Broeck, J. (1977). Efficiency estimation from Cobb-Douglas production functions with composed error. *Int. Econ. Rev.* **18**, 435–444.

Meinhold, R. J. and Singpurwalla, N. D. (1983). Understanding the Kalman filter. *Amer. Statist.* **37**, 123–127.

Meinhold, R. J. and Singpurwalla, N. D. (1989). Robustification of Kalman filter models. *J. Amer. Statist. Assoc.* **84**, 479–486.

Merton, R. C. (1973). Theory of rational option pricing. *Bell J. Econ.* **4**, 141–183.

Middleton, J. F. and Thompson, K. R. (1986). Extreme sea levels from short records. *J. Geophysical Research* **91**, 11707–11716.

Mills, T. (1995). Modelling skewness and kurtosis in the London stock exchange FT-SE index return distributions. *J. R. Statist. Soc. D* **44**, 323–332.

Monahan, J. F. (2001). *Numerical Methods of Statistics*. Cambridge, UK: Cambridge University Press.

Morgan, B. J. T. (1983). Observations on quantitative analysis. *Biometrics* **39**, 879–886.

Mossin, J. (1966). Equilibrium in a capital asset market. *Econometrica* **34**, 768–783.

Mudholkar, G. S. and Hutson, A. D. (2000). The epsilon-skew-normal distribution for analyzing nearly normal data. *J. Statist. Plann. Inference* **83**, 291–309.

Mukhopadhyay, S. and Gelfand, A. E. (1997). Dirichlet process mixed generalized linear models. *J. Amer. Statist. Assoc.* **92**, 633–639.

Mukhopadhyay, S. and Vidakovic, B. (1995). Efficiency of linear Bayes rules for a normal mean: skewed prior class. *J. R. Statist. Soc. D* **44**, 389–397.

Nadarajah, S. and Kotz, S. (2003). Skewed distributions generated by the normal kernel. *Statist. Probab. Let.* **65**, 269–277.

Nandram, B. and Chen, M.-H. (1996). Accelerating Gibbs sampler convergence in the generalized linear models via a reparameterization. *J. Statist. Comput. Simulation* **54**, 129–144.

Naveau, P., Ammann, C., Oh, H. S. and Guo, W. (2003). An automatic statistical methodology to extract pulse like forcing factors in climatic time series: Application to volcanic events. In *Volcanism and the Earth's Atmosphere*. A. Robock, C. Oppenheimer (eds.), Geophysical Monograph **139**, Washington, DC: American Geophysical Union, 177–186.

Naveau, P., Genton, M. G. and Shen, X. (2004). A skewed Kalman filter, *J. Multivariate Anal.*, to appear.

Nelder, J. A. and Mead, R. (1965). A simplex method for function minimization. *Comput. J.* **7**, 308–313.

Nelson, D. (1991). Conditional heteroskedasticity in asset returns: a new approach. *Econometrica* **59**, 347–370.

Nelson, L. S. (1964). The sum of values from a normal and a truncated normal distribution. *Technometrics* **6**, 469–471.

Neter, J., Kutner, H. K., Nachtsheim, C. J. and Wasserman, W. (1996). *Applied Linear Statistical Models.* 4th edition, Boston: IRWIN.

Newey, W. K. (1990). Semiparametric efficiency bounds. *J. Appl. Econ.* **5**, 99–135.

Newton, M. A., Czado, C. and Chappell, R. (1996). Bayesian inference for semiparametric binary regression. *J. Amer. Statist. Assoc.* **91**, 142–153.

Newton, M. A. and Raftery, A. E. (1994). Approximate Bayesian inference with the weighted likelihood bootstrap (with discussion). *J. Roy. Statist. Soc. B* **56**, 3–48.

O'Hagan, A. and Leonard, T. (1976). Bayes estimation subject to uncertainty about parameter constraints. *Biometrika* **63**, 201–202.

Owen, D.B. (1956). Tables for computing bivariate Normal probabilities. *Ann. Math. Statist.* **27**, 1075–1090.

Paczynski, B. (2004). A problem of distance. *Nature* **427**, 299–300.

Parsons, L. J. (2002). Using stochastic frontier analysis for performance measurement and benchmarking. *Econometric Models in Marketing* **16**, 317–350.

Pawlowicz, R., Beardsley, B. and Lentz, S. (2002). Classical tidal harmonic analysis including error estimates in MATLAB using TTIDE. *Computers and Geosciences* **28**, 929–937.

Pearson, K. (1893). Asymmetrical frequency curves. *Nature* **48**, 615–616.

Pearson, K. (1895). On skew probability curves. *Nature* **52**, 317.

Peiró, A. (1999). Skewness in financial returns. *J. Banking Finance* **23**, 847–862.

Pewsey, A. (2000). Problems of inference for Azzalini's skew-normal distribution. *J. Appl. Statist.* **27**, 859–870.

Pinheiro, J. C. and Bates, D. M (2000). *Mixed-Effects Models in S and S-PLUS.* New York: Springer.

Polachek, S. W. and Robst, J. (1998). Employee Labor Market Information: comparing direct world of work measures of workers' knowledge to stochastic frontier estimates. *Lab. Econ.* **5**, 231–242.

Poliannikov, O. V. and Krim, H. (2003). On sampling closed planar curves and surfaces. *J. Sampling Theory Image Signal Proc.* **2**, 53–82.

Pont, F. (1999). The Cepheid distance scale after Hipparcos. *ASP Conference Series* **167**, 113–128.

Praetz, P. (1972). The distribution of share price changes. *J. Bus.* **45**, 49–55.

Pregibon, D. (1980). Goodness of link tests for generalized linear models. *Appl. Statist.* **29**, 15–24.

Prentice, R. L. (1976). A generalization of the probit and logit methods for dose response curves. *Biometrics* **32**, 761–768.

Prentice, R. L. (1988). Correlated binary regression with covariates specific to each binary observation. *Biometrics* **44**, 1033–1048.

Pugh, D. T. (1987). *Tides, Surges and Mean Sea Level: A Handbook for Engineers and Scientists.* Chichester: Wiley.

Pugh, D. T. and Vassie, J. M. (1980). Applications of the joint probability method for extreme sea level computations. *Proceedings of the Institute of Civil Engineers Part 2* **69**, 959–975.

Qiou, Z. Q., Ravishanker, N. and Dey, D. K. (1999). Multivariate survival analysis with positive stable frailties. *Biometrics* **55**, 637–644.

Racine-Poon, A. (1992). SAGA: Sample Assisted Graphical Analysis. In *Bayesian Statistics* **4**, J. M. Bernardo, J. O. Berger, A. P. Dawid, A. F. M. Smith. (eds.), London: Oxford University Press.

Raftery, A. E., Madigan, D. and Hoeting, J. A. (1997). Bayesian model averaging for linear regression models. *J. Amer. Statist. Assoc.* **92**, 179–191.

Raftery, A. E., Madigan, D. and Volinsky, C. T. (1995). Accounting for model uncertainty in survival analysis improves predictive performance. In *Bayesian Statistics* **5**, J. M. Bernardo, J. O. Berger, A. P. Dawid, A. F. M. Smith (eds.), Oxford: Oxford University Press, 323–350.

Rao, C. R. (1985). Weighted distributions arising out of methods of ascertainment: What populations does a sample represent? In *A Celebration of Statistics: The ISI Centenary Volume.* A. G. Atkinson, S. E. Fienberg (eds.), New York: Springer-Verlag, 543–569.

Ravishanker, N. and Dey, D. K. (2000). Multivariate survival models with a mixture of positive stable frailties. *Methodology and Computing in Applied Probability* **2**, 293–308.

Reid, N. (1995). The roles of conditioning in inference. *Statist. Sci.* **10**, 138–157.

Roberts, C. (1966). A correlation model useful in the study of twins. *J. Amer. Statist. Assoc.* **61**, 1184–1190.

Rydberg, T. H. (2000). Realistic statistical modelling of financial data. *Internat. Statist. Rev.* **68**, 233–258.

Sahu, S. K., Dey, D. K., Aslanidou, H. and Sinha, D. (1997). A Weibull regression model with Gamma frailties for multivariate survival data. *Lifetime Data Anal.* **3**, 123–137.

Sahu, S. K., Dey, D. K. and Branco, M. (2003). A new class of multivariate skew distributions with application to Bayesian regression models. *Canad. J. Statist.* **31**, 129–150.

Salvan, A. (1986). Locally most powerful invariant tests of normality (in Italian). In *Atti della XXXIII Riunione Scientifica della Società Italiana di Statistica* **2**, 173–179. Bari: Cacucci.

Sansó, B. and Guenni, L. (2000). A nonstationary multisite model for rainfall. *J. Amer. Statist. Assoc.* **95**, 1089–1100.

Sartori, N. (2003). Bias prevention of maximum likelihood estimates: skew-normal and skew-t distributions. *Working Paper 2003.1*, Department of Statistics, University of Padua, Padua.

Schick, I. C. (1989). Robust recursive estimation of a discrete-time stochastic linear dynamic system in the presence of heavy-tailed observation noise. Ph.D. Dissertation, Massachusetts Institute of Technology, Cambridge, MA.

Schwert, G. W. (1989). Why does stock market volatility change over time? *J.*

Finance **44**, 1115–1153.

Scott, C. and Nowak, R. D. (2001). Template learning from atomic representations: A wavelet-based approach to pattern analysis. *2nd International Workshop on Statistical and Computational Theories of Vision – Modeling, Learning, Computing, and Sampling.*

Sharpe, W. F. (1964). Capital asset prices: a theory of market equilibrium under conditions of risk. *J. Finance* **19**, 425–442.

Shephard, N. (1994). Partially non-Gaussian state-space models. *Biometrika* **81**, 115–131.

Shumway, R. H. and Stoffer, D. S. (1991). Dynamic linear models with switching. *J. Amer. Statist. Assoc.* **86**, 763–769.

Simaan, Y. (1993). Portfolio selection and asset pricing – three parameter framework. *Management Science* **39**, 568–587.

Sinha, D. and Dey, D. K. (1997). Semiparametric Bayesian analysis of survival data. *J. Amer. Statist. Assoc.* **92**, 1192–1212.

Small, C. G. (1996). *The Statistical Theory of Shape.* New York: Springer.

Smith, R. L. and Miller, J. E. (1986). A non-Gaussian state space model and application to prediction of records. *J. R. Stat. Soc. Ser. B* **48**, 79–88.

Smith, R. L. and Robinson, P. J. (1997). A Bayesian approach to the modelling of spatial-temporal precipitation data. In *Case Studies in Bayesian Statistics III*, G. Gatsonis, J. S. Hodges, R. E. Kass, R. McCulloch, P. Rossi, N. D. Singpurwalla (eds.). New York: Springer, 237–269.

Spiegelhalter, D. J., Best, N. G., Carlin, B. P. and van der Linde, A. (2002). Bayesian measures of model complexity and fit (with discussion). *J. R. Stat. Soc. Ser. B* **64**, 583–639.

Stefanski, L. A. and Boos, D. D. (2002). The calculus of M-estimation. *Amer. Statist.* **56**, 29–38.

Stevenson, R. E. (1980). Likelihood function for generalized stochastic frontier estimation. *J. Econometrics* **13**, 57–66.

Stidd, C. K. (1973). Estimating the precipitation climate. *Water Resources Res.* **9**, 1235–1241.

Stigler, S. M. (1978). Francis Ysidro Edgeworth, statistician (with discussion). *J. R. Stat. Soc. Ser. A* **141**, 287–322.

Stigler, S. M. (1986). *The History of Statistics: The Measurement of Uncertainty before 1900.* Cambridge and London: The Belknap Press of Harvard University Press.

Stukel, T. (1988). Generalized logistic models. *J. Amer. Statist. Assoc.* **83**, 426–431.

Sutton, J. (1997). Gibrat's legacy. *J. Econ. Lit.* **35**, 40–59.

Tancredi, A. (2003). Accounting for heavy tails in stochastic frontier models. *Technical Report, Dip. Scienze Statistiche, Università di Padova,* Italy.

Tanner, M. A. (1996). *Tools for Statistical Inference.* New York: Springer.

Tao, H., Palta, M., Yandell, B. S. and Newton, M. A. (1999). An estimation method for the semiparametric mixed effects model. *Biometrics* **55**, 102–110.

Tawn, J. A. and Vassie, J. M. (1989). Extreme sea levels – the joint probabilities method revisited and revised. *Proceedings of the Institute of Civil Engineers Part 2 – Research and Theory* **87**, 429–442.

Terzopoulos, D. (1986). Regularization of inverse problems involving discontinuities. *IEEE Trans. on Pattern Analysis and Machine Intelligence* **8**, 413–424.

Theodossiou, P. (1998). Financial data and the skewed generalized T distribution. *Management Science* **44**, 1650–1661.

Tukey, J. W. (1977). *Modern Techniques in Data Analysis.* NSF-sponsored regional research conference at Southeastern Massachusetts University, North Dartmouth, MA.

Udalski, A., Soszynski, I., Szymanski, M., Kubiak, M., Pietrzynski, G., Wozniak, P. and Zebrun, K. (1999). The optical gravitational lensing experiment. Cepheids in the Magellanic clouds. V. Catalog of Cepheids from the small Magellanic cloud. *Acta Astronomica* **49**, 437–520.

Verbeke, G. and Lesaffre, E. (1996). A linear mixed-effects model with heterogeneity in the random-effects population. *J. Amer. Statist. Assoc.* **91**, 217–221.

Verbeke, G. and Molenberghs, G. (2000). *Linear Mixed Models for Longitudinal Data.* New York: Springer.

Wacholder, S. and Weinberg, C. R. (1994). Flexible maximum likelihood methods for assessing joint effects in case-control studies with complex sampling. *Biometrics* **50**, 350–357.

Wahba, G. (1978). Improper priors, spline smoothing and the problem of guarding against model errors in regression, *J. R. Stat. Soc. Ser. B* **40**, 364–372.

Wahed, A. and Ali, M. M. (2001). The skew-logistic distribution. *J. Statist. Res.* **35**, 71–80.

Wakefield, J. (1996). The Baysian analysis of population pharmacokinetic models. *J. Amer. Statist. Assoc.* **91**, 62–75.

Wakefield, J., Smith, A. F. M., Racine-Poon, A. and Gelfand, A. E. (1994). Bayesian analysis of linear and non-linear propulation models using the Gibbs sampler. *Appl. Statist.* **43**, 201–221.

Walker, S. G. and Mallick, B. K. (1997). Hierarchical generalized linear models and frailty models with Bayesian non-parametric mixing. *J. Roy. Statist. Soc. Ser. B* **59**, 845–860.

Wang, J., Boyer, J. and Genton, M. G. (2004a). A skew-symmetric representation of multivariate distributions. *Statist. Sinica,* to appear.

Wang, J., Boyer, J. and Genton, M. G. (2004b). A note on an equivalence between chi-square and generalized skew-normal distributions. *Statist. Probab. Lett.* **66**, 395–398.

Wang, J. and Genton, M. G. (2004). The multivariate skew-slash distribution. *J. Statist. Plann. Inference,* to appear.

Wecker, W. E. and Ansley, C. F. (1983). The signal extraction approach to nonlinear regression and spline smoothing. *J. Amer. Statist. Assoc.* **78**, 81–89.

Weinstein, M. A. (1964). The sum of values from a normal and a truncated normal distribution. *Technometrics* **6**, 104–105.

Weisstein, E. W. (2002). *CRC Concise Encyclopedia of Mathematics.* Second Edition, Boca Raton, FL: Chapman & Hall.

West, M. and Harrison, J. (1997). *Bayesian Forecasting and Dynamic Models.* New York: Springer.

Whittemore, A. S. (1983). Transformations to linearity in binary regression. *SIAM J. Appl. Math.* **43**, 703–710.

Winsten, C. B. (1957). Discussion on Mr. Farrell's Paper. *J. R. Stat. Soc. Ser. A* **120**, 282–284.

WOCE Data Products Committee (2002). WOCE Global Data, Version 3.0, WOCE International Project Office, WOCE Report No. 180/02, UK.

Yaglom, A. M. (1987). *Correlation Theory of Stationary and Related Random Functions I. Basic Results.* New York: Springer.

Yang, R. and Berger, J. O. (1994). Estimation of a covariance matrix using reference priors. *Ann. Statist.* **22**, 1195–1211.

Ye, J. C., Bresler, Y. and Moulin, P. (2000). Asymptotic global confidence regions in parametric shape estimation problems. *IEEE Trans. Information Theory* **46**, 1881–1895.

Ye, J. C., Bresler, Y. and Moulin, P. (2001). Cramér-Rao bounds for 2-D target shape estimation in nonlinear inverse scattering problems with application to passive radar. *IEEE Trans. Antennas Propagation* **49**, 771–783.

Ye, J., Bresler, Y. and Moulin, P. (2003). Cramér-Rao bounds for parametric shape estimation in inverse problems. *IEEE Trans. Image Proc.* **12**, 71–84.

Zakoian, J. M. (1994). Threshold heteroskedastic models. *J. Econ. Dynam. Control* **18**, 931–935.

Zellner, A. (1976). Bayesian and non-Bayesian analysis of the regression model with multivariate Student-t error terms. *J. Amer. Statist. Assoc.* **71**, 400–405.

Zhang, B. (2000). A goodness of fit test for multiplicative-intercept risk models based on case-control data. *Statist. Sinica* **10**, 839–865.

Zhang, D. and Davidian, M. (2001). Linear mixed models with flexible distributions of random effects for longitudinal data. *Biometrics* **57**, 795–802.

Zhou, G. (1993). Asset pricing tests under alternative distributions. *J. Finance* **48**, 1927–1942.

Index

A

Absolute values, asymmetric distributions, 115–117
Adcock and Clark studies, 196
Adcock and Shutes studies, 4, 193–194
Adcock studies, *vi,* 191–204
Additive component, 108
Agresti studies, 136
Ahn studies, 181
Aigner, Lovell and Schmidt studies
  GARCH model, 209
  skew-normal distributions, 4, 7, 76
  stochastic frontier analysis, 223, 232, 235
Aisbett, McGilchrist and, studies, 332
Albert and Chib studies, 131–132, 149
Ali, Wahed and, studies, 89
van Altena, Shelus, Franz, Wasserman, Hemenway, Duncombe, Story, Whipple and Bradley, Benedict, McArthur, Fredrick, Harrison, Lee, Slesnick, Rhee, Patterson, Nelan, Jefferys, studies, 314
Amemiya studies, 131
Ammann, Kiehl, Zender, Otto-Bliesner and Bradley studies, 270, 273
Ammann, Naveau, Genton and, studies, *vi,* 259–278
Ammann, Oh and Guo, Naveau, studies, 269
Andel, Netuka and Zvara studies, 7
Anderson-Darling test, 14
Andrews and Mallows studies, 52
Ansley, Wecker and, studies, 269
Aparicio Acosta and Estrada studies, 193
Aranda-Ordaz studies, 132
Arellano-Valle, del Pino and San Martín studies
  asymmetric distributions, 115, 119–120, 122
  closed skew-normal distributions, 27, 40
  skew-elliptical distributions, 63
Arellano-Valle, Gómez and Quintana studies, 116–118, 124
Arellano-Valle and del Pino studies, *vi,* 113–130
Arellano-Valle and Genton studies, 118, 128
Arenou and Luri studies, 318
Arjas and Gasbarra studies, 322, 326–327
Arjas and Heikkinen studies, 327
Arnold, Beaver, Groeneveld and Meeker studies
  closed skew-normal distributions, 25
  hidden truncation, 103
  skew-normal distributions, 3
Arnold, Castillo and Sarabia studies, 47
Arnold and Beaver studies
  Bayesian techniques, 155, 175
  capital asset pricing, 194–195
  closed skew-normal distributions, 25, 38–39
  hidden truncation, *vi,* 101–112
  linear mixed effects models, 345
  skew-elliptical distributions, 44, 46–47, 62–64, 89
  skew-normal distributions, 3, 66, 74
  stochastic frontier analysis, 224–225
Aslanidou and Sinha, Sahu, Dey, studies, 322, 330, 333
A-SN type, skew-elliptical distributions, 53
Astrometric satellites, 311
Astronomical distance determination
  astrometric satellites, 311
  basics, 309–310
  Cepheids, 312–314
  fixed slope, 316–317
  free slope, 317–318
  parent distributions, 312

period-luminosity relation calibration, 314–319
regression, 315–316
sample formation, 312
skew-normal errors, 315–316
trigonometric parallax, 310–312
Asymmetric distributions, *see also* Skew-symmetric distributions
absolute values, 115–117
basics, 113–115, 118
canonical form, 128–129
conditional representatives, 119–120
density formula, 120–121, 123–124
GARCH model, 211
invariance, 118–121
latent variables, 117–118
maximal invariant, 120–121
moments, 125–126
multivariate case, 125–128
probability function, 120–121
random variables, 116–117, 126
real line, 115–118
selection models, 117–118
sign invariant class, 121–126
signs, 115–117
skewed distribution, 123–124
skew-elliptical distributions, 128–129
skewness, 118–121
skew-normal distribution, 126–128
stochastic representation, 124–126
symmetry, 118–121
transformations, 119
Athletes data, 74–76, 93–95
Australian Bureau of Statistics studies, 74
Autocovariance function, skew-symmetric distributions, 83–85
Azzalini and Capitanio studies
asymmetric distributions, 116, 124, 127–129
Bayesian techniques, 157–160, 164
closed skew-normal distributions, 25–26, 42
hidden truncation, 109
multivariate skew-*t* distributions, 244, 246–247, 251–252
skew-elliptical distributions, 47, 81, 85, 89–90
skew-normal distributions, 3, 10, 12, 16, 20–21, 65, 67–68
spatial prediction, 279, 281
time series analysis, 260, 263
Azzalini and Dalla Valle studies
asymmetric distributions, 126–128
Bayesian techniques, 155, 166, 175
capital asset pricing, 194–195, 197, 204
closed skew-normal distributions, 25, 28, 37–38
hidden truncation, 105
linear mixed effects models, 341
skew-elliptical distributions, 43, 45, 53, 57, 62, 81, 85, 89, 94
skew-normal distributions, 3, 14, 20, 65, 68–69, 71, 74, 79
spatial prediction, 279
time series analysis, 260
Azzalini and Stanghellini, Capitanio, studies, 244, 247, 252
Azzalini studies
Bayesian techniques, 158, 160, 164–165
capital asset pricing, 194
closed skew-normal distributions, 25, 38
GARCH model, 214, 216–217
hidden truncation, 101–104
skew-elliptical distributions, 43, 62, 65, 81, 89
skew-normal distributions, *v,* 3, 8, 11, 14, 23
stochastic frontier analysis, 223–224
time series analysis, 260

B

Balakrishnan and Johnson, Kotz, studies, 226, 236
Ball and Mankiw studies, 4
Baloch, Krim and Genton studies, *vi,* 291–308
Bartolucci, De Luca and Loperfido studies, 209
Baseline hazard function, 326–327
Basu and Mukhopadhyay studies, 132, 139
Basu and Pereira studies, 72
Bates, Pinheiro and, studies, 339
Bayarri and De Groot studies, 66, 95
Bayesian techniques
application to firm size, 181–189
asymmetric distributions, 129
basics, 175–176, 189–190
categorical response data, 133, 139, 142–148
FS skewed distributions, 177
linear mixed effects models, 346–350

multivariate skewed distributions, 176–179
multivariate survival model, 322, 330
numerical implementation, 180–181
prior distributions, 179–180
regression models, 179–181
SDB skewed distributions, 178–179
skew-elliptical distributions, 63
skew-normal distributions, 4
spatial prediction, 282–284
Beardsley and Lentz, Pawlowicz, studies, 248
Beaver, Arnold and, studies
Bayesian techniques, 155, 175
capital asset pricing, 194–195
closed skew-normal distributions, 25, 38–39
hidden truncation, *vi*, 101–112
linear mixed effects models, 345
skew-elliptical distributions, 44, 46–47, 62–64, 89
skew-normal distributions, 3, 66, 74
stochastic frontier analysis, 224–225
Beaver, Groeneveld and Meeker, Arnold, studies
closed skew-normal distributions, 25
hidden truncation, 103
skew-normal distributions, 3
Benedict, McArthur, Fredrick, Harrison, Lee, Slesnick, Rhee, Patterson, Nelan, Jefferys, van Altena, Shelus, Franz, Wasserman, Hemenway, Duncombe, Story, Whipple and Bradley studies, 314
Berger, De Oliveira and Sansó studies, 153, 282
Berger, Lee and, studies, 97
Berger, Rios Insua and Ruggeri studies, 157
Berger, Yang and, studies, 168
Berger and Bernardo studies, 164
Bernardo, Berger and, studies, 164
Bernardo and Smith studies, 133
Bernoulli properties, 136, 283
Bessel, F., 310
Best, Carlin and van der Linde, Spiegelhalter, studies, 144, 148
Bhaumik, Chen and Dey studies, 139, 151
Bibby, Mardia, Kent and, studies
skew-elliptical distributions, 87
skew-normal distributions, 68, 79
stochastic frontier analysis, 227
Bickel, Klaassen, Ritov and Wellner studies, 96
Binomial Theorem, 221
Birnbaum studies, 101, 224
Black and Scholes studies, 191
Blake and Isard studies, 291
Blattberg and Gonedes studies, 193
Blenman and Thatcher, Louis, studies, 4
Blood pressure, 87
Bohidar and Ciminera, Mantel, studies, 334
Bollerslev, Chou and Kroner studies, 206
Bollerslev and Engle studies, 206
Bollerslev studies, 193, 208, 211
Boos, Stefanski and, studies
astronomical distance determination, 316
linear mixed effects models, 346
skew-elliptical distributions, 99
Booth and Hobert studies, 345–346
Boville and Gent studies, 270, 273
Boyer and Genton, Wang, studies
asymmetric distributions, 116–117, 124
flexible skew-symmetric distributions, 292, 296
skew-elliptical distributions, 47–48, 61–63, 81–82, 85–86
skew-normal distributions, 3, 22, 66–67
Bradley, Ammann, Kiehl, Zender, Otto-Bliesner and, studies, 270, 273
Bradley, Benedict, McArthur, Fredrick, Harrison, Lee, Slesnick, Rhee, Patterson, Nelan, Jefferys, van Altena, Shelus, Franz, Wasserman, Hemenway, Duncombe, Story, Whipple and, studies, 314
Brain, *see* Flexible skew-symmetric (FSS) distributions
Branco, Sahu, Dey and, studies
astronomical distance determination, 315
asymmetric distributions, 128–129
Bayesian techniques, 153, 157, 160–162, 175, 189
categorical response data, 142
multivariate survival data, 322–323
skew-elliptical distributions, 55, 89
Branco and Dey, Sahu, studies, 50, 64
Branco and Dey studies
asymmetric distributions, 128
Bayesian techniques, 155, 175
categorical response data, 142
closed skew-normal distributions, 25
hidden truncation, 109–110
skew-elliptical distributions, 44–47, 55–56, 61–62, 89
skew-normal distributions, 65–66
Bresler and Moulin, Ye, studies, 292

Breslow studies, 327
Brockwell and Davis studies, 83
Van den Broeck, Koop, Osiewalski, Steel and, studies, 162
Van der Broeck, Meeusen and, studies, 223, 235
Brown, Guo, Wang and, studies, 260
Brown, Gupta and, studies, 23, 62
Brown, Harlow and Tinic studies, 205
Brown studies, 156
B-SN type, skew-elliptical distributions, 54
Buckle studies, 323, 329

C

Cain studies, 103
Campbell and Hentschel studies, 205
Canonical form, asymmetric distributions, 128–129
Capital asset pricing
- basics, 191–194, 204
- data, 198–199
- empirical study, 199–204
- estimation method, 198–199
- market model, 196–198
- multivariate case, 194–196
- skew-normal model, 194–196

Capitanio, Azzalini and, studies
- asymmetric distributions, 116, 124, 127–129
- Bayesian techniques, 157–160, 164
- closed skew-normal distributions, 25, 42
- hidden truncation, 109
- multivariate skew-*t* distributions, 244, 246–247, 251–252
- skew-elliptical distributions, 47, 81, 85, 89–90
- skew-normal distributions, 3, 10, 12, 16, 20–21, 65, 67–68
- spatial prediction, 279, 281
- time series analysis, 260, 263

Capitanio, Azzalini and Stanghellini studies, 244, 247, 252
Carlin, Polson and Stoffer studies, 259, 275
Carlin and van der Linde, Spiegelhalter, Best, studies, 144, 148
Cartesian coordinates, 294
Casella, Lehmann and, studies, 114, 118
Castillo and Sarabia, Arnold, studies, 47
Catchpole, Feast and, studies, 314, 316
Catchpole and Morgan studies, 103
Categorical response data models
- basics, 131–133, 151
- Bayesian techniques, 143–148
- conditional predictive ordinate, 147–148
- correlated binary and/or ordinal models, 140–142
- CPO-based residuals, 149–150
- deviance information criterion, 148
- diagnostics, 148–151
- discrete choice models, 142–143
- general skewed link models, 137–143
- independent binary and/or ordinal models, 137–140
- latent residuals, 149
- links, 134–137
- models, 144–151
- observationwise weighted L measure, 150–151
- outlier detection, 148–151
- prediction, 135–137
- regression coefficients, 134–135
- residuals, 149–150
- weighted L measure, 144–147, 150–151

Cauchy/skew-Cauchy distributions
- astronomical distance determination, 311
- hidden truncation, 107, 111
- skew-elliptical distributions, 52
- skew-normal distributions, 3, 66

Caudill studies, 4
Cepheids, 310, 312–315
Chamberlain studies, 193
Chang, Dey, Chen and, studies, 147
Chang, Gelfand, Dey and, studies, 144, 147, 286
Chen, Dey and, studies, 140
Chen, Dey and Ibrahim studies, 144–147
Chen, Dey and Shao studies, 3, 131, 151
Chen, Gupta and, studies, 13
Chen, Ibrahim and, studies, 143
Chen, Ibrahim and Yiannoutsos studies, 133
Chen, Nandram and, studies, 137
Chen, Shao and Ibrahim studies, 148
Chen, Zhang and Davidian studies, 346
Chen and Chang, Dey, studies, 147
Chen and Deely studies, 150
Chen and Dey, Bhaumik, studies, 139, 151
Chen and Dey studies, 132, 137, 140, 148–149
Chen and Shao studies, 132, 143–144, 148
Chen and Sinha, Ibrahim, studies, 133, 145, 322

Chen studies, *vi,* 131–151
Chib, Albert and, studies, 131–132, 149
Chib and Greenberg studies, 140, 284
Chib and Jeliazkov studies, 166
Chiogna studies, 11, 218–219
Choleski decomposition, 91, 344
Cholesterol data, 356–357
Chou and Kroner, Bollerslev, studies, 206
Choy studies, 51–52
Christie studies, 205
Ciminera, Mantel, Bohidar and, studies, 334
Clark, Adcock and, studies, 196
Cliff and Ord studies, 84
Climatic impacts, *see* Time series analysis
Closed skew-normal (CSN) distributions, *see also* Skew-normal distributions
  Arnold and Beaver studies, 38–39
  Azzalini and Dalla Valle studies, 37–38
  basics, 25–26, 41–42
  characterization, 31
  independent random vectors, 33–37
  joint distributions, 33–35
  linear transformations, 29–32
  Liseo and Loperfido studies, 39–40
  marginal density, 31
  multivariate case, 26–29, 40–41
  random vectors, 33–40
  singular skew-normal distributions, 32
  skew-elliptical distributions, 40–41
  skew-normal random vectors, 37–40
  time series analysis, 262–263
Clyde studies, 133
Coastal flooding
  basics, 243–244, 258
  estimating flooding risk, 255–258
  observations, 247–250, 253–254
  quality control, 253–254
  seasonal variance, 244–247
  secular changes, 254
  skew-*t* distributions, 244–247, 250–258
COLS, *see* Corrected ordinary least squares (COLS) method
Compound errors, stochastic frontier analysis, 230–234, 236–238
Conditional closure property, 58
Conditional distributions, 22–23
Conditional predictive ordinate (CPO), 147–150
Conditional representations, 119–120
Conditioning method, 6
Construction of regions, 23–24
Cook and Weisberg studies, 74, 93
Cooper and Graham, Cootes, Taylor, studies, 291
Coordinate reflections, 119
Cootes, Edwards and Taylor studies, 291
Cootes, Taylor, Cooper and Graham studies, 291
Copas and Li studies
  closed skew-normal distributions, 26–27
  skew-elliptical distributions, 89
  skew-normal distributions, 3, 66
Corcoran and Schneider studies, 265
Corrected ordinary least squares (COLS) method, 233–234
Correlated binary and/or ordinal models, 140–142
Correlated errors, stochastic frontier analysis, 227–228, 239–241
CPO, *see* Conditional predictive ordinate (CPO)
Cressie studies, 84, 279
CSN, *see* Closed skew-normal (CSN) distributions
Cumulative distributions function, 16–17
Czado and Chappell, Netwon, studies, 139
Czado and Santner studies, 132
Czado studies, 132

D

Dalla Valle, Azzalini and, studies
  asymmetric distributions, 126–128
  Bayesian techniques, 155, 166, 175
  capital asset pricing, 194–195, 204
  closed skew-normal distributions, 25, 28, 37–38
  hidden truncation, 105
  linear mixed effects models, 341
  skew-elliptical distributions, 43, 45, 53, 57, 62, 81, 85, 89, 94
  skew-normal distributions, 3, 14, 20, 65, 68–69, 71, 74, 79
  spatial prediction, 279
  time series analysis, 260
Dalla Valle studies, *vi,* 3–24
Data
  capital asset pricing, 198–199
  GARCH model, 211–213
  spatial prediction, 280–288
Davidian, Chen, Zhang and, studies, 346

Davidian, Ma, Genton and, studies, *vii,* 339–358
Davidian, Zhang and, studies, 339, 342, 352, 355–357
David studies, 7
Davis, Brockwell and, studies, 83
Deely, Chen and, studies, 150
De Forest studies, *v*
De Groot, Bayarri and, studies, 66, 95
del Pino, Arellano-Valle and, studies, *vi,* 113–130
del Pino and San Martín, Arellano-Valle, studies
  asymmetric distributions, 115, 119–120, 122
  closed skew-normal distributions, 27, 40
  skew-elliptical distributions, 63
De Luca and Loperfido, Bartolucci, studies, 209
De Luca and Loperfido studies, *vi,* 205–222
Density formula, asymmetric distributions, 120–121, 123–124
Density shape, skew-elliptical distributions, 62
De Oliveira, Kedem and Short studies, 279, 282, 284, 286
De Oliveira and Sansó, Berger, studies, 153, 282
Deviance information criterion, 148
Dey, Aslanidou and Sinha, Sahu, studies, 322, 330, 333
Dey, Bhaumik, Chen and, studies, 139, 151
Dey, Branco and, studies
  asymmetric distributions, 128
  Bayesian techniques, 155, 175
  categorical response data, 142
  closed skew-normal distributions, 25
  hidden truncation, 109–110
  skew-elliptical distributions, 44–47, 56, 61–62, 89
  skew-normal distributions, 65–66
Dey, Chen and, studies, 132, 137, 140, 148–149
Dey, Chen and Chang studies, 147
Dey, Gelfand and, studies, 329
Dey, Liu and, studies, *vi,* 43–64
Dey, Qiou, Ravishanker and, studies, 322
Dey, Ravishanker and, studies, 323, 337
Dey, Sahu, Branco and, studies, 50, 64
Dey, Sahu and, studies, *vi,* 321–338
Dey, Sinha and, studies, 322, 326
Dey and Branco, Sahu, studies
  astronomical distance determination, 315
  asymmetric distributions, 128–129
  Bayesian techniques, 153, 157, 160–162, 175, 189
  categorical response data, 142
  multivariate survival data, 322–323
  skew-elliptical distributions, 55, 89
Dey and Chang, Gelfand, studies, 144, 147, 286
Dey and Chen studies, 140
Dey and Ibrahim, Chen, studies, 144–147
Dey and Lin studies, 48–50
Dey and Liu studies, 44, 48–50, 58–59, 63
Dey and Shao, Chen, studies, 3, 131, 151
Diagnostics, categorical response data models, 148–151
Dickey studies, 161
Dirichlet distributions, 140
Discrete choice models, 142–143
Dixon, Hill and, studies, 65
Domínguez-Molina, González-Farías and Gupta studies
  Bayesian techniques, 155
  skew-elliptical distributions, 54–55
  stochastic frontier analysis, 239
  time series analysis, 260, 262, 264
Domínguez-Molina, González-Farías and Ramos-Quiroga studies, *vi,* 25, 28, 223–241
Domínguez-Molina, Gupta, González-Farías and, studies, 28, 54
Domínguez-Molina and Gupta, González-Farías, studies
  closed skew-normal distributions, *vi,* 25–42
  stochastic frontier analysis, 226, 228
  time series analysis, 262–263
Dudley, Ludbrook and, studies, 66
Duncombe, Story, Whipple and Bradley, Benedict, McArthur, Fredrick, Harrison, Lee, Slesnick, Rhee, Patterson, Nelan, Jefferys, van Altena, Shelus, Franz, Wasserman, Hemenway, studies, 314

E

Eaton studies, 118, 121
Econometrics, skew-normal distributions, 4
Eddy, Geisser and, studies, 144, 147, 322, 329
Edgeworth studies, *v*
Edwards and Taylor, Cootes, studies, 291
Efficiencies/inefficiencies, stochastic frontier analysis, 229–230

Efficient semiparametric estimators, 95–100
EM algorithm, 343–346
Empirical study, 199–204
Endogenous shocks, 207
Engle, Bollerslev and, studies, 206
Engle and Gonzalez-Rivera studies, 211
Engle and Ito, Ling, studies, 207
Engle and Ng studies, 209
Engle studies, 193
Erkanli, Stangl and Müller studies, 139
Estimation
 coastal flooding, 255–258
 GARCH model, 213–215
 stochastic frontier analysis, 225–230
Estimation and inference, 9–13
Estimation method, 198–199
Estrada, Aparicio Acosta and, studies, 193
Euclidean properties, 69, 233
Eun and Shim studies, 207
European financial markets, *see* GARCH model
Exogenous shocks, 207
Extended skew-elliptical distributions, 40–41, *see also* Skew-elliptical distributions
Eyer and Genton studies, *vi,* 309–319

F

Fahrmeir and Tutz studies, 151
Failure rate, skew-normal distributions, 23
Fang, Kotz and Ng studies, 40, 44, 88
Fang and Zhang studies, 61
Feast and Catchpole studies, 314, 316
Ferguson studies, 139
Fernández, Osiewalsky and Steel studies, 114, 116, 120
Fernández and Steel studies, 114, 116, 175, 177
Ferreira and Steel studies, *vi,* 142, 175–190
FGSE (flexible generalized skew-elliptical) class, *see* Linear mixed effects models
Financial markets, *see* GARCH model
Firth studies, 12
Fisher, Ronald (Sir), 114
Fisher-Cochran theorem, 21
Fisher information matrix, 11, 164–165
Fixed slope, 316–317
Flexible generalized skew-elliptical (FGSE) class, *see* Linear mixed effects models
Flexible skew-symmetric (FSS) distributions, *see also* Shape representation
 athletes data, 93–95
 basics, 90–91
 efficient semiparametric estimators, 95–100
 multimodality, 91–93
Foreign markets' returns, 208
De Forest studies, *v*
Forsythe, Malcolm and Moler studies, 330
Frailty models, 323–326, *see also* Multivariate survival model
Franz, Wasserman, Hemenway, Duncombe, Story, Whipple and Bradley, Benedict, McArthur, Fredrick, Harrison, Lee, Slesnick, Rhee, Patterson, Nelan, Jefferys, van Altena, Shelus, studies, 314
Fredrick, Harrison, Lee, Slesnick, Rhee, Patterson, Nelan, Jefferys, van Altena, Shelus, Franz, Wasserman, Hemenway, Duncombe, Story, Whipple and Bradley, Benedict, McArthur, studies, 314
Freedman, Madore and, studies, 315
Free slope, 317–318
French, Schwert and Stambaugh studies, 205
Frontier data, skew-normal distributions, 10
FSS, *see* Flexible skew-symmetric (FSS) distributions
FS skewed distributions, 177
Future research, skew-normal distributions, 24

G

Gallant, Hsieh and Tauchen studies, 211
Gallant and Nychka studies, 339
Gamerman studies, 322, 326–327
GARCH model
 assumptions, 207–208
 basics, 205–206, 215
 data, 211–213
 estimation, 213–215
 news, 208–210
 proofs, 216–222
 returns, 207–208, 210–211

skew-elliptical distributions, 163
Gasbarra, Arjas and, studies, 322, 326–327
Gaussian case, 261–262
Gautschy studies, 313
Geary index, 84
Geisser and Eddy studies, 144, 147, 322, 329
Geisser studies, 133, 147
Gelfand, Dey and Chang studies, 144, 147, 286
Gelfand, Mukhopadhyay and, studies, 140
Gelfand, Wakefield, Smith, Racine-Poon and, studies, 341, 348
Gelfand and Dey studies, 329
Gelfand and Ghosh studies, 133, 144
Gelfand and Smith studies, 284, 322
Gelfand studies, 285
Gelman, Meng and Stern studies, 133
Generalized skew-elliptical distributions, *see* Skew-elliptical distributions
General skewed link models, 137–143
Genesis, skew-normal distributions, 15–16
Gent, Boville and, studies, 270, 273
Genton, Arellano-Valle and, studies, 118, 128
Genton, Baloch, Krim and, studies, *vi*, 291–308
Genton, Eyer and, studies, *vi*, 309–319
Genton, He and Liu studies
asymmetric distributions, 126
skew-elliptical distributions, 83, 85
skew-normal distributions, 20, 22, 71
Genton, Ma and, studies
asymmetric distributions, 117
flexible skew-symmetric distributions, 92, 292, 297
linear mixed effects models, 343
multivariate survival data, 337
skew-elliptical distributions, 48, 81, 84, 90, 93, 98–99
skew-normal distributions, 67
Genton, Wang, Boyer and, studies
asymmetric distributions, 116–117, 124
flexible skew-symmetric distributions, 292, 296
skew-elliptical distributions, 47–48, 61–63, 81–82, 85–86
skew-normal distributions, 3, 22, 66–67
Genton, Wang and, studies, 89
Genton and Ammann, Naveau, studies, *vi*, 259–278
Genton and Davidian, Ma, studies, *vii*, 339–358
Genton and Loperfido studies
asymmetric distributions, 116–117, 120, 124
skew-elliptical distributions, 81, 88–89
skew-normal distributions, 3, 22, 65, 68, 72–75
time series analysis, 260
Genton and Shen, Naveau, studies, 260, 262
Genton and Strang, Gorsich, studies, 84
Genton and Thompson studies, 244, 246, 250, 255, 258
Genton and Tsiatis, Ma, studies, 81, 96, 98
Genton studies, *vi*, 81–100
Genz studies, 232
George, McCullagh and Tsay studies, 133
George and McCullagh studies, 133
Geroski, Lazarova, Urga and Walters studies, 181
Gersch, Kitagawa and, studies, 259–260
Ghosh, Gelfand and, studies, 133, 144
Gibbs sampling
Bayesian skewed regression, 181
categorical response data, 139, 142
skew-elliptical distributions, 50, 157
spatial prediction, 284
Gibrat studies, 176, 182
Gilks, Richardson and Spiegelhalter studies, 283
Gómez and Quintana, Arellano-Valle, studies, 116–118, 124
Gonedes, Blattberg and, studies, 193
González-Farías, Domínguez-Molina and Gupta studies
closed skew-normal distributions, 25–42
stochastic frontier analysis, 226, 228
time series analysis, 262–263
González-Farías and Domínguez-Molina, Gupta studies, 28, 54
González-Farías and Gupta, Domínguez-Molina, studies
Bayesian techniques, 155
skew-elliptical distributions, 54–55
stochastic frontier analysis, 239
time series analysis, 260, 262, 264
González-Farías and Ramos-Quiroga, Domínguez-Molina, studies, *vi*, 25, 28, 223–241
Gonzalez-Rivera, Engle and, studies, 211
Gorsich, Genton and Strang studies, 84
Graham, Cootes, Taylor, Cooper and, studies, 291

Greenberg, Chib and, studies, 140, 284
Green studies, 322, 330
Grenander studies, 291
Groeneveld and Meeker, Arnold, Beaver, studies
closed skew-normal distributions, 25
hidden truncation, 103
skew-normal distributions, 3
De Groot, Bayarri and, studies, 66, 95
Guenni, Sansó and, studies, 279
Guerrero and Johnson studies, 132
Guo, Naveau, Ammann, Oh and, studies, 269
Guo, Wang and Brown studies, 260
Gupta, Domínguez-Molina, González-Farías and studies
Bayesian techniques, 155
skew-elliptical distributions, 54–55
stochastic frontier analysis, 239
time series analysis, 260, 262, 264
Gupta, González-Farías, Domínguez-Molina and, studies
closed skew-normal distributions, 25–42
stochastic frontier analysis, 226, 228
time series analysis, 262–263
Gupta, González-Farías and Domínguez-Molina studies, 28, 54
Gupta and Brown studies, 23, 62
Gupta and Chen studies, 13

H

Ha and Mallick, Kim, studies, *vi*, 279–289
Hadamard product, 345
Halifax, Canada, *see* Coastal flooding
Hall, Bronwyn, 181
Halley, E., 309
Hall studies, 181
Handcock and Stein studies, 284
Harlow and Tinic, Brown, studies, 205
Harrison, Lee, Slesnick, Rhee, Patterson, Nelan, Jefferys, van Altena, Shelus, Franz, Wasserman, Hemenway, Duncombe, Story, Whipple and Bradley, Benedict, McArthur, Fredrick, studies, 314
Harrison, West and, studies, 260
Harrison and Stevens studies, 260
Harvey and Renault studies, 206
Harvey and Siddique studies, 206
Healy's plot, 70
He and Liu, Genton, studies
asymmetric distributions, 126
skew-elliptical distributions, 83, 85
skew-normal distributions, 20, 22, 71
Heart, *see* Flexible skew-symmetric (FSS) distributions
Heckman studies, 209
Heikkinen, Arjas and, studies, 327
Hemenway, Duncombe, Story, Whipple and Bradley, Benedict, McArthur, Fredrick, Harrison, Lee, Slesnick, Rhee, Patterson, Nelan, Jefferys, van Altena, Shelus, Franz, Wasserman, studies, 314
Hentschel, Campbell and, studies, 205
Henze studies, 7–8, 224
Heteroscedastic errors, 227–228
Hidden truncation and selective sampling
additive component, 108
basics, 101, 112
Jones construction, 108
multivariate case, 105–108
skew-elliptical distributions, 63, 109–112
skew-normal estimation, 103–104
univariate case, 101–105
Hill and Dixon studies, 65
Hipparcos mission, 311
Hobert, Booth and, studies, 345–346
Hoeting, Raftery, Madigan and, studies, 133
Homoscedastic errors, 227
Hong studies, 211
Horn and Johnson studies, 33
Hotelling statistic, 72
Hougaard studies, 322–323, 338
Hsieh and Tauchen, Gallant, studies, 211
Hsieh studies, 210–211
Human brain and heart, *see* Flexible skew-symmetric (FSS) distributions
Hutson, Mudholkar and, studies, 115–116
Hyper-parameter values, 329–330
Hypothesis, skew-normal distributions, 13–14

I

Ibrahim, Chen, Dey and, studies, 144–147
Ibrahim, Chen, Shao and, studies, 148
Ibrahim, Chen and Sinha studies, 133, 145, 322

Ibrahim, Laud and, studies, 133, 144
Ibrahim and Chen studies, 143
Ibrahim and Laud studies, 133, 144
Ibrahim and Yiannoutsos, Chen, studies, 133
Independent binary and/or ordinal models, 137–140
Independent random vectors, 33–37
Intensive care unit data, 86–88
Invariance, 82–85, 118–121
Isard, Blake and, studies, 291
Ishwaran and James studies, 140
Ito, Ling, Engle and, studies, 207

J

Jacobian properties, 121
James, Ishwaran and, studies, 140
Jeffreys, van Altena, Shelus, Franz, Wasserman, Hemenway, Duncombe, Story, Whipple and Bradley, Benedict, McArthur, Fredrick, Harrison, Lee, Slesnick, Rhee, Patterson, Nelan, studies, 314
Jeffreys' prior, 164–166
Jeliazkov, Chib and, studies, 166
Johnson, Guerrero and, studies, 132
Johnson, Kotz, Balakrishnan and, studies, 226, 236
Joint distributions, 33–35
Jones construction, 108
Jones studies, 105–107, 112
Jorissen, Knapp, Pourbaix, Platais and, studies, 318

K

Kalbfleish and Prentice studies, 327
Kalbfleish studies, 233
Kalman filter, 260–262, *see also* Skewed-Kalman filter
Kaplan and Meier studies, 332
Karolyi studies, 207
Kass, Witkin and Terzopoulos studies, 291
Kass and Raftery studies, 133
Kedem and Short, De Oliveira, studies, 279, 282, 284, 286
Kendall school, 291
Kendall studies, 291
Kent and Bibby, Mardia, studies
  skew-elliptical distributions, 87
  skew-normal distributions, 68, 79
  stochastic frontier analysis, 227
Kidney infection data, 332–334
Kiehl, Zender, Otto-Bliesner and Bradley, Ammann, studies, 270, 273
Kim, Ha and Mallick studies, *vi,* 279–289
Kim and Mallick studies
  Bayesian techniques, 163
  skew-elliptical distributions, 84
  spatial prediction, 280, 282, 286, 289
Kitagawa and Gersch studies, 259–260
Klaassen, Ritov and Wellner, Bickel, studies, 96
Knapp, Pourbaix, Platais and Jorissen studies, 318
Kolmogorov distributions, 51
Kolmogorov-Smirnov test, 13–14
Koop, Osiewalski, Steel and Van den Broeck, studies, 162
Korea, *see* Spatial predictions
Kotz, Balakrishnan and Johnson studies, 226, 236
Kotz, Nadarajah and, studies, 67
Kotz and Ng, Fang, studies, 40, 44, 88
Kriging methods, 286
Krim, Poliannikov and, studies, 294
Krim and Genton, Baloch, studies, *vi,* 291–308
Kronecker matrix product, 33
Kroner, Bollerslev, Chou and, studies, 206
Kubiak, Pietrzynski, Wozniak and Zebrun, Udalski, Soszynski, Szymanski, studies, 313
Kumbhakar and Lovell studies, 225
Kurtosis, GARCH model, 210–211
Kutner, Nachtsheim and Wasserman, Neter, studies, 134

L

Latent residuals, 149
Latent variables, 117–118
Laud, Ibrahim and, studies, 133, 144
Laud and Ibrahim studies, 133, 144
Lazarova, Urga and Walters, Geroski studies, 181

LCLC1/LCLC2, *See* Linear constraint and linear combination (LCLC1/LCLC2)
Leadbetter, Lindgren and Rootzén studies, 255
Leavitt, Henrietta, 312–313
Lee, Slesnick, Rhee, Patterson, Nelan, Jefferys, van Altena, Shelus, Franz, Wasserman, Hemenway, Duncombe, Story, Whipple and Bradley, Benedict, McArthur, Fredrick, Harrison, studies, 314
Lee and Berger studies, 97
Lehmann and Casella studies, 114, 118
Lentz, Pawlowicz, Beardsley and, studies, 248
Leonard, O'Hagan and, studies, *v*
  Bayesian techniques, 153–154
  capital asset pricing, 194
  closed skew-normal distributions, 25
  skew-normal distributions, 4
Lesaffre, Verbeke and, studies, 339
Li, Copas and, studies
  closed skew-normal distributions, 26–27
  skew-elliptical distributions, 89
  skew-normal distributions, 3, 66
Lietchy studies, 162
Likelihood specification, 228, 327
van der Linde, Spiegelhalter, Best, Carlin and, studies, 144, 148
Lindgren and Rootzén, Leadbetter, studies, 255
Linear constraint and linear combination (LCLC1/LCLC2), 44, 49–50, 53–54, 58–64
Linear Gaussian state-space model, 263–265
Linear mixed effects models
  applications, 356–357
  basics, 339–340, 358
  Bayesian inference, 346–350
  cholesterol data, 356–357
  EM algorithm, 343–346
  FGSE distributions, 340–343
  implementation and interface, 343–350
  maximum likelihood, 343–346
  MCMC simulation, 346–350
  simulation results, 350–355
Linear transformations, 29–32
Ling, Engle and Ito studies, 207
Links, categorical response data models, 134–137
Lintner studies, 191
Liseo and Loperfido studies, 25, 39–40, 153–156, 164–168
Liseo and Macaro studies, 163
Liseo studies
  Bayesian techniques, *vi*, 153–171
  hidden truncation, 103
  skew-normal distributions, 4, 9
Littell, Milliken, Stroup and Wolfinger studies, 339
Litters, multivariate survival model, 334–336
Liu, Dey and, studies, 44, 48, 50, 58–59, 63
Liu, Genton, He and, studies
  asymmetric distributions, 126
  skew-elliptical distributions, 83, 85
  skew-normal distributions, 20, 22, 71
Liu and Dey studies, *vi*, 43–64
Logarithm of the pseudo-marginal likelihood (LPML), 147–148
Loperfido, Bartolucci, De Luca and, studies, 209
Loperfido, Genton and, studies
  asymmetric distributions, 116–117, 120, 124
  skew-elliptical distributions, 81, 88–89
  skew-normal distributions, 3, 22, 65, 68, 72–75
  time series analysis, 260
Loperfido, Liseo and, studies
  Bayesian techniques, 153, 155–156, 164–168
  closed skew-normal distributions, 25, 39–40
  skew-elliptical distributions, 54
Loperfido, De Luca and, studies, *vi*, 205–222
Loperfido studies
  Bayesian techniques, 157
  hidden truncation, 102
  skew-elliptical distributions, *vi*, 65–80
  skew-normal distributions, 7, 20–22, 24, 72, 76
Lo studies, 140
Louis, Blenman and Thatcher studies, 4
Lovell, Kumbhakar and, studies, 225
Lovell and Schmidt, Aigner, studies
  GARCH model, 209
  skew-normal distributions, 4, 7, 76
  stochastic frontier analysis, 223, 232, 235
LPML, *see* Logarithm of the pseudo-marginal likelihood (LPML)
De Luca and Loperfido, Bartolucci, studies, 209
De Luca and Loperfido studies, *vi*, 205–222

Ludbrook and Dudley studies, 66
Luri, Arenou and, studies, 318

M

Ma, Genton and Davidian studies, *vii,* 339–358
Ma, Genton and Tsiatis studies, 81, 96, 98
Ma and Genton studies
  asymmetric distributions, 117
  flexible skew-symmetric distributions, 92, 292, 297
  linear mixed effects models, 343
  multivariate survival data, 337
  skew-elliptical distributions, 48, 81, 85, 90, 93, 98–99
  skew-normal distributions, 67
Macaro, Liseo and, studies, 163
Madigan and Hoeting, Raftery, studies, 133
Madigan and Raftery studies, 133
Madigan and Volinsky, Raftery, studies, 133
Madore and Freedman studies, 315
Magder and Zeger studies, 340
Mahalanobis distances, 70, 82
Malcolm and Moler, Forsythe, studies, 330
Mallick, Kim, Ha and, studies, *vi,* 279–289
Mallick, Kim and, studies
  Bayesian techniques, 163
  skew-elliptical distributions, 84
  spatial prediction, 280, 282, 286, 289
Mallick, Walker and, studies, 337
Mallows, Andrews and, studies, 52
Mankiw, Ball and, studies, 4
Mantel, Bohidar and Ciminera studies, 334
Mardia, Kent and Bibby studies
  skew-elliptical distributions, 87
  skew-normal distributions, 68, 79
  stochastic frontier analysis, 227
Mardia's tests
  multivariate skewed regression modeling, 175
  skew-elliptical distributions, 163
  skew-normal distributions, 71
  skew-symmetric distributions, 87
Mardia studies, 71
Marginal closure property, 58
Marginal density, 31
Market model, 196–198, *see also* Capital asset pricing
Markov Chain Monte Carlo (MCMC) simulation
  Bayesian skewed regression, 180
  linear mixed effects models, 340, 346–351, 356
  multivariate survival model, 322, 330, 337–338
  skew-elliptical distributions, 153–154, 156, 159, 161, 163
  spatial prediction, 284
Markov process, 293
Markowitz studies, 191
San Martín, Arellano-Valle, del Pino and, studies
  asymmetric distributions, 115, 119–120, 122
  closed skew-normal distributions, 27, 40
  skew-elliptical distributions, 63
Matérn class, 288
Maximal invariant, 120–121
Maximum likelihood function
  capital asset pricing, 194, 199
  categorical response data, 134, 143
  linear mixed effects models, 341–346
  skew-elliptical distributions, 159, 164
  skew-normal distributions, 10, 24, 72
  skew-symmetric distributions, 93–94
  stochastic frontier analysis, 230, 232
McArthur, Fredrick, Harrison, Lee, Slesnick, Rhee, Patterson, Nelan, Jefferys, van Altena, Shelus, Franz, Wasserman, Hemenway, Duncombe, Story, Whipple and Bradley, Benedict, studies, 314
McCullagh, George and, studies, 133
McCullagh and Nelder studies, 114, 132, 137
McCullagh and Tsay, George, studies, 133
McCullagh studies, 132
McDonald and Nelson studies, 193
McDonald and Newey studies, 193, 206
McDonald and Xu studies, 193
McDonald studies, 206
McGilchrist and Aisbett studies, 332
McKeague and Tighiouart studies, 322, 327, 330
MCMC, *see* Markov Chain Monte Carlo (MCMC) simulation
Mead, Nelder and, studies, 12
Mean life function, 23

Meeker, Arnold, Beaver, Groeneveld and, studies
  closed skew-normal distributions, 25
  hidden truncation, 103
  skew-normal distributions, 3
Meeusen and Van den Broeck studies, 223, 235
Meier, Kaplan and, studies, 332
Meinhold and Singpurwalla studies, 259, 261
Meng and Stern, Gelman, studies, 133
Merton studies, 191
Metropolis Hastings algorithm
  linear mixed effects models, 348
  skew-elliptical distributions, 166
  spatial prediction, 284
Miller, Smith and, studies, 259
Milliken, Stroup and Wolfinger, Littell, studies, 339
Mills studies, 163
Mixed effects models, *see* Linear mixed effects models
Models, *see also* Multivariate survival model
  categorical response data models, 144–151
  spatial prediction, 280–284
  stochastic frontier analysis, 226–228, 238
Molenberghs, Verbeke and, studies, 340
Moler, Forsythe, Malcolm and, studies, 330
Moments
  asymmetric distributions, 125–126
  skew-elliptical distributions, 56–58
  skew-normal distributions, 7–8, 17–19
Monahan studies, 345
Monte Carlo simulation and techniques, *see also* Markov Chain Monte Carlo (MCMC) simulation
  astronomical distance determination, 314, 316
  categorical response data, 142
  linear mixed effects models, 343, 345
  Markov Chain Monte Carlo (MCMC) simulation, 346–350
  skew-normal distributions, 13
  skew-symmetric distributions, 99
Moran index, 84
Morgan, Catchpole and, studies, 103
Morgan studies, 132
Mossin studies, 191
Moulin, Ye, Bresler and, studies, 292
Mount Pinatubo, *see* Time series analysis
Mudholkar and Hutson studies, 115–116
Mukhopadhyay, Basu and, studies, 132, 139
Mukhopadhyay and Gelfand studies, 140
Mukhopadhyay and Vidakovic studies, 4, 156–157
Müller, Erkanli, Stangl and, studies, 139
Multimodality, skew-symmetric distributions, 91–93
Multi-process linear models, 269
Multivariate case
  asymmetric distributions, 125–128
  Bayesian techniques, 176–179
  capital asset pricing, 194–196
  closed skew-normal distributions, 26–29
  extended skew-elliptical distributions, 40–41
  hidden truncation and selective sampling, 105–108
  skewed regression modeling, 175–190
  skew-elliptical distributions, 44–50, 53–56
  skew-normal distributions, 14–19
  skew-symmetric distributions, 86
  stochastic frontier analysis, 236–239
Multivariate skew-*t* distributions, *see* Coastal flooding
Multivariate survival model
  baseline hazard function, 326–327
  basics, 321–323, 336–338
  examples, 332–336
  frailty models, 323–326
  hyper-parameter values, 329–330
  kidney infection data, 332–334
  likelihood specification, 327
  litters example, 334–336
  reversible jump steps, 330–331
  sensitivity, 329–330
  specification, 327–330

N

Nachtsheim and Wasserman, Neter, Kutner, studies, 134
Nadarajah and Kotz studies, 67
Nandram and Chen studies, 137
Naveau, Ammann, Oh and Guo studies, 269
Naveau, Genton and Ammann studies, *vi*, 259–278
Naveau, Genton and Shen studies, 260, 262
Negative skewness, 210

Nelan, Jefferys, van Altena, Shelus, Franz, Wasserman, Hemenway, Duncombe, Story, Whipple and Bradley, Benedict, McArthur, Fredrick, Harrison, Lee, Slesnick, Rhee, Patterson, studies, 314
Nelder, McCullagh and, studies, 114, 132, 137
Nelder and Mead studies, 12
Nelson, MacDonald and, studies, 193
Nelson studies, *v*
Neter, Kutner, Nachtsheim and Wasserman studies, 134
Netuka and Zvara, Andel, studies, 7
Newey, MacDonald and, studies, 193, 206
Newey studies, 96
News, GARCH model, 208–210
Newton, Czado and Chappell studies, 139
Newton, Tao, Palta, Yandell and, studies, 340, 352
Newton and Raftery studies, 183182
Newton-Raphson methods, 12, 24
Ng, Engle and, studies, 209
Ng, Fang, Kotz and, studies, 40, 44, 88
Nowak, Scott and, studies, 292
Numerical implementation, Bayesian techniques, 180–181
Nychka, Gallant and, studies, 339

O

Observationwise weighted L measure, 150–151
O'Hagan and Leonard studies
  Bayesian techniques, 153–154
  capital asset pricing, 194
  closed skew-normal distributions, 25
  skew-normal distributions, *v*, 4
Oh and Guo, Naveau, Ammann, studies, 269
De Oliveira, Kedem and Short studies, 279
De Oliveira and Sansó, Berger, studies, 153, 282
Ord, Cliff and, studies, 84
Osiewalski, Steel and Van den Broeck, Koop, studies, 162
Osiewalsky and Steel, Fernández, studies, 114, 116, 120
Otto-Bliesner and Bradley, Ammann, Kiehl, Zender, studies, 270, 273
Outlier detection, 148–151
Owen studies, 8

P

Paczynski studies, 311
Paleoclimate time series, *see* Time series analysis
Palta, Yandell and Newton, Tao, studies, 340, 352
Parent distributions, astronomical distance determination, 312
Parsons studies, 223
Patterson, Nelan, Jefferys, van Altena, Shelus, Franz, Wasserman, Hemenway, Duncombe, Story, Whipple and Bradley, Benedict, McArthur, Fredrick, Harrison, Lee, Slesnick, Rhee, studies, 314
Pawlowicz, Beardsley and Lentz studies, 248
Pearson's test, 13–14, 136
Pearson studies, *v*
Peiró studies, 163, 206
Pereira, Basu and, studies, 72
Period-luminosity relation calibration, 314–319
Pewsey studies
  Bayesian techniques, 158, 160
  closed skew-normal distributions, 42
  hidden truncation, 103
  skew-normal distributions, 12–13
Pietrzynski, Wozniak and Zebrun, Udalski, Soszynski, Szymanski, Kubiak, studies, 313
Pinatubo, Mount, *see* Time series analysis
Pinheiro and Bates studies, 339
del Pino, Arellano-Valle and, studies, *vi*, 113–130
del Pino and San Martín, Arellano-Valle, studies
  asymmetric distributions, 115, 119–120, 122
  closed skew-normal distributions, 27, 40
  skew-elliptical distributions, 63
Platais and Jorissen, Knapp, Pourbaix, studies, 318
Poisson distributions, 328, 337
Polachek and Robst studies, 4
Polar symmetry, 119
Poliannikov and Krim studies, 294
Polson and Stoffer, Carlin, studies, 259, 275
Pont studies, 314, 316
Pourbaix, Platais and Jorissen, Knapp, studies, 318
Praetz studies, 193

Predictions, categorical response data models, 135–137, *see also* Spatial predictions
Predictive asymmetry, GARCH model, 211
Pregibon studies, 132
Prentice, Kalbfleish and, studies, 327
Prentice studies, 132, 140
Prior distributions, 179–180
Probability function, asymmetric distributions, 120–121
Pseudo-Bayes factor, 333, 336, 338
Pugh and Vassie studies, 248, 256
Pugh studies, 243

Q

Qiou, Ravishanker and Dey studies, 322
Quadratic forms, 20–22, 59–61
Quality control, coastal flooding, 253–254
Quasi-Newton methods, 12
Quintana, Arellano-Valle, Gómez and, studies, 116–118, 124

R

Racine-Poon and Gelfand, Wakefield, Smith, studies, 341, 348
Racine-Poon studies, 348
Raftery, Kass and, studies, 133
Raftery, Madigan and, studies, 133
Raftery, Madigan and Hoeting studies, 133
Raftery, Madigan and Volinsky studies, 133
Raftery, Newton and, studies, 183
Rainfall, *see* Spatial predictions
RAL, *see* Regular asymptotically linear (RAL) estimators
Ramos-Quiroga, Domínguez-Molina, González-Farías and, studies, *vi*, 25, 28, 223–241
Random variables, asymmetric distributions, 116–117, 126
Random vectors, 33–40
Rao studies, 95
Ravishanker and Dey, Qiou, studies, 322
Ravishanker and Dey studies, 323, 337
Real line, asymmetric distributions, 115–118
Regression models and coefficients
  astronomical distance determination, 315–316
  Bayesian techniques, 179–181
  categorical response data models, 134–135
Regular asymptotically linear (RAL) estimators, 97
Reid studies, 72
Reliability function, 23
Renault, Harvey and, studies, 206
Residuals, categorical response data models, 149–150
Returns, GARCH model, 207–208, 210–211
Reversible jump steps, 330–331
Rhee, Patterson, Nelan, Jefferys, van Altena, Shelus, Franz, Wasserman, Hemenway, Duncombe, Story, Whipple and Bradley, Benedict, McArthur, Fredrick, Harrison, Lee, Slesnick, studies, 314
Richardson and Spiegelhalter, Gilks, studies, 283
Rios Insua and Ruggeri, Berger, studies, 157
Ritov and Wellner, Bickel, Klaassen, studies, 96
Roberts studies, *v*
  closed skew-normal distributions, 25
  hidden truncation, 102
  skew-elliptical distributions, 62
  skew-normal distributions, 7
Robinson, Smith and, studies, 279
Robst, Polachek and, studies, 4
Rootzén, Leadbetter, Lindgren and, studies, 255
Ruggeri, Berger, Rios Insua and, studies, 157
Rydberg studies, 205

S

Sahu, Dey, Aslanidou and Sinha studies, 322, 330, 333
Sahu, Dey and Branco studies
  astronomical distance determination, 315
  asymmetric distributions, 128–129
  Bayesian techniques, 153, 157, 160–162, 175, 189
  categorical response data, 142
  multivariate survival data, 234, 322
  skew-elliptical distributions, 50, 55, 64, 89
Sahu and Dey studies, *vi*, 321–338

Salvan studies, 13
Sample formation, astronomical distance determination, 312
Sampling, *see* Hidden truncation and selective sampling
San Martín, Arellano-Valle, del Pino and, studies
  asymmetric distributions, 115, 119–120, 122
  closed skew-normal distributions, 27, 40
  skew-elliptical distributions, 63
Sansó, Berger, De Oliveira and, studies, 153, 282
Sansó and Guenni studies, 279
Santner, Czado and, studies, 132
Sarabia, Arnold, Castillo and, studies, 47
Sartori studies, 12–13, 160
Scalar properties, 19–20
Schick studies, 259
Schmidt, Aigner, Lovell and, studies
  GARCH model, 209
  skew-normal distributions, 4, 7, 76
  stochastic frontier analysis, 223, 232, 235
Schneider, Corcoran and, studies, 265
Scholes, Black and, studies, 191
Schwert and Stambaugh, French, studies, 205
Schwert studies, 205
Scott and Nowak studies, 292
SDB skewed distributions, 178–179
Seasonal variance, coastal flooding, 244–247
Secular changes, coastal flooding, 254
SE_k, skew-elliptical distributions, 55–56
Selection models, asymmetric distributions, 117–118
Selective sampling, *see* Hidden truncation and selective sampling
Sensitivity, multivariate survival model, 329–330
SGARCH, *see* GARCH model
Shao, Chen, Dey and, studies, 3, 131, 151
Shao, Chen and, studies, 132, 143–144, 148
Shao and Ibrahim, Chen, studies, 148
Shape representation, *see also* Flexible skew-symmetric (FSS) distributions
  analysis, 294–302
  angles, 299
  basics, 291–292, 307
  case studies, 303–305
  distributions, 300–301
  experimental results, 301–307
  performance assessment, 301–302
  posterior learning, 296–298
  problem statement, 292–294
  proofs, 308
  sampling, 306–307
Sharpe studies, 191
Shelus, Franz, Wasserman, Hemenway, Duncombe, Story, Whipple and Bradley, Benedict, McArthur, Fredrick, Harrison, Lee, Slesnick, Rhee, Patterson, Nelan, Jefferys, van Altena, studies, 314
Shen, Naveau, Genton and, studies, 260, 262
Shen, Thompson and, studies, *vi*, 243–258
Shephard studies, 260
Shim, Eun and, studies, 207
Shocks, GARCH model, 207
Short, De Oliveira, Kedem and, studies, 279, 282, 284, 286
Shumway and Stoffer studies, 260
Shutes, Adcock and, studies, 4, 193–194
Siddique, Harvey and, studies, 206
Sign invariant class, 121–126
Signs, 115–117
Simaan studies, 195
Simulations, time series analysis, 273, *see also* Monte Carlo simulation and techniques
Singpurwalla, Meinhold and, studies, 259, 261
Singular skew-normal (SSN) distributions, 32
Sinha, Ibrahim, Chen and, studies, 133, 145, 322
Sinha, Sahu, Dey, Aslanidou and, studies, 322, 330, 333
Sinha and Dey studies, 322, 326
Skewed components, time series analysis, 270–272
Skewed distribution, asymmetric distributions, 123–124
Skewed frailty, *see* Multivariate survival model
Skewed-Kalman filter, 262–268, 275–278, *see also* Kalman filter
Skewed regression modeling, *see* Bayesian techniques
Skew-elliptical distributions, *see also* Bayesian techniques
  Arnold and Beaver approach, 46–47
  A-SN type, 53
  asymmetric distributions, 128–129
  athletes data, 74–76
  basics, 43–44, 56–62, 65–66, 76–77

Branco and Dey approach, 45–46, 55, 61–62
B-SN type, 54
characterization, 66–67, 76
closed skew-normal distribution, 40–41
conditional closure property, 58
density shape, 62
Dey and Liu approach, 48–50
diagnostics, 69–71, 76
Domínguez-Molina, González-Farías and Gupta class, 54–55
estimation, 76–77
examples, 50–56
exponential distribution, 52
finite mixture, 51
flexible generalized skew-elliptical (FGSE), 81–100
generalized skew-elliptical (GSE), 81–100
hidden truncation and selective sampling, 109–112
LCLC1/LCLC2, 49–50, 58–61
Liseo and Loperfido class, 54
logistic distributions, 51
marginal closure property, 58
moments, 56–58, 68–69
multivariate cases, 44–50, 53–56
multivariate extended distribution, 40–41
parametric inference, 72–73
Pearson type-II, 53
proofs, 77–80
properties, 56–62
quadratic forms, 59–61
scale mixture, 50–53, 56–58
SE_k, 55–56
skew-*t* distribution, 52–53
stable distributions, 51–52
stochastic frontier analysis, 234–235
symmetric distributions, 88–89
transformations, 68–69
univariate case, 62
Wang, Boyer and Genton approach, 47–48
Skew-exponential distributions, 52
Skew-finite mixture, 51
Skewing function, 63, 82
Skew-in-mean GARCH model, *see* GARCH model
Skew-logistic distributions, 51
Skewness, asymmetric distributions, 118–121
Skew-normal distributions, *see also* Closed skew-normal (CSN) distributions
astronomical distance determination, 315–316
asymmetric distributions, 126–128
basics, 3–4
capital asset pricing, 194–196
closed skew-normal distribution, 37–40
conditional distribution, 22–23
conditioning method, 6
construction of regions, 23–24
cumulative distribution function, 16–17
estimation and inference, 9–13
failure rate, 23
future research, 24
generalized skew-normal (GSN), 65–80
genesis, 15–16
hidden truncation and selective sampling, 103–104
hypothesis, 13–14
mean life function, 23
moments, 7–8, 17–19
multivariate case, 14–19
quadratic forms, 20–22
reliability function, 23
scalar properties, 19–20
strength-stress models, 23
transformation method, 6–7
univariate case, 4–9
Skew-Pearson type-II, 53
Skew-scale mixture, 50–53, 56–58
Skew-stable distributions, 51–52
Skew-symmetric distributions, *see also* Asymmetric distributions
athletes data, 93–95
basics, 81–82, 100
efficient semiparametric estimators, 95–100
flexible class, 90–100
intensive care unit data, 86–88
invariance, 82–85
multimodality, 91–93
multivariate case, 86
skew-elliptical distributions, 88–89
skewing function, 82
stochastic representations and simulations, 85
Skew-*t* distributions
coastal flooding, 244–247, 250–258
heavy tail priors, 157
skew-elliptical distributions, 52–53
Slesnick, Rhee, Patterson, Nelan, Jefferys, van Altena, Shelus, Franz, Wasserman,

Hemenway, Duncombe, Story, Whipple and Bradley, Benedict, McArthur, Fredrick, Harrison, Lee, studies, 314
Small studies, 291
Smith, Bernardo and, studies, 133
Smith, Gelfand and, studies, 284, 322
Smith, Racine-Poon and Gelfand, Wakefield, studies, 341, 348
Smith and Miller studies, 259
Smith and Robinson studies, 279
Smoothing spline model, 269–270
Soszynski, Szymanski, Kubiak, Pietrzynski, Wozniak and Zebrun, Udalski, studies, 313
Spatial predictions
 analysis of data, 284–288
 basics, 279, 288–289
 Bayesian analysis, 282–284
 data, 280–288
 model, 280–284
Specification, multivariate survival model, 327–330
Spherical symmetry, 119
Spiegelhalter, Best, Carlin and van der Linde studies, 144, 148
Spiegelhalter, Gilks, Richardson and, studies, 283
SSN, *see* Singular skew-normal (SSN) distributions
Stambaugh, French, Schwert and, studies, 205
Stanghellini, Capitanio, Azzalini and, studies, 244, 247, 252
Stangl and Müller, Erkanli, studies, 139
State-space model, 260–261
Steel, Fernández, Osiewalsky and, studies, 114, 116, 120
Steel, Fernández and, studies, 114, 116, 175, 177
Steel, Ferreira and, studies, *vi*, 142, 175–190
Steel and Van den Broeck, Koop, Osiewalski, studies, 162
Stefanski and Boos studies
 astronomical distance determination, 316
 linear mixed effects models, 346
 skew-elliptical distributions, 99
Stein, Handcock and, studies, 284
Stern, Gelman, Meng and, studies, 133
Stevens, Harrison and, studies, 260
Stevenson studies, 232
Stidd studies, 279
Stigler studies, *v*
Stochastic frontier analysis
 assumptions, 226–228
 basics, 223–225, 235
 compound errors, 230–234, 236–238
 correlated errors, 227–228, 239–241
 efficiencies/inefficiencies, 229–230
 estimation, 225–230
 heteroscedastic errors, 227–228
 homoscedastic errors, 227
 likelihood, 228
 model, 226–228, 238
 multivariate case, 236–239
 skew-elliptical distributions, 162, 234–235
 truncated multivariate normal distributions, 229–230, 238–239
 uncorrelated errors, 227–228
Stochastic representation and simulations, 85, 124–126, *see also* Stochastic frontier analysis
Stoffer, Carlin, Polson and, studies, 259, 275
Stoffer, Shumway and, studies, 260
Story, Whipple and Bradley, Benedict, McArthur, Fredrick, Harrison, Lee, Slesnick, Rhee, Patterson, Nelan, Jefferys, van Altena, Shelus, Franz, Wasserman, Hemenway, Duncombe, studies, 314
Strang, Gorsich, Genton and, studies, 84
Strength-stress models, skew-normal distributions, 23
Stroup and Wolfinger, Littell, Milliken, studies, 339
Stukel studies, 132
Survival model, *see* Multivariate survival model
Sutton studies, 182
Symmetry, *see* Asymmetric distributions; Flexible skew-symmetric (FSS) distributions; Skew-symmetric distributions
Szymanski, Kubiak, Pietrzynski, Wozniak and Zebrun, Udalski, Soszynski, studies, 313

T

Tancredi studies, 224
Tanner studies, 285
Tao, Palta, Yandell and Newton studies, 340, 352
Tauchen, Gallant, Hsieh and, studies, 211
Tawn and Vassie studies, 248
Taylor, Cooper and Graham, Cootes, studies, 291
Taylor, Cootes, Edwards and, studies, 291
Taylor expansion, 134–135
Temperature, *see* Time series analysis
Terzopoulos, Kass, Witkin and, studies, 291
Thatcher, Louis, Blenman and, studies, 4
Theodossiou studies, 193
Thompson, Genton and, studies, 244, 246, 250, 255, 258
Thompson and Shen studies, *vi*, 243–258
Tighiouart, McKeague and, studies, 322, 327, 330
Time series analysis
  basics, 259–260, 274–275
  closed skew-normal distributions, 262–263
  Gaussian case, 261–262
  Kalman filter, 260–262
  linear Gaussian state-space model, 263–265
  multi-process linear models, 269
  paleoclimate time series, 268–273
  simulations, 273
  skewed components, 270–272
  skewed Kalman filter, 262–268, 275–278
  smoothing spline model, 269–270
  state-space model, 260–261, 263–265
Tinic, Brown, Harlow and, studies, 205
Trancredi studies, 162
Transformations, 6–7, 119
Trigonometric parallax, 310–312
Truncated multivariate normal distributions, 229–230, 238–239
Truncation, *see* Hidden truncation and selective sampling
Tsay, George, McCullagh and, studies, 133
Tsiatis, Ma, Genton and, studies, 81, 96, 98
Tukey studies, *v*
Tutz, Fahrmeir and, studies, 151

U

Udalski, Soszynski, Szymanski, Kubiak, Pietrzynski, Wozniak and Zebrun studies, 313
Uncorrelated errors, 227–228
Univariate case
  hidden truncation and selective sampling, 101–105
  skew-elliptical distributions, 62
  skew-normal distributions, 4–9
Urga and Walters, Geroski, Lazarova, studies, 181

V

van Altena, Shelus, Franz, Wasserman, Hemenway, Duncombe, Story, Whipple and Bradley, Benedict, McArthur, Fredrick, Harrison, Lee, Slesnick, Rhee, Patterson, Nelan, Jefferys, studies, 314
Van den Broeck, Koop, Osiewalski, Steel and, studies, 162
Van den Broeck, Meeusen and, studies, 223, 235
van der Linde, Spiegelhalter, Best, Carlin and, studies, 144, 148
Vassie, Pugh and, studies, 248, 256
Vassie, Tawn and, studies, 248
Verbeke and Lesaffre studies, 339
Verbeke and Molenberghs studies, 340
Vidakovic, Mukhopadhyay and, studies, 4, 156–157
Volatility clustering, 211
Volcanic eruptions, *see* Time series analysis
Volinsky, Raftery, Madigan and, studies, 133

W

Wacholder and Weinberg studies, 95
Wahba studies, 269
Wahed and Ali studies, 89
Wakefield, Smith, Racine-Poon and Gelfand studies, 341, 348
Wakefield studies, 348
Walker and Mallick studies, 337

Walters, Geroski, Lazarova, Urga and, studies, 181
Wang, Boyer and Genton studies
asymmetric distributions, 116–117, 124
flexible skew-symmetric distributions, 292, 296
skew-elliptical distributions, 47–48, 61–63, 81–82, 86, 8584
skew-normal distributions, 3, 22, 66–67
Wang and Brown, Guo, studies, 260
Wang and Genton studies, 89
Wasserman, Hemenway, Duncombe, Story, Whipple and Bradley, Benedict, McArthur, Fredrick, Harrison, Lee, Slesnick, Rhee, Patterson, Nelan, Jefferys, van Altena, Shelus, Franz, studies, 314
Wasserman, Neter, Kutner, Nachtsheim and, studies, 134
Wecker and Ansley studies, 269
Weighted L measure, 144–147, 150–151
Weinberg, Wacholder and, studies, 95
Weinstein studies, *v*
Weisberg, Cook and, studies, 74, 93
Weisstein studies, 294
Wellner, Bickel, Klaassen, Ritov and, studies, 96
West and Harrison studies, 260
Whipple and Bradley, Benedict, McArthur, Fredrick, Harrison, Lee, Slesnick, Rhee, Patterson, Nelan, Jefferys, van Altena, Shelus, Franz, Wasserman, Hemenway, Duncombe, Story, studies, 314
Whittmore studies, 132
Wishart distributions and prior, 72, 167
Witkin and Terzopoulos, Kass, studies, 291
Wolfinger, Littell, Milliken, Stroup and, studies, 339
Wozniak and Zebrun, Udalski, Soszynski, Szymanski, Pietrzynski, studies, 313

X

Xu, MacDonald and, studies, 193

Y

Yaglom studies, 284
Yandell and Newton, Tao, Palta, studies, 340, 352
Yang and Berger studies, 168
Ye, Bresler and Moulin studies, 292
Yiannoutsos, Chen, Ibrahim and, studies, 133

Z

Zebrun, Udalski, Soszynski, Szymanski, Kubiak, Pietrzynski, Wozniak and, studies, 313
Zeger, Magder and, studies, 340
Zellner studies, 279
Zender, Otto-Bliesner and Bradley, Ammann, Kiehl, studies, 270, 273
Zero correlation, 211
Zero expectation, 210
Zhang, Fang and, studies, 61
Zhang and Davidian, Chen, studies, 346
Zhang and Davidian studies, 339, 342, 352, 355–357
Zhang studies, 96
Zhou studies, 193
Zvara, Andel, Netuka and, studies, 7

Printed in the United States
by Baker & Taylor Publisher Services

Printed in the United States
by Baker & Taylor Publisher Services